Desheng Dash Wu • Yong Zhou
Editors

# Modeling Risk Management for Resources and Environment in China



Springer

*Editors*
Desheng Dash Wu
University of Toronto
Risklab
Spadina Crescent 1
M5S 3G3 Toronto Ontario
Canada
dash@risklab.ca

Yong Zhou
Chinese Academy of Sciences
Academy of Mathematics and
Systems Sciences
Beijing
China, People's Republic

ISSN 2191-1436 e-ISSN 2191-1444
ISBN 978-3-642-27083-3 ISBN 978-3-642-18387-4 (eBook)
DOI 10.1007/978-3-642-18387-4
Springer Heidelberg Dordrecht London New York


Softcover reprint of the hardcover 1st edition 2011


*Cover design:* SPI Publisher Services

Printed on acid-free paper

Springer is part of Springer Science+Business Media (www.springer.com)

# Computational Risk Management

*Series Editors*



For further volumes:
http://www.springer.com/series/8827

# Preface

We are living in a risky world, and it is getting riskier and riskier. Prediction of extreme risk events is almost unlikely, as being expressed by many scholars such as in Taleb's 2007 book titled *Black Swan*. Effective risk management needs integration of various risks facing the organization (Wu and Johnson 2009, 2010 HERA). Enterprise risk management has been defined as a process that uses integrated, systematic approaches to manage risks that face the organization. Therefore, enterprise risk management has been deemed as an effective risk management philosophy.

In the past, we have tried to discuss different aspects of risk, to include finance, information systems, disaster management, and supply chain perspectives (Olson and Wu 2008a, b, 2010). In this edited volume, we present the state-of-the-art views of the perspective of Risk Management in Sustainable Economy, Risk Management in Engineering Projects, Risk Management in Sustainable Enterprise, Environmental Risk Management, Energy Risk Management, Risk Management Modeling, Case Study of Risk Management

The bulk of this volume is devoted to presenting a number of modeling approaches and case studies that have been (or could be) applied to risk management in sustainability from the International Symposium on Statistics and Management Science in Nanjing, P. R. China. We include dynamic particle swarm optimization models, regime-Switching models to assist risk managers to capture risk dynamics in their decisions. Rough Set and fuzzy set theories are employed in non-probabilistic uncertain environment. Dynamic models such as GARCH family models are used to handle risky project management when achieving sustainable development purpose. Failure Mode models are used in lots of engineering risk management problems. We hope that this book provides some view of how models can be applied by more readers aiming to achieve sustainable development in face of risks.

October 2010

Desheng Dash Wu
Yong Zhou

## References

Wu D, Olson DL (2009) Introduction to the special section on optimizing risk management. Methods and Tools, Human and Ecological Risk Assessment 15(2): 220–226
Wu D, Johnson BL (2010) HERA emphasizing risk management: an editorial. Human and Ecological Risk Assessment accepted and in press
Olson DL, Wu D (2008a) Enterprise risk management. World Scientific Publishing, Singapore
Olson DL, Wu D (2008b) New frontiers in risk management. Springer, Heidelberg
Olson DL, Wu D (2010) Enterprise risk management models. Springer, Heidelberg

# Contents

# Part I
# Risk Management in Sustainable Economy

# Research on Recycling Mining System and Its Risk Analysis in Eastern China

**Lili Bian, Min Zhou, Aibin Li, and Shitong Ge**

**Abstract** It is the most economically developed region in Eastern China of the country. But the resources are relatively scarce there. As economic development, many mining areas in Eastern China are facing resource depletion problems, and the resulting is series of social problems and environmental issues. Recycling mining system is the solution using the principle of circular economy for sustainable mining development. The construction of recycling mining system is the future trends of coal mining. In this paper, Datun coal is as an example, to explore a useful model for the development of recycling mining system in Eastern China, and a brief analysis of economic risk is made in the paper.

**Keywords** Circular economy · Recycling mining system · Risk analysis

## 1 Introduction

Since the late 1990s, the development of circular economy (CE) in China has experienced there stages (Ren and Zhou 2009): the introduction of the concept by scholars, accepted by and relevant decisions made by the national government, and national pilot and demonstration started in 2006 (Ren and Zhou 2009). Recycling mining system is guided by circular economy theory, through reduction, reuse and recycling (that is 3R) principle, to achieve mine economic benefits, environmental benefits, social benefits together.

L. Bian (✉), M. Zhou, A. Li, and S. Ge
Management School, China University of Mining & Technology, Jiangsu Province 221008, People's Republic of China
e-mail: 1207104492@qq.com

Y. Zhou and D.D. Wu (eds.), *Modeling Risk Management for Resources and Environment in China*, Computational Risk Management,
DOI 10.1007/978-3-642-18387-4_1, © Springer-Verlag Berlin Heidelberg 2011

Eastern China, as China's most dynamic economic growth region, is the most developed area in the country. At the same time, the resources and energy are also relatively scarce. Many of the current mining area in Eastern China, have changed from the previous resource-based to be resource-exhausted, left behind a series of social and environmental problems. How to get a better solution to get rid of this development cycle, and to change the traditional extensive mode of economic growth? How to be sustainable use of resources, improve energy efficiency? Development of circular economy based on mining is an urgent requirement and the only realistic choice. In this paper, a kind of recycling mining model is developed based on case studies.

## 2 Overview of the Study Area

Datun mining area is located at the junction of Jiangsu and Shandong province, Weishan Lake Peixian territory. It is located in the high developed eastern coastal areas, close to one of the three economic circles Yangtze River Delta. Across two provinces, two cities and three counties, Datun mining area is the Economic radiation intersection of five cities of Xuzhou, Zaozhuang, Jining, Shangqiu and Huaibei. At the same time it is near the main port of Lianyungang, Jiangsu China. The total mining area of it is 245 $km^2$. The distance from mining center to Peixian county is 10 km, Xuzhou City 72 km, Lianyungang City, 305 km, and Shanghai 833 km. It has geographical advantages of both consumer and sales market.

## 3 Construction Conditions of Recycling Mining System

Datun coal is a resource-based economy. As one of the large industrial enterprises, its core businesses Shanghai Datun Energy Co., Ltd., is engaged in collection of coal production, power generation, aluminum processing, and rail transport. With the development of mining industry, electric power, coal washing, transport, aluminum and other subsidiary companies in the region, the electricity load and water demand have been increased dramatically, water, coal, electricity, land and other resources are increasingly inadequate, and eco-construction and environmental protection are facing an increasingly grim situation. The traditional "three highs" (high consumption, high energy consumption and high pollution) extensive mode of economic development can no longer meet the need for sustainable development in mining. After 30 years of development, the resource situation has not optimistic, taking into account of the coal to be suppressed by "three under", especially under villages and Ordovician limestone, the current recoverable reserves of coal left only 180 million t. How to achieve sustainable development in Datun mining, has to be considered in the next 20 years.

# 4 Planning for Recycling Mining System

## *4.1 Design Ideas*

### 4.1.1 Circular Economy, Eco-industrial-Oriented (Zhou 2005)

The coal mining area based on comprehensive utilization of resources, in accordance with the "reduce, reuse, recycle" concept, the industry level, the optimal structure of industrial chains, through material substitution, process innovation and product upgrades and other cleaner production measures to improve resource utilization rate, reducing waste emissions, reduce pollution, increase economic efficiency. The traditional model of industrial development towards ecological direction.

### 4.1.2 Industry Horizontal Symbiotic, Resource Vertical Coupling (Qi 2009; Li and Luo 2006; Niu 2004)

In accordance with the "non-coal set optimization, industrial ecology" requirement, optimizing industrial structure and layout of non-coal mining area, the lateral coupling among different industries, and gradually reduce the proportion of resources to primary processed products, to promote the vertical chain, network-based development. Establishment of the exploitation of resources, basic products and fine high-end products, functional products, the common development of multi-level model of industrial to wider areas and more space for internal and external resources and links to related enterprises, expansion and integration efforts for non-coal industrial chain of organic integration, optimization and upgrading, a "coal–electricity–Aluminum–transport– building" an integrated development system, the formation of multi-lateral expansion of industrial coupling, the vertical extension of processing resources, as well as by-products and waste utilization pattern of three-dimensional structure.

### 4.1.3 Production–Science–Research Combined, Technology–Industry–Trade Integrated

## 4.2 Structural Framework

Based on Datun coal industrials, the planning involved in five industries, that is, coal industry, power industry, aluminum, building materials and transport. These five industries formed the coal-power-building materials, and coal-power-aluminum two vertical main industrial chain, and formed gangue materials, waste heat recycling, mine water reuse, reclamation, four lateral collapse vice industry chain. Datun coal by region is divided into six industrial park area and a Vein Industrial Park (Zhu and Zhang 2007; Meng and Yu 2008), that is, the Central Area, Datun Power Plant, Yao Qiao Mine, Xu Zhang Mine, Long Dong Mine, Kong Zhang Mine Planning framework shown in Fig. 1.

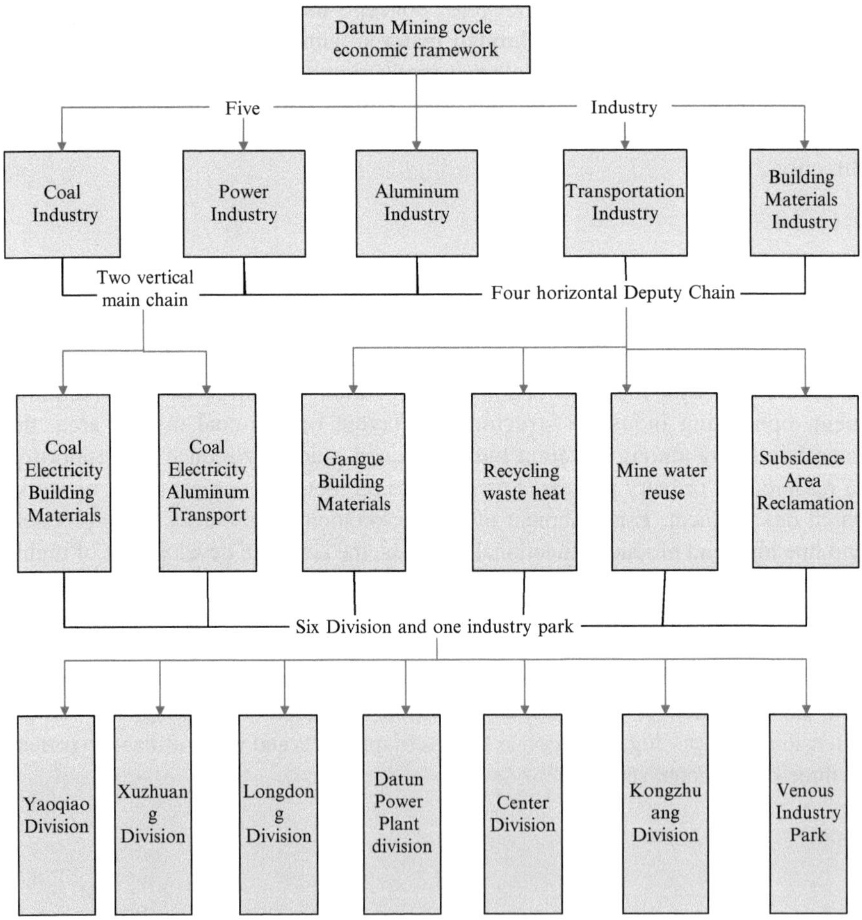


**Fig. 1** Recycling Datun coal structure frame diagram

### 4.3 *Overall Model*

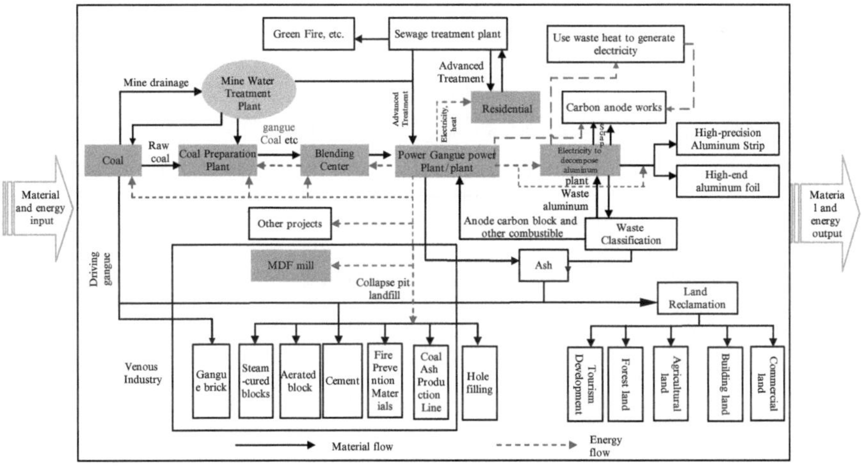


**Fig. 2** Datun recycling mining overall pattern

## 5 Logistics Plan of Recycling Mining

Analysis shows that the establishment of a symbiotic relationship, resulting in solid waste coal mines, power plants and other companies to reduce the waste piled up and disposal of low pay compensation for pollution; use of waste deliveries of the raw material cost is a party and enjoy the country the use of preferential policies to encourage waste (Figs. 2 and 3).

## 6 Cycle-Benefit Analysis and Risk Analysis

### 6.1 *Cycle-Benefit Analysis*

The added value distribution of the planning is shown in Fig. 4 as follows. As can be seen, the proportion of the aluminum processing project reached 41% first; project to improve recovery rate is 25%; power generation has reached 15%; coal separation process is about 11%, 10% for fly ash utilization.

The added profit distribution of the planning is shown in Fig. 5 as follows. The aluminum processing is the highest too, reaching 31%. The power generation and resource recovery projects improve the profit contribution rate of 26% and 15%, respectively. The coal separation process up is to 11%.

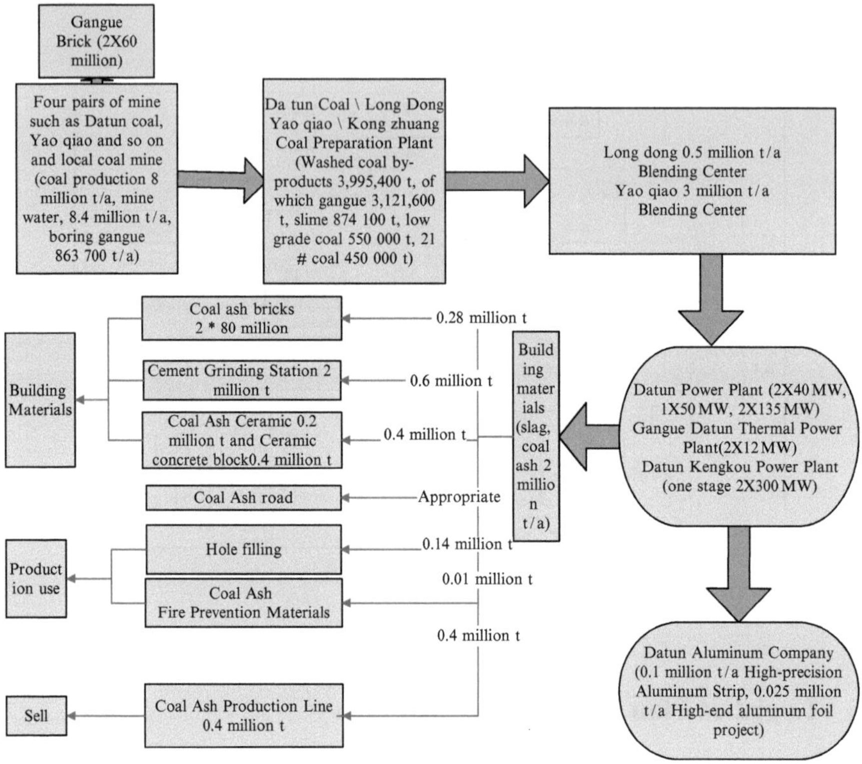


**Fig. 3** Datun mining 2010 flow diagram of material recycling economy

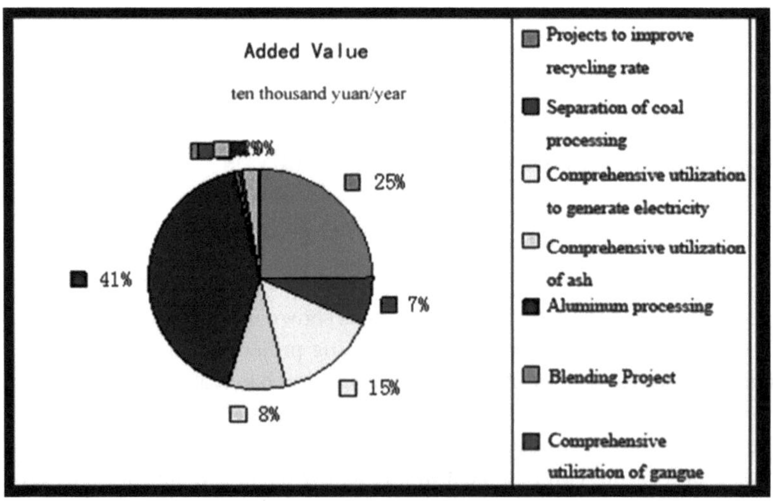


**Fig. 4** The added value distribution of the projects planning

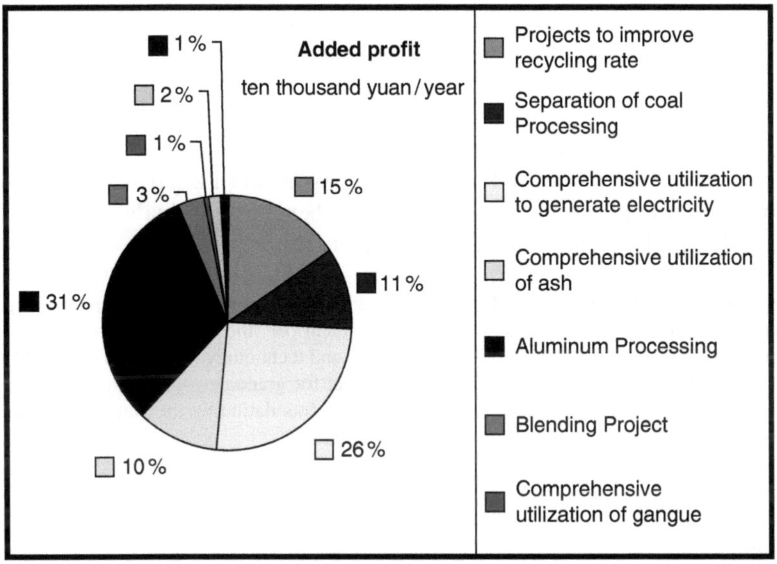


**Fig. 5** The added profit distribution of the project planning

Therefore, further extend the industry chain, to generate electricity using coal mine waste, coal aluminum joint development of circular economy is to conduct deep processing of aluminum, used to carry out solid waste, especially for power coal blending for the boiler combustion to create more substantial profits; gangue brick and fly ash brick, low cost, direct economic benefits may be a loss, but the analysis of indirect economic, environmental and social benefits, is still very substantial.

### 6.2 *Risk Analysis*

From the above analysis, the Datun coal Circular Economy is the main source of profits: First, extend the industrial chain, especially coal–electricity–aluminum industry chain extension; second vein industrial Chain. From the perspective of sustainable development, this model there is a resource risk, technology risk, economic risk and policy risk (Lu and Zhao 2005).

## 7 Conclusion

Under the guidance of theory of the circular economy and ecological industrial theory, the direction of the development of Datun Mining is to stabilize the coal industry, develop the electric power, extend the aluminum industry, and expansion of trade. And at the same time, Implementation utilization of mine water, sewage, coal gangue, coal slime, power plant fly ash, and other wastes, the Datun coal into a technologically advanced, knowledge-intensive, management, civilization, environment friendly, harmonious cyclic structure mining, achieving economic, social and environmental benefits of win-win.

Research on the Mode of Recycling mining in the Eastern China will help smooth transition of the mining area, and has benefit on China's resource-based sustainable development as model and reference.

**Acknowledgement** We are grateful for the financial support from the key discipline development project 'Coal industry strategy and security management' of 'the Third 211 Project' of China University of Mining & Technology(CUMT), science and technology foundation of CUMT (NO. 2007B026), 2008 scientific research innovation project for graduates of Jiangsu provincial universities (NO. CX08B-042R) and 2008 opening research foundation for the State Key Laboratory of Coal Resources and Mine Safety of CUMT(NO. 08KF05).

## References

Li W, Luo N (2006) Based on circular economy of coal industry chain construct [J]. Coal Econ Res (11) (in Chinese)
Lu T, Zhao Z (2005) Guard against the coal mining industry chain extension risk [J]. China Coal 5:24–25, (in Chinese)
Meng Y, Yu J (2008) Venous industry circular economy and energy conservation [J]. Dongbei Univ Finance (4) (in Chinese)
Niu W (2004) Circular economy: achieving sustainable development of the ideal economic model [J]. Chinese Acad Sci (6) (in Chinese)
Qi Z et al (2009) Research summary on symbiosis and coupling analysis of eco-industrial chains based on circular economy [J]. Ecol Econ (2) (in Chinese)
Ren Y, Zhou G et al (2009) Circular economy development in China: approaches and policies [M]. China Environmental Science Press, Beijing (in Chinese)
Zhou H et al (2005) The circular economy [M]. China Development Press, Beijing (in Chinese)
Zhu S, Zhang L (2007) Discussion of venous industry development pattern [J]. Ecol Econ (11) (in Chinese)

# Information Entropy of a Rainfall Network in China

**Jiguo Zhang, Huimin Wang, and Vijay P. Singh**

**Abstract** In this paper, we use the directional information transmission index which is based on information entropy to investigate the spatial nonuniformity of rainfall in the Huaihe River basin in China, and propose methods of information transfer function, information distance and information area which are employed to determine the area where a rain gauge is capable of representing. The analysis also involves transformation of the rainfall field from discrete state to continuous state.

**Keywords** Distribution nonuniformity · Information entropy · Information field · Information transmission · Rainfall network · Uncertainty

## 1 Introduction

Entropy, as a measure of information and uncertainty, has been employed for evaluation and design of rainfall network as well as for expansion of networks. Caselton and Husain (1980) used the maximum information transmission to select

J. Zhang (✉)
Business School, Hohai University at Changzhou, Changzhou 213022, People's Republic of China
and
State Key Laboratory of Hydrology-Water Resources and Hydraulic Engineering, Hohai University, Nanjing 210098, People's Republic of China
e-mail: zhangjg@hhuc.edu.cn

H. Wang
State Key Laboratory of Hydrology-Water Resources and Hydraulic Engineering, Hohai University, Nanjing 210098, People's Republic of China

V.P. Singh
Department of Biological and Agricultural Engineering, Texas A&M University, College Station, TX 77843, USA

Y. Zhou and D.D. Wu (eds.), *Modeling Risk Management for Resources and Environment in China*, Computational Risk Management,
DOI 10.1007/978-3-642-18387-4_2, 

stations in a hydromeric network. Husain (1989) applied an entropy-based methodology to (a) selecting the optimum number of stations from a dense network, and (b) expanding a network using data from an existing sparse network by interpolation of information and identification of zones with minimum hydrological information. Harmancioglu and Yevjeich (1987) used entropy to measure the transmission of information among gauging stations on a river. Krstanovic and Singh (1992a, b) developed an entropy-based approach for space and time evaluation of rainfall networks in Louisian. Employing a measure of information flow, called directional information transfer index (DITI), between gauging stations in a network, Yang and Burn (1994) presented an entropy-based methodology for design of data collection networks. DITI is both a measure of the information transmission capacity and an indicator of the dependence of station pairs. The concept of DITI can be applied to regionalization of networks.

Mutual information and DITI can be applied to describe the degree of relationship between stations in a rainfall network. DITI between station pairs and area pairs can be used to show the spatial distribution of rainfall in a river basin from one direction to the other. Similarly, a graph of DITI and distance can show the difference in information transmission from one direction to the other, e.g. from east (E) to west (W) or from south (S) to north (N).

Huaihe River basin is located in east of China and between Yangtse River basin and Yellow River basin from which the 99 gauging stations, lying between N 31°–35° and E 112°–118°, upper Benbu Station (No. 86, E 117°23′, N 32°56′), as shown in Fig. 1, are selected by this paper. The objective of this paper is to investigate the spatial distribution nonuniform of rainfall in the area.

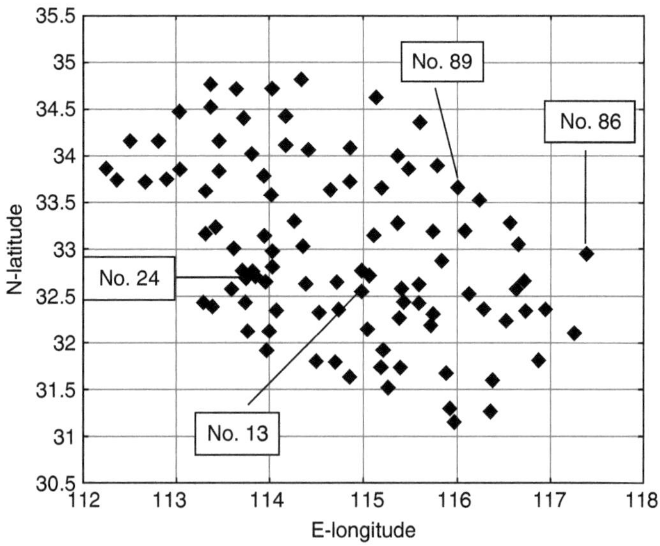


**Fig. 1** Sketch map on 99 stations and 4 special stations

## 2 Preliminary Concepts

The Shannon information entropy is defined as:

$$H(X) = H(p_i, p_2 \cdots, p_n) = -\sum_{i=1}^{n} p_i \log p_i. \tag{1}$$

where $X$ is a discrete random variable, $p_i$ is the probability that $X$ assumes a value $X = x_i$, and $n$ is the number of values (sample size) that $X$ can take on. Equation (1) defines entropy in real time or space.

To measure the transmission of information and indicate the dependence between two stations $X$ and $Y$, a directional information transfer index (*DITI*) is defined as:

$$DITI_{XY} = I(X,Y) \Big/ H(Y) = \frac{H(Y) - H(Y/X)}{H(Y)}. \tag{2}$$

where $DITI_{XY}$ is *DITI* of $X$ about station $Y$, $H(Y/X)$ is the entropy of $Y$ conditioned on $X$, and $I(X,Y)$ is the mutual information defined as:

$$I(X,Y) = H(Y) - H(Y/X) = H(X) - H(X/Y) = I(Y,X). \tag{3}$$

Equation (3) is a measure of information transmission. In $DITI_{XY}$, $X$ is called the basic point (basic station) and $Y$ the auxillary point (auxillary station). $DITI_{XY}$ varies from 0 to 1; a large value indicates large capability of information transmission from $X$ to $Y$. The concept of *DITI* can be extended to any number of stations and measures inherent dependence among stations of a set as:

$$SDITI = \sum_{i=1}^{m} \sum_{j=1, j \neq i}^{m} DITI_{ij}. \tag{4}$$

where $m$ is the number of stations in the set.

## 3 Data Analysis

Using (1), which eliminates random uncertainty of the annual rainfall series and shows spatial characteristics of precipitation distribution, the entropy values of 99 stations were calculated. Using (3), the mutual information values between station pairs were obtained.

Generally, the majority of Huaihe River basin is influenced by plum rains from July to September because of the southeast monsoon. The plum rains starts in Yangtse River basin at first, and then Huaihe River basin, finally moves to Yellow

River basin. For the sake of observing this phenomenon and demonstrating that *DITI* is a correlative index between station pairs, the area mentioned was partitioned according to latitude as well as longitude.

It is obvious that *DITI* was associated with distance. Thus, when information transmission is combined, the distance must be accounted for at the same time. Two stations, No. 24 and No. 89 (see Fig. 1), are selected to show the information transfer graph with distance. There are two reasons for choosing these stations. One is that these two stations are much different in location, weather and topography, for station No. 24 is located in the mountainous area and station No. 89 in the northeast plain area. The other is that they are located in the border of the region. Therefore, if a direction is extended from the basic station along E-longitude or N-latitude, then more auxiliary stations can be located in order to analyze the relationship between *DITI* and distance.

Now let stations No. 24 and No. 89 (see Fig. 1) be basic stations, some auxiliary station points are chosen separately.

①Station No. 24 as a basic station, 7 auxiliary stations to be chosen from S to N
②Station No. 24 as a basic station, 8 auxiliary stations to be chosen from W to E
③Station No. 89 as a basic station, 7 auxiliary stations to be chosen from E to W
④Station No. 89 as a basic station, 7 auxiliary stations to be chosen from N to S

The relation between DITIs and distances is illustrated separately in Figs. 2 and 3.

These figures display the following properties: (1) *DITI* decreases with increasing distance; (2) *DITI* decreases faster in the S-N direction than in the E-W direction. In other words, the information transmission is greater in the E-W direction than in the S-N direction; and (3) the longer the distance is, the smaller the *DITI* is. This shows that the rainfall field of the Huaihe River basin spreads out

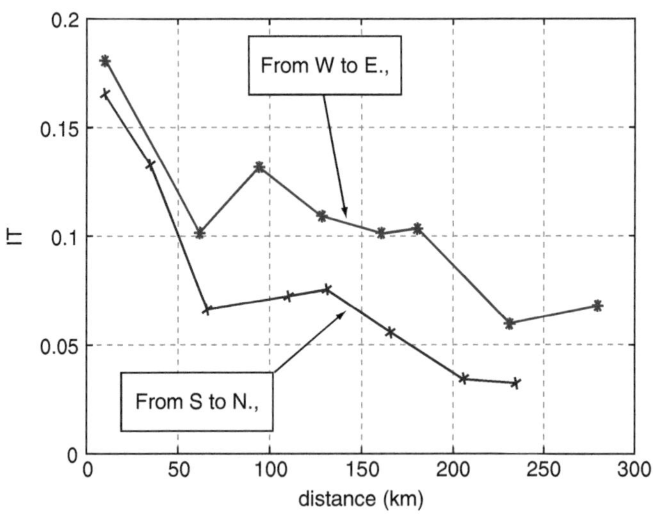


**Fig. 2** Information transfer graph of basic station No. 24

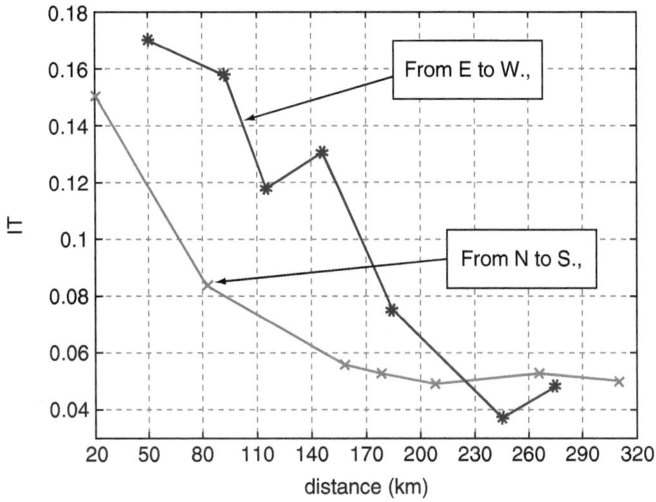


**Fig. 3** Information transfer graph of basic station No. 89

in the E-W direction and moves from south to north. Thus, *DITI* reflects the relationship of a gauging station to its nearby region, and its value reduces with increasing distance. Based on a large number of stations studied, it was observed that this characteristic not only exists for each station in the Huaihe River basin, but also the functional curves have the same properties. This phenomenon will be further researched in the following section.

## 4 Theoretical Models

The use of entropy requires defining the spatial structure of rainfall by a multivariate probability distribution. In general, a continuous distribution function is assumed (Husain 1989) or a non-parametric distribution is applied (Yang and Burn 1994). This study extends the discrete field to the continuous field. In this case, each point in the area is a gauging station. In the continuous field, an information transfer function model is derived, and two information indexes, information distance and information field which measure the capability of a station with respect to its nearby regions, are defined. Then, an algorithm in the frequency domain is employed following the principle of maximum entropy.

### *4.1 Information Transfer Function*

The rainfall field is characterized by a finite number of stations and is called a discrete field. It may, however, be more convenient to extend the discrete field to the continuous field for developing a theoretical model for information transmission. In this case, every point in an area on a spatial coordinate surface is considered a rainfall gauging station. Therefore, the transmission of information of a given station successively changes and be in progress in any of the directions.

Let $X$ be a basic point in the rainfall field and $Y$ an auxillary station being in a certain direction. Let $s_{XY}$ be the geometric distance between $X$ and $Y$. Then the information transfer function (*ITF*) of $X$ about $Y$ can be defined as:

$$ITF_X = \frac{1}{(1+as)^b}. \tag{5}$$

where $a, b > 0$ are parameters.

When $a, b$ are smaller in (5), $ITF_X$ is greater, suggesting greater information from $X$ in a given direction. Although different river basins have probably different information transmission function, two general properties can be easily summarized:

1. $ITF_X(0) = 1 \qquad ITF_X \to 0(s \to \infty)$
2. $ITF_X$ is a differential function, and $ITI = \frac{d(ITF_X)}{ds} \le 0$

where *ITI* is the abbreviation for information transfer intensity.

Thus, using a nonlinear regression analysis, the simulated curves representing 4 broken lines in Figs. 2 and 3 can be expressed analytically as:

$$\text{Station No.24} \quad \text{from W to E } ITF = 1/(1+2.784\text{s})^{0.471} \quad \text{R} = 0.80. \tag{6}$$

$$\text{from S to N } ITF = 1/(1+0.126\text{s})^{1.289} \quad \text{R} = 0.94. \tag{7}$$

$$\text{Station No.89} \quad \text{from E to W } ITF = 1/(1+0.092\text{s})^{1.074} \quad \text{R} = 0.94. \tag{8}$$

$$\text{from N to S } ITF = 1/(1+1.181\text{s})^{0.633} \quad \text{R} = 0.98. \tag{9}$$

Equations (6), (7), (8) and (9) are called information transfer functions which are represented with curves as in Figs. 4 and 5 (*IT* is *ITF* in following figures).

### 4.2 Information Distance

Let $X$ be a station and $N$ the length from $X$ to the border of a basin along a given direction, then the information distance ($ID_X$) of $X$ in this direction can be defined as:

$$ID_X = \left(2\int_0^N sITF_X ds\right)^{\frac{1}{2}} = \frac{1}{a^2}\left[\frac{(1+aN)^{2-b}-1}{2-b} - \frac{(1+aN)^{1-b}-1}{1-b}\right]. \tag{10}$$

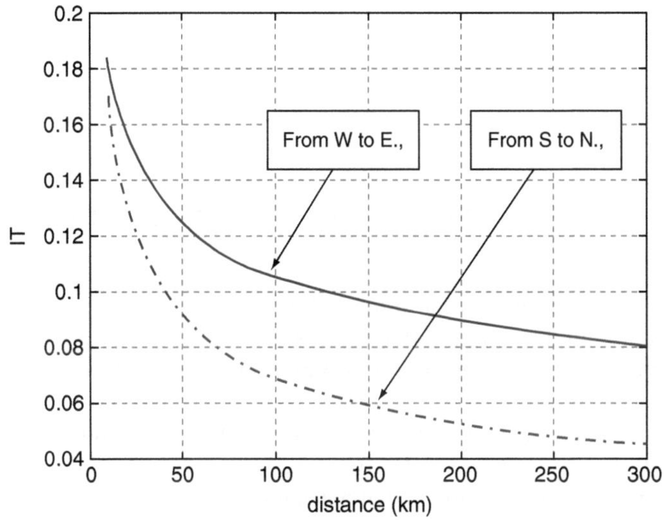


**Fig. 4** Sketch map of ITF on No. 24 in W-E and S-N

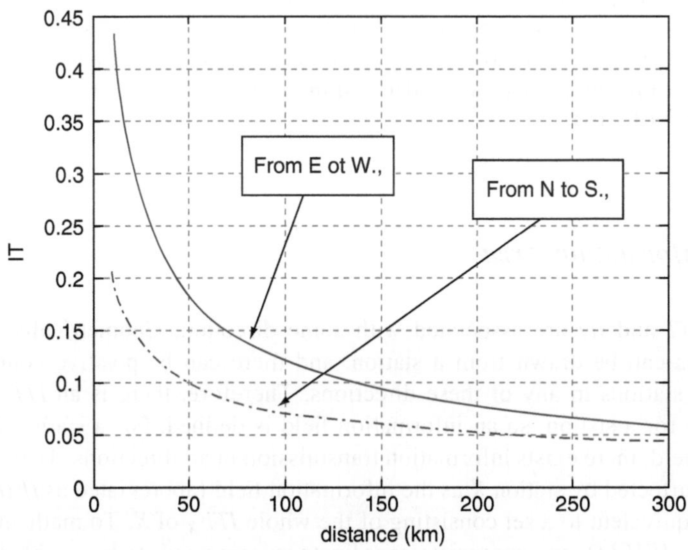


**Fig. 5** Sketch map of ITF on No. 89 in E-W and N-S

In light of the principle of fuzzy mathematics, $ITF_X$ may be considered a possibility distribution, that means the membership function of station $Y$ belongs to the basic station $X$. Therefore, $ID_X$ measures the average information transfer of $X$ in the above-mentioned direction in a statistical sense. For example, if $ITF_X = 1$

**Table 1** ID's of stations No. 24 and 89 with distances (km)

| Dis. | W→E (No. 24) | E→W (No. 89) | S→N (No. 24) | N→S (No. 89) |
|---|---|---|---|---|
| 50 | 8.47 | 16.84 | 7.47 | 8.83 |
| 100 | 14.16 | 24.23 | 11.38 | 13.13 |
| 150 | 19.16 | 29.60 | 14.57 | 16.52 |
| 200 | 23.77 | 33.96 | 17.35 | 19.44 |
| 250 | 28.11 | 37.68 | 19.87 | 22.04 |
| 300 | 32.24 | 40.97 | 22.20 | 24.41 |

then $ID_X = \mathrm{N}$, which means the information may be entirely transferred to the basin boundary; on the contrary, if, except $\mathrm{N} = 0$, $ITF_X = 0$, then $ID_X = 0$, in this case, station *X* does not transfer any information outward. Thus, $ID_X$ defined is reasonable.

Consider the information distance of stations No. 3 and 89. The geometric distances of station 3 to the border from west to east and from south to north separately are 325 and 270 km. And about station 89, the distances separately are 305 km (from E to W) and 335 km (from N to S). Then, the *ID*'s of stations No. 24 and 89 with distances separately were graphed in Table 1.

From Table 1, *ID* in S-N is even larger than that in E-W within about 65 km. Thus, on average information transfer capacity, there is no obvious difference between E-W and S-N no more than 70 km. But it is also seen that increasing velocity of *ID* in S-N is more slowly than that in E-W. It is shown that time and intensity of precipitation are very identical in the zone of N 32°–33°, but not in the zone of E 113°–114°.

## 4.3 Information Area

Both *DITI* and *ID* are associated with some direction. Geometrically, infinite directions can be drawn from a station, and there can be positive connections between stations in any of these directions. Therefore, there is an *ITF* in each direction for a ststion, so an information field is defined. For a station *X* in the rainfall field, there exists information transmission in all directions. Then, we call the area affected by station *X* as the information field (abbreviated as *IFIELD*) of *X*. It is equivalent to a set consisting of the whole $ITF_X$ of *X*. To mathematically formulate *IFIELD*, an appropriate coordinate system needs to be established. Let the geometric location of *X* be a pole and one direction (e.g., selecting the direction from east to west) be a polar axis; thus, polar coordinates are obtained. Let $\theta$ be the polar angle. Then, one can consider parameters *a* and *b* in ITF relating to $\theta$, that is, $a = a(\theta)$, $b = b(\theta)$, and $ITF = ITF(\theta)$. Therefore, information field of *X* is defined as:

$$IFIELD_X = \{ITF_X(\theta),\ 0 \le \theta \le 2\pi\}. \tag{11}$$

Suppose that *ITF* is a continuous function of s and θ. Then, we designate:

$$IA_X = \frac{1}{2}\int_0^{2\pi} (ID_x)^2 d\theta. \tag{12}$$

as the information area (*IA*) of *X*, which measures the total region of influence of *X* with respect to its nearby region.

In practice, all *ITFs* of *X* can not be obtained due to the limitation on the number of stations. One can obtain only finite *ITFs*. For illustration, station No. 13 (E 114°59′ N 32°33′) (see Fig. 1) is considered, because it is located almost in center of the area. Beginning with $\theta = 0$, and at a set interval of $\pi / 4$, 8 *ITFs* are computed and given as:

$$\mathrm{ITF} = 1/(1+0.578\mathrm{s})^{0.542} \quad \mathrm{ITF} = 1/(1+1016.441\mathrm{s})^{0.212}$$

$$\mathrm{ITF} = 1/(1+2.501\mathrm{s})^{0.500} \quad \mathrm{ITF} = 1/(1+0.874\mathrm{s})^{0.593}$$

$$\mathrm{ITF} = 1/(1+6.454\mathrm{s})^{0.407} \quad \mathrm{ITF} = 1/(1+0.914)^{0.532}$$

$$\mathrm{ITF} = 1/(1+0.052)^{1.602} \quad \mathrm{ITF} = 1/(1+0.477\mathrm{s})^{0.689}$$

There are 8 distance values from station No. 13 (see Fig. 1) to border of the area along 8 different directions. The minimum of them is 145 (km) which is considered as a integral upper limit while computing *ID*. It is seen that *IA* is 5761 ($km^2$) if distance is 90 (km), and *IA* is 13256 ($km^2$) if distance is 145 (km).

## 5 Conclusion

The following conclusions are drawn from this study: (1) The Information transfer graph shows a decline in Information of a station in the east-west as well as in the south-north direction; (2) The cluster graphs show spatial nonuniformity of the rainfall field; (3) In section of Data Analysis, the values of SDITI of station-sets separately from N 32°–33° and E 115°–116° are relatively larger, the reason is, the most of stations chosen come from the region where the distribution of precipitation is uniformity, thus, relative *DITI* values are larger than others; (4) *ID* is an index used to measure average information transmission capacity of a station in the given direction; and *IA* is an influence indicator of a station to its nearby region in information.

**Acknowledgments** This work is supported by the Public-interest Industry Project of Ministry of Water Resources of China (No. 200801027).

## References

Caselton WF, Husain T (1980) Hydrologic network information transmission. J Water Resour Plan Manag 106:503–520

Harmancioglu N, Yevjevich V (1987) Transfer hydrologic information among river points. J Hydrol 191:103–118

Husain T (1989) Hydrologic uncertainty measure and network design. Water Resour Bull 25(3):527–534

Krstanovic PF, Singh VP (1992a) Evaluation of rainfall network using entropy: I. theoretical development. Water Resour Manag 6:279–293

Krstanovic PF, Singh VP (1992b) Evaluation of rainfall network using entropy: II. application. Water Resour Manag 6:295–314

Yang Y, Burn DH (1994) An entropy approach to data collection network design. J Hydrol 157:307–324

# Research on Industry-Related and Policy Risk in Provincial Level Based on SAM: Case Study for Yunnan

**Lu Qicheng and Li Yijia**

**Abstract** In this paper, structural path analysis in SAM has been explained and also its application in provincial region in China has been discussed. By setting up several policies simulation scenarios, this method has been used to study industry-related and policy risk in Yunnan province base on detailed SAM of Yunnan province. It concludes that it's a useful way to examine how an endogenous accounting will be affected by exogenous variables and to reveal its uncertainty and risk of industry policy.

**Keywords** Industry-related · Policy risk · SAM · Structural path analysis

## 1 Introduction

The Social Accounting Matrix (SAM) models are extensions of the input-output model that account for income generation from non-industrial demand source. The SAM approach to modeling is a very flexible one and is a basic element in the tool kit of the general equilibrium economist. SAMs have been used to study: (1) growth strategies in developing economies, (2) income distribution and redistribution, (3) fiscal policy in national or regional settings, and (4) decomposition of activity multipliers. In SAM multiplier analysis, it distinguishes the national

L. Qicheng (✉)
Faculty of Management and Economics, Kunming University of Science and Technology, 650031 Kunming, People's Republic of China
and
School of Business Administration, Yunnan University of Finance and Economics, 650221 Kunming, People's Republic of China
e-mail: Qicheng_lu@yahoo.com.cn

L. Yijia
Faculty of Management and Economics, Kunming University of Science and Technology, 650031 Kunming, People's Republic of China
e-mail: tilburg@sohu.com

Y. Zhou and D.D. Wu (eds.), *Modeling Risk Management for Resources and Environment in China*, Computational Risk Management,
DOI 10.1007/978-3-642-18387-4_3, 

accounting into endogenous accounting and exogenous accounting, and reveals related affects for the whole economic systems via researching certain endogenous accounting impacted by additional exogenous variables. Within a general equilibrium frameworks not sub-equilibrium framework, multiplier decomposition shows internal and external relationships among three types endogenous accounting through multiplier effects analysis. However, in view of policy-making, results from multiplier analysis are limited in practice. Meanwhile, it is a black box to decision-maker how additional endogenous variables play its role. To utilize structural path analysis, the black box will be opened, and we can find how additional endogenous variables affect specific accounting along its path from its starting point to its different ending points. Thus, decision-maker will know all kinds of multiplier affects but also clarify the affecting paths. However, in current research, it hardly is found that the studies on industry-related association in provincial level based on SAM structural path analysis.

In our paper, we generally discuss SAM structural path analysis principles, and apply it in provincial level industry-related association characteristics study. Using data from Yunnan provincial SAM in 2002, we examine Yunnan provincial industry-related association and its policy risk in empirical ways.

## 2 Multiplier Effects via Social Accounting Matrix (SAM)

A standard SAM gives a panoramic view of social-economic development in certain period. It is not only to reflect direct relationships among main parts of economy and other institutions but also to reveal indirect relationships among them clearly. Table 1 is a partially simplified macro SAM including three endogenous accountings: production activities, factors and institutions (firms and residents involved) and one exogenous accounting (government, capital and foreign accountings involved). It meets all requirements for SAM:

- The column represents outlay and the row represents income.
- Whole income will be equal to whole outlay. In other words, sum of columns equals to sum of rows in the matrix.

On the field of endogenous accounting which presented as a 3×3 block matrix, $T_{11}$ represents mid-outlay needs for producing activities; in fact it is the flow part in

**Table 1** Simplified macro SAM schematic diagram

| Income outlay | I | II | III | IV | V |
|---|---|---|---|---|---|
| Endogenous accounting | | | | | |
| I Production activities | $T_{11}$ | O | $T_{13}$ | $X_1$ | $Y_1$ |
| II Factors | $T_{21}$ | O | O | $X_2$ | $Y_2$ |
| III Institutions | O | $T_{32}$ | $T_{33}$ | X3 | $Y_3$ |
| IV Exogenous accounting | $L_1$ | $L_2$ | $L_3$ | $LX$ | $Y_4$ |
| V Total | $Y_1$ | $Y_2$ | $Y_3$ | $Y_4$ | |

input-output table. $T_{13}$ describes output model of residents and firms which are two accountings from institutions. $T_{21}$ presents added-value allocation among factors which created by production activities; in other words, it shows factors' income from productions activities. $T_{32}$ presents factors' income allocation among residents and firms. $T_{33}$ shows internal income transfer payment among residents and firms.

Generally in SAM, we assume price as constant, outcome directed by needing, all economic entities with linear relationship. In SAM, we define expenditure propensity matrix as $A_n$, in which matrix elements come from elements of endogenous accounting divided by total of each column. According to balance relation of rows and columns in SAM, total income of endogenous accounting ($y_n$) can be presented as:

$$y_n = A_n y_n + x_n \tag{1}$$

With further transformation, we can get:

$$y_n = (I - A_n)^{-1} x_n = M_a x_n \tag{2}$$

We call $(I - A_n)^{-1}$ as accounting multiplier matrix which represented with $M_a$. It can combine endogenous income ($y_n$) with exogenous variables ($x_n$).

## 3 SAM-Based Structural Path Analysis

Under structural path analysis framework, we define the direct influence of accounting $i$ on accounting $j$ as the element $a_{ji}$ in expenditure propensity matrix $A_n$.

Thus,

$$I^D_{(i \to j)} = a_{ji} \tag{3}$$

When $i$ and $j$ as two destination of one path linked with many accounting points, multiplier influence can be defined as the multiplication of each influence value in that path.

Thus

$$I^D_{(i \ldots j)} = a_{jn} \cdots a_{mi} \tag{4}$$

In complicated social & economic systems, there are not only an elementary path linked with two points but also many causal feedback loops. In order to clarify all influence covered those paths and loops; we define the concept of total influence as sum of direct influence yielded by elementary path and indirect influence yielded by causal feedback loops. It can be expressed as multiplication of direct influence and path multiplier.

Thus

$$I^{T}_{(i\to j)p} = I^{D}_{(i\to j)p} M_p \tag{5}$$

Within (5), $M_p$ is division of two determinates $\Delta_p/\Delta$. $\Delta$ is $|I - A_n|$ and $\Delta_p$ is sub-determinates of $|I - A_n|$ expected all points on the elementary path.

Path multiplier depends on path length and feedback strength. If more points on elementary path, more feedback loops will be found and path multiplier will be larger. To transform (5) yield:

$$1\Big/M_p = I^{D}_{(i\to j)p}\Big/I^{T}_{(i\to j)p} \tag{6}$$

Path multiplier reciprocal shows direct influence proportion in complete influence in an elementary path. Also, it shows exogenous variables conducting time in a degree. With more $1/M_p$, direct influence proportion will be bigger and exogenous variables influence will conduct faster. On the contrary, with larger path multiplier, direct influence will be amplified much and influence conduction will be slowly. So, to decision maker, it is an important reference when facing a dilemma between exogenous variables influence strength and it's conducting time.

In complicated economic system, because of countless paths between two destination of elementary path, it need to consider global influence instead of a specific path influence along in order to examine effects caused by all paths (elementary path and feedback loop involved). Owning to complete influence abstracted from each path influence, it can be expressed with elements in accounting multiplier matrix $M_a$. The element from ***i***th row and ***j***th column in $M_a$ shows exogenous variable $x_i$ influence on endogenous variable $y_j$.

Thus

$$I^{G}_{(i\to j)} = M_{aji} \tag{7}$$

In the light of global influence composed of each complete influence from all points linked with fundamental path destination, assume that there are ***p***th paths between destination ***i***th and ***j***th, and three type influences can be expressed as:

$$I^{G}_{(i\to j)} = M_{aji} = \sum_{p=1}^{n} I^{T}_{(i\to j)p} = \sum_{p=1}^{n} I^{D}_{(i\to j)p} M_p \tag{8}$$

Compared with direct influence, global influence summarizes effects from all paths linked with two ends (elementary path and feedback loop involved). Due to countless elementary paths in a complicated economic system, to find out those paths with comparatively large global influence and examine them, it will help us to understand economic characteristic in a region. Meanwhile, it will help us to comprehend exogenous variables conducting mechanism and to grasp critical policy fields in course of policy implementation.

# 4 Empirical Study Based on Yunnan SAM

In our empirical study, we choose detailed SAM of Yunnan province in 2002 as data resource. It includes 42 industry sections, 5 type production factors, 2 type residents, 1 firm accounting and macro accounting (government subsidies, production tax, import-export, and capital changes involved). In detail, producing activities comprise with 1 agriculture section, 24 industry sections, 1 construction industry sections and 16 service sections. Labours can be divided into rural labour, workers and technicians. Residents can be divided into rural and non-rural household. In light of production tax, two major taxes: value added tax and business tax can be considered and other taxes can be considered as other types. We construct 1 exogenous accounting including government, tax, import-export and capital changes. Therefore, we yield a $51 \times 51$ SAM constituted with 3 endogenous accountings: production activities (42 industry section involved), factors (5 type factors involved) and institution (two kinds of residents and one kinds of firm involved), and one exogenous accounting.

In Table 2, we assume six kinds of scenario to examine internal and mutual path influence and its characteristics among production activities accounting, factors accounting and resident's accountings. Also, to examine the capability of path conducting, we calculate proportion of total influence for each path in global influence.

For the purpose of examine different industry sections affecting mutually, we choose restaurant section and agriculture in scenario I and construction industry and metal mining industry in scenario II. In view of the restaurant section affecting agriculture, the major influence path depends on "Restaurant →agriculture" which is the shortest and most direct path. Within this path, the proportion of global influence achieves 47.22%. Taking food and tobacco manufacture section as middle section into account, the proportion of global influence achieves 11.42% along the path "Restaurant →food and tobacco manufacture → agriculture". Surprisingly, if tourism section considered, the proportion is zero. In scenario II, it yields contrary characteristics. In the direct path "construction industry → metal mining industry", the proportion of global influence is zero. The path with largest proportion is "construction industry →metal extension and smelting industry →metal mining industry" in which the proportion of global influence is 67.4%. In other path, within two industries as middle section, "construction industry → metalwork industry →metal extension and smelting industry →metal mining industry" yields 10.31% in the proportion and reflects significantly conducting capability. This shows that path with stronger conducting capability will not always be the direct path and path with longer length will not mean weaker conducting capability. The strength of conducting capability reveals correlation among those industry sections linked by the path.

We consider relation between producing activities accounting and factors accounting from scenario III to scenario V. In scenario III, to investigate how increasing output from production section affects labor, we examine which kind

**Table 2** Empirical study of structural path decomposition by SAM

| (0) | (1) Start point $(i) \rightarrow$ | (2) End point $(j)^+_+$ | (3) Global influence $I^G_{(i \rightarrow j)} = m_{aji}$ | (4) Elementary path $(i \rightarrow j)^+_{p^+}$ | (5) Direct influence $I^D_{(i \rightarrow j)p}$ | (6) Path multiplier $\times M_p =$ | (7) Total influence $I^T_{(i \rightarrow j)p}$ | (8) $\frac{I^T_{(i \rightarrow j)p}}{I^G_{(i \rightarrow j)}}$ (%) |
|---|---|---|---|---|---|---|---|---|
| I | Restaurant (Rst) | Agriculture (Agr) | 0.3303 | Rst→Agr | 0.0971 | 1.6068 | 0.1559 | 47.22 |
| | | | | Rst→Manu→Agr | 0.0214 | 1.7637 | 0.0377 | 11.42 |
| | | | | Rst→Tour→Manu→Agr | 0.0000 | 1.8641 | 0.0000 | 0 |
| II | Construction (Con) | Metal mining (Mine) | 0.0463 | Con→Mine | 0.0000 | 1.1558 | 0.0000 | 0 |
| | | | | Con→Metsp→Mine | 0.0234 | 1.3375 | 0.0312 | 67.40 |
| | | | | Con→Metp→Metsp→Mine | 0.0035 | 1.3688 | 0.0048 | 10.31 |
| III | Metalworking industry (Metp) | Production worker (Pw) | 0.1064 | Metp→Pw | 0.0312 | 1.1197 | 0.0349 | 32.78 |
| | | | | Metp→Metsp→Pw | 0.0090 | 1.3144 | 0.0118 | 11.06 |
| | | | | Metp→Power→Pw | 0.0011 | 1.4269 | 0.0016 | 1.53 |
| | | | | Metp→Metsp→Mine→Pw | 0.0047 | 1.4799 | 0.0069 | 6.53 |
| IV | Real estate (Est) | Capital (Cap) | 0.8731 | Est→Cap | 0.7144 | 1.1355 | 0.8111 | 92.91 |
| | | | | Est→Fin→Cap | 0.0096 | 1.1564 | 0.0111 | 1.27 |
| | | | | Est→Nmet→Cap | 0.0014 | 1.2296 | 0.0017 | 0.20 |
| V | Metal mining (Mine) | Production worker (Pw) | 0.2532 | Mine→Pw | 0.1066 | 1.2526 | 0.1335 | 52.75 |
| | Nonmetal (Nmet) | | 0.1832 | Nmet→Mine→Pw | 0.0006 | 1.3531 | 0.0009 | 0.47 |
| | Metal extension (Metsp) | | 0.1656 | Metsp→Mine→Pw | 0.0196 | 1.4470 | 0.0284 | 17.17 |
| | Metalworking (Metp) | | 0.1064 | Metp→Mine→Pw | 0.0001 | 1.2838 | 0.0001 | 0.08 |
| VI | Food and tobacco manufacturing (Manu) | Rural household (RHou) | 0.2414 | Manu→Pw→Rhou | 0.0238 | 1.5734 | 0.0375 | 15.54 |
| | | Urban household (UHou) | 0.1764 | Manu→Pw→Uhou | 0.0264 | 1.3190 | 0.0349 | 19.77 |
| VII | Transportation and warehouse industry (Ware) | Rural household (RHou) | 0.2475 | Ware→Pw→Rhou | 0.0934 | 1.5728 | 0.1469 | 59.35 |
| | | Urban household (UHou) | 0.3053 | Ware→Pw→Uhou | 0.1036 | 1.2954 | 0.1342 | 43.94 |
| VII | Rural household (RHou) | agriculture (Agr) | 0.6150 | Rhou→Agr | 0.3281 | 1.6447 | 0.5396 | 87.74 |
| | | | | Rhou→Manu→Agr | 0.0217 | 1.7861 | 0.0387 | 6.29 |
| | | | | Rhou→Saw→Agr | 0.0037 | 1.8036 | 0.0067 | 1.09 |
| IX | Urban household (UHou) | agriculture (Agr) | 0.1797 | Uhou→Agr | 0.0582 | 1.7445 | 0.1015 | 56.52 |
| | | | | Uhou→Manu→Agr | 0.0147 | 1.8876 | 0.0277 | 15.44 |
| | | | | Uhou→Saw→Agr | 0.0008 | 1.9151 | 0.0016 | 0.88 |

of jobs will be affected by the growing needs for metalwork. According to our calculating, jobs for metalwork will be benefit from it mostly with the direct path "metalwork industry → worker". The proportion of global influence is 32.78%. In scenario IV, to investigate how real estate industry induce needs for capital, we compare different paths and find out that exogenous variables affecting business is mainly through face to face conducting for example "real estate → capital" and conducting capability is limited in the path with middle section. In scenario V, choosing worker from metal mining industry as objects and assuming different starting points to investigate which industry will increase jobs for metal mining with its exogenous needs increased, we yield result that increasing needs for metal mining will create more job for this industry with the conducting path "metal mining industry → worker" and its proportion of global influence is 52.75%. In addition to this, workers will be benefit from increasing needs for related industry of metal mining such as metal extension and melting industry.

In scenario VI and VII, we investigate correlation among food and tobacco manufacture industry, transportation and warehouse industry and resident accounting. Due to resident's income from factors, there are no flows in the column of resident accounting and the row of production activities accounting. Thus, no direct influence exists among them. Considering of this, we use worker as middle section to examine how production activities affect residents. To rural residents, food and tobacco manufacture industry affects them more than urban ones. On the contrary, to urban residents, transportation & warehouse industry and tourism industry affects them more than rural ones. Through the influence path and comparing different industries, we find out that transportation & warehouse industry depends more on direct path than food & tobacco industry or tourism industry. Two of those industries affect resident's income via worker with more general ways.

In scenario VIII and IX, we investigate how residents affect producing activities accounting. Assuming that government gives transfer payment to rural and urban residents respectively, how will agriculture be benefit from this payment? It shows that the proportion of global influence in rural household is 61.50% much larger than urban household's 17.97%. Meanwhile, through the direct path "rural resident → agriculture", increasing income for rural household affects agriculture much with a proportion 87.74%. However, in same direct path, increasing income for urban household affects agriculture with a proportion 56.52%. It is because of more influence scattered into other paths.

## 5 Conclusions

Accountings in SAM cover all sections in national economy. With adding income re-allocation accounting in SAM, it reflects comprehensive social economic accounts. By structural path analysis in SAM, it reveals exogenous variables affecting ways and mechanism and clarifies the path and degree that exogenous variables influencing other economic entities. These results will give useful decision references to policy maker. In further studies, with the 2007 Input-output table of national and Yunnan published, a new SAM will be established; we can get more accurate information from it by structural path analysis in SAM.

# Pro-poor Growth Measurement: A Survey

**Shengyun Wang and Shujuan Yu**

**Abstract** Based on the perspective of change in the growth rate of income or poverty index, seven types of measurement methods in pro-poor growth are analyzed. The multi-dimensional perspective of pro-poor growth can still go further, and could study the growth and distributional effects from a multi-dimensional perspective. In addition, dealing with the non-anonymous nature of pro-poor growth is a serious problem.

**Keywords** Inequality · Poverty · Pro-poor growth · Sensitivity

## 1 Introduction

There are two views on the relationship between economic growth and poverty. One is that economic growth helps reduce poverty. Economic growth can realize a trickle-down effect to the poor through capital accumulation. Despite the absence of other poverty reduction measures, economic growth can increase employment opportunities and benefit the poor by increasing financial income through transfer payments. Another idea is that economic growth cannot reduce poverty. The poor do not spontaneously benefit through economic growth in low-income countries. These views focus on whether economic growth is pro-rich or pro-poor. The concept and measurement of pro-poor growth vary in many studies. According to concepts of pro-poor growth, this article puts forward seven measurement methods, namely, pro-poor growth rate (PPGR), poverty growth curve (PGC), and growth

S. Wang (✉)
China Academy for Rural Development, Zhejiang University, Hangzhou 310029, China
and
Department of Management, Zhejiang Water Conservancy and Hydropower College, Hangzhou 310018, China

S. Yu
Department of Management, Zhejiang Water Conservancy and Hydropower College, Hangzhou 310018, China

Y. Zhou and D.D. Wu (eds.), *Modeling Risk Management for Resources and Environment in China*, Computational Risk Management,
DOI 10.1007/978-3-642-18387-4_4, 

elasticity of poverty (GEP), pro-poor growth index (PPGI), poverty-equivalent growth rate (PEGR), inequality growth trade-off index (IGTI), and improved IGTI (IIGTI).

## 2 Concept of Pro-poor Growth

The concept of pro-poor growth has two meanings. The first is that economic growth is pro-poor (Kakwani and Pernia 2000), with poor income growth faster than rich income growth. The second is if poverty index reduces, then it is pro-poor growth (Kraay 2004). For example, if the poverty rate decreases, then economic growth is pro-poor. (Duclos and Wodon 2004) defined pro-poor growth as either relative pro-poor growth or absolute pro-poor growth. Economic growth is relative pro-poor if the income growth rate of the poor is higher than the growth rate of per capita income. This is because relative pro-poor growth will reduce poverty more quickly if average income growth rate remains unchanged (Ravallion and Datt 1999), and declining inequality has intrinsic value for the poor. The welfare of the poor depends not only on their income level, but also on the income gap between the rich and the poor (Klasen 2008).

Absolute pro-poor growth focuses not just on the decline in relative inequality, but also on the decline in absolute inequality. There are two meanings of absolute pro-poor growth. The first is strong absolute pro-poor growth. In this case, income growth among the poor is greater than that among the rich. Kakwani and Son (2003) also called this super pro-poor economic growth. (Kakwani and Son 2003). Despite the difficulty of obtaining this type of economic growth, many scholars still tend to hold this view. They argue that if we pay too much attention to relative income inequality, it is easy to overlook that although relative inequality is decreasing, absolute inequality is still expanding (Amiel and Cowell 1999; Atkinson and Brandolini 2004; Ravallion 2005).The second is weakly absolute pro-poor growth, which comes from a policy point of view. As long as the income growth rate of the poor is greater than zero, economic growth is pro-poor (OECD 2006). It views all poverty reduction problems as income growth of the poor, not the comparison between the poor and the rich in income growth.

## 3 Measurement of Pro-poor Growth

According to the definition of pro-poor growth, its measurement is based on the change of income growth rate or poverty index. Of the seven measure methods of pro-poor growth, two are based on the change of income growth rate, which are PPGR and PGC. The other five are based on the change of poverty index, which include GEP, PPGI, PEGR, IGTI, and improved IGTI.

## 3.1 *Measurement of Pro-poor Growth Based on Income Growth Rate*

**PPGR.** Ravallion and Chen (2003) offered the growth incidence curve (GIC) to measure whether economic growth is pro-poor. Let $y_t(p) = L_t'(p)\mu_t$ represents the income of pth quantile, where $L_t(p)$ is the Lorenz curve, and $\mu_t$ is the mean. Then, the income growth rate of pth quantile is as follows

$$g_t(p) = \frac{y_t(p)}{y_{t-1}(p)} - 1 = \frac{L'_t(p)}{L'_{t-1}(p)}\left(\frac{\mu_t}{\mu_{t-1}} - 1 + 1\right) - 1 = \frac{L'_t(p)}{L'_{t-1}(p)}(GRIM + 1) - 1 \quad (1)$$

where $GRIM = \mu_t/\mu_{t-1} - 1$ means average income growth rate. If the Lorenz curve does not change, inequality has not changed, and any person's income growth rate is equal to the average income growth rate, namely $g_t(p) = GRIM$.

Ravallion and Chen showed that the change in the Watts poverty index is the area enclosed by the poverty incidence curve below the poverty rate. The Watts poverty index can be expressed as $W_t = \int_0^{H_T} \log[z/y_t(p)]dp$, where $H_t$ is a poverty rate of t period. Generally, the change rate of Watts's poverty index is $-\frac{dW_t}{dt} = \int_0^{H_t} g_t(p)dp$. Ravallion and Chen put forward the PPGR to measure whether economic growth is pro-poor. Here, PPGR is

$$PPGR = \frac{\int_0^{H_T} g_t(p)dp}{H_t} \quad (2)$$

If PPGR>0, economic growth is weak absolute pro-poor; if PPGR<0, economic growth is not pro-poor (Grimm 2007). If PPGR>GRIM, economic growth is relative pro-poor, while income inequality decreases; if PPGR<GRIM, economic growth is not relative pro-poor, and income inequality deepens; if PPGR=GRIM, income inequality remains unchanged (Grosse et al. 2006).

PPGR is based on the change of income growth rate. It can reflect the average income growth rate of the poor and is easily understood. However, there are two drawbacks. First, it ignores the poor income distribution, and provides the same weight to each income growth rate; the income growth rate of poorer people should be given greater weight. Second, this indicator ignores income growth of the rich. The average income growth rate of the poor may be positive, but the amount of income growth and rate of income growth of the rich are likely to greater than the poor's are, in which case economic development may worsen inequality.

**PGC.** Son (2004) proposed PGC as an alternative measure approach to pro-poor growth. PGC is similar to the GIC in that they investigate the income growth rate of the pth quantile population. The obvious difference is that the former investigates the average income or consumption share of p percent of the total population, while the latter investigates the share in the p percentile.

Let the Lorenz curve is $L(p) = \frac{1}{\mu}\int_0^x yf(y)dy$, where p is the proportion of people below income x, $\mu$ is per capita income, y is personal income, and $f(y)$ is the income distribution density function. If the Lorenz curve shifts upward for any p, there is $\Delta L(p) \geq 0$; thus, economic growth will be accompanied by a decline in inequality and economic growth is pro-poor. Based on the relationship between second-order dominance and poverty reduction to the general additive measure methods of poverty, for any p, if $\Delta(\mu L(p)) \geq 0$, poverty indicators are falling. From the definition of the Lorenz curve, we can refer to $L(p) = \mu_{\mathrm{p}} p/\mu$, where $p$ is the cumulative percentage of population and $\mu_{\mathrm{p}}$ is the corresponding average income. Logarithmic on both sides are $L(\mu_p) = Ln(\mu L(p)) - Ln(p)$. After the difference, we can get $g(p) = \Delta Ln(\mu L(p))$, where $g(p) = \Delta Ln(\mu_p)$ is the corresponding income growth rate. When p moves from 0 to 100%, $g(p)$ is the PGC. According to Atkinson (1987), as axioms for any p, if there is $g(p) > 0$(or $g(p) < 0$) during the two periods of economic development, then poverty declined (or increased). We can get:

$$PGC = g(p) = g + \Delta Ln(L(p)) \tag{3}$$

where $g = \Delta Ln(\mu)$ is the average income growth of the entire society. When $\mathrm{p} = 100\%$, $g(p) = g$, there is $\Delta L(p) = 0$. Thus, when $\mathrm{p} < 100\%$ and $g(p) > g$, economic growth is pro-poor and inequality is falling because the Lorenz curve shifts upwards (for any p, there is $\Delta L(p) \geq 0$). When $\mathrm{p} < 100\%$ and $0 < g(p) < g$, economic growth will reduce poverty, but increase inequality (for any p, there is $\Delta L(p) < 0$). In this case, economic growth might be called trickle-down growth. If $g(p) < 0$ and $g > 0$, economic growth will lead to an increase in poverty (Bhagwati 1988).

Let $x_p = \mu L'(p)$ represents the pth percentile income, where $L'(p)$ is the first order derivative of the Lorenz curve and $\mu$ is per capita income growth rate. Then the pth percentile income growth rate is: $r(p) = \Delta Ln(x_p)) = g + \Delta Ln(L'(p))$.

The difference between GIC and PGC is the calculation method used. PGC is expressed as a total income growth rate of the p percentage population, while GIC represents the average income growth rate of the pth percentile people. Let $(x_1^{(1)}, x_2^{(1)}, \ldots, x_p^{(1)})$ and $(x_1^{(2)}, x_2^{(2)}, \ldots, x_p^{(2)})$ represent two periods income of p percentage population, then there are

$$GIC = \frac{\frac{x_1^{(2)}}{x_1^{(1)}} + \frac{x_2^{(2)}}{x_2^{(1)}} + \ldots + \frac{x_p^{(2)}}{x_p^{(1)}}}{p} - 1 \quad PGC = \frac{x_1^{(2)} + x_2^{(2)} + \ldots + x_p^{(2)}}{x_1^{(1)} + x_2^{(1)} + \ldots + x_p^{(1)}} - 1 \tag{4}$$

PGC has an advantage over GIC because it focuses on the entire income growth rate of the poor. The drawback of giving each person's income equal weight is that the index is more sensitive to the income growth of high-income groups and incompletely reflects the change in inequality.

## 3.2 *Measurement of Pro-poor Growth Based on Change in Poverty Index*

Sen (1976) pointed out that a good poverty index should be consistent with 12 axioms; the core axioms include focus, weak transfer sensitivity, continuity, replication invariance, symmetry, subgroup consistency, and increasing poverty line (Sen 1976). However, no poverty index can meet all these axioms. Therefore, all defects in poverty indicators are embodied in measuring pro-poor growth. In the following, we will review five measure indices of pro-poor growth.

**GEP.** GEP refers to the percentage reduction in poverty rate associated with a percentage change in mean (per capita) income, namely

$$GEP = \frac{dH/H}{d\mu/\mu} \tag{5}$$

where H is poverty rate and $\mu$ is per capita income. When GEP is greater than 1, economic growth is pro-poor. This method is simple and not sensitive to data. However, it can be misleading, because the response of poverty to income growth is affected with initial income levels and income distributions. GEP is more sensitive to low-income changes. The same elasticity may suggest pro-poor growth in low-income countries, but it may not mean the same in the middle-income countries. Similarly, under a high degree of inequality, GEP will not be sensitive to increasing income. GEP is also sensitive to the position of the poverty line on income distribution (Bourguignon and Morrisson 2002; Bourguignon and Chakravarty 2003; Ravallion 1997; Ravallion 2001).

In addition, there are defects in the poverty indicators, such as less information, violation of monotonicity and metastatic axiom, misleading the anti-poverty policy, not being used in household surveys (Watts 1968; World Bank 2007), and so on.

**PPGI.** Kakwani and Pernia (2000) proposed PPGI by accounting for economic growth impact on income distribution and poverty. This index is designed on the assumption that poverty reduction is the result of economic growth and change in income distribution. That is, if inequality increases, even with economic growth, poverty reduction is limited. If the rate of increase in inequality is small, the poverty-reducing effect of economic growth is significant.

Therefore, measuring pro-poor growth must take the estimation error of the poverty reduction effect caused by inequality (overestimate poverty reduction effect under the circumstance of increasing inequality) into account.

Let $\theta = \theta(z, \mu, L(p))$ represents the poverty index, where z is the poverty line, $\mu$ is per capita income, and L(p) is the Lorenz curve. The change in poverty can be expressed as

$$P_{12} = Ln[\theta(z, \mu_2, L_2(p))] - Ln[\theta(z, \mu_1, L_1(p))] \tag{6}$$

Let

$$G_{12} = 1/2[Ln[\theta(z,\mu_2,L_1(p))] - Ln[\theta(z,\mu_1,L_1(p))] + Ln[\theta(z,\mu_2,L_2(p))] - Ln[\theta(z,\mu_1,L_2(p))]] \quad (7)$$

$$I_{12} = 1/2[Ln[\theta(z,\mu_1,L_2(p))] - Ln[\theta(z,\mu_1,L_1(p))] + Ln[\theta(z,\mu_2,L_2(p))] - Ln[\theta(z,\mu_2,L_1(p))]] \quad (8)$$

then $G_{12}$ means purely economic growth effect (the income distribution remains unchanged, while per capita income changes), and $I_{12}$ means income distribution effect (per capita income unchanged, income distribution changes), and the poverty reduction effect of economic growth ($P_{12}$) can be separated into purely economic growth effect ($G_{12}$), income distribution effect ($I_{12}$), and $P_{12} = G_{12} + I_{12}$.

Pure economic growth effect is negative. However the income distribution effect can be positive (inequality increasing) or negative (inequality reducing). Let $GRIM = \mu_2/\mu_1 - 1$ represents the average income growth, then total GEP is $\eta = \frac{P_{12}}{\frac{\mu_2}{\mu_1}-1} = \frac{P_{12}}{GRIM}$, where $\eta$ means the percentage reduction in poverty associated with a percentage change in mean (per capita) income.

Purely economic GEP is $\eta_g = \frac{G_{12}}{\frac{\mu_2}{\mu_1}-1} = \frac{G_{12}}{GRIM}$, where $\eta_g$ is the percentage reduction in purely economic growth effect, which is associated with a percentage change in mean (per capita) income.

Purely income distribution elasticity of poverty is $\eta_I = \frac{I_{12}}{\frac{\mu_2}{\mu_1}-1} = \frac{I_{12}}{GRIM}$, where $\eta_I$ is the percentage reduction in pure income distribution effect. It associates with a percentage change in mean (per capita) income. Kakwani and Pernia (2000) proposed the PPGI as

$$PPGI = \frac{\eta}{\eta_g} \quad (9)$$

If PPGI<0, economic growth is pro-rich; if 0<PPGI≤0.33, economic growth is mildly pro-poor; if 0.33<PPGI≤0.66, economic growth is moderately pro-poor; if 0.66<PPGI≤1, economic growth is pro-poor; and if PPGI≥1, economic growth is highly pro-poor.

Therefore, when 0<PPGI<1, although economic growth can reduce poverty, the income distribution effect is negative because the benefit the rich get from the economic growth is greater than the poor, and inequality is increasing. This type of economic growth is relative to pro-poor. Only if PPGI≥1 is economic growth absolute pro-poor. Under this condition, the economy grows, poverty is reduced, and inequality declines.

**PEGR.** According to PPGI, increasing income leads to poverty reduction, while increasing inequality leads to increased poverty. Then Kakwani and Son (2003) considered the condition of constant inequality, and created a consuming PPGI, a method that measured poverty-equivalent growth rate. PEGR can be expressed as:

$PEGR = PPGI * GRIM$. The index, a monotone increasing function of PEGR, is characterized by the ratio of poverty reduction. If $PEGR>GRIM$, economic growth is pro-poor; if $0<PEGR<GRIM$, economic growth is accompanied by rising inequality and poverty reduction; and if $PEGR<0$, economic growth means worsening poverty, namely, the benefits of economic growth are offset by increasing inequality. Kakwani and Son argued that the poverty reduction performance of a country should be measured by PEGR rather than economic growth.

**IGTI.** Kakwani (2000) pointed out that poverty depends on two factors: the average income level and the degree of inequality. Poverty can be expressed as: $\theta = \theta(\mu, G)$, where $\mu$ is per capita income, G is the Gini coefficient. Then

$$\frac{d\theta}{\theta} = \frac{\partial\theta}{\partial\mu} \cdot \frac{\mu}{\theta} \cdot \frac{d\mu}{\mu} + \frac{\partial\theta}{\partial G} \cdot \frac{G}{\theta} \cdot \frac{dG}{G} \tag{10}$$

The impact of economic growth on poverty can be decomposed into growth effect and distribution effect, namely

$$\frac{d\theta}{\theta} = \eta_\theta \cdot \frac{d\mu}{\mu} + \varepsilon_\theta \cdot \frac{dG}{G} \tag{11}$$

where $\eta_\theta = \frac{\partial\theta}{\partial\mu} \cdot \frac{\mu}{\theta}$ and $\varepsilon_\theta = \frac{\partial\theta}{\partial\mu} \cdot \frac{\mu}{\theta}$. Here $\eta_\theta$ means growth effect, and $\varepsilon_\theta$ is distribution effect. On this basis, Kakwani (2000) constructed IGTI as

$$IGIT = \frac{\partial\mu}{\partial G} \times \frac{G}{\mu} = -\frac{\varepsilon_\theta}{\eta_\theta} \tag{12}$$

The index shows how much economic growth is needed to make up for an increase in poverty when the Gini coefficient increased by 1%. For example, if IGIT is 3, it needs the average income growth rate increased by 3 percent points to make up for increasing poverty when the Gini coefficient increased by 1%. Whether PPGI or IGTI, the poverty index is expressed as a function of per capita income and inequality. The difference is that PPGI uses Lorenz curves to represent the inequality, but IGTI uses Gini coefficients. However, IGTI can give more policy guidance.

**IIGTI.** Ji Hong and Ruan Jing (2007) improved on the IGTI. Economic growth leads to changes in income distribution (curve shift and shape variation). To measure the impact of economic growth on poverty reduction, it needs to take income growth and distribution effects, change in the poverty line, and their interactions into account. Assumed income distribution in time t and t+1 are shown in Fig. 1.

The chart shows that the change in total welfare of poverty groups can be represented as area change surrounded by the income distribution curve and poverty line, namely

$$\Delta S = \int_0^{Z_{t+1}} f_{t+1} dx - \int_0^{Z_t} f_t dx \tag{13}$$

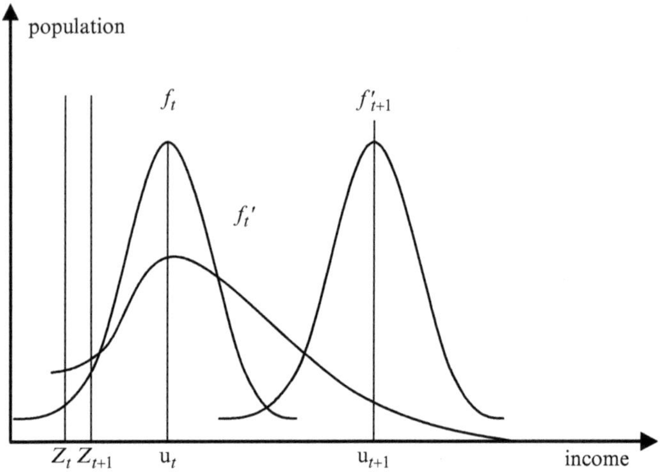


**Fig. 1** Income distribution curve

Let the coefficient of determination of pro-poor growth be

$$\phi = \frac{\int_0^{Z_{t+1}} f_{t+1} dx}{\int_0^{Z_t} f_t dx} \tag{14}$$

If $\phi>1$, economic growth is pro-poor; if $\phi<1$, economic growth is pro-rich. Furthermore, we can decompose the change in poverty into growth effect and distribution effect as

$$\text{Growth effect} : \Delta G = \int_0^{Z_{t+1}} f_t dx - \int_0^{Z_t} f'_{t+1} dx \tag{15}$$

$$\text{Distribution effect} : \Delta I = \int_0^{Z_{t+1}} f_t dx - \int_0^{Z_t} f'_t dx \tag{16}$$

where $f'_{t+1}$ and $f_t$ have the shape distribution shape, but different distribution center. $f_t$ and $f'_t$ have the same distribution center, but different distribution shape of distribution. Thus, IIGTI can be represented as

$$IIGIT = -\frac{\Delta I}{\Delta G} \tag{17}$$

PPGI decomposes the change in poverty rate into growth effect and distribution effect. Therefore, this index is significant and more intuitive.

# 4 Summary and Outlook

This article mainly discusses the construction of each PPGI and divides the measurement of pro-poor growth into two categories, namely, measurement of pro-poor growth based on change in income growth rate and on change in poverty index. We also explore various measure methods, and describe their characteristics and shortcomings. The measurement of pro-poor growth is closely related to the constructions of poverty index. However, we only discuss the measurement of pro-poor growth from the perspective of income. The poverty problem involves multi-dimensional poverty; therefore, building PPGI from a multi-dimensional perspective is significant to research, such as education poverty, nutrition poverty, human capital poverty, and so on. Furthermore, if we can decompose poverty changes into growth effect and distribution effect from the multi-dimensional perspective, it could provide additional practical guidance.

# References


Amiel Y, Cowell F (1999) Thinking about inequality: personal judgement and income distribution. Cambridge University Press, Cambridge

Atkinson AB (1987) On the measurement of poverty. Econometrica 55:749–764

Atkinson A, Brandolini A (2004) Global world inequality: Absolute, relative or intermediate. Presented at the 28th general conference of the International Association in Income and Wealth, Cork, Ireland, August, pp 22–28

Bhagwati JN (1988) Poverty and public policy. World Dev 16(5):539–654

Bourguignon F, Chakravarty S (2003) Measurement of multidimensional poverty. J Econ Inequality 1:25–49

Bourguignon F, Morrisson C (2002) Inequality among world citizens: 1820–1992. Am Econ Rev 92(4):727–744

Duclos J, Wodon Q (2004) What is pro poor? Working Paper, CIRPEE, pp 04–25

Grimm M (2007) Removing the anonymity axiom in assessing pro-poor growth. J Econ Inequality 5:179–197

Grosse M, Harttgen K, Klasen S (2006) Measuring pro-poor progress toward the non-income millennium development goals. WIDER Research Paper (2006-38), WIDER, Helsinki

Ji Hong, Ruan Jing (2007) Measurement and decomposition of pro-poor growth based on income distribution (in Chinese). Res Econ Manag 8:38–43

Kakwani N (2000) Poverty and Well-being in the Philippines with a Focus on Mindanao. Asian Development Bank Technical Report

Kakwani N, Pernia E (2000) What is pro-poor growth? Asian Dev Rev 18(1):1–16

Kakwani N, Son HH (2003) Pro-poor growth: concepts and measurement with country case studies. Pak Dev Rev 42(4):417–444

Klasen S (2008) Economic growth and poverty reduction: measurement issues using income and nonincome indicators. World Dev 36:420–445

Kraay A (2004) When is growth pro-poor? Cross-country evidence. World Bank Policy Research Working Paper, The World Bank, Washington, DC

OECD (2006) Promoting pro-poor growth: key policy messages. OECD, Paris

Ravallion M (1997) Can high inequality developing countries escape absolute poverty? Econ Lett 56:51–57

Ravallion M (2001) Growth inequality and poverty: looking beyond averages. World Dev 29:1803–1815
Ravallion M (2005) A poverty-inequality trade-off? J Econ Inequality 3(2):169–181
Ravallion M, Chen S (2003) Measuring pro-poor growth. Econ Lett 78:93–99
Ravallion M, Datt G (1999) When is growth pro-poor? Evidence from the diverse experience of India's States. Policy Research Paper 2263, World Bank
Sen AK (1976) Poverty: an ordinal approach to measurement. Econometrica 44:219–231
Son HH (2004) A note on pro-poor growth. Econ Lett 82:307–314
Watts HW (1968) An economic definition of poverty. In: Moynihan DP (ed) On understanding poverty. Basic Books, New York
World Bank (2007) The world development index. The World Bank, Washington DC

# The Dynamic Economic Equilibrium Model and Uncertainty Applied Study About Forest Resources Sustainable Utilization

**Wenhui Chen, Junchang Liu, and Aili Yao**

**Abstract** In this paper, we regarded the forest resources as research objects, and used the dynamic economic model about renewable resources utilization based of the Logistic model to create the forest resources dynamic economic equilibrium model with the many restriction conditions about ecological production supply, taxation, economic composition and so on. Then we used the mathematic method about normal differential equation and maximum theory to obtain the optimal forest accumulation stock quantity and the optimal harvest quantity. After, we selected the North-east of China as the case site of applied research, and used the dynamic economic equilibrium theoretical model and the data from forest resources investigation, forestry statistics yearbook and typical survey to carry the applied analysis to obtain the optimal harvest quantity dynamic function about forest resources in North-east of China. Then we used the function to obtain the development trend of the equilibrium after 2008, and the optimal harvest quantity should be 4.26 million $m^3$.

**Keywords** Dynamic economic model · Forest resources · Sustainable harvest quantity · Uncertainty applied study

## 1 Introduction

The forest resources not only are one of the important renewable nature resources, but also is the substance base of forestry sustainable development. The rational development and optimal management of forest resources is the key of harmonious development between forestry industry and economic and social

---

W. Chen (✉), J. Liu, and A. Yao
School of Economics and Management, Beijing Forestry University, Beijing 100083, People's Republic of China
e-mail: wenhui@bjfu.edu.cn; chenwenhui77@163.com

Y. Zhou and D.D. Wu (eds.), *Modeling Risk Management for Resources and Environment in China*, Computational Risk Management,
DOI 10.1007/978-3-642-18387-4_5, 

sustainable development. Some research results were about the qualitative analysis of forest sustainable management, the technologies and the single quantitative analysis of forest resources sustainable management. And a few researchers regarded the forest resources as the whole to take quantitative economic analysis. It introduced the Faustmann model and optimal selected deforestation model in the book of *Mathematical bioeconomics: the optimal management of renewable resources* by C.W. Clark in 1976. Kilkki and Vaisanan used the dynamic layout method to design the growing and selected deforestation model of Scotland pine in Finland in 1969. Sethi and Thompson excogitated the optimal management model about reforestation in 1981. ZHAO created the forest development equation and designed the object function of forest optimal management with restriction conditions on the cost and benefit in 2003. ZHANG used the change differential coefficient method and optimal control theory to obtain the optimal control strategy about the selected deforestation degree and accumulation density of the manual forest passing the dynamic mathematical model in 2006. CHEN created the dynamic economic model about sustainable deforestation with the restriction condition on ensuring the ecological services for society and obtained the optimal strategy about forest resources accumulation stock quantity and harvest quantity.

In a word, the optimal management was considered in the known research results, but the optimal management with many restriction conditions on nature, social development, and economic policies still is a key problem of forest resource economics and management. So the paper study the dynamic economic model with many restriction conditions about forest growing rule, ecological services supply for society, taxation and economic compensation by government to obtain the optimal management strategy.

## 2 Research Method and Object

In the paper, it systematically analyzes the dynamic equilibrium with many restriction conditions about forest growing rule, ecological services supply for society, taxation and economic compensation by government and looks for the optimal forest resources accumulation stock quantity and the optimal harvest quantity about the whole forest resources. It is the basal of constituting rational forest resources management policy by the government.

It uses the cost and benefit analysis to create the dynamic economic equilibrium model of forest resources with the many restriction conditions on the Logistic model of forest growing, ecological production supply, taxation, and economic compensation, and uses the mathematic method about normal differential equation and maximum theory to solve the optimal forest accumulation stock quantity and the optimal harvest quantity.

# 3 The Theoretical Analysis of the Dynamic Economic Model About Forest Resources Sustainable Utilization

## *3.1 The Basal Hypothesis of the Model*

For satisfied the request of model and the characteristic of forest resources, some hypothesis are as follow:

1. The forest resources utilization market is completely competitive market which is based on the request of the dynamic equilibrium model about renewable resources sustainable utilization.
2. The foremost status is on the increase course.
3. The forest resources production includes two parts, one part is timber product, its price is the average price of different tree categories and different age classes; others is byproduct, its price is the average price of all kinds of byproducts.
4. The influence to forest resources is only the influence to forest accumulation quantity.

## *3.2 The Restriction Conditions of the Model*

### 3.2.1 The Natural Growing Model of Forest Resources

In natural condition, the forest growing accord with the logistic growing model of biology population. It is as follow:

$$\frac{dR(t)}{dt} = F(R) - h(t) \tag{1}$$

According to formula (1), defined as

R = R(t) as forest resources population at time t;

F(x) as the known forest natural population growth rate functions;

h (t) as the harvest rate.

This model shows the sustainable management method which is based on the biological characteristic of forest resources. Its aim is for obtaining the Maximum sustainable yield (MSY). In this model, if the forest resources stock quantity is less than the load quantity of environment, there is the surplus quantity to harvest for ever and not change the forest resources stock quantity. And if it is not harvested, the forest resources stock quantity will increase and consume more and more environmental elements to make the forest resources system to ruin. So it shows as like as the U shape between the cure growth rate of forest resources and environmental load capacity. The common model form about forest resources population growth is as follow:

$$\frac{dR(t)}{dt} = rR\left(1 - \frac{R}{K}\right) \tag{2}$$

According to formula (2), defined as
R = R (t) as forest resources population at time t;
K as the environmental load capacity;
r as the cure growth rate of forest resources; $r = F(R) - h(t) - \beta_0$, in this formula,
F (x) as the known forest natural population growth rate functions;
h (t) as the harvest rate.
$\beta_0$ as the withered rate

### 3.2.2 The Least Forest Resources Accumulation Quantity Satisfied the Request of Ecological Services for Society

Forest is characteristic renewable natural resources. It not only supplies the forest production, but also supplies the public ecological services. The forest resources accumulation quantity must keep the level which is able to supply ecological services. In the paper, A(t) defines as the least forest resources accumulation quantity satisfied the request of ecological services for society. A(t) is the exogenous variable, which don't change with the change of forest resources accumulation quantity. Its change is only determined by social request which includes the social development level, population and environmental pollution and so on. So the restriction on A(t) is that the forest resources accumulation stock quantity after deforestation in every time t should be more than the least forest resources accumulation quantity satisfied the request of ecological services for society in the same time. It is as follow:

$$R(t) - h(t) \geq A(t) \tag{3}$$

### 3.2.3 Taxation

In the paper, the tax variable is the function of the forest resources harvest quantity and price variable. The tax sum is equal to the product of the tax rate, the harvest quantity and the price. So the average tax quantity $\tau(t)$ is equal to the product of the tax rate $\alpha$ and the price p(t). It is as follow:

$$\tau(t) = \alpha P(t) \tag{4}$$

So the restriction on $\tau(t)$ is that the average tax quantity $\tau(t)$ of the every unit accumulation quantity of forest resources in every time t should be less than the surplus of the price p(t) minus the cost C(R(t),W(t)) in the same time. It is as follow:

$$P(t) - C(R(t), W(t)) \geq \tau(h(t), P(t)) \tag{5}$$

### 3.2.4 Economic Compensation

In the paper, the economic compensation is the compensation which forest manager obtain because they decrease deforestation quantity for the ecological services supply. The economic compensation B(t) is less than the cost of forest management. It is as follows:

$$C(R(t), W(t)) \geq B(h(t)) \tag{6}$$

## 3.3 *Creating the Model and Solving the Optimal Conditions*

As forest resources manager, they take forestation, arboriculture and deforestation activities to obtain the maximization benefit in a long term. The aim function of the model is created which is based on cost-benefit analysis in economics. It is as follows:

$$\max PV = \int_0^T [p(t) - C(t) - \tau(t) + B(t)]h(t)e^{-\delta t}dt + g(R(t))e^{-\delta t} \tag{7}$$

According to formula (7), defined as
PV as the current value of net benefit of managing forest resources;
R = R (t) as forest resources population at time t;
h (t) as the harvest rate;
p(t) as the average price of forest resources at time t;
C(t) as the average cost of every unit harvest of forest resources at time t;
$\tau(t)$ as the average tax quantity of every unit harvest of forest resources at time t;
B(t) as the economic compensation of every unit harvest of forest resources at time t;
$\delta$ as discount rate, is a constant more than zero;
$g(R(t))$ as residual function in theoretical;
T as the term of the whole forest resources utilization.

The restriction conditions of the model includes formula (2), (3), (5), (6) and the other two conditions from the basal assumption, it is as follows:

$$R(0) = R_0 > 0(\text{known}) \tag{8}$$

$$h(t) > 0 \tag{9}$$

The dynamic economic model of forest resources sustainable utilization is compose of aim function formula (7) and the restriction conditions of the formula (2), (3), (5), (6), (8) and (9).

According as extremum principle, it uses the Lagrange function method to solve the optimal conditions. First necessary conditions are as follows:

$$[B'(\bullet) - \tau_h]h(t) - \lambda_1 e^{\delta t} - \lambda_2 e^{\delta t} \\ + P(t) - C(\bullet) - \tau(\bullet) + B(\bullet) - \lambda_3(t)\tau_h e^{\delta t} - \lambda_4(t)B'(\bullet)e^{\delta t} = 0 \quad (10)$$

$$e^{-\delta t}[-C_W]h(t) + \lambda_1(t)F_W - \lambda_3(t)C_W + \lambda_4(t)C_W = 0 \quad (11)$$

$$C_R\left[\lambda_4(t)e^{\delta t} - \lambda_3(t)e^{\delta t} - h(t)\right] + e^{\delta t}\left[\lambda_1(t)F_R + \lambda_2(t) + \dot{\lambda}_1\right] = 0 \quad (12)$$

$$\lambda_2(t) = 0 \quad (13)$$

$$(1 - \tau_P)\left[\lambda_3(t) - h(t)e^{-\delta t}\right] = 0 \quad (14)$$

$$g'(\bullet) = \lambda_1(T)e^{\delta t} = 0 \quad (15)$$

According to formula (7), defined as $F(\bullet)$ as $F(R(t), W(t))$; $\dot{\lambda}_1$ as partial derivative of λ1; $F_R$ as partial derivative of function F(R(t)) to R(t), next is same in the paper.

Solving the equations formula (10), (11), (12), (13), (14) and (15) to obtain the optimal harvest quantity of forest resources $h^*(t)$. $h^*(t)$ is as follows:

$$h^*(t) = \frac{1}{F_R} \quad (16)$$

It uses formula (16), (1) and (8) to obtain the optimal forest resources accumulation stock quantity $R^*(t)$.

## 4 The Applied Research About Forest Resources Sustainable Utilization Dynamic Economic Analysis in North-East of China

According as the theoretical model, the total accumulation quantity of forest resources in China doesn't satisfy the request of forest resources least accumulation quantity for ecological services supply. So in the paper, it selects the North-east of China which has the abundant forest resources and high market degree as the case to create the applied model.

## 4.1 The Basal Data in the Model

The basal data in the model come from china forestry statistics yearbook, the published paper and the other open data. The accumulation quantity data in the model are from the fourth, fifth and sixth forest resources investigation results in North-east of China. They are as shown in Table 1.

The discount rate is 5% which is the average of capital yield in current three years used by State Asset Regulatory Commission in forestry enterprise evaluation.

The total cost of forest resources includes the forestation cost, manage cost and deforestation cost. The cost data are from representative survey in North-east of China. It is as shown in Table 2.

The data of labor effort W(t) is the labor productivity of forestation from the typical survey in North-east of China. It is as shown in Table 3.

The forest resource price data are the average price of timber which is that the total sales amount divides by the total sales quantity in North-east of China. The sales amount and sale quantity are from China forestry statistics yearbook (Table 4).

**Table 1** The fourth, fifth and sixth forest resources investigation results in North-east of China. Unit: $10^4$ ha, %, $10^4$ $m^3$ (From China forestry statistics yearbook)

| Sort | Woodland area | Forest area | Forest cover rate | Total accumulation quantity of stand tree | Forest storage quantity |
|---|---|---|---|---|---|
| Sixth forest resources investigation results (1999–2003) | 3466.46 | 2998.15 | 38.05 | 254058.59 | 236624.39 |
| Fifth forest resources investigation results (1994–1998) | 3528.41 | 2918.34 | 37.03 | 256731.60 | 235863.01 |
| Fourth forest resources investigation results (1988–1993) | 3561.3 | 2642.62 | 33.48 | 250485.5 | 224111.10 |

**Table 2** The cost table of every unit accumulation quantity. Unit: Yuan (From representative survey in North-east of China)

| Year | 2003 | 2004 | 2005 | 2006 | 2007 | 2008 |
|---|---|---|---|---|---|---|
| Forestation and management cost | 156 | 169 | 189 | 158 | 165 | 173 |
| Deforestation cost | 137 | 158 | 182 | 206 | 244 | 295 |
| Total cost | 293 | 327 | 371 | 364 | 409 | 468 |

**Table 3** The labor productivity of forestation in North-east of China. Unit: Yuan

| Year | 2002 | 2003 | 2004 | 2005 | 2006 | 2007 |
|---|---|---|---|---|---|---|
| Labor productivity | 12877 | 6056 | 8436 | 7894 | 15767 | 26231 |

**Table 4** The average price of timber in North-east of China. Unit: Yuan

| Year | 2002 | 2003 | 2004 | 2005 | 2006 | 2007 |
|---|---|---|---|---|---|---|
| Average price | 537 | 516 | 556 | 642 | 747 | 895 |

It adopts the reverse calculation method of market price to obtain forest resources price. The total value method of forest resources is as follows:

$$V = \frac{\sum_{t=0}^{r} \bar{R}_t (1+\delta)^{r-t} - \sum_{t=0}^{r} \bar{C}_t (1+\delta)^{r-t}}{(1+\delta)^r - 1} \tag{17}$$

According to formula (17), defined as

V as the total value of forest resources;

$\bar{R}_t$ as the total income of sale forest resources production at time t;

$\bar{C}_t$ as the total cost of forest resources harvest at time t;

$\delta$ as discount rate, is a constant more than zero;

According with formula (17), the price p of every unit accumulation quantity of forest resources is equal to the quotient that the total value of forest resources divided the total harvest quantity. The account formula is as follow:

$$p = \frac{\sum_{t=0}^{r} \rho_t (1+\delta)^{r-t} - \sum_{t=0}^{r} C_t (1+\delta)^{r-t}}{(1+\delta)^r - 1} \tag{18}$$

According to formula (18), defined as

p as the average price of forest resources;

$\rho_t$ as the market price of timber at time t;

$C_t$ as the average cost of forest resources harvest at time t;

$\delta$ as discount rate, is a constant more than zero;

According with formula (18), it obtains the price of forest resources. It shows in Table 5.

The basal data which obtain the least accumulation quantity of forest resources for satisfied the request of ecological services for society are from the relative data of necessary forest area in different typical area in the book of *China forest resources and sustainable development in 2007*. It obtains the least forest cover rate of 37.7% by the weighted arithmetic mean method. The least accumulation quantity of forest resources for satisfied the request of ecological services for society is the continued product of the total land area, the least forest cover rate and every unit accumulation quantity. The account formula is follow:

$$78.92 \text{ million ha} \times 37.7\% \times 78.92\,\text{m}^3/\text{ha} = 2.34809 \text{ billion m}^3$$

The least accumulation quantity of forest resources for satisfied the request of ecological services for society is 2.34809 billion $m^3$.

**Table 5** The price of forest resources in North-east of China. Unit: Yuan

| Year | 2003 | 2004 | 2005 | 2006 | 2007 |
|---|---|---|---|---|---|
| The price of forest resources | 223 | 229 | 271 | 364 | 409 |

The average taxation rate is 5% which is from the known paper.

The economic compensation is not considered in the applied model because there aren't compensation activities in North-east of China.

## 4.2 Solving the Model

### 4.2.1 The Fitting About the Growth Model of Forest Resources

According with formula (2), it obtains the solution that is as follow:

$$R(t) = \frac{K}{1 + (K/R_0 - 1)e^{-rt}} \tag{19}$$

According to formula (19), defined as

R = R (t) as forest resources population at time t;

K as the environmental load capacity;

r as the cure growth rate of forest resources;

$R_0$ as the initial values of forest resources.

The account formula of parameters K and r in the model is as follow:

$$\begin{cases} K = \dfrac{2R_0R_1R_2 - R_1^2(R_0 + R_2)}{R_0R_2 - R_1^2} \\ r = \dfrac{1}{T}\ln\left[\dfrac{R_2(K - R_1)}{R_1(K - R_2)}\right] \end{cases} \tag{20}$$

According to formula (20), defined as

$R_0$, $R_1$ and $R_2$ as forest resources population at different time;

K as the environmental load capacity;

r as the cure growth rate of forest resources;

T as the interval between different terms.

In the paper, $R_0$, $R_1$ and $R_2$ are in turn the accumulation quantity of forest resources in the fourth, fifth and sixth forest resources investigation in North-east of China. So $R_0$ is equal to 2241.1110 million $m^3$, $R_1$ is equal to 2357.6301 million $m^3$, $R_2$ is equal to 2366.2439 million $m^3$. It obtains the value of parameters K and r according with formula (20) and the data of $R_0$, $R_1$, $R_2$ and T. The value of K is equal to 236674.33, and the value of r is equal to 0.5582.

According with the formula (19) and the data of K and r, the growth function of forest resources in North-east of China is as follows:

$$R(t) = \frac{236674.33}{1 + 0.0561e^{-0.5582\,t}} \tag{21}$$

The initial time is in 1993, the t is equal to zero when the time is in 1993, and the interval is 1 year.

### 4.2.2 The Fitting About the Price Function Model of Forest Resources

According with the accumulation quantity of forest resources by formula (21) and the forest resources price from the Table 4, the price function model of forest resources is as follow:

$$P(t) = -824133 + 3.485\ R(t) \tag{22}$$

The fitting goodness is equal to 0.652; the F-statistics value is equal to 5.618. The price function model of forest resources has passed the hypothesis test.

### 4.2.3 The Fitting About the Harvest Cost Function Model of Timber

According with the total cost from the Table 2, the accumulation quantity of forest resources by formula (21) and the labor productivity from the Table 3, the harvest cost function model of forest resources is as follows:

$$C(R(t), W(t)) = -563562 + 2.383\ R(t) + 0.004\ W(t) \tag{23}$$

$$\text{T-value}: \quad -8.042 \qquad 8.046 \qquad 5.722$$

The fitting goodness is equal to 0.981; the F-statistics value is equal to 78.449. The price function model of forest resources has passed the hypothesis test.

### 4.2.4 The Fitting About the Taxation Function Model of Forest Resources

According with the taxation rate of 5% and the price function model as formula (22), the taxation function model of forest resources is as follow:

$$\tau(t) = \alpha P(t) = -40206.65 + 0.17425\ R(t) \tag{24}$$

### 4.2.5 The Applied Model Form

It put the fitting functions which include the growth model, the price function model, the harvest cost function model and the taxation function model of forest resources into the theoretical model to obtain the applied dynamic economic model about forest resources sustainable utilization in North-east of China. It is as follow:

$$\max PV = \int_0^T [-220364.35 + 0.928\ R(t) - 0.004\ W(t)]h(t)e^{-0.05t}dt$$

Restriction condition:

$$\frac{236674.33}{1+0.0561e^{-0.5582\,t}} - h(t) \geq 234787 \tag{25}$$

$$-55091087.5 + 231.9375\ R(t) \geq W(t)$$

$$F(R(t), W(t)) = \frac{7411.461377e^{-0.5582t}}{(1+0.0561e^{-0.5582t})^2} + h(t)$$

According with the theoretical model, the optimal dynamic harvest quantity function is as follow:

$$h(t) = 118762.7863 - \frac{118337.165e^{0.5582t}}{0.0561 + e^{0.5582t}} \tag{26}$$

## 4.3 The Result of the Model

The result of the dynamic economic model with the restriction conditions about forest growing rule, ecological production supply, and taxation is as follow:

The optimal accumulation stockage quantity function is $R(t) = \frac{236674.33}{1+0.0561e^{-0.5582t}}$, and the optimal harvest quantity is $h(t) = 118762.7863 - \frac{118337.165e^{0.5582t}}{0.0561+e^{0.5582t}}$ (t is time).

# 5 Conclusions

According with the model, the results are as follow:

1. It is growth equilibrium status that the forest resources accumulation quantity change in North-east of China.
2. In creating the dynamic economic model, they are exogenous variables which there are simplified hypothesis that they are the least forest resources accumulation quantity for ecological services supply for society, taxation rate, and natural growth rate of the forest resources. In fact, the values of these exogenous variables are change in the long periods. So the applied model and its results, to a great extent, affect the policy establishment of forest resources management.
3. According with the model to forecast, when the time t is equal to 15(namely the time is in 2008), the optimal harvest quantity converge 4.26 million $m^3$. After 2008, the forest resources will be equilibrium status, and the maximum harvest quantity is 4.26 million $m^3$.
4. According with the account method of the least forest resources accumulation quantity for ecological services supply for society, the necessary forest cover of

China should reach 32.64%, but according with the seventh forest resources investigation, the factual forest cover is only 20.36%, can't satisfy the least forest resources accumulation quantity for ecological services supply for China. So in the whole country, it should be speedily forestation, increase the forest resources accumulation quantity.

**Acknowledgement** This research was supported by "the Fundamental Research Funds for the Central Universities" (Grant RW2010-4).

## References

Clark CW (1976) Mathematical bioeconomics: the optimal management of renewable resources. ZHOU qinxue, QIU zhaofu (trans: Beijing) Agriculture Publishing House, 1984:10, 43, 277–278

Kilkki P, Vaisanan U (1969) Determination of optimal policy for forest stands by means of dynamic programming [J]. Acta Forestalia Fennica 102(1):100–112

Sethi SP, Thompson GL (1981) Optimal control theory, applications to management science [M]. Martinus Nijhoff Publishing

# Sensitive Analysis of Sustainable Economy Growth Based on Resources Consumption, Technology Progress and Human Capital

**Zheng-xia He**

**Abstract** This paper constructs a dynamic economy growth model with the elements of resource consumption, technology progress and endogenous human capital to discuss the optimal economy growth path applying the optimized control methods. Through sensitive analysis the research demonstrates that consumer's intertemporal substitute elasticity, human capital accumulation efficiency, and time discount rate determine the optimal economy growth path. Consumer's intertemporal substitute elasticity, time discount rate, and factor output elasticity of product have effects on stable economy growth rate and resource consumption velocity.

**Keywords** Human capital · Resources consumption · Sensitive analysis · Sustainable growth · Technology progress

## 1 Introduction

With the rapid development of economy, human beings are also facing various serious problems such as the lack of resources. How to coordinate the relationship between economy and resources calls for quick solution. As early as the 1970s the Rome club puts forward the growth extreme which advocates that the economy is not able to develop steadily in the long term due to the restriction of the limited natural resources, and advocates that it be necessary to deliberately slow down economy development in order to protect the nature resources. After that time, Dasgupta and Heal (1979), D'Arge and Kogiku (1973), Keeler et al. (1971), Forster (1973), Garg and Sweeney (1978), Solow (1974), and Stiglitz (1974) analyze the compact of resources and environment on economy. Then in the mid and late

This paper is partial research result of social science fund of Xuzhou Normal University: the environmental regulation and green innovations of enterprises (09XWA14) and national social science fund: research on environmental policy of long-term mechanism of perfecting energy saving and pollutants abatement (10CJY028).

Z.-x. He
School of Economics, Xuzhou Normal University, Xuzhou 221116, People's Republic of China

Y. Zhou and D.D. Wu (eds.), *Modeling Risk Management for Resources and Environment in China*, Computational Risk Management,
DOI 10.1007/978-3-642-18387-4_6, © Springer-Verlag Berlin Heidelberg 2011

1980s, with the emergence of the endogenous economy growth model represented by Romer (1990), Aghion and Howitt (1998), Bovenberg and Smulder (1995) regard technology progress as one of endogenous variables in the economy growth model, and discuss the relationship between ecological environment and long term economy growth. Hereafter, Grimaud and Rouge (2003) introduce limited natural resources to economic model and analyze the compact of resources on sustainable development.

This paper, based on the previous research, builds a simple endogenous growth model and makes research on the sustainable best growth path.

# 2 Construction of the Economic Model

## 2.1 Consumer Utility

The representative consumers' utility not only comes from the consumption of the product, but also from the natural resources of their living conditions. This paper adopts the addable equal elasticity utility function as following:

$$U(C_t, N_t) = \int_0^\infty \left( \frac{C_t^{1-\varepsilon} - 1}{1-\varepsilon} + \frac{N_t^{1+\delta}}{1+\delta} \right) e^{-\rho t} dt \tag{1}$$

In the equation, $C_t$ and $N_t$ stand for respectively the quantity of the material products and natural resources consumption at time t. $\rho$ is the subjective time discount rate, parameter $\varepsilon$, $\delta$ is the positive constant, evaluating respectively the effects of material products and natural resources on consumer utility.

## 2.2 Input and Output of the Material Product Department

Setting that the production function of material product department is Cobb-Douglas production function and the quantity of labor force is 1, and the production of material product department is constant scale of economies return, the detailed production function is:

$$Y_t = A_{Yt} K_t^\alpha (\mu H_t)^\beta R_t^{1-\alpha-\beta}, \quad 0 < \alpha, \beta, 1 - \alpha - \beta < 1 \tag{2}$$

$Y_t$ stands for per capita output during t time period, $A_{Yt}$ stands for the production efficiency of material product department during t time period( here we set that the production efficiency of material product department mainly depends on the production technology and the quantity of knowledge); $K_t$ stands for per capita physical capital during t time period; $H_t$ stands for per capita human capital during t time period; $\mu$ stands for per capita human capital ratio that is put into the material products production department; $R_t$ stands for per capita natural resources during t time period; Based on the hypothesis that the remainder of output of material product department is used as physical capital and invested in the reproduction process besides that one part of output is used for consumption and setting the discounted rate of physical capital is $\psi$, the change over time of per capital physical capital at random time point can be showed as:

$$\dot{K}_t = Y_t - C_t - \psi K_t \quad (3)$$

## 2.3 Human Capital Accumulation

Following the foreign scholars' presumption for human capital accumulation, the accumulation of human capital is mainly determined by the production efficiency of human capital and the preset available accumulation level of human capital, so the production function of the human capital department can be written as:

$$\dot{H}_t = A_H v H_t \quad (4)$$

In this function, $\dot{H}_t$ is the added quantity of human capital accumulation per person in the time of t; $A_H$ is the coefficient of the human capital accumulation efficiency, which reflects a parameter for the production efficiency of the human capital department; $v$ is the ratio of human capital per person put into the process of the human capital department production; $H_t$ stands for the already available human capital level per person in the time of t in the economic system.

## 2.4 The Research and Development of Technology

Assuming that material production department production efficiency $A_Y$ mainly depends on the level of production technological knowledge. According to the model of Jones (1995), the production function of R&D department is:

$$\dot{A}_{Yt} = A_T(1 - \mu - v)H_t A_{Yt} \quad (5)$$

$A_T$ stands for production efficiency parameter of R&D department; $(1 - \mu - v) H_t$ stands for the quantity of per capita human capital put into the R&D department in the time of t; $A_{Yt}$ stands for present available level of production technological knowledge in the economy in the time of t; $\dot{A}_{Yt}$ stands for the added quantity of technological knowledge in the time of t.

### 2.5 Natural Resources Consumption

Supposed natural resources are all not regenerative, initialization is $N_0$, and natural resources department exploits $R_t$ and sells them to material product department at time t. According to Stiglitz (1974), the existed quantity of natural resources at time t is used the following to denote:

$$N_t = N_0 - \int_0^t R_t(v)dv \tag{6}$$

derivating t yields:

$$\overset{\bullet}{N}_t = R_t \tag{7}$$

This denotes diversification of natural resources during the time of t.

## 3 Equilibrium Analysis

### 3.1 The Necessary Condition of Optimal Equilibrium Growth Path

Based on above economic models, Maximize (1) with constraints (2), (3), (4), (5), (7), and review the social optimization problem (the time t is omitted). Construct the current-valued Hamiltonian:

$$J = \frac{C^{1-\varepsilon} - 1}{1-\varepsilon} + \lambda_1(Y - C - \psi K) + \lambda_2 A_H vH + \lambda_3(-R) + \lambda_4 A_T(1-\mu-v)HA_Y \tag{8}$$

From above equation, $g_R$, $g_H$, $g_{A_Y}$ and $g_Y$ can yield in the steady economy growth path ($g_i$ denotes the growth rate $\overset{\bullet}{i}/i$):

$$g_R = \frac{1-\varepsilon}{1+\delta} g_Y \tag{9}$$

$$g_H = (1-\varepsilon)g_Y + A_H - \rho \tag{10}$$

$$g_{A_Y} = \frac{(1-\mu-v)\beta A_H}{\mu} \tag{11}$$

$$g_Y = \beta\left(\frac{1-v}{\mu}A_H - \rho\right)\left[(1-\alpha) - \beta(1-\varepsilon) - \frac{(1-\alpha-\beta)(1-\varepsilon)}{1+\delta}\right]^{-1} \tag{12}$$

From the yields of model, we can judge in the steady optimal growth path, economy growth rate should be positive and natural resources consumption rate should be negative. On this condition, the sustainable economy growth will come true. With (9) and (12), we can conclude that:

$$\varepsilon > 1 \tag{13}$$

$$A_H > \mu\rho/(1 - v) \tag{14}$$

$\varepsilon > 1$ suggests that consumer's intertemporal substitute elasticity $1/\varepsilon$ should be $0 < 1/\varepsilon < 1$. Because the variable $\varepsilon$ determines consumer's choice of transferring consumption present to future, the greater $\varepsilon$ is, the little consumption at present. This constraint can make consume at present not too much, and induce rational consumption. $A_H > \mu\rho/(1 - v)$ demonstrates that if the ratio between human capital accumulation efficiency and time discount is greater than the ratio between the rate of human resources in material production department and material production department together with R&D department, economy growth will be sustainable. The reason is that sufficient human capital can make economy grow and conquer the constraint of consumer's impatience.

## 3.2 *The Comparative Static Analysis of Optimal Equilibrium Growth Path*

We will go further by sensitive analysis to examine the effects of $\rho$, $\varepsilon$, $\delta$, $\alpha$, $\beta$, $A_H$ on $g_Y$ and $g_R$. We can easily get Table 1 by computing the partial derivative of (9) and (12).

We can see the sensitivities of $g_Y$ and $g_R$ from Table 1. The larger the consumers' time discount rate is, the lower the steady state per capita economy growth rate is, and the higher the growth rate of per capita natural resources consumption. This is because the larger $\rho$ means consumers lack of patience, the current consumption brings consumers higher utility compared to future consumption(consumers have relative low evaluation of future consumption utility), consumers will increase the current consumption which will enhance the current output and ultimately add more pressure of natural resources. At the same time, the larger $\rho$ indicates that consumers have weaker awareness of sustainable development, which is not beneficial for long term economy growth.

**Table 1** Influencing of parameters on sustainable growth in steady state

| $\xi$ | $\rho$ | $\varepsilon$ | $\delta$ | $\alpha$ | $\beta$ | $A_H$ |
|---|---|---|---|---|---|---|
| $\frac{\partial g_Y}{\partial \xi}$ | $< 0$ | $< 0$ | $> 0$ | $> 0$ | $> 0$ | $> 0$ |
| $\frac{\partial g_R}{\partial \xi}$ | $> 0$ | $< 0$ | $> 0$ | $< 0$ | $< 0$ | $< 0$ |

The larger $\varepsilon$ indicates that the substitution elasticity of products consumption at random two time points is lower, the curvier the utility function form of consumers' material products consumption is, the quicker the rate of marginal utility decreasing is, the more consumers do not prefer the current consumption, the lower the steady state economy growth rate is. At the same time, in the consumption function, when the consumption of material products and natural resources demand are supplementary, the consumption growth lead to the improvement of environment quality, and it is easier for the economy to achieve sustainable optimal growth path; On the contrary, when the consumption of material products and natural resources demand are substitutes, the growth of consumption is on the cost of sacrificing the natural resources and environment, alternatively we have to limit consumption to protect environment, therefore, it is harder for us to achieve the optimal sustainable growth path. Herein we find that there is certain substitution relationship between the consumption of material products and natural resources demand, therefore, with the rise of $\varepsilon$, steady state economy growth rate has tendency to decline and so does the growth rate of natural resources consumption, which is beneficial for the improvement of natural resources and environment quality.

The larger $\delta$ indicates consumers emphasize the current environment quality more and more. The stronger consumers' awareness of sustainable development, the more beneficial it is for the economy growth.

As can be seen from Table 1, enhancing the output elasticity of physical capital and human capital factor of material products department will enhance per capita economy growth rate in steady state and reduce per capita natural resources consumption growth rate. It can easily be understood that the enhancing of production factor output capacity is beneficial for the transformation of extensive economy into intensive economy, which is ultimately beneficial for long term sustainable economy growth.

Table 1 indicates that the rise of human capital accumulation efficiency parameter $A_H$ means the rise of marginal returns of human capital which is beneficial for the rise of steady state per capita economy growth rate and the drop of per capita natural resources consumption growth rate.

## 4 Conclusion

This paper is trying to build a dynamic growth model to discuss the optimal growth path restricted by natural resources consumption. From this paper we can get the following conclusions:

Firstly, the analysis of the dynamic growth model shows that the optimal sustainable economy growth path can be realized on the condition that intertemporal substitution elasticity is between zero and one and the ratio between human capital accumulation efficiency and time discount is greater than the ratio between the rate of human resources in material production department and material production department together with R&D department.

Secondly, as far as the typical consumers are concerned, in the optimal growth path the larger the consumption time discount rate $\rho$ is, the less patience the consumers have, the weaker awareness of sustainable development the consumers have, the lower the steady state per capita economy growth rate is, the faster the speed of natural resources consumption is, the more unbeneficial it is for the long term economy growth. The smaller material products consumption substitution elasticity of typical consumers is, per capita growth rate in the steady state economy has tendency to decline while the natural resources environment has tendency to be improved, which means there is certain substitution relationship between consumers' consumption of material products and the demand for natural resources environment. The more attention consumers pay to the current natural resources environment quality, and the stronger awareness of sustainable development consumers have, the more beneficial it is for economy growth. These demonstrate that government should pay attention to inducing consumers to consume rationally and enhance awareness of sustainable development.

Thirdly, as far as physical production department is concerned, the larger output elasticity of physical capita and human capital factors means the stronger production capacity of production factors, and the more beneficial it is for the economy to transform from extensive economy to intensive economy and sustainable economy growth. As far as human capital accumulation department is concerned, the higher human capital accumulation efficiency, the higher marginal returns of human capital, the more beneficial it is for long term sustainable economy growth. These demonstrate that government should pay attention to improving human capital accumulation efficiency.

## References


Aghion P, Howitt P (1998) Howitt, endogenous growth theory [M]. MIT Press, Cambridge, MA, pp 3–105

Bovenberg A, Smulders S (1995) Environmental quality and pollution-augmenting technological change in a two-sector endogenous growth model [J]. J Public Econ 57:368–389

Dasgupta PS, Heal G (1979) Economic theory and exhaustible resources [M]. Cambridge University Press, Cambridge, UK

D'Arge RC, Kogiku KC (1973) Economic growth and the environment [J]. Rev Econ Stud 61–77

Forster BA (1973) Optimal capital accumulation in a polluted environment [J]. Rev Econ Stud 39:544–547

Garg PC, Sweeney JL (1978) Optimal growth and depletable resources [J]. Resour Energy 43–56

Grimaud A, Rouge L (2003) Non-renewable resources and growth with vertical innovations: optimum, equilibrium and economic policies [J]. J Environ Econ Manage 45:433–453

Jones C (1995) R&D-based models of economic growth [J]. J Polit Econ 103:769–785
Keeler E, Spence M, Zeckhauser R (1971) The optimal control of pollution [J]. J Econ Theory 4:19–34
Romer P (1990) Endogenous technological change [J]. J Polit Econ 98:71–102
Solow R (1974) Intergenerational equity and exhaustible resources [J]. Rev Econ Stud (Symposium Issue):29–46
Stiglitz J (1974) Growth with exhaustible natural resources: efficient and optimal growth paths [J]. Rev Econ Stud (Symposium) 41:123–137

# The Sensitive Analysis of Industry-University Collaboration Mode on Regional Innovation Capacity

**Hui Yin, Yun Jun, and Fang Xie**

**Abstract** The research of the Impacts of Industry–University (I/U) collaboration mode on regional innovation capacity is vital which has theoretical value and practical significance for building the national innovation system, improving national independent innovation capability and economic development. Some researchers have studied the topic, but they had different opinions of the impact. Based on the statistical data of 31 provinces of China, we do a regression analysis of this issue with the modified Jaffe knowledge production function. Analysis showed that the enterprise leading mode of I/U collaboration has significant impact on regional innovation capabilities, and the impact on output quantity is slightly greater than the output quality. At the same time, universities and research institutions leading mode have no significant effect on regional innovation capabilities. Finally, we put forward some policy recommendations to promote regional innovation.

**Keywords** Industry-University (I/U) collaboration mode · Knowledge production function · Patent data · Regional innovation capacity · Regression · Sensitivity

## 1 Introduction

Reported by OECD, in the era of knowledge economy, the strength and quality of the relationship of industry and science (ISRs) is playing an increasingly important role. In the process of the implementation of independent innovation and intellectual property strategy, the university-Industry collaboration is one of the

H. Yin (✉) and F. Xie
School of Management, Zhejiang University, YuHangTang Road No. 388, 310058, HangZhou, People's Republic of China

Y. Jun
School of Business, Hohai University, XiKang Road No. 1, 210098, Nanjing, People's Republic of China

Y. Zhou and D.D. Wu (eds.), *Modeling Risk Management for Resources and Environment in China*, Computational Risk Management,
DOI 10.1007/978-3-642-18387-4_7, 

key issues, but also is the weakest link. Therefore, the research of the Impacts of Industry-University (I/U) collaboration mode on regional innovation capacity is vital which has theoretical value and practical significance for building the national innovation system, improving national independent innovation capability and economic development.

Industry-University (I/U) collaboration means enterprises, universities and research institutes take various economic cooperation including R&D, marketing, consulting through negotiation for the market needs and mutual benefits based on the market economy mechanism.

In the perspective of cooperation type, I/U collaboration mode can be divided into technology transfer, contract research, joint research, internal integration, building a science facility, setting up R&D entities, joint training of talents and talents exchanges, industrial technology Union and other types. In the perspective of cooperation body, I/U collaboration mode can be divided into three kinds which are point to point, point to chain and cooperation network. In the perspective of the role of government, I/U collaboration mode can be divided into the spontaneous market, government guidance and the government leading.

There are two different systems of I/U collaboration in China. One is university-leading & technology-driven mode, and the other is enterprises-leading & market pull mode. In the latter mode, the subject of cooperation project is decided by enterprises, which provide the essential collaboration resources, manage the cooperation projects, commercialize the technology. I/U collaboration have been dominated by the former pattern for nearly 20 years. More and more challenges aroused in the recent years. According to the reality, the cooperation funding of contract research is increasing, which means that the second mode has been paid more attention to.

Many researchers have studied I/U collaboration and regional innovation ability respectively. For instance, Michael D. Santoro, Alok K. Chakrabarti studied the benefits for the enterprises through I/U collaboration. This study indicates that large more mechanistic firms especially those in resource intense industrial sectors use knowledge transfer and research support relationships to build competencies in non-core technological areas. In contrast, smaller more organic firms particularly those in high tech industrial sectors focus more on problem solving in core technological areas through technology transfer and cooperative research relationships. Sun analyzed theoretically the factors which cause regional income gap in the perspective of regional innovation and new economic geography. The result showed that the difference of the regional innovation capability is one of the main reasons of regional income gap.

But there are few studies on the relationship between I/U collaboration and regional innovation ability. And there are big rift on this topic in academic circles: 1. some scholars believe that the I/U collaboration play an important role in regional innovation ability such as Gibbons and Johnston (1974), Salter and Martin. Etzkowitz stated I/U collaboration has played a critical role in regional dynamics by increasing knowledge and human resource reserves, stimulating byproduct of technology or methodology, and affecting the formation of the network. The case study done by Varga, Rip and other scholars verified the importance of I/U collaboration (to regional innovation), and emphasized the interaction between the universities and other regional members. Based on the use of the patents by I/U collaboration, Deng and Zhu (2009) ranked the top eight provinces in regional innovation ability according to their I/U collaboration situation. Comparing to the rank about regional innovation ability, they found that provinces having the higher ability are also stronger in I/U collaboration. 2. Some scholars believe that I/U collaboration have little effect on the regional innovation ability. Yu Mingnan, Zhu Shuai and Bao Lin-Lin studied the impact of I/U collaboration on regional innovation of Beijing and Liaoning Province with modified knowledge production function. They found that the integration of enterprises and universities had not effectively prompted the regional innovation output, and R & D combination did not play its role in strengthening the regional innovation capability.

In short, the related research on I/U collaboration is quite mature in the world, but still being relatively preliminary stage in china. Most of the discussions are qualitative analysis, lacking of detailed quantitative model and empirical analysis of China actual situation. At the same time there are big rift on the relationship between I/U collaboration and regional innovation ability in academic circles.

This paper attempts to do in-depth analysis on the Impacts of I/U Collaboration on regional innovation capacity based on the statistical data of 31 provinces of China with the modified Jaffe knowledge production function. On above study, we do a further analysis on the role of different modes of I/U Collaboration.

The paper proceeds as follows: the second part after the introduction is about the variable design and theoretical model. Then the third part is detailed analysis of the empirical results. The fourth part is the main conclusions and countermeasures.

## 2 Methodology

The research on the relationship between I/U collaboration and regional innovation ability involve the dependent variable of regional innovation ability, and the independent variable of I/U collaboration and other control variables. In the previous study, measurement of these indicators can be classified into following categories.

In the report of Regional Innovation Capability of China (2006–2007), the measurement of I/U collaboration is mainly by two indicators: One is the he proportion of funds from the enterprise in all raised funds of universities and research institutes; another is joint application data of three types of patent.

Deng and Zhu (2009) analyzed the changing rate of the I/U collaboration patent and the proportion of invention patent in the total patents of several years. They used I/U collaboration patent for the measurement of I/U collaboration. In addition, Shen (2005) said that the technical content of invention patents is highest among the three forms of patents (inventions, utility models, industrial designs). Therefore, the number of invention patents and its proportion in total patents are core indicators for national or regional innovation ability. The trend of patent structure also evidenced the strategic orientation of regional innovation and the accumulation degree of the original innovation capacity.

In view of the scholar's opinions above, we use the number of patent applications as a regional innovation output indicator, while using the number of invention patent applications as the core indicator of regional innovation capability which reflects the quality of regional innovation. The joint application of three types of patent is used as an indicator of I/U collaboration. This indicator can be subdivided into three kinds: enterprises leading, universities leading and research institutions leading.

Knowledge Production Function (abbreviation KPF) is a useful tool for analysis of regional innovation (Uallachain and Leslie 2007), for example, Michael Fritsch used a KPF approach to measure the quality of regional innovation systems of 11 Euroupe cities, such as Barcelona, Rotterdam, Vienna and so on.

The basic assumption of KPF is that the output of the innovation process is viewed as a function of R&D investment. Griliches and Jaffe viewed patents as tangible manifestations of the investment firms and other institutions make in generating new knowledge. In regional analysis, this approach assumes that there exists some function:

$$CP = f(I, U, F, \varepsilon). \tag{1}$$

where CP is commercial patent production and I, U and F are levels of R&D investments performed by industry, universities and federal agencies.

Given commercial spillovers from university R&D in the KPF, Jaffe (1989) and others used a modified Cobb-Douglas model with two inputs as new form of KPF:

$$Ln(P_{ikt}) = \beta_{1k}\ln(I_{ikt}) + \beta_{2k}\ln(U_{ikt}) + \beta_{3k}[\ln(U_{ikt})\ln(C_{ikt})] + \varepsilon_{ikt}. \tag{2}$$

where i indexes the unit of observation (states, in this case), k indexes technological areas, and t indexes time. P is corporate patents, a proxy for new economically useful knowledge; I is R&D performed by industry and U is university research. $\varepsilon_{ikt}$ is a stochastic error. The variable C is a measure of the geographic coincidence of university and industrial research activity within the state. Allowing the potency of university spillovers to depend on C is intended to mitigate the arbitrariness of states as observation units.

According to the principle modeling extended KPF by Jaffe (1989), combining conditions of R&D investment subjects in China, we modified (2) again in order to easily analyze impacts of three kinds of I/U collaboration.

$$LnP_i = \beta_1 LnS_i + \beta_2 LnU_i + \beta_3 LnI_i + \beta_4 LnSUI_i + \beta_5 LnUSI_i + \beta_6 LnISU_i + \varepsilon_i \tag{3}$$

Here, (3) represents the quantity model of regional innovation capability, i indexes the unit of observation (provinces, in this case), P is the number of patent applications accepted; I, S,U is R&D investment performed by industry, research institutes and university, respectively. SUI,USI,ISU is research institutions leading, universities leading, enterprises leading I/U collaboration respectively. I/U collaboration is a measure of coincidence of university and industrial research activity within the province. Different I/U collaboration can be measured by the number of joint application patents with different first applicant accordingly.

$$\begin{aligned}\mathrm{Ln\,IP_i} = {} & \beta_1' \,\mathrm{Ln\,S_i} + \beta_2' \,\mathrm{Ln\,U_i} + \beta_3' \,\mathrm{Ln\,I_i} + \beta_4' \,\mathrm{Ln\,SUI_i} + \beta_5' \,\mathrm{Ln\,USI_i} \\ & + \beta_6' \,\mathrm{Ln\,ISU_i} + \varepsilon_i' \end{aligned} \tag{4}$$

Here, (4) represents the quality model of regional innovation capability. IP is the number of invention patent applications accepted, other variables have same meaning as these of (3).

## 3 The Results of Empirical Analysis

The empirical data come from two sources published in China, that is, China and Foreign patent database service platform (online) and China statistical yearbook on science and technology 2009. The number of patent and Invention patent applications accepted directly from the former, The number of joint applications patent and Invention patent indirectly from the former by identifying field of first joint applicant.

In order to collect data of R&D investment of universities, research institutes and large & medium-sized enterprises in 31 provinces except Hong Kong, Macao, Taiwan, we refer to items of the latter data source, included Intramural Expenditures on S&T Activities in Institutions of Higher Education by Region (2008), Intramural Expenditures on S&T Activities in Large and Medium-sized Industrial Enterprises by Region (2008), Intramural Expenditures on S&T Activities in R&D Institutions by Region (2008) et al.

Given different unit between dependent variables and independent variables, These variable were standardized at first, then, we make the stepwise regression analysis on (3) and (4) by sending data into SPSS.

Table 1 showed the stepwise regression part results of the quantity model (3) using SPSS16.

**Table 1** Quantity model 1 summary

| Model | R | R square | Adjusted R square | Std. error of the estimate | Durbin-Watson |
|---|---|---|---|---|---|
| 1 | .928[a] | .861 | .854 | .28140221 | |
| 2 | .967[b] | .935 | .928 | .19759181 | 1.808 |

[a]Predictors: (Constant), ZLn

[b]Predictors: (Constant), ZLnI, ZLnISU

**Table 2** Quantity model summary

| Model | R | R square | Adjusted R square | Std. error of the estimate | Durbin-Watson |
|---|---|---|---|---|---|
| 1 | .883[a] | .780 | .769 | .31674120 | |
| 2 | .943[b] | .888 | .877 | .23097885 | |
| 3 | .963[c] | .927 | .915 | .19216845 | 2.015 |

[a]Predictors: (Constant), ZLnISU
[b]Predictors: (Constant), ZLnISU, ZLnI
[c]Predictors: (Constant), ZLnISU, ZLnI, ZLnU

The analysis on (3) indicates that the two variables (ZLnI – R&D investment of large & medium-sized enterprises and ZLnISU – enterprises leading I/U collaboration) stand out and enter the regression model in a specific order, having the positively significant impacts on the quality of regional innovation (Coefficients were 0.850, 0.346, respectively; all $p < 0.01$), Results of the one-way ANOVA indicated that there are a significant difference between two variables ($F = 142.706$, $p < 0.001$). The model fit well (adjusted $R^2 = 0.935$).

Table 2 showed the stepwise regression part results of the quantity model (4) using SPSS16.

The analysis on (4) indicates that the three variables stand out and enter the regression model in a specific order, including ZLnISU – enterprises leading I/U collaboration, ZLnI – R&D investment of large & medium-sized enterprises and ZLnU – R&D investment of universities, having the positively significant impacts on the quality of regional innovation (Coefficients were 0.278, 0.486, 0.340, respectively; all $p < 0.01$), Results of the one-way ANOVA indicated that there are a significant difference between two variables($F = 79.946$, $p < 0.001$). the model also fit well (adjusted $R^2 = 0.927$).

## 4 Conclusions and Suggestions

We can conclude from above third part of the analysis that (1) there are most significant impacts of R&D investment of large and medium-sized enterprises on quantity (patents number) and quality (invention patents number) of regional innovation capability; (2) there are significant impacts of enterprises leading I/U collaboration on quantity and quality of regional innovation capability; and (3) there is significant impact of R&D investment of universities on quality of regional innovation capability.

Jaffe (1989) reported a statistically significant positive relationship between commercial patenting and industrial and university R&D expenditures. This conclusion can be evidenced by the contribution of R&D investment of large and medium-sized enterprises and universities to regional innovation. Moreover, our findings further shows that R&D investment of large and medium-sized enterprises play a vital role in regional innovation, especially for quantity of regional innovation, while R&D investment of universities only has significant impacts on quality.

For enterprises leading I/U collaboration, It has significant impacts on regional innovation. Moreover, its impacts on the quantity are slightly greater than that on the quality of regional innovation. This result shows the enterprises as the main subject effectively enhance regional innovation capabilities through actively conducting I/U collaboration driven by the market mechanism.

As for research institutes in China, its reform have been carried out since 1985, included use of the market mechanism, alteration of internal and external environment and improvement of the efficiency & effectiveness of scientific researches. However, empirical studies have found that research institutions leading I/U collaboration doesn't significantly affect regional innovation capacity.

Meanwhile, universities leading I/U collaboration also doesn't significantly affect regional innovation capacity in despite of universities R&D investment effectively enhancing regional innovation quality. Parity reason is poor evaluation index system of universities in China. Nie Ye, Wupei Kai, Guo-Gang Sun ( 2008) think that our current evaluation system of research universities is unfavorable to cooperation among industries and universities and research institutes. Only number of papers and funds, no practical value of papers are considered in evaluating teachers in many universities. To some extent, it hindered process of translation from scientific research achievements into productive force.

According to this empirical research, we also propose some policy suggestions to improve the regional innovation capability as follows:

1. In view of a vital role played by enterprises, Government should encourage and guide enterprises in R&D investment. Simultaneously, actively promote enterprises leading I/U collaboration in the whole society.
2. Deep reforms of research institutes need be continued. National research agents should be run as enterprises, on the contrary. some research institutes established by large and medium companies have played an essential role in regional innovation, for example, 97 patents (invention, designs and practical new patents were 3, 32 and 62, respectively) are joint applied by Zhejiang Geely Automobile Research Institute Company Limited and Zhejiang Geely Holding Group Co., Ltd. in 2008, its proportion in total joint application patents in Zhejiang Province is about 20%.
3. Evaluation system in Colleges and Universities need be further reformed in china, especially for colleges of science and engineering. As an indispensable link in technology innovation system, cooperation between industry-university-research should be took into account in evaluation system to encourage teachers to apply for more patents of practical value.

This study makes a quantitative analysis of I/U collaboration mode and regional Innovation capability using patent data. Limitations lie in that a single indicator is easy to understand and analyze, however, it might not be comprehensive. In future research, we will increase the time series data or other indicators in order to conduct further research on this issue.

## References

Deng Y, Zhu G (2009) The patent analysis research on the industry and university collaboration in China. Sci Sci Manag S&T 12:16–22

Gibbons M, Johnston R (1974) The economic benefits from science. Res Policy 22:220–242

Jaffe AB (1989) Real affects of academic research. Am Econ Rev 79(5):957–970

Nie Ye, Wu PeiKai, Sun GuoGang (2008) A study on quantitative assessment of university rankings and of industry-university (I/U) collaboration's effects. Sci Technol Manag Res 11: 76–81

Research Group on Development and Strategy of Science and Technology of China (2006–2007) Annual Report of Regional Innovation Capability of China. . . Intellectual Property Publishing House, Beijing, 2007, pp 27–28

Shen N (2005) A study on divergence of regional innovation capacity in china based on patent structure. In: Proceedings of the 1'st Symposium of S&T Policy and Management, China, pp 700–710

Uallachain BO, Leslie TF (2007) Rethinking the regional knowledge production function. J Econ Geogr Adv Access 7:1–16

# Statistical Investigation into the Accelerating Cyclical Dynamics Among Gold, Dollar and U.S. Interest Rate

**Nan Yang and Jieyi He**

**Abstract** Based on time-series decomposition, HP filter and cycle locating procedure, the paper extracts the concrete cycle information from gold, US$ exchange rate, and US interest rate time series, on all four durations of the past 30 years. (In this paper the US$ exchange rate is reflected by U.S. Dollar Index, and the interest rate refers to the US$ interest rate.) It shows that the cycle length has shortened and a more sensitive interaction among the three variables appeared since 2008 financial crisis. Further, application of VAR model indicates not only autoregressive relationship among the three variables, but also gold price to be the leading indicator, with 1-month ahead of the dollar exchange rate and 2-months ahead of the interest rate. This significant counter-cyclical pattern may be suggestive of government's reserve portfolio allocation, especially when facing the uncertainty of the economy and the underlying inflation expectation.

**Keywords** Counter-cyclical pattern · Cross correlation · Gold price · HP filter · VAR

## 1 Introduction

During the recent US recession, there are obvious winners in terms of asset performance, like fixed-income assets, and obvious losers, like cyclical stocks. However, among all the assets, gold has aroused substantial interest for its role as a diversifying asset. Attention has been focused on gold's superior role of early warning indicator of inflation, the relationship between gold and each exchange rate, the structural change in government's reserve asset management, and so on

N. Yang (✉) and J. He
School of Statistics and Management, Shanghai University of Finance and Economics, Shanghai, China
e-mail: yangnan@shufe.edu.cn; hejieyixs@yahoo.cn

Y. Zhou and D.D. Wu (eds.), *Modeling Risk Management for Resources and Environment in China*, Computational Risk Management,
DOI 10.1007/978-3-642-18387-4_8, © Springer-Verlag Berlin Heidelberg 2011

(Dempster 2008; Aritgas 2010; Capie et al. 2004; Dempster and Milling-Stanley 2010).

Actually, the study of gold market behaviour and its relationship with the spot market has long been concentrated. I.S. Gulati and Ashoka Mody (1982) have proposed that higher inflationary expectations and exchange rate fluctuations may create greater investment demand for gold. They also believe that since gold is a non-interest bearing asset, large increase in interest rates may discourage holding of gold. Eric J. Levin and Robert E. Wright (2006) found the negative relationship between changes in the gold price and changes in US dollar trade-weighted exchange rate, as well as gold's role as a long-run hedge against inflation. Further researches were down by M. W. Luke Chan and Dean C. Mountain (1988) to discover that change in a precious metal price generally can be a function of both present and lagged values of a substituted precious metal price, the interest rate, and the lagged values of its own price.

Now given the close relationship and dramatic changing behaviour, in contrast of the gold price, US dollar and interest rate, this paper focuses on solving the following questions quantitatively: how the dynamic among them changed before and after the recent financial crisis, and what does it imply?

To investigate this, the paper is, therefore, organized as follows: examine the cycles of gold price movement and related variables; diagnose the cycle properties of the involving variables; test the leading and lagging quality among them and especially, explore the unique and correspondingly predicting characteristic of gold. If the mapping of US dollars and gold price's moving trends has accelerated, the political implication may be valuable to China's government.

## 2 Exploratory Analysis

History has witnessed large fluctuations in gold price, especially after the collapse of the Bretton Woods regime of fixed exchange rates in the early 1970s. Many researchers are concerned with the length and size of gold-price cycles, for improving the understanding of these cycles will be a key input in efforts to stabilize the macroeconomic effects of movements in prices (Cashin and McDermott 2002). Among the macroeconomic factors, inflationary expectations, exchange rate fluctuations and changes in interest rate are the main forces to be thought to correlate closely to gold prices (Gulati and Mody 1982; Mahdavi and Zhou 1997).

The paper is concerned with the empirical behaviour of gold price, US$ exchange rates and interest rate before and after the 2008 financial crisis, particularly changes in the periodicity and endogenous lead-lag relationship to gold. Firstly, the long-run trend movements indicate the counter-cyclical fluctuation between Gold Price and USDX, Gold Price and US$ interest rate, which is shown in Fig. 1.

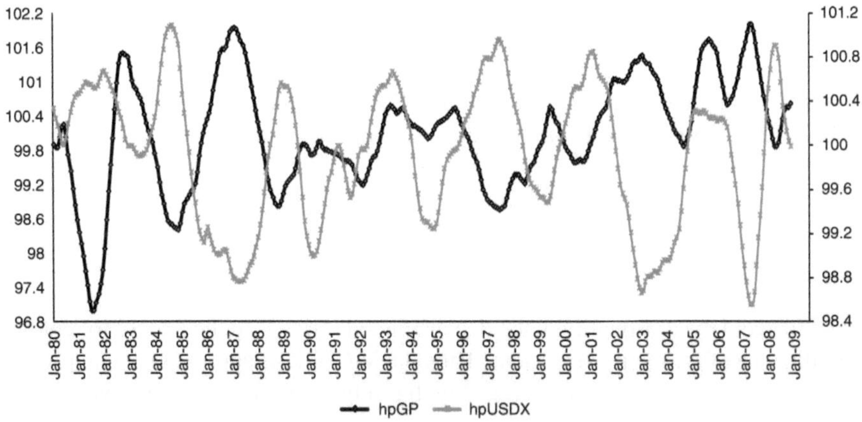


**Fig. 1** Linear graph of gold price and USDX movements (Data of gold price from http://www.research.gold.org, data of USDX from http://www.federalreserve.gov)

# 3 Empirical Investigation

## *3.1 Data Description*

Empirical monthly series of gold price, USDX and interest rate from 1980–2010 are used in this paper. To distinguish any different characteristic of the series after 2008, we focus on the movement divergences of four durations in these 30 years, which are:

1980–01~1986–12, the period after the disintegration of Bretton Woods system
1987–01~1996–12, the period covering the great crash of the world stock market
1997–01~2007–12, the period after the Asia Financial Crisis
2008–01~2010–04, the period after the recent financial shock

## *3.2 Cyclical Component Series Calculation*

To extract the cyclical component from the original series, the decomposition technique is applied in cyclical analysis, enabling the cycle to be defined in terms of deviation of the series from trend, with proper modification (Cuddington 1992; Reinhart & Wickham 1994).

The cyclical component series for Gold Price/USDX/Interest Rate is calculated by time-series decomposition and the monthly relative returns for $\{X_t\}$ are used:

$$X_t = T_t \times S_t \times C_t \times I_t \tag{1}$$

$$X_t^{GP} = \ln R^{GP} \tag{2}$$

Here $R^{GP}$ is the real price of gold in dollar. So do USDX and Interest Rate.

In order to observe the behaviour of $C_t$, the decomposition is done as follows:

$$Y_t = \frac{1}{n}\sum_{k=t}^{t+n-1} X_k, t = 1, 2, \cdots, n = 12 \tag{3}$$

As to avoid the seasonality in the deconstructed series $Y_t$, define n=12 to remove the influence of seasonal component. Then the MA of $Y_t$ is applied in decomposing the irregular component of the series, highlighting long-term trends or cycles:

$$MA_t = TC = \frac{1}{2}\sum_{k=t}^{t+1} Y_k,\ t = 1, 2, \cdots \tag{4}$$

Next, use HP Filter to simulate the long term progressing-$T_t$:

$$\sum_{t=1}^{T} (y_t - s_t)^2 + \lambda \sum_{t=2}^{T-1} ((s_{t+1} - s_t) - (s_t - s_{t-1}))^2 \tag{5}$$

Where $y_t$ represents $X_t$ and $s_t$ represents $T_t$.

HP filter, first used by Hodrick and Prescott (1997) to analyze post-war U.S. business cycles, is a two-sided linear filter that computes the smoothed series $S_t$ of $y_t$ by minimizing the variance of $y_t$ around $S_t$, subject to a penalty that constrains the second difference of $S_t$. That is, the HP filter chooses $S_t$ to minimize. The parameter $\lambda$ controls the smoothness of the series, the larger the $\lambda$, the smoother the $S_t$. Here we specify $\lambda$ directly, yielding the original HP values for $\lambda$, which is 14400 in monthly series. Finally, $C_t$ is defined as:

$$C_t = MA_t/T_t \tag{6}$$

## 3.3 Cycle Locating Procedure

Here the commonly used definition of a cycle is used. A boom is defined as a sequence of absolute increases in the level of prices. A slump is defined as a sequence of absolute decrease in the level of prices. A cycle includes one boom and one slump (Cashin and McDermott 2002). Besides, the following algorithm is designed to locate cycles:

Step1 (S1 in short): Search for $C_{t^*} = \max\{C_t, t = 1, 2, \cdots, k\}$, and use it to separate the left and right sides of the whole duration, denoted by $A_L$ and $A_R$ correspondingly.

S2: Search for all the peaks and valleys:

$$P_i = \{C_i : C_{i-1}<C_i, C_{i+1}<C_i\}, V_j = \{C_j : C_{j-1}>C_j, C_{j+1}>C_j\} \tag{7}$$

S3: Locate the booms and slumps on $A_L$:

Define $B_{L,1}$ to be the first span on which all the peaks and valleys are increasing, and is the first boom from the left side of $C_{t^*}$, $S_{L,1}$ to be the first span on which all the peaks and valleys are decreasing, and is the first slump to the left of $B_{L,1}$.

$$B_{L,1} = [l_{1,L}, l_{1,U}], S_{L,1} = [h_{1,L}, h_{1,U}] \tag{8}$$

$$l_{1,U} = C_{t^*} l_{1,L} = V_m \; m = \min\{v : P_i \geq V_v, V_i \geq V_v, i = m+1, \cdots, t^*\} \tag{9}$$

$$h_{1,U} = V_m \; h_{1,L} = V_n \; n = \min\{v : P_i \leq V_v, V_i \leq V_v, i = n+1, \cdots, m\} \tag{10}$$

Starting from $S_{L,1}$ and searching to the left, repeat the same procedure to detect all the booms and slumps on $A_L$.

S4: Symmetrically use the procedure in S3 to locate the booms and slumps on $A_R$.

S5: Search on $A_L$from $C_{t^*}$, define a cycle to be the union of a slump and its succeeding boom, and detect all the cycles. Symmetric algorithm applies to $A_R$.

The above algorithm is applied on series of gold price, USDX and interest rate and all the cycles are detected, except when the first and last ones are round up if not complete. Accordingly, cycle properties of the three variables, including numbers, length, maximum amplitudes, can be recognized and are shown in Table 1.

**Table 1** Detected cycle properties

| Duration | Variables | Cycle numbers | Cycle length (year) | Max amplitude (%) |
|---|---|---|---|---|
| 1980–01 ~ 1986–12 | GP | 3 | 2.33 | 4.58 |
| | USDX | 2 | 3.5 | 2.1 |
| | IR | 2 | 3.5 | 18.15 |
| 1987–01 ~ 1996–12 | GP | 2.5 | 4 | 3.18 |
| | USDX | 2.5 | 4 | 1.90 |
| | IR | 2 | 5 | 16.35 |
| 1997–01 ~ 2007–12 | GP | 3 | 3.67 | 3.23 |
| | USDX | 3 | 3.67 | 2.44 |
| | IR | 2.5 | 4.4 | 15.8 |
| 2008–01 ~ 2010–04 | GP | 1 | 2.34 | 2.08 |
| | USDX | 1 | 2.34 | 1.72 |
| | IR | 1 | 2.34 | 23.64 |

Information on the average duration and amplitude of commodity-price cycles can be used in designing domestic counter-cyclical policies, in examining whether it is useful to borrow externally in presence of a temporary adverse shock, or in determining the efficacy of national commodity stabilization and international market sharing agreement (Deaton 1999; Deaton and Miller 1996).

Focusing on the properties of cycles above, it is clear that the three variables keep shorter cycles after 2008, and the cycle lengths turn out to conform to each other. The amplitude presented between gold price and USDX over the four durations suggests that they are contrary in the trend, but conform in absolute value over 2008. A high frequency with declining amplitude series of the real gold price is exactly what we expect of a perfect inflation hedge, theoretically. Meanwhile, the conforming, after 2008, between gold price and USDX probably indicates an immediate interaction mechanism since the recent financial crisis.

## 3.4 Lag Selection with Cross Correlation

The price of gold bears a relation to the growth of money supply and may be a leading indicator of a recovery in velocity and a rise in inflation pressure (Aritgas 2010). On the other hand, fluctuations of USDX and interest rate usually accompanied with adjustment of money supply. Therefore, a lead/lag relationship is highly possible between gold and USDX/interest rate. In this section, Cross Correlation analysis is applied to figure out the proper lead/lag relationship among these three variables. The data is transformed as following:

$$x_{t\pm i} = \frac{C\_X^{GP}_{t\pm i}}{X^{GP}_{t\pm i}\_trend}, \; i = 0, 1, 2, \cdots \tag{11}$$

$$y_t = \frac{C\_Y^{USDX}_t}{Y^{USDX}_t\_trend} \tag{12}$$

Where

$C\_X^{GP}_{t\pm i}$ = the cyclical component of i months prior or delay.
$X^{GP}_{t\pm i}\_trend$ = the long-term trend of i months prior or delay.[1]
$C\_Y^{USDX}_t$ = the cyclical component of USDX.
$Y^{USDX}_t\_trend$ = the long-term trend of USDX. So do the other two variables.

Correlation coefficients are calculated as follows:

$$r(\pm i) = \sum x_{t\pm i} \cdot y_t \Big/ \sqrt{\sum x^2_{t\pm i} \cdot \sum y^2_t} \tag{13}$$

[1]Here $C\_X^{GP}_{t\pm i} = X^{GP}_{t\pm i} - X^{GP}_{t\pm i}\_trend$, $X^{GP}_{t\pm i}\_trend = T^{GP}_{t\pm i}$, and so does $Y_t$.

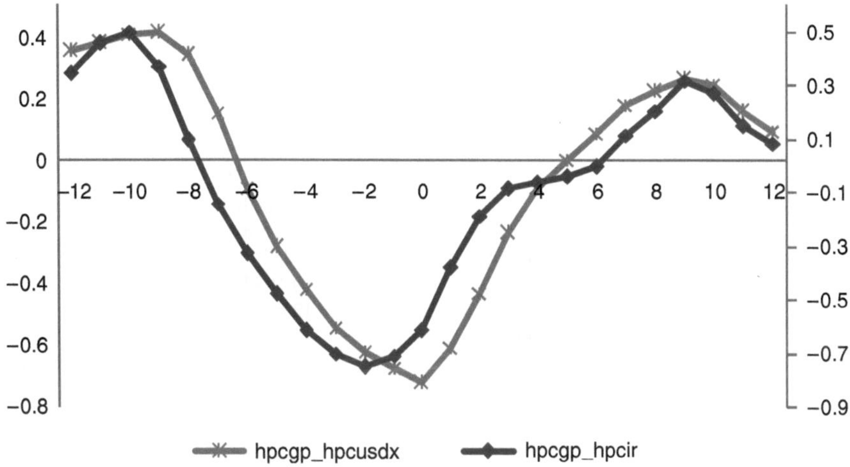


**Fig. 2** Graph of correlation coefficients during 2008~2010

The results show that the peak absolute value of correlation coefficients is 2-months lag to 1-month prior in gold price in 2008–01~2010–04, and 1-month lag to 2-month prior in other three periods. Figure 2 shows the case in 2008–01~2010–04. Thus [−2, 2] is concluded as the most proper lead / lag. In other words, the gold price movement has the most significant impact on the USDX, or interest rate, which is 2-months lag to 2-months prior.

## 3.5 VAR Model Application

We use VAR model to specify the accurate leading/lagging order of the gold price, and the dynamic interaction among the three variables. The model is to excavate the interactive mechanism among gold price, USDX, interest rate, differing only in the lags. Namely:

$$Y_t = A_t + \Gamma_1 Y_{t-1} + \Gamma_2 Y_{t-2} + \ldots.. + \Gamma_k Y_{t-k} + \varepsilon_t \tag{14}$$

Where $Y_{t-i} = 3^*1$ matrix, $i = 0, 1, 2$ $\quad A_t = 3^*1$ parameter matrix

$\Gamma_j = 3^*3$ parameter matrix, $j = 1, 2$ $\quad \varepsilon_t \sim N(0, \Omega)$

As results of cross correlation have indicated the comparatively higher correlation between the gold price and USDX, interest rate with an upper limit order of 2, we adjust the model like this way:

$$Y_t = A_t + \Gamma_0 Y_t + \Gamma_1 Y_{t-1} + \Gamma_2 Y_{t-2} + \varepsilon_t \tag{15}$$

**Table 2** Empirical result of VAR model in 2008–2010

| | Y=lnUSDX | Y=lnIR |
|---|---|---|
| Variables | Coefficient | Coefficient |
| C | −0.026407<br>(0.493034)<br>[−0.053560] | −12.16408<br>(10.49636)<br>[−1.158885] |
| lnGP | −0.242769<br>(0.096864)<br>[−2.506284] | – |
| lnGP(−1) | 0.265467<br>(0.103731)<br>[2.559171] | – |
| lnGP(−2) | – | 1.498295<br>(0.710083)<br>[2.110026] |
| lnUSDX | – | – |
| LnUSDX(−1) | 0.971777<br>(0.082298)<br>[11.80805] | −5.446411<br>(2.009567)<br>[−2.710241] |
| lnUSDX(−2) | – | 5.857539<br>(2.551067)<br>[2.296114] |
| lnIR | – | – |
| LnIR(−1) | – | 0.990015<br>(0.075043)<br>[13.19259] |
| lnIR(−2) | – | – |
| R-squared | 0.859725 | 0.963293 |
| Adjusted R-squared | 0.841428 | 0.956301 |
| F-statistic | 46.98782 | 137.7753 |
| Prob (F-statistic) | 0.000000 | 0.000000 |

Standard error in ( ); t-statistics in [ ]

All the possible leading/lagging relationships among the three variables are tried and the best simulation is presented in Table 2.

The analysis suggests that a 1% increase in current gold price, on average, will cause a 0.24% decline in USDX, assuming the other variables remain unchanged. Similarly, a 1% increase in gold price 1 month ahead will cause a 0.27% increase in USDX and a 1% increase in USDX itself 1 month ahead will cause the current increase by 0.97%. As to the interest rate, a 1% increase in gold price 2 months ahead tends to increase the current interest rate by 1.50%.

## 4 Conclusion

The first conclusion, therefore, there is a significant counter-cyclical pattern between the gold price and the dollar exchange rates, and also such a pattern between the gold price and the US interest rate through all the past 30 years. That is, if the dollar and the interest rate appreciate, the dollar gold price falls and vice versa. It is also noticeable that the pattern among the three series has accelerated since the recent financial crisis. The cycle length has shortened and the amplitude of gold price and USDX has conformed, both indicating an even sensitive corresponding system among the three variables since 2008.

Gold is indeed a hedge against fluctuations of the US$ on the foreign exchanges, and more than it. It has been accepted by previous researches that microeconomic variables such as exchange rates are autoregressive to some extent, and our paper provides clear statistical support for this relationship after the impulse of 2008 financial crisis. The empirical result of VAR model indicates not only the autoregressive relationship among the three variables, but also gold price to be the leading indicator. Gold exhibits a negative current correlation to dollar, which is consistent with their long-run behaviour, and 1-month leading property for the dollar exchange rate and 2-months for interest rate. Thus, there is evidence that gold price has an impact on future dollar exchange rate and US$ interest rate.

Because the dollar remains the mainstay of the reserve portfolio of both advanced and developing economies, and the US economy's exceptionally loose monetary and fiscal policy will lead to renewed dollar depreciation, policy makers of China have even called into question the dollar's role as a reserve currency. When turning to the policy implications of this analysis, US wealth holders should profit from holding gold, especially after this recession. Gold has a long-standing negative relationship with the dollar, which makes it an effective hedge against any further dollar weakness. The matching of US dollars and gold price has accelerated, and this point is valuable to China's government. According to the data from the World Gold Council in June 2010, China is now the fifth on gold holdings and its occupation of gold is 1.6% among the total foreign reserves. Adjustment on the government's reserve portfolio allocation seems to be necessary, especially when facing the uncertainty of the economy and the underlying inflation expectation.

**Acknowledgments** This work was partially supported by Shanghai University of Finance and Economics through Project 211 Phase III and Shanghai Leading Academic Discipline Project, Project Number: B803.

## References

Aritgas JC (2010) Linking global money supply to gold and to future inflation. Technical report, World Gold Council, Feb 2010

Capie F, Mills TC, Wood G (2004) Gold as a hedge against the US dollar. Technical report, World Gold Council, Sept 2004

Cashin P, McDermott CJ (2002) The long-run behavior of commodity prices: small trends and big variability. IMF Staff Pap 49(2):175–199

Cuddington JT (1992) Long-run trends in 26 primary commodity prices. J Dev Econ 39:207–227

Deaton A (1999) Commodity prices and growth in Africa. J Econ Perspect 13:23–40

Deaton A, Miller R (1996) International commodity prices, macroeconomic performance and politics in sub-Saharan Africa. J Afr Econ 5:99–191

Dempster N (2008) What does a US recession imply for the gold price? Technical report, World Gold Council, Apr 2008

Dempster N, Milling-Stanley G (2010) The importance of gold in reserve asset management. Technical report, World Gold Council, June 2010

Gulati IS, Mody A (1982) International gold price movements, 1972–1982. Econ Polit Wkly 17 (46/47):1861–1870

Hodrick RJ, Prescott EC (1997) Postwar U.S. business CYCLES: an empirical investigation. J Money Credit Bank 29:1–16

Levin EJ, Wright RE (2006) Short-run and long-run determinants of the price of gold. In: World Gold Council, June 2006

Luke Chan MW, Mountain DC (1988) The interactive and causal relationships involving precious metal price movements: a analysis of the gold and silver market. J Bus Econ Stat 6(1):69–77

Mahdavi S, Zhou S (1997) Gold and commodity prices as leading Indicators of inflation: tests of long-run relationship and predictive performance. J Econ Bus 49:475–489

Reinhart CM, Wickham P (1994) Commodity prices: cyclical weakness or secular decline? Staff Pap Int Monet Fund 41:175–213

# An Index Tracking Model: One Application of Integer Programming

**Dexiang Wu**

**Abstract** Index tracking, as a popular form of investment strategy, has wildly accepted by portfolio managers in industry, i.e. constructing your own portfolio to reproduce the target index. Traditionally, reproducing stock market index requires large amount of money to invest into all stocks, but this seems impossible when stock number up to thousands, one alternative way to capture the market movements is to invest in a few representative stocks that mimic the index. Portfolio managers often face a problem of figuring out the portfolio that will optimally perform this task. In this project we adopt the index tracking model from Cornuejols and Tutuncu (2007) and apply it to track the S&P100 and compare the computational results of the target index and the portfolio. Matlab and Gurobi V4.0 are used as solvers in the project.

**Keywords** Index tracking · Integer programming · Portfolio optimization

## 1 Introduction

Since the first Mean-Variance Optimization (MVO) model is introduced by Markowitz (1952), the financial products get an exponential increase every year. Portfolio management involves investing large amount of money into various equities in the world's market. But how to figure out which combination of stocks to invest in to achieve the highest possible return is a challenge for portfolio managers. Basic strategies adopted by fund managers are always can be classified as active management or passive management (Beasley et al. 2003), here we focus on the second strategy, i.e. passive management.

One passive investing strategy is index tracking – to emulate market movement, because historical data and research experience has shown market always has a

D. Wu (✉)
Department of Mechanical and Industrial Engineering, University of Toronto, Toronto, ON, Canada
e-mail: dextrewu@gmail.com; dextre.wu@utoronto.ca

Y. Zhou and D.D. Wu (eds.), *Modeling Risk Management for Resources and Environment in China*, Computational Risk Management,
DOI 10.1007/978-3-642-18387-4_9, 

positive expected return over a long period of time. Some famous indexes, e.g. S&P500, are always have a better performance than the most single stock in stock market because the index offset the market fluctuation, so tracking the target index become a nature action for investor, another reason is that index tracking freed investor from the frequent trading to other investment activities, which also save huge transaction cost. To achieve this passive investing strategy, it requires one to monitor the market activity. Fortunately, thousands of indexes have been created to keep track of the particular market of a particular industry, or as a whole. Give the assumption that the target has a good performance, we just truck these indexes. The simplest investment strategy is to replicate all stocks to your portfolio. However, total replication yields many disadvantageous, for example insufficient budget, transaction costs, number of stocks available for holding, and small fraction of investment and so on (Canakgoz and Beasley 2008).

To accomplish index tracking in an elegant manner, one possible objective is to find out most representative stocks to invest to achieve highest possible return, and this can be challenging, but many researchers are involving in portfolio construction for many years. Dimitris Bertsimas et al. (1999) used a mixed integer programming to build the multi-objective portfolio, they minimum the downside risk between portfolio and target index. Li et al., related index tracking as minimizing a return weighted squared difference between stock weights in the portfolio and stock weights in target (Li et al. 2005).

The rest of the chapter is organized as follows: Sect. 2 formulates the Portfolio Selection model and develops the model with sector limit. Section 3 presents the computation results of validation experiments, and Sect. 4 concludes the chapter.

## 2 Model Formulation

### *2.1 Basic Index Tracking Model*

The model selects stocks with highest correlation to track the index shows as following pure Zero–one integer model,

$$\max \sum_{i=1}^{n} \sum_{j=1}^{n} \rho_{ij} x_{ij}$$

$$\begin{aligned} \text{s.t.} \quad & \sum_{j=1}^{n} y_j = q \\ & \sum_{j=1}^{n} x_{ij} = 1 \quad \forall i = 1, \ldots, n \\ & x_{ij} \le y_j \quad \forall i = 1, \ldots, n; \quad j = 1, \ldots, n \\ & x_{ij}, y_j \in \{0, 1\} \end{aligned}$$

Where $\rho_{ij}$ represent the similarity between asset $i$ and asset $j$; $q$ denotes portfolio size. The first constraint denotes portfolio size; the second constraint means each stock has exactly one representative in the portfolio; the third constraint means any stock must be in the portfolio to be a representative.

### 2.1.1 Decision Variables

Let $x_{ij}$ represent whether stock $j$ is a representative of stock $i$, $x_{ij}$ is 1 if $j$ is the most similar asset in the portfolio to $i$, 0 otherwise. Let $y_j$ represent if asset $j$ is selected to be in the portfolio (1 if true, 0 otherwise). i.e.

$$x_{ij} = \begin{cases} 1 & if \quad stock\ j\ is\ most\ similar\ to\ stock\ i \\ 0 & o/w \end{cases}$$

$$y_j = \begin{cases} 1 & if \quad stock\ j\ is\ selected\ to\ represent\ any\ stock \\ 0 & o/w \end{cases}$$

Having solved the model, a weight $w_j$ is calculated for each selected asset $j$ using sum of the market value, $V_i$, of each stock from the index it is representing by

$$w_j = \sum_{i=1}^{n} V_i x_{ij}$$

For example, if stock 1 represents stock 2 and 3 in the portfolio, we sum the market value of stock 2 and 3, and then divide the whole market value; the ratio $w_1$ is the proportion we should invest in stock 1.

## 2.2 *An Simple Example*

Assume that stocks pool $i = \{1,2,3,4,5,6,8,9,0\}$ set $q = 5$.

We randomly generate correlation of return between stock $i$ and $j$, i.e. 10*10 matrix. Then compare the solution time under Matlab and Gurobi.

We can easily see how the portfolio is built in Table 1, we also find that Gurobi is much quick than Matlab, especially stocks number over 50. So we use Gurobi for large scale computation.

**Table 1** Compute time (CT: second)

| | Variables # | Constraints # | Opt. solution | Opt. value |
|---|---|---|---|---|
| Portfolio selection model | 110 | 111 | {1,2,3,5,9} | 5.8706 |
| | Matlab(bintprog) | | Gurobi_mex | |
| Portfolio selection model | 0.5086 | | 0.0704 | |

# 3 Experimental Results

## *3.1 Model Testing*

We randomly generate the correlation and test the capacity of our laptop. All results are computed by Gurobi with 1.5 GB virtual memory available, we find that our laptop can cover 1,200*1,200 at most, with means we can track a target index which including 1,200 stocks. Test result show as follow:

The last column of Table 2 show the general number of variable and constraint and tell us how long we can solve the problem, we can get a picture about the problem, for example, if we want track 400 stocks then there are 160,400 variables and 160,401 constraints, this size can be solve in 5 min with Gurobi. Table 3 shows the results.

## *3.2 Tracking S&P100*

We obtained historical data for S&P100 from 2000 to 2010 using Bloomberg. The data is taken in into Matlab as Excel file. The data is available from internet, and is also provided as soft copy when the project is handed in. Daily returns from January 3, 2000 until December 31, 2008 were used to generate portfolio by using the model; Performance was tracked out of sample from January 2, 2009 to December 31, 2010. (2,767 days). We set portfolio size $q$ from 10 to 100 per 2, and solve the problem one by one. We main show four Performance metrics, i.e. the Mean return of the portfolio; standard deviation (SD); risk/return; Objective value. Result show as follow:

Figure 1 tell us that the difference when portfolio size between 10 and 40 is bigger than portfolio size between 40 and 100, Fig. 2 give a drop tendency of standard deviation with portfolio size go to large, Fig. 3 shows that the Sharp ratio changes and it better than the situation that portfolio size is small, Fig. 4 tell us that the tracking result of the portfolio selection model, actually when portfolio size greater than 70, it almost tracking the whole market capitalization of target index.

**Table 2** Decision variable number and constraint number

| | | 100*100 | 400*400 | 800*800 | 1,200*1,200 | i*j |
|---|---|---|---|---|---|---|
| Time of yield data (s) | | 6.2994 | 237.0881 | 452.1953 | 676.6974 | – |
| Selection model | CT(s) | 8.1170 | 58.2257 | 78.7544 | 129.0825 | – |
| | Variable# ($10^5$) | 0.101 | 1.604 | 6.408 | 14.412 | i*j + j |
| | Constraint# ($10^5$) | 0.10101 | 1.60401 | 6.40801 | 14.41201 | i*j + i + 1 |

**Table 3** Comparison between the two models

| | | 100*100 | 400*400 | 800*800 | 1,200*1,200 |
|---|---|---|---|---|---|
| Selection model | q | 40 | 180 | 380 | 580 |
| | zmax1 | 65.8970 | 248.0347 | 502.0956 | 797.2367 |

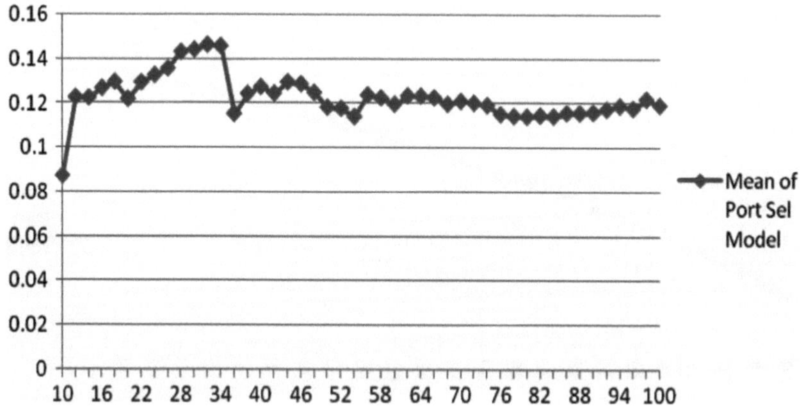


**Fig. 1** Mean return of portfolio

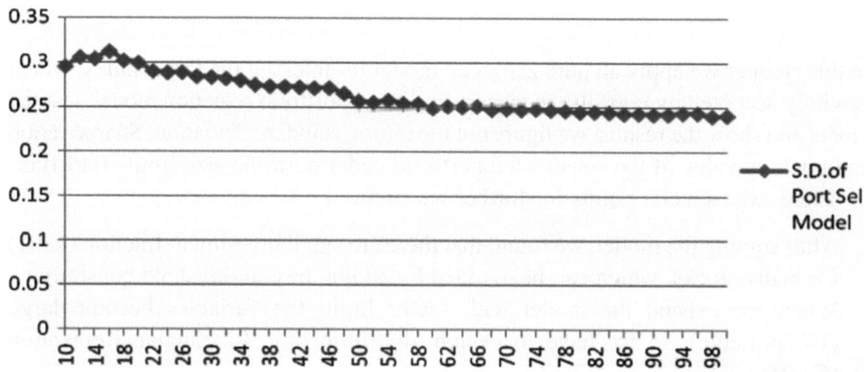


**Fig. 2** Standard deviation of the portfolio

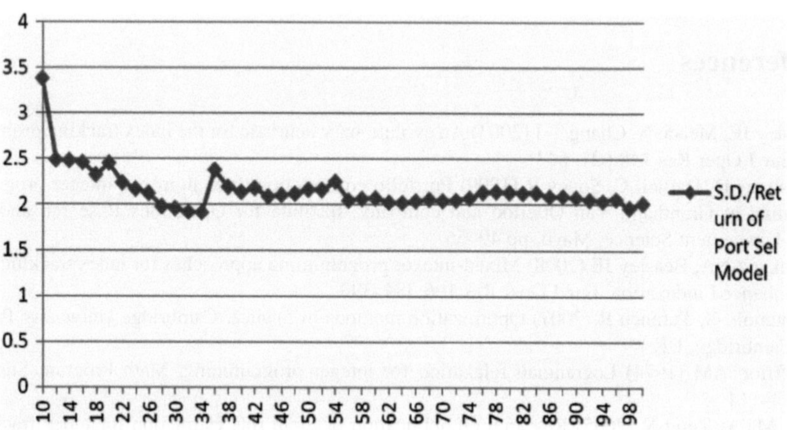


**Fig. 3** Sharp ratio (Std/Mean) of the portfolio

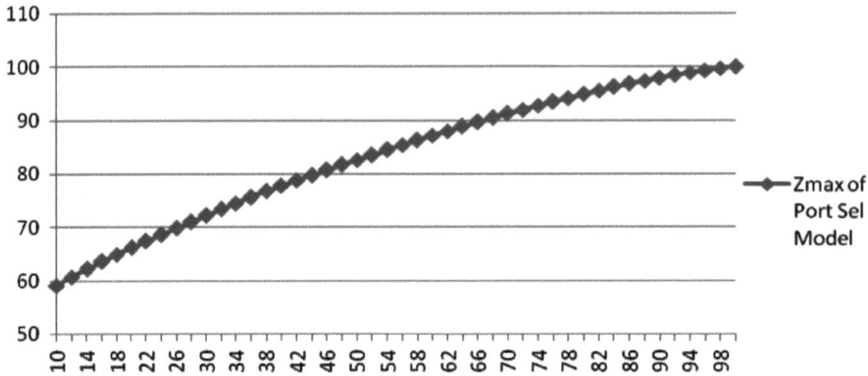


**Fig. 4** Tracking result for each portfolio

# 4 Conclusion

In this chapter we apply an pure zero–one model for tracking the target index, which is wildly accepted by portfolio managers, we use a portfolio selection model, test the model and show the results, we figure out the return, standard deviation, Sharpe ration and tracking value of the constructed portfolio under portfolio size from 10 to 100.

There exist several points for further research:

1. After solving the model, we found that there are small investment fraction (<.01) for many stocks, which can be avoided by adding buy-in threshold constraint;
2. When we expand the model with sector limit, the variables become large ($10^7$ possible), so we have to design algorithm (e.g. Lagrangian Relaxation [Geoffrion 1974]) to solve it;
3. The objective value will be sensitive since the similarity will vary over time, so we must consider about robust mechanism to overcome it.

# References

Beasley JE, Meade N, Chang T-J (2003) An evolutionary heuristic for the index tracking problem. Eur J Oper Res 148:621–643

Bertsimas D, Darnell C, Soucy R (1999) Portfolio construction through mixed-integer programming at Grantham. Van Otterloo and company, Institute for Operations Research and the Management Science, Mayo, pp 49–66

Canakgoz NA, Beasley JE (2008) Mixed-integer programming approaches for index tracking and enhanced indexation. Eur J Oper Res 196:384–399

Cornuejols G, Tutuncu R (2007) Optimization methods in finance. Cambridge University Press, Cambridge, UK

Geoffrion AM (1974) Lagrangian relaxation for integer programming. Math Program Stud 2: 82–114

Li J, Ma Y, Zeng Y (2005) Research on application of Stein rule estimation for index tracking problem. In: Proceedings of the 2005 international conference on management science and engineering, 12th (1–3): 434–439

Markowitz HM (1952) Portfolio selection. J Financ 7(1):77–91

# Part II
# Risk Management in Engineering Projects

# Shallow Groundwater Quality Evaluation in Huaibei Based on the Uncertainty Theory

**Haifeng Lu and Baoyuan Yuan**

**Abstract** In order to understand the risk level of the shallow groundwater pollution in Huaibei, a method of the support vector machine (SVM) based on the uncertainty theory was proposed to evaluate the water quality. First, the principle of SVM was introduced, and a multi-class classifier of the SVM model was established based on it. Data from some water samples were used to train in the SVM model secondly. And then the model was used to predict the unknown samples, which were fit to the results calculated by fuzzy comprehensive evaluation (FCE). The results show that the SVM model of groundwater quality evaluation is reliable. It is suitable for the data processing based on finite number of training samples, with special technique to restrict overfitting. And method to establish the SVM model is simple.

**Keywords** Stochastic · SVM · Uncertainty · Water quality evaluation

## 1 Introduction

Huaibei is a comprehensive city which subjects with coal, power and chemical industry in Anhui province of China. Due to the industrial and mining enterprises wastewater, domestic sewage, mining solid wastes (coal gangue, fly ash, etc.) and farmland sewage irrigation, the urban shallow groundwater environment suffers great influences. So, to understand the condition of groundwater pollution, gradation and evaluation for water pollution is the primary task at present.

There are some traditional methods for water quality evaluation, such as scoring method and statistical method, etc. (Cao 1989). In the recent years, some new evaluation methods (such as clustering method, economic analysis method, fuzzy

H. Lu (✉) and B. Yuan
College of Earth Science & Engineering, Hohai University, Nanjing 210098, China
e-mail: luhaifeng7571@126.com

Y. Zhou and D.D. Wu (eds.), *Modeling Risk Management for Resources and Environment in China*, Computational Risk Management,
DOI 10.1007/978-3-642-18387-4_10, © Springer-Verlag Berlin Heidelberg 2011 

evaluation method, gray correlation method and artificial neural network method, etc.) were proposed by many scholars (Feng et al. 2002). These methods can basically reflect the water quality condition. However, because of the complex nonlinear relationships between the evaluation factors and the water quality, they have exposed some defects, particularly the modeling difficult, calculation complexity and poor general availability.

The support vector machine (SVM) is a small sample studying method with solid theoretical foundation based on statistical theory. SVM has many special advantages in solving the problem of small sample learning, and nonlinear and high dimensional pattern recognition, which makes the SVM become an excellent machine learning algorithm (Yu and Yang 2008; Geng and Zhong 1980). In view of this, in order to increase the accuracy of classification, SVM was used to evaluate the shallow groundwater quality in Huaibei.

## 2 Basic Principle of the SVM

Construction of optimal hyperplane is the core technology of SVM. Its basic idea can be explained as shown in Fig. 1 (in two dimensional case) (Deng and Ti 2004).

In Fig. 1, a series data points for two different classes of data are shown, black squares for negative class and white circles for positive class. The SVM tries to place a linear boundary between the two different classes, and orientate it in such way that the margin represented by the dotted line is maximized. Furthermore, SVM attempts to orientate the boundary to ensure that the distance between the boundary and the nearest data point in each class is maximal. Then, the boundary is placed in the middle of this margin between two points. The nearest data points that used to define the margin are called support vectors, represented by the grey circles and squares. When the support vectors have been selected, the rest of the feature set is not required, as the support vectors can contain all the information-based need to define the classifier. For mathematical description of the SVM see reference (Deng and Ti 2004).

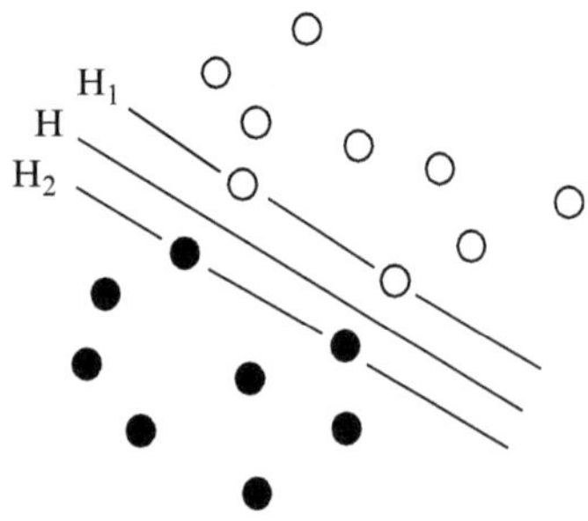


**Fig. 1** Schematic diagram showing SVM

# 3 Model of Water Quality Evaluation

## 3.1 Determination of Model Parameters

We choose 40 groundwater sampling points in Huaibei (Fig. 2). Sampling time was in April, 2007. Sampling points basically located in the main groundwater exploitation area. Therefore, they were representative groundwater samples. Test items in the laboratory included salinity, total hardness, nitrate, sulfate, chloride, fluoride, cyanide, arsenic, hexavalent chromium, ammonia nitrogen, volatile phenol and escherichia coli. Indoor detection method of national standard was used to test the samples. Based on the water quality characteristics in Huaibei, measured values of 6 main influential factors were used as evaluation indexes of water quality

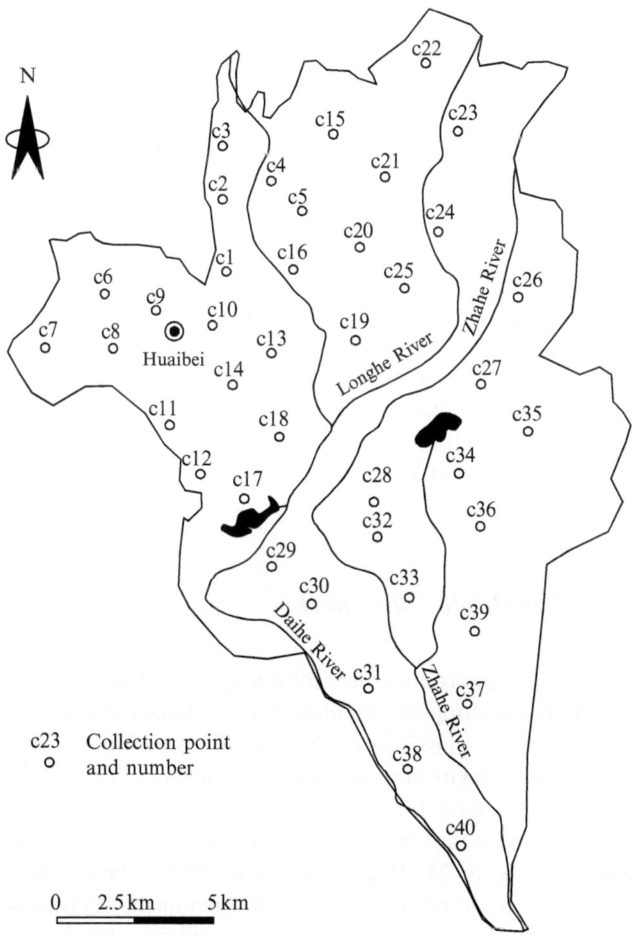


**Fig. 2** Location of groundwater samples

**Table 1** Analysis data of 15 groups of groudwater samples in Huaibei (mg/l)

| Number | Total hardness | Chloride | Sulfate | Total phosphorus | Total nitrogen | Nitrate nitrogen |
|---|---|---|---|---|---|---|
| c2 | 543.21 | 34.21 | 112.34 | 0.10 | 0.66 | 0.31 |
| c4 | 248.33 | 41.34 | 37.5 | 0.17 | 11.32 | 3.27 |
| c6 | 348.37 | 54.27 | 59.74 | 0.12 | 6.13 | 1.89 |
| c8 | 313.25 | 66.72 | 141.83 | 0.19 | 6.33 | 1.13 |
| c12 | 488.21 | 169.53 | 99.78 | 0.07 | 2.05 | 0.91 |
| c14 | 475.48 | 89.62 | 47.23 | 0.18 | 7.15 | 1.73 |
| c17 | 364.27 | 89.6 | 170.9 | 0.33 | 5.32 | 0.76 |
| c19 | 373.32 | 7.19 | 26.92 | 0.18 | 0.01 | 0.31 |
| c21 | 732.7 | 279.9 | 196.03 | 0.11 | 23.73 | 15.37 |
| c23 | 316.35 | 78.21 | 31.44 | 0.25 | 5.56 | 1.59 |
| c28 | 588.33 | 227.38 | 62.15 | 0.11 | 3.61 | 1.23 |
| c31 | 342.31 | 151.21 | 32.38 | 0.13 | 0.14 | 0.35 |
| c33 | 374.41 | 13.45 | 75.45 | 0.13 | 5.78 | 1.14 |
| c36 | 285.14 | 23.36 | 41.25 | 0.12 | 0.10 | 0.25 |
| c40 | 483.22 | 314.72 | 79.72 | 0.21 | 0.16 | 0.30 |

**Table 2** Comparison of algorithm of the multi-class classifier of SVM

| Classification algorithm | Correct rate | Cost of training | Cost of studying | Samples of repetitive training |
|---|---|---|---|---|
| One to many | Moderate | Moderate | Greater | More |
| One to one | Moderate | Less | Moderate | More |
| Decision tree | Higher | Less | Moderate | Less |

(Table 1). Taken the groundwater quality standard (GB/T14848-93) and the drinking water quality standard (GB5749-85) of China as the evaluation standard, combined with the environmental hydrogeology condition and pollution source distribution in this area, evaluation grade of water quality was divided into five levels: level I (clean), level II (relatively clean), level III (slight pollution), level IV (relatively pollution) and level V (serious pollution).

## 3.2 *Choice of Multiple Classifiers*

Algorithm of SVM is actually classifiers for binary classification. Due to the water quality evaluation has multiple levels, classifiers for binary classification does not meet the requirements of classified evaluation. So, algorithm of the multi-class classifiers are a research focus in recent years. Algorithm of the multi-class classifiers of SVM mainly has three types at present (Table 2).

Through comparison on the classification effects of three kinds of the multi-class classifier patterns of the SVM (Wang and Wang 2004), classification pattern of binary decision tree was adopted in the model of water quality evaluation. Based on the structure of the binary tree, the arrangement order of levels of quality was carried out according to the decreasing order of different kinds of samples. And it

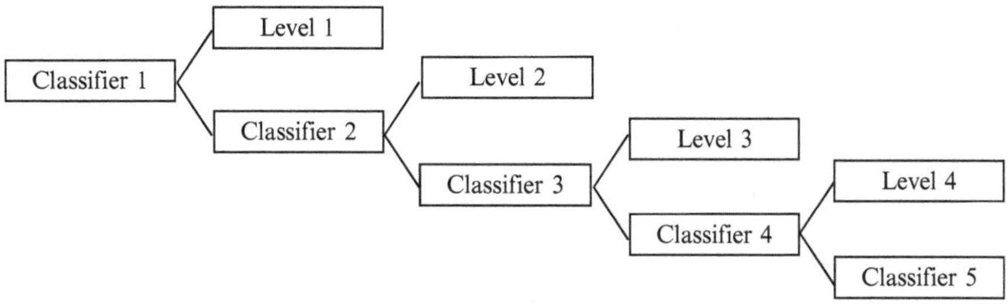


**Fig. 3** Model of the binary tree of five classifications

was numbered I, II, III, IV and V in sequence. The model of the binary tree was finished as shown in Fig. 3. It indicated that four classifiers should be established.

## *3.3 Sample Learning*

Based on the selection principle of the samples and water quality standard, 15 sets of data were used as training samples. They were as follows: c2, c4, c6, c8, cl2, c14, c17, c19, c22, c24, c26, c28, c30, c32, c45, including samples of five levels (I, II, III, IV and V). And the other 25 sets of data were used as testing samples. In addition, to avoid the dominance of a few large descriptors on the diversity score, scaling becomes necessary if the chemical descriptors use different units and vary significantly in their magnitudes. It is customary to standardize data entries by "minimum-maximum transformation".

$$x_{std} = \frac{x - x_{\min}}{x_{\max} - x_{\min}}. \tag{1}$$

where $x$ is the original data, $x_{std}$ is the standardized data, $x_{max}$ and $x_{min}$ are the maximum and minimum value for descriptors, respectively.

LIBSVM library file was used to establish the multi-class classifier of the SVM (Yan et al. 2006). The LIBSVM which uses MATLAB 6.5 software as platform is the current popular library of algorithm of SVM. It can determine the kernel function and penalty factor ($C$), etc. through changing the parameters. Performance of SVM often depends on the selection of kernel function and its width coefficient, penalty factor and insensitive loss factor. To establish the model of groundwater quality evaluation need to select those parameters rightly. Radial basis function (RBF) is taken as a kernel function in the paper, and in order to improve the accuracy of regression model, the insensitive loss factor is taken as 0.01. The penalty factor $C$ and the Gauss RBF kernel's deviation $\sigma$ were determined 100 and 1, respectively.

# 4 Evaluation Results

During binary decision tree is used for classification, object attribute is substituted into classification function to test value from the tree root. When the test value is 1, it means that this attempt arrives at the leaf node and stops advancing. This indicates that the class where the data is situated is representative class of this leaf node. When the test value is $-1$, this attempt goes down along the branch until it arrives at a given leaf node. Trained model of groundwater quality evaluation was applied in identification and classification of 40 water samples in Huaibei. In addition, in order to check the performance of this method, 15 water samples which were randomly selected were evaluated by fuzzy comprehensive evaluation (FCE). The results were as shown in Table 3 (Yan et al. 2006; Xu and Yang 2006; Guo and Li 2003).

The results indicated that, evaluation results by SVM coincided well with results by fuzzy comprehensive evaluation. Discrimination model was confirmed according to the water quality standard when FCE was used to evaluate the water quality. Then category of water quality in the whole area was defined. But compared with the fuzzy comprehensive evaluation, water samples were trained by the SVM based on the distribution characteristics of water quality. And then water quality in the whole area was defined corresponding category. From the above, the method of the SVM can reflect more the actual situation than the method of the fuzzy comprehensive evaluation. 40 water samples were evaluated through using this method. The results were as follows: 4 level I samples, 9 level II samples; 21 level III samples, 4 level IV samples and 2 level V samples.

**Table 3** Comparison of the test results

| No. | FCE | SVM | | | | | |
|---|---|---|---|---|---|---|---|
| | Level | SVM1 | SVM2 | SVM3 | SVM4 | SVM5 | Level |
| c2 | II | −1 | 1 | −1 | −1 | −1 | II |
| c4 | II | −1 | 1 | −1 | −1 | −1 | II |
| c6 | III | −1 | −1 | 1 | −1 | −1 | III |
| c8 | III | −1 | −1 | 1 | −1 | −1 | III |
| c12 | II | −1 | 1 | −1 | −1 | −1 | II |
| c14 | IV | −1 | −1 | −1 | 1 | −1 | IV |
| c17 | III | −1 | −1 | 1 | −1 | −1 | III |
| c19 | III | −1 | −1 | 1 | −1 | −1 | III |
| c21 | II | −1 | 1 | −1 | −1 | −1 | II |
| c23 | III | −1 | −1 | 1 | −1 | −1 | III |
| c28 | III | −1 | −1 | 1 | −1 | −1 | III |
| c31 | II | −1 | 1 | −1 | −1 | −1 | II |
| c33 | III | −1 | −1 | 1 | −1 | −1 | III |
| c36 | V | −1 | −1 | −1 | −1 | 1 | V |
| c40 | I | 1 | −1 | −1 | −1 | −1 | I |

## 5 Conclusion

- It is suitable for the data processing based on finite number of training samples, with special technique to restrict overfitting. And method to establish the SVM model is simple.
- Data from some samples was used to train and test in the SVM model, the results calculated by SVM model were fit to the results calculated by fuzzy comprehensive evaluation. The study results prove that it is reliable to evaluate shallow groundwater quality by SVM model.
- Many effect factors related to groundwater quality can be comprehensively considered in SVM model Therefore, a new method to evaluate shallow groundwater quality is provided for future research.

**Acknowledgments** The authors gratefully acknowledge Mr. Liang of Nan jing University and Mr. Li for they meaningful comments on this manuscript and support.

## References

Cao WJ (1989) Water resource calculation, evaluation and management. Hohai University Press, Nanjing (in Chinese)

Deng NY, Ti YJ (2004) A novel method of data mining-support vector machine. Science Press, Beijing (in Chinese)

Feng LJ, Li JS, Shao GQ (2002) The application of adaline in recognition of mine water quality types. Coal Geol L Explor 30:35–37 (in Chinese)

Geng D, Zhong W (1980) Determination of basic parameters in ground movement using comprehensive factor for rock character evaluation. J China Coal Soc 4:13–25 (in Chinese)

Guo GD, Li SZ (2003) Content-based audio classification and retrieval by support vector machine. IEEE TMNS Neural Netw 14:209–215

Wang XD, Wang JQ (2004) Research and application of support vector machine. J Air Force Eng Univ (Natural Science Edition) 5:49–55 (in Chinese)

Xu HM, Yang TX (2006) Evaluation of lake water quality based on classification algorithms of support vector machines. J Jilin Univ (Earth Science Edition) 4:570–573 (in Chinese)

Yan ZG, Zhang HR, Du PJ (2006) Application of SVM in analyzing the headstream of gushing water in coal mine. J China Univ Min Technol 16:433–438

Yu N, Yang H (2008) Calculation of ground subsidence coefficient in mining areas using support vector machine regression. J Liaoning Tech Univ (Natural Science) 27:365–367 (in Chinese)

## 5 Conclusion

# The Economic Hazard Analysis of Three Major Regions in China

**Ma Hailiang and Huang Dechun**

**Abstract** The level of China's overall economic development is determined by that of three major economic regions (the Yangtze River Delta, the Bohai Zone and Pearl River Delta). Using 1996–2008 panel data and production function including knowledge stock, the Malmquist method is adopted to measure technical Efficiency, technological Progress and total-factor productivity (TFP). The results are listed as follows: (1) The TFP growth rates of Yangtze River Delta and Pearl River Delta regional were 4.2% and 3.1%, which were higher than that of Bohai Rim region. (2) The TFP growth rates of three major economic regions moved back and forth in 13 years, and negative growth rates were mainly caused by poor pure technical efficiency and technological progress, which shows China's economic development still have greater uncertainty.

**Keywords** Economic hazard · Malmquist index · Total factor productivity · Uncertainty

## 1 Introduction

Since reform and opening, China has become one of the fastest growing economies, so economic performance and growth reason are focused by applying Solow's growth model in many papers. Solow's growth model attributed economic growth to factor inputs increases and total factor productivity (TFP). The technological progress and technical efficiency are usually reflected by TFP without considering factor inputs. Due to not requiring a specific functional form and behaviour assumptions and effectively avoiding the mistakes of production

M. Hailiang (✉)
Business Department, Hohai University, Changzhou, China 213022,
e-mail: hilima@vip.sina.com

H. Dechun
Business Department, Hohai University, Nanjing, China 210098,
e-mail: huangdechun@hhu.edu.cn

Y. Zhou and D.D. Wu (eds.), *Modeling Risk Management for Resources and Environment in China*, Computational Risk Management,
DOI 10.1007/978-3-642-18387-4_11, © Springer-Verlag Berlin Heidelberg 2011

function (Li Gucheng 2009), Non-parametric methods, represented by DEA (Data Envelopment Analysis), are used widely to calculate TFP by constructing the production frontier.

Three major economic regions play significant role in China. In 2008, the sum of GDP in three major economic regions reached 19.67 trillion, accounting for 60.56% of total GDP in China. At the same time, three major economic regions are most efficient in production technology and fastest in economic growth. Therefore, it is important to reach "three-step" strategic objective whether three major economic regions could adjust the industrial structure by technological progress and improve technical efficiency.

Using panel data to analysis empirically TFP of China's three major economic regions, the existing literature will be expanded in the following areas: (1) most papers are related to TFP of a province or an industry's of China, while TFP of the most developed economic region of China is ignored; (2) the input and output factors of Hong Kong and the Macao Special Administrative Region are considered in exploring TFP of the Pearl River Delta; (3) using 1995–2008 panel data of three major economic region, The Malmquist index will be adopted to measure technological progress, pure technical efficiency and scale efficiency; (4) the impact about knowledge stock on TFP growth will be considered.

## 2 Literature Review

Firstly, the TFP growth rates, especially in the agricultural and industrial sectors, were discussed widely. Lambert and Parker, Meng Jie, Chen Weiping used DEA to measure China's agricultural TFP growth rates, and Li GuCheng analyzed technological progress and technical efficiency using micro-data of Jiangsu and Hubei provinces. It is more similar in findings of the agricultural sector. From 1978 to 1985, the TFP growth rate was greatly improved, but in the late 80 it slowed down, and in the early 90s the growth rate was high. The TFP growth of agricultural sector was mainly contributed by technological progress, not technical efficiency. But research results are quite different in industrial sector because of different methods and data applied.

The DEA and Malmquist methods were used by Zheng Jinghai and Arne Bigsten to measure technical efficiency of 600 state-owned enterprises by exploring data of 1980–1994. It was showed that technological progress played important impact on productivity growth and productivity growth rate in large corporations was higher than that in small businesses. Using data of manufacture sector in 28 regions of china in 1987–1995, Gong Jun Tao though that TFP growth came from technological progress while technical efficiency had negative effects on growth. Tu Zhengge and Xiao Geng believed that the increasing gap between the technical efficiencies had challenged the transformation mode of economic growth. Using DEA and SFA methods, the TFP growth of three major economic regions of china was studied by Tang Jie and Meng Yajiang, it was explained that the efficiency improvement was the only way to narrow the gap between regions and the convergence effect showed "first decrease and then increase" trend with the passage of time.

Secondly, the factors that affect total factor productivity were also studied. Using 1960–1990 data of 92 countries, Edwards analyzed impacts had by nine open indicates on TFP growth. It was discovered that opening had significant positive correlation with TFP growth and open country grew faster than protectionist countries. Tu Zhengge pointed out technological progress forefront and resource-allocated efficiency of China's industry were promoted by economic globalization, FDI, ownership changes and orderly competition, which also promoted China's industry from extensive growth to intensive growth. Based on 12 years data in 29 regions, Wang Yanli researched influences exerted by the degree of opening up on technical efficiency difference by SFA method. The results were as explained that from 1992 to 2003 the technical efficiency showed a rising trend, and the technical efficiency of the eastern region is much higher than that of western regions. He Feng believed that the economic openness had an extremely significant impact on the growth of technical efficiency in China, and the impact of FDI was higher than that of international trade. In addition, HeYuanqing, Yao Weifeng, Zai Bin Wu, etc. also analyzed the influence exerted by human capital on TFP growth.

## 3 Research Method and Data Sources

In essence, the ability and effort to shake off poverty and develop economic for a country (region) in a certain period would be reflected by FTP, which is a comprehensive reflection exerted by technological progress on economic development. TFP had three sources: technical efficiency, technological progress and scale efficiency. As Fig. 1 shows, when the $X2$ input factor remains unchanged, by improving management and operational efficiency, the production set moved from point A to point B. That is to say, the production could reach the production frontier by increasing technical efficiency. But the point B is not the best production set. The point C is clearly more efficient than B, and C reached the production frontier at the same time. The movement from B point to C point reflects the scale efficiency. Under the condition of certain input factors, such as X1, $X2$, the production could increase much by moving production frontier (from F to F,). That is the impact of technological progress. Therefore, TFP is the result of interaction of technical efficiency, technological progress and scale efficiency.

### 3.1 *Malmquist Method*

A convenient tool is provided by Malmquist index to analyze TFP changes of all regions. According to this method, not only technological progress change (Techch)

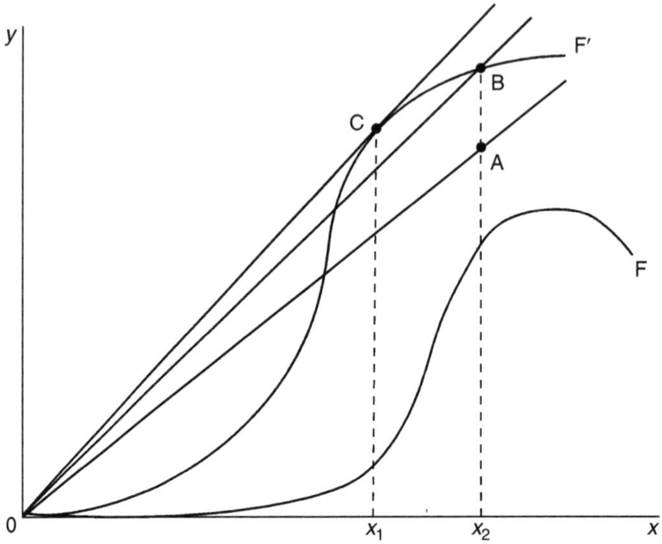


**Fig. 1** The decomposition of total factor productivity

but also technical efficiency change (Effch) could be examined and Effch could be divided into pure technical efficiency change (Pech) and scale efficiency change (Sech), so TFP change (Tfpch) could also be calculated and could show more practical value information. As Arcelus and Arozena pointed out, comparing Fisher index and Tomqvist index, Malmquist index could be divided into technical change and efficiency change without considering price information, thus avoiding the price problems caused by asymmetric information; and it could use the majority of inputs and output factors without considering the conditions of cost minimization and profit maximization.

RD and Fare thought Malmquist index should be defined on the base technology. Supposing (Xt,Yt) and (Xt+1,Yt+1) representing respectively input– output relationship in t and t +1 periods, the change of input–output relationship from (Xt,Yt) to (Xt+1,Yt+1) is the change of productivity change. Productivity change is not only from technological progress, but also from technical efficiency change. Supposing

$D_c^t(x_t, y_t)$ and $D_c^{t+1}(x_{t+1}, y_{t+1})$ being distance functions, the Malmquist index based on t and t +1 periods are as follows:

$$\begin{aligned} M_t(x_t, y_t, x_{t+1}, y_{t+1}) &= \frac{D_c^t(x_{t+1}, y_{t+1})}{D_c^t(x_t, y_t)} \\ M_{t+1}(x_t, y_t, x_{t+1}, y_{t+1}) &= \frac{D_c^{t+1}(x_{t+1}, y_{t+1})}{D_c^{t+1}(x_t, y_t)} \end{aligned} \quad (1)$$

The above index is symmetrical in economic implications. According to Fisher ideal index, their geometric average of them is defined as the integrated productivity index, which is Malmquist index.

$$M(x_t, y_t, x_{t+1}, y_{t+1}) = (M_t \times M_{t+1})^{1/2} = \left[\frac{D_c^t(x_{t+1}, y_{t+1})}{D_c^t(x_t, y_t)} \frac{D_c^{t+1}(x_{t+1}, y_{t+1})}{D_c^{t+1}(x_t, y_t)}\right]^{1/2} \quad (2)$$

How to decompose Malmquist index? Ray and Desli did not agree with Fare. Because most scholars recognized Ray and Desli decomposition (referred as RD model), the RD model is also used in this paper.

$$\begin{aligned} M(x_t,y_t,x_{t+1},y_{t+1}) &= \frac{D_v^{t+1}(x_{t+1},y_{t+1})}{D_v^t(x_t,y_t)} \times \left[\frac{D_v^t(x_t,y_t)}{D_v^{t+1}(x_t,y_t)} \frac{D_v^t(x_{t+1},y_{t+1})}{D_v^{t+1}(x_{t+1},y_{t+1})}\right]^{1/2} \\ &\times \left[\frac{D_c^t(x_{t+1},y_{t+1})/D_v^t(x_{t+1},y_{t+1})}{D_c^t(x_t,y_t)/D_v^t(x_t,y_t)} \frac{D_c^{t+1}(x_{t+1},y_{t+1})/D_v^{t+1}(x_{t+1},y_{t+1})}{D_c^{t+1}(x_t,y_t)/D_v^{t+1}(x_t,y_t)}\right]^{1/2} \\ &= Pech \times Techch \times \mathrm{Sech} \end{aligned} \quad (3)$$

Pech (pure efficiency change) stands for technical efficiency change under the assumption of variable returns to scale. Techch (technical change) reflects the contribution of production frontier movement to productivity changes. Sech (scale efficiency change) indicates the influence of scale economic on productivity change. If TFP index is bigger than 1, the integrated productivity increased, while if less than 1, it means that productivity deteriorated. When one of changes constituting TFP index is bigger than 1, it is explained that it is the root causes of productivity increase. On the contrary, it leads to lower productivity levels.

## 3.2 Data Sources

The input–output data of three major regions in 1995–2008 are adopted. Three major rejoins includes six provinces, three municipalities and two special administrative regions. In economic growth model, capital stock, labor and knowledge stock are as three inputs. The specific input and output data are as follows:

(1) Capital stock. Zhang Jun made systematic research on estimating capital stocks of China using perpetual inventory method. Many researches were made according to capital stock data estimated by him. The period 1995–2008 capital stock data was acquired by using the same method. In addition, the original data is expressed in 1995 price, the corresponding data expressed in 2000 price is adopted by GDP deflator index. The data of Hong Kong and Macao are required by annual exchange rate in yearbook. The unit is million yuan. (2) Labor employment. Labor employment is calculated using average numbers of beginning year employment

and end year employment. The unit is million people. (3) Knowledge stock. Granted patent amounts can be used as knowledge stock. In this paper, the amounts or three granted patent (inventions, utility models, and designs) are regarded as knowledge stock according to this view. (4) GDP. The annual GDP of three regions are used as output factor. The unit is million yuan. GDP are also expressed in 2000 price in consistent with the capital stock.

# 4 Empirical Analysis

Using software Deap 2.1, technical Efficiency, technological progress and total-factor productivity (TFP) of three major economic regions during 1996–2008 are measured. The results are listed as follows:

1. The TFP changes (Tfpch) of three major economic regions showed somewhat fluctuations. During 1996–2008, The Tfpch growth rate of Yangtze River Delta and Pearl River Delta Regional reached 4.2% and 3.1%, which were higher than that of Bohai Rim region. It showed that comparing to the other two regions, Bohai Zone should implement "catch up" strategy and there was a large potential to develop economy. In addition, technological progress and technical efficiency had played important roles to promote regional economic growth for Yangtze River Delta and Bohai Zone. Effch and Techch of Yangtze River Delta were up 3% and 1.2% and those of Bohai Zone reached 2.3% and 6% respectively. It was surprising that while Tfpch reached 3.1%, Techch was −0.2% for Pearl River Delta region during 13 years, which was somewhat similar with Wang Bing's conclusion. It should be pointed out that the outbreak of the financial crisis (2007) lead to negative growth for three regions. Especially, The Tfpch of Pearl River Delta suffered a lot, which was subject to its industrial structure.
2. The average technical efficiency change (Effch) of Pearl River Delta was fastest in three regions and reached 3.3%, while Bohai Zone reached 2.3%. This showed Pearl River Delta was a leading position in technology digestion and production management. Negative Effch growth rate of Yangtze River Delta region appeared three times during 1996–2008, accounting for 23.08% and the lowest one appeared during 2006–2007. The main reason was that scale efficiency was only 0.981. It indicated that pure technical efficiency and scale efficiency were not well coordinated in the same year. Pearl River Delta had negative Effch growth rate for two times and the lowest one was during 2006–2007. Before 2001, Bohai Zone region always had negative pure technical efficiency growth, but Effch growth was always positive for its scale efficiency (Table 1).
3. The technical change (Techch) of three major economic regions almost kept simultaneous trend. During 1996–1997, the lowest Techch appeared in Yangtze River Delta and Pearl River Delta and Techch growth reached −3.7% and −7.9%. The Bohai Zone Techch growth rate also decreased, but the rate

**Table 1** TFP decomposition of three major regions

| | Yangtze River Delta<br>Effch | Techch | Pech | Sech | Tfpch |
|---|---|---|---|---|---|
| 1995–1998 | 1.080 | 1.005 | 1.001 | 1.079 | 1.084 |
| 1998–2001 | 1.013 | 1.042 | 1.000 | 1.014 | 1.055 |
| 2001–2004 | 1.012 | 1.020 | 0.978 | 1.036 | 1.033 |
| 2004–2008 | 1.017 | 0.989 | 0.999 | 1.019 | 1.006 |
| 1995–2008 | 1.030 | 1.012 | 0.995 | 1.036 | 1.042 |
| | Bohai Zone<br>Effch | Techch | Pech | Sech | Tfpch |
| 1995–1998 | 1.004 | 1.041 | 0.976 | 1.028 | 1.045 |
| 1998–2001 | 1.007 | 1.018 | 0.984 | 1.023 | 1.024 |
| 2001–2004 | 1.058 | 0.984 | 1.044 | 1.013 | 1.039 |
| 2004–2008 | 1.024 | 0.987 | 1.008 | 1.015 | 1.011 |
| 1995–2008 | 1.023 | 1.006 | 1.004 | 1.019 | 1.029 |
| | Pearl River Delta<br>Effch | Techch | Pech | Sech | Tfpch |
| 1995–1998 | 1.072 | 0.973 | 1.010 | 1.062 | 1.044 |
| 1998–2001 | 1.008 | 1.016 | 1.003 | 1.005 | 1.024 |
| 2001–2004 | 1.032 | 1.001 | 0.997 | 1.036 | 1.035 |
| 2004–2008 | 1.024 | 1.000 | 1.011 | 1.013 | 1.024 |
| 1995–2008 | 1.033 | 0.998 | 1.005 | 1.028 | 1.031 |

was not obvious. The reason was that flood disaster appeared in 1998 posed a great negative impact on production of two major regions, but Bohai Zone was less affected. In addition, during 2007–2008, there appeared relatively greater Techch grow rate increase in Pearl River Delta. Facing shrinking market, promoting technical level and speeding up industrial transition was the right choice, which led to a more significant growth rate in Techch.

## 5 Conclusion

Using 1995–2008 panel data and the Malmquist method, technical efficiency, technological progress and TFP were measured. The conclusions are listed as follows:

1. The TFP growth of Yangtze River Delta and Pearl River Delta reached 4.2% and 3.1% in 14 years, which was higher than that of Bohai Zone region. The TFP growth of Yangtze River Delta was mainly due to technological progress and scale efficiency, while pure technical efficiency had a negative effect on TFP growth. Technological progress, pure technical efficiency and scale efficiency all played role on TFP growth in Bohai Zone. However, in the Pearl River Delta, the technical had a greater negative impact on TFP growth.
2. The TFP growth rates of three major economic regions moved back and forth in 14 years, and negative growth rates were mainly caused by poor pure technical

efficiency and technological progress. This shows there are still considerable risks for economy growth in three major economic regions. Therefore, it is urgent to improve technical efficiency, promote reforming market economy system in China.

It should be explained due to different data, in particular different input–output factors, different conclusions of TFP may be drawn by different researchers. The impact of human capital on production and factors affecting TFP were not considered because of time, which are research directions in future.

## References

Li Gucheng (2009) Technical efficiency, technological progress and agricultural productivity in china [J]. Econ Rev 1:60–69

# Research on Estimation Method of Gross Registered Tonnage of Ships in Grand Canal

**Xu Peng, Hou Lei, and Wu Zhong**

**Abstract** In this paper, the author uses the method of mathematical statistics to analyze several statistical samples, finds the relationship between the Gross Registered Tonnage of ships with the size of them and finally gets a fitted equation. Subsequently, according to the specification of "The standard ships' main dimensions in Grand Canal", the equation is used to point out the relationship between the Gross Registered Tonnage of common type Grand Canal ships with the size of them.

**Keywords** Gross registered tonnage · regression analysis · size of ships · SPSS

## 1 Introduction

Gross Registered Tonnage (shorted for GRT) is commonly used in counting lockage in toll administration of Waterway Department. The definition of GRT is a quotient that stands for a cubage of ship, and gets from the total enclosed internal space of ship multiplied a certain factor and divide 100 $ft^3$ (2.83 $m^3$).The GRT is commonly used with a unit of Ton, though it is a dimensionless unit.

By the definition of GRT, the GRT has relationship with the size of ships, so the gross tonnage of standard ship can be directly gotten from relevant documents. However, a lot of ships are built in a nonregular shipyard and get no standard dimensions, so there are some difficulties in calculating GRT. At the same time, the ship often uses a smaller certificate of registry instead of an official one for the sake

Foundation Program: The chapter is funded by the natural science funds of Hohai University (2008429911)

X. Peng (✉), H. Lei, and W. Zhong
College of Civil Engineering and Transportation, Hohai University, Nanjing 210098, China
e-mail: robinxp@sina.com

Y. Zhou and D.D. Wu (eds.), *Modeling Risk Management for Resources and Environment in China*, Computational Risk Management,
DOI 10.1007/978-3-642-18387-4_12, 

of evasion of duty when registering (Li 2009). This phenomenon would misguide ships dispatchers and cause dispatching accident which delays ships' flowing through the locks, so it is the most concerned focus for the Waterway Department. On the other hand, it is difficult to check all the ships because it may take much time. Thus, a scientific method is required to support the staff to confirm a factual gross tonnage of ships so as to protect the shipping market environment.

Currently, the computing studies of GRT have caught much attention from the relevant administration and researchers; however there are not sufficient related research achievements. Some references come as follows:

*Primary exploration on the simplified calculational method of minimal ships' maximum capacity* by Shan (2000) analyzes and discusses the relationship between the deadweight tonnage of a ship and its length, width, least freeboard, block coefficient, and provides suggested calculation methods for calculating minimal ships' maximum capacity

*Statistic analysis on net tonnage measurement of ships in domestic voyage* by Sun (2002) talks over the gaps of net tonnages between domestic voyage ships by comparing the calculated value with the standard value of actual ships.

*Suggestions on modification of domestic ships tonnage measurement method* by Lei (2002) probes into the disparities of net tonnages between domestic voyage ships and keelboat, as well as the disparities between voyage ships norm and *1969 International Convention on Tonnage Measurement of Ships.*

*Research on keelboat net tonnage calculation* by Meng (2003) , according to the definition of gross tonnage and net tonnage, establishes the net tonnage coefficient of inland river deck-cargo barge by analyzing and comparing the calculation of typical ships. Furthermore, it presents the net tonnage coefficient of inland river barge by distinguish its different functions and purposes.

*The Regression Analysis about Ship Displacement and Ship Deadweight* by Yang (2008) collects several statistical samples of four ship types of container, bulk cargo, oil and general cargo, takes a relevant analysis on displacement and deadweight tonnage of ships by the method of mathematical analysis, and establishes a regression equation

According to the latest survey data from the Waterway Department, the paper uses the method of mathematical statistics (Lei 2009) to find the function relationship between the size and GRT of ships. It will be helpful for the relevant work staff to estimate the gross tonnage of ships.

## 2 Analysis

From 2008, the author who was entrusted by the Waterway Department of Grand Canal in North Jiangsu, selected passing registration data of vessels in North Jiangsu Grand Canal as a statistical analysis object in a certain period in hopes of getting exact relationship between the size and GRT of ships and supporting the Waterway Department staff's work.

## 2.1 Data Analysis and Introduction to Sample Selection

During the process of resorting to and paying lockage toll in locks, the size and GRT of ships are required to be filled in and stored into toll data base of waterway which also needs to be checked by the staff from the remote dispatch station. Due to the perfection of information system, the lockage information of all ships each time when passing through locks would be preserved. For ships of North Jiangsu Canal, though the phenomenon of misdeclaration and false declaration still exists, ships data are still available and believable widespread by repeated manual checking and system revision.

Lockage data of four locks in North Jiangsu Canal in July, 2008, has been selected to analyze the relationship between the tonnage (gross tonnage, permitted load tonnage) and the size of ships. The result shows that the gross tonnage and the size of ships are highly related with each other, so a function relation can be considered to be established. According to the above analysis, the gross tonnage distribution frequency and the size (length, width, and depth)of ships (fleet, single vessel) distribution frequency are analyzed which provides bases for type distribution research as well as charges approval of ships.

According to the statistics, the selected fleet samples are totally 1,850, the single vessel 6,314. For the sake of error data in ships declaration, by analyzing and communicating with professional staff, two types are to be rejected, that is, type of fleet of length $> 520$ m, width $> 23$ m, depth $= 0$ or depth $> 5$ m, and type of single vessel of 20 m $<$ length $<$ 260 m, 5 m $<$ width $<$ 23 m, depth $= 0$ or depth $> 5$ m.

Statistics are respectively completed in group of fleer and single vessel which are very common in inland waterway transportation. This thesis takes investigation from three indexes: the relationship between tonnage and size, tonnage frequency distribution, and size frequency distribution. By the analysis of the relationship between tonnage and size, regression simulation formula is to be established. By reflecting tonnage frequency distribution and size frequency distribution to several common ship types, actual tonnage of ships in different sizes is to be confirmed, and analogy of ships' tonnage and size are formed finally.

## 2.2 Data Analysis and Fitting

According to the statistics of available ship types in North Jiangsu Canal, by a functional fitting equation, the fitting relationships between ships' tonnage and size come as follows (see Fig. 1 and 2):

### 2.2.1 The Relationship Between Tonnage and Size of Single Vessel

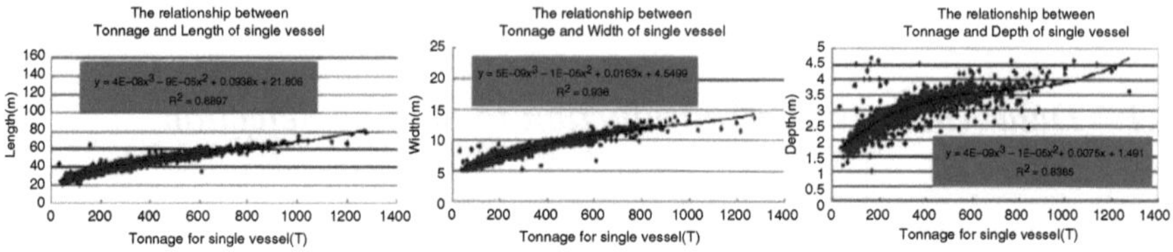


**Fig. 1** Statistical graph of the relationship between tonnage and size of single vessel

### 2.2.2 The Relationship Between Tonnage and Size of Fleet

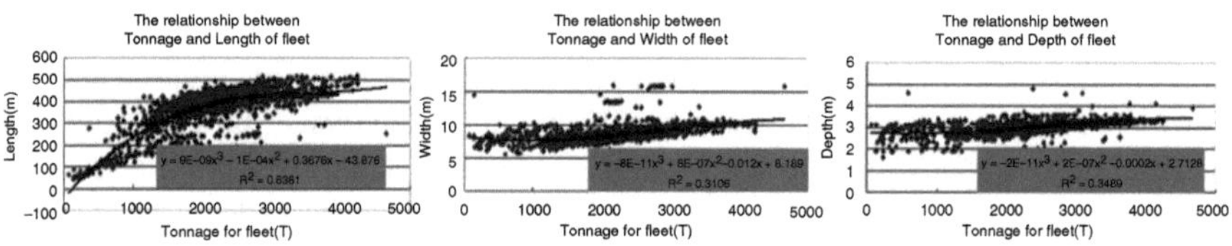


**Fig. 2** Statistical graph of the relationship between tonnage and size of fleet

By the relationship among the different sizes, it's found that the length of sample fleet is mainly in the range of 300–500 m, the gross tonnage of 1,500– 3,500 t, the width of 7–10 m, while the length of about half of single vessels is mainly in the range of 30–50 m, the gross tonnage of 100–200 t, the width of 6–10 m, (as shown in Figs. 3 and 4).

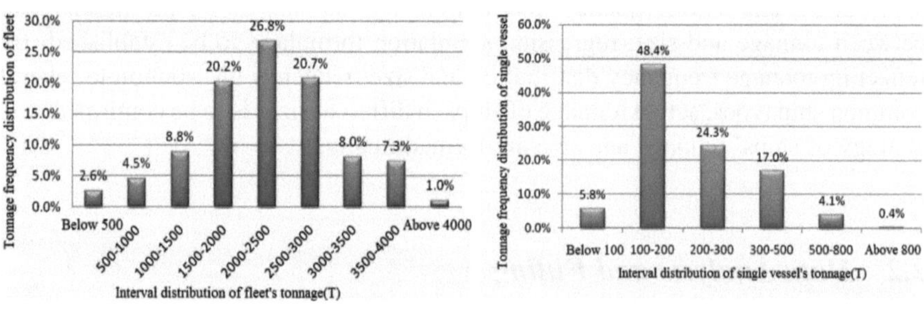


**Fig. 3** Tonnage frequency distribution of available ship types in North Jiangsu Canal

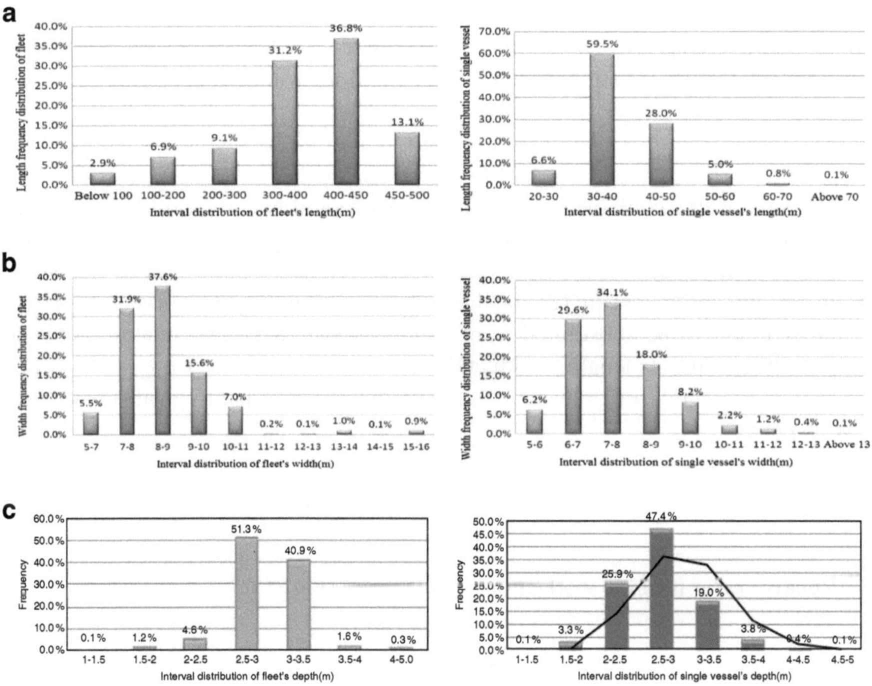


**Fig. 4** Size frequency distribution of available ship types in North Jiangsu Canal. (**a**) Chart of length frequency distribution of fleet and single vessel. (**b**) Chart of width frequency distribution of fleet and single vessel. (**c**) Chart of depth frequency distribution of fleet and single vessel

# 3 Analysis of Statistical Results

According to the data distribution chart, this thesis research aims to find the functional relation between tonnage and size of ships with a view to laying a foundation of reasonable estimation of ship's tonnage.

## *3.1 A Suggested Model*

According to the above simulation chart of single vessel, it's inferred that gross tonnage is closely related with length, width and depth respectively, and formed a power functional relation, which is used generally.

Combined with a mass formula $m = \rho v$ ($m$ stands for mass, $\rho$ for density, $v$ for volume), its functional relation can be expressed as:

$$T = \alpha L^a W^b D^c \tag{1}$$

Here: T stands for gross tonnage of single vessel, L for length of single vessel, W for width of single vessel, D for depth of single vessel, α, a, b, c are parameters.

Both sides of the equal mark in (1) get natural logarithm at one time:

$$LnT = Ln\alpha + aLnL + bLnW + cLnD \tag{2}$$

In terms of the available data of gross tonnage and length, width and depth of single vessel, corresponding transformation (getting natural logarithm respectively) is taken. Then, by getting multiple linear regression analysis from SPSS statistical software, any coefficients in (2) and data-related coefficients are worked out. These parameters are to be substituted in (1), thus get the functional relation between ship's tonnage and its length, width and depth.

Moreover, relative error mean between predicted value and true value is worked out by (3):

$$e = \frac{|T' - T|}{T} \times 100\% \ (T' \text{ is predicted value}) \tag{3}$$

Relative error mean is worked out by (4):

$$\overline{e} = \frac{1}{n}\sum_{i=1}^{n} e \tag{4}$$

## 3.2 SPSS Error Analysis

Through the research, it's found that the length of ships is the most related element with gross tonnage. The relationship between single vessel and its gross tonnage is part of the relation between fleet and its tonnage, thus by dividing the length of ships, the functional relation between the length and gross tonnage can be got.

According to length distribution, and taking the standard size of inland waterway ships into consideration, the section are divided as Table 1 in the thesis:

The Table 2–3 below shows the result of statistical analysis, in which the related coefficient and the relative error mean are checked and ratified by SPSS statistical software.

**Table 1** Length distribution of single vessel

| Total size | Section (m) | | | |
|---|---|---|---|---|
| 20–80 m | 20–42 | 42–50 | 50–64 | 64–80 |

**Table 2** Model Summary

| Model | R | R square | Adjusted R square | Std. error of the estimate |
|---|---|---|---|---|
| 1 | 0.984[a] | 0.968 | 0.968 | 0.066411 |

[a]Predictors: (Constant), LN Depth, LN Length, LN Width

**Table 3** Coefficients

| Model | | Unstandardized coefficients | | Standardized coefficients | t | Sig. |
|---|---|---|---|---|---|---|
| | | B | Std. error | Beta | | |
| 1 | (Constant) | −1.419 | 0.040 | | −35.079 | 0.000 |
| | LN Length | 0.851 | 0.022 | 0.304 | 39.442 | 0.000 |
| | LN Width | 1.502 | 0.024 | 0.513 | 61.309 | 0.000 |
| | LN Depth | 0.645 | 0.015 | 0.204 | 42.643 | 0.000 |

**Table 4** The relation function of ship's tonnage and its size

| Scope of length | Predictor formula | Related coefficient | Mean value of relative error |
|---|---|---|---|
| 20–42 m | $T = 0.348 \times L^{0.690} \times W^{1.570} \times D^{0.730}$ | 0.963 | 4.90% |
| 42–50 m | $T = 0.159 \times L^{1.036} \times W^{1.415} \times D^{0.567}$ | 0.936 | 4.00% |
| 50–64 m | $T = 0.367 \times L^{0.983} \times W^{1.241} \times D^{0.383}$ | 0.937 | 5.70% |
| 64–80 m | $T = 0.004 \times L^{2.347} \times W^{0.823} \times D^{0.204}$ | 0.838 | 10.10% |
| 20–80 m | $T = 0.242 \times L^{0.851} \times W^{1.502} \times D^{0.649}$ | 0.984 | 4.80% |

In the table, T stands for the gross tonnage of ships, L for the length, W for the width, D for the depth

## 3.3 *Establishment of Regression Formula*

The sectional predicted functional relation of each brake cable in Table 4 is worked out from the analysis model and its statistical analysis results in the Part 3.1 of the thesis.

# 4 Conclusion

By the regression formula in the thesis and compared with the standard ships, the figure of relationship between ship type and tonnage in North Jiangsu Canal is finally determined (Wang 2001) .the administrative staff in ship lock can infer that which type and fleet formation of ships or fleet is in accordance with the size of ships or fleet, and further estimate the gross tonnage. When it's found that the reported data are definitely not consistent with the estimated data, a further field survey is required so as to avoid national tax loss effectively.

The establishment of GRT data estimation method defines the basis of rational charging for the passing ships in locks in order that it endures the administrative and executive seriousness, waterway management, and the good order of waterway transportation. Moreover, this kind of research finding also assists in other aspects as follows.

1. The computative data for the relation between ship's size and its type help know well of the waterway requirements such as safety water depth and beveling radius so as to arrange the waterway transportation timely and improve its safety.
2. The current developmental tendency of waterway transportation can also be analyzed. Since the North Jiangsu Canal has prohibited against earthenware ships and paddle vessel passage, the navigation environment has been greatly improved, and the tendency of large-scale vessel has been gradually developed.
3. It also provides data for the advancement of shipbuilding standardization. Because individual shipping carriage takes full consideration of developing the use ratio of navigation facilities of future waterways and locks, as well as enhancing economic benefits in shipping business, the shipping standardization has been greatly improved and the ship's size has been gradually unified.
4. It provides data for new administrative policy in waterway transportation, so as to reinforce supervision of shipping market, and promotes the scientific development of inland waterway transportation.

## References

Lei H (2009) SPSS method for forecasting inland waterway cargo traffic in the north of Jiangsu [J]. China Water Transp 8

Lei J-T (2002) Suggestions on modification of domestic ships tonnage measurement method [J]. China Ship Surv 9:57–58

Li G-c (2009) Research on the casual and its correspondent strategy of the phenomenon: "unmatched ship and its' registration" [J]. Transp Constr Manage 02:49–51

Meng X-Q (2003) Research on keelboat net tonnage calculation [J]. Ship Ocean Eng 06:49–51

Shan G-Y (2000) Primary exploration on the simplified calculational method of minimal ships' maximum capacity [J]. Fish Econ Res 02:40–42, 45

Sun C-F (2002) Statistic analysis on net tonnage measurement of ships in domestic voyage [J]. China Ship Surv 1

Wang L-z (2001) Research on principle dimension series of Grand Canal cargo ship [J]. 2

Yang X (2008) The regression analysis about ship displacement and ship deadweight [J]. Port Eng Technol 01:17–18

# Sensitive Analysis of the Cargo Transported by Ports Contributes to Hinterland Economy Based on Input–Output Model

**Guiliang Tian, Yi Zhuang, Jianzhong Hu, and Xiujuan Guo**

**Abstract** This paper analyzes the sensitivity of cargo transported by ports contributes to regional economic development from inside of port industry making specific port transport cargo as its research objective. Firstly, membership relation between port transport cargo and port industry is clarified before analyzing forward connections effect of port entry cargo to regional economy and backward connections effect of port leaving cargo to regional economy based on input–output model in this paper. Since ripple effect of various cargoes to economy is completed in many years, Taylor Expansion is used to peel off the current contribution of port transport cargo to regional economic development. Finally, an empirical research is conducted with data from Fujian Province. Conclusions draw from this research can serve for decision department of government to diagnose economic operation by ports promptly.

**Keywords** Backward connections effect · Cargo transported by ports · Forward connections effect · Fujian · Input–output model · Port contribution rate · Sensitive analysis

## 1 Introduction

As the transfer station of product imported and exported for hinterland for economic development, port's operation reflects the regional economic development to a certain extent. Therefore, the port has become the "barometer" of hinterland's economic development. Profound research on the relation between port operation and regional economic development is foundation for decision department to conduct economic diagnosis through port operation.

G. Tian (✉), Y. Zhuang, and X. Guo
Investment Institute, Hohai University, Jiangsu Nanjing 210098, China
e-mail: tianguiliang@hhu.edu.cn

J. Hu
Traffic Planning Office of Fujian Province, Fujian Fuzhou 350001, China

Y. Zhou and D.D. Wu (eds.), *Modeling Risk Management for Resources and Environment in China*, Computational Risk Management,
DOI 10.1007/978-3-642-18387-4_13, 

Seldom literatures researched contribution of port economy since high data requirement for research on contribution of port economy. Usually, it requires questionnaires conducting and data collection to ten thousand of units which is hardly realised by common researchers. Dickey (1999) investigated contribution of port when studying port industry of Netherlands. Waters (1977) suggested that analyzing direct economic contribution by cost-benefit analysis and studying subsequent effect by regional input–output model. Bingliang (2001) divided port economic contribution into direct port economic contribution and indirect port economic contribution. He and Wang (2005) divided port economic contribution into direct contribution, indirect contribution and spread contribution and studied it by input–output table and multiple models. Sun et al. (2004) divided port activities into transportation-related activities, commercial activities and industry-related activities. He analyzed characteristics of them respectively and studied the influence of port economy by input–output methods and multiple models. Xiujuan (2010) put forward the measuring indexes and the models to estimate the economic contribution of several ports in a same area are given based on the input–output table of 45 sectors.

However, current researches mainly study contribution of port industry to regional economy regarding port as an independent industrial sector. We will analyze the contribution of port transport cargo to regional economic development from inside of port industry making specific port transport cargo as its research objective. Therefore, this research establishes input–output research to measure contribution of port transport cargo to regional economy and makes an empirical research based on data from Fujian Province so as to serve for decision department of government to make port economy analysis.

# 2 Analyses on Contribution Mechanism of Port Transport Cargo to Hinterland Economy

## *2.1 Contribution Mechanism of Port Entry Cargo to Hinterland Economy*

Port is the transfer station of product imported and exported for hinterland for economic development. Port entry cargo provides substantial raw material for regional economic development. When the port entry volume of a certain cargo increases, this means the increase of raw material which can be used as input by other industry. In order to guarantee convenient availability to raw material by other industry, other industry scoters which use such cargo as input will expand production. Therefore, the contribution of port entry cargo to regional industrial development reflects as forward connections effect. Since a certain cargo belongs to relative industry which produces the goods, every unit weight increase of cargo in port means contribution of expanding production scale by local relative industry to regional economy.

### 2.2 *Contribution Mechanism of Port Leaving Cargo to Hinterland Economy*

Increase of port leaving cargo means increase of export of relative industry. However, export is a vital composition of final demand of society. Thus, export increase must induce production expansion of industry sectors. Therefore, contribution of change in the volume of port leaving cargo to hinterland economic development reflects as backward connections effect. For example, increase of entry volume of chemicals raw materials and products means regional chemical industry products not only meet the needs of its own development, but also provide chemical raw material and products for other domestic provinces or even foreign countries. Therefore, the chemical industry in the region must expand production scale. Furthermore, production scale expansion of chemical industry not only increase the production of this industry, but also will induce other related industries expand scale and increase output such as electronic power, petroleum processing, machinery and equipment manufacturing.

## 3 Input–Output Model for Measuring Contribution Rate of Port Transport Cargo to Hinterland Economy

Input–output method was put forward by American economist Leontief. The method reflects technical and economic connection between production and distribution, input and output of products in every sector of economic system. Then input–output table is worked out as Table 1. Then, input–output mathematic model

**Table 1** Input–output table of value type

| Input \ Output | | Interim Utilization: Sector 1 ... Sector n | Final Consumption: Export, Consumption, Accumulate, major repair and reconstruction expense, fixed assets | Total Output |
|---|---|---|---|---|
| Interim Input | Sector 1 | $X_{11}$ ··· ··· ··· ··· $X_{1n}$ | $Y_1$ | $X_1$ |
| | ⋮ | ... $X_{ij}$ ... ⋮ | ⋮ | ⋮ |
| | Sector n | $X_{n1}$ ··· ··· ··· $X_{nn}$ | $Y_n$ | $X_n$ |
| Initial Input | Fixed Assets Depreciation | $D_1$ ··· ··· $D_n$ | | |
| | Income of Laborer | $V_1$ ··· ··· $V_n$ | | |
| | Net Income | $M_1$ ··· ··· $M_n$ | | |
| Total Input | | $X_1$ ··· ··· $X_n$ | | |

is established according to balance relationship in input–output table. Finally, works such as balance program, economic forecast and economic analysis can be conducted by input–output model (Zhang and Zhang 1999).

## *3.1 Membership Relation Between Port Transport Cargo and Industry*

Input–output method is way reflecting consumption relationship between industries while statistical data of port operation is mainly the volume of port entry and leaving cargo. Therefore, membership relation between port transport cargo and industry should be clarified first. It can be found in Table 2.

**Table 2** The industry of port transport cargo subjected

| Number of port transport cargo | Port transport cargo | Number of industry | Industry |
|---|---|---|---|
| ① | Coal and related products | I | Coal mining and dressing |
| ② | Petroleum, natural gas and related products | II | Petroleum and natural gas extraction, processing of petroleum, coking, processing of nuclear fuel |
| ③ | Metallic ore | III | Nonmetal minerals mining and dressing |
| ④ | Steel | IV | Metal smelt and pressing for processing |
| ⑤ | Mineral building materials | V | Construction industry |
| ⑥ | Cement | VI | Nonmetal mineral products industry |
| ⑦ | Timber | VII | Timber processing and related products and furniture manufacture |
| ⑧ | Nonmetal ores | VII | Nonmetal minerals mining and dressing |
| ⑨ | Fertilizer and pesticide | IX | Chemical industry |
| ⑩ | Salt | IX | Chemical industry |
| ⑪ | Grain | X | Agriculture |
| ⑫ | Machinery, equipment and electric appliance | XI | Electric equipment and machinery industry |
| ⑬ | Chemical raw materials and related products | IX | Chemical industry |
| ⑭ | Nonferrous metal | XII | Metal products industry |
| ⑮ | Light industry medical products | IX | Chemical industry |
| ⑯ | Agriculture, forestry, husbandry and fishery products | X | Agriculture |
| ⑰ | Others | XIII | Other industry |

### 3.2 Forward Connections Effect of Port Entry Cargo to Industrial Development

Connection effect of port entry cargo to industrial development of hinterland is forward connection effect because a certain cargo belongs to the industry which produces the cargo (Table 2). Therefore, its membership industry is used to conduct connection analysis when calculating forward connections effect of port entry cargo to industrial development by input–output method.

The impact of cargo volume change of port entry to industrial development is:

$$FL_i = \sum_{j=1}^{m} x_{ij} \quad (i = 1, 2, \ldots n)$$

$FL_i$ means forward connections effect of volume change of port entry of cargo $i$ to industrial development is. That is, total output values of each industrial change caused by per unit $i$ port entry cargo (value form: 10,000 Yuan). $x_{ij}$ is complete consuming coefficient to cargo $i$ by development of $j$ industry. That is, the element in row $i$ and line $j$ of complete consuming coefficient table.

### 3.3 Backward Connections Effect of Port Leaving Cargo to Industrial Development

Connection effect of port leaving cargo to industrial development of hinterland is backward connections effect.

The impact of cargo volume change of port leaving to industrial development is:

$$BD_j = \sum_{i=1}^{n} x_{ij} \quad (j = 1, 2, \ldots n)$$

$BD_j$ means backward connections effect of volume change of port leaving of cargo $j$ to industrial development is. That is, total output values of each industrial change caused by per unit $j$ port leaving cargo (value form: 10,000 Yuan). $x_{ij}$ is complete consuming coefficient to cargo $i$ by development of $j$ industry. That is, the element in row $i$ and line $j$ of complete consuming coefficient table.

### 3.4 Peeling off of Contribution in Current Year

Calculate contribution of certain cargo of per unit value to social total output according to volume of port entry and leaving cargo considering the phenomena

**Table 3** The total output growth types and their calculations

| Direct consumption coefficient matrix | Growth matrix $H^{(n)}$ | Growth types |
|---|---|---|
| $A$ | $A\Delta X$ | Direct growth |
| $A^2$ | $A^2\Delta X$ | The 1st round of the secondary growth |
| $A^3$ | $A^3\Delta X$ | The 2nd round of the secondary growth |
| $\vdots$ | $\vdots$ | $\vdots$ |
| $A^k$ | $A^k\Delta X$ | The (k−1)th round of the secondary growth |

various kinds of cargo may enter and leave port at the same time. Since GDP is the part of total social output with the exclusion of intermediate input, contribution of a certain cargo to GDP is 1 subtracted from value of contribution rate of port transport cargo to total output.

This paper calculates the contribution of unit port transport cargo to GDP of present year. That is, contribution of the current year. Since sectors which expand production under radiation of port will need production from other sectors as its intermediate input, direct increase will then lead to the first time indirect increase. And the first time indirect increase will cause the second time indirect increase and so on. It is shown as formula: complete increase = direct increase + 1st indirect increase + 2nd indirect increase + …, such as Table 3.

In line with the abovementioned equality relation of total social output increase, the following recurrence formula can be established:

That is, volume of complete increase of total output in all industrial sectors caused by port transport cargo is:

$$H = H^{(1)} + H^{(2)} + H^{(3)} + \cdots + H^{(n)} = A\Delta X + A^2\Delta X + A^3\Delta X + \cdots + A^k\Delta X$$
$$= (A + A^2 + A^3 + \cdots + A^k) \bullet \Delta X = [(I - A)^{-1} - I] \bullet \Delta X$$

In accordance with many years' experience of industrial analysis, generally, contribution in the current year of port operation to GDP can be reflected by two to three times ripple effect. To be secure, we suppose contribution of per unit port transport cargo increase to social total output and GDP is $H^{(1)} + H^{(2)}$, that is, two times ripple effect. $H$ is matrix of ripple effect of port transport cargo to social total output.

## 4 Empirical Research

Input–output table is basis tool of input–output analysis. Chinese government and every province has formed the system that establishing input–output table one time 5 years. Forward connections effect of per unit Fujian port entry cargo to Fujian's economy can be calculated with the latest <2002 Fujian Input–output Table>, and the results shown in Table 4.

**Table 4** The forward promoted effect of cargo imported to hinterland economy in Fujian unit:1

| Port transport cargo | ① | ② | ③ | ④ | ⑤ | ⑥ | ⑦ |
|---|---|---|---|---|---|---|---|
| Forward connections effect | 1.680 | 2.001 | 2.446 | 2.574 | 1.624 | 1.379 | 2.213 |
| Port transport cargo | ⑧ | ⑨, ⑩, ⑬, ⑮ | ⑪, ⑯ | ⑫ | ⑭ | ⑰ | |
| Forward connections effect | 0.181 | 3.522 | 4.672 | 1.124 | 1.628 | 0.932 | |

Notes: The numbers in the table correspond to the cargo in Table 2, the following tables are so.

**Table 5** The backward pulled effect of cargo exported to hinterland economy in Fujian unit: 1

| Port transport cargo | ① | ② | ③ | ④ | ⑤ | ⑥ | ⑦ |
|---|---|---|---|---|---|---|---|
| Backward connections effect | 1.017 | 2.330 | 1.855 | 2.061 | 1.759 | 1.793 | 1.318 |
| Port transport cargo | ⑧ | ⑨, ⑩, ⑬, ⑮ | ⑪, ⑯ | ⑫ | ⑭ | ⑰ | |
| Backward connections effect | 1.387 | 2.124 | 1.011 | 2.202 | 2.166 | 1.860 | |

**Table 6** The contribution of unit value cargo port transported to Fijian's GDP in the same year unit. 1

| Port transport cargo | ① | ② | ③ | ④ | ⑤ | ⑥ | ⑦ |
|---|---|---|---|---|---|---|---|
| The contribution to GDP | 0.263 | 0.439 | 0.569 | 0.589 | 0.274 | 0.112 | 0.405 |
| Port transport cargo | ⑧ | ⑨, ⑩, ⑬, ⑮ | ⑪, ⑯ | ⑫ | ⑭ | ⑰ | |
| The contribution to GDP | 0.037 | 0.708 | 1.026 | 0.368 | 0.411 | 0.157 | |

**Table 7** The average market prices of cargo transported by Fijian ports in 2002 unit: Yuan/t

| Port transport cargo | ① | ② | ③ | ④ | ⑤ | ⑥ | ⑦ |
|---|---|---|---|---|---|---|---|
| The average market prices | 184 | 747 | 120 | 2,012 | 10 | 39 | 166 |
| Port transport cargo | ⑧ | ⑨, ⑩, ⑬, ⑮ | ⑪, ⑯ | ⑫ | ⑭ | ⑰ | |
| The average market prices | 14 | 1,917 | 3,175 | 1,841 | 1,231 | 157 | |

Backward connections effect of per unit Fujian port entry cargo to Fujian's economy is shown in Table 5.

Contribution of per unit port transport cargo in the present year to Fujian's GDP can be peeled off by methods of calculating Contribution of per unit port transport cargo in the present year to regional economy. The results can be found in Table 6.

We transfer measurement unit of money and products into physical measurement unit (ton) in line with average market price in 2002 in order to further reflect contribution of per unit weight Fujian port transport cargo to GDP. Then, contribution of per unit port transport cargo to GDP in Fujian Province is calculated.

Contribution of per unit port transport cargo to GDP in Fujian Province is 157.56 Yuan per ton after calculation with the results from Table 7 and combination of products throughput of different kinds in 2002, Fujian Province.

**Table 8** The proportion of each cargo's contributions to GDP in the same year unit:%

| Port transport cargo | ① | ② | ③ | ④ | ⑤ | ⑥ | ⑦ |
|---|---|---|---|---|---|---|---|
| The proportion | 4.8 | 17.7 | 0.3 | 15.6 | 0.9 | 0.1 | 0.2 |
| Port transport cargo | ⑧ | ⑨, ⑩, ⑬, ⑮ | ⑪, ⑯ | ⑫ | ⑭ | ⑰ | |
| The proportion | 0.1 | 20.7 | 29.9 | 1.2 | 0.1 | 8.5 | |

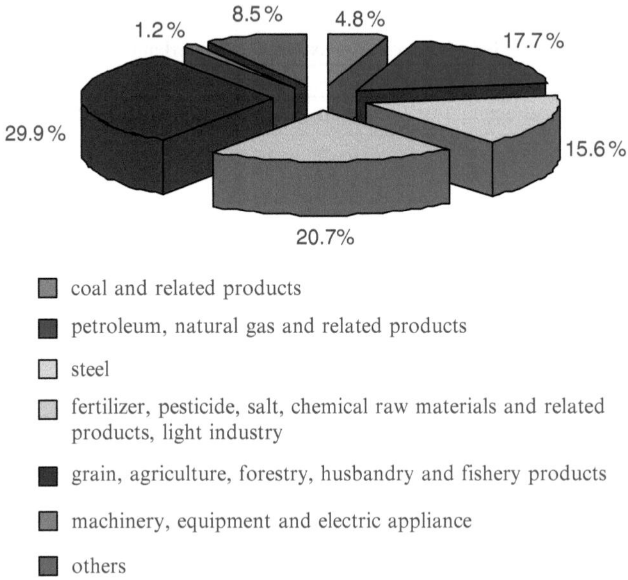


**Fig. 1** The proportion of each cargo's contributions to GDP in the same year

The proportion of contribution of each type of products is shown in Table 8.

Some products with high economic contribution rate in Fujian are selected from Table 8 and shown as pie chart in Fig. 1:

**Acknowledgement** We would like to show our deepest gratitude each member of research project (Research on Economic Contribution of Ports in Fujian province) group. This research was supported by the National Natural Science Foundation of China(Grant No. 41001377), Social Sciences Programming Fund of Ministry of Education of China(Grant No. 09YJC790067) and "211" project key disciplines (Technology Economy and Management) projects of Hohai University, All remaining shortcomings are ours.

## References

Bingliang S (2001) On the evaluation of the overall contributions of Shanghai port. J Shanghai Marit Univ 22(4):26–29

Dickey A (1999) Dutch maritime cluster: a network of excellence. Lloyd's List, London

He X, Wang W (2005) Application of input-output model in research on contribution of port to regional economy. Shipping Manage 27(3):16–18
Sun J, Shan M, Jia D, Ning T (2004) Analysis on comparison of port economy contribution. China Ports 2:26–28
Waters RC (1977) Port economic impact studies: practice and assessment. Transportation Journal, Spring, pp 14–18
Xiujuan G (2010) Economic contribution estimation of several ports in a same area based on input-output model. Port Eng Technol 47(1):25–29
Zhang Q, Zhang H (1999) Application of input-output model in discussion on relationship between transportation and social economic system. Shipping Manage 3:23–26

# Scenario Construction of Flood Emergency Management in River Basin Based on Scene Perception

**Gao-feng Liu, Hui-min Wang, and Jin-ping Tong**

**Abstract** In view of the prominent characteristics of complexity, uncertainty, inevitability and relief of river basin floods, this paper presents a research of "scene perception – scenario construction – scenario analysis" on flood emergency management in river basin by applying the related theories of scenarios, that is, perceiving the situations of key factors (rain, flooding, etc.), acquiring information from the perception and forming the possible future scenarios of flood emergency management. A case study of Huaihe River Basin is researched, in which the future scenarios of flood emergency management are described and corresponding recommendations of emergency measures are made. This method is also suitable for flood emergency management in other river basins.

**Keywords** Emergency management · Flood disasters · River basin · Scenario analysis · Scene perception · Uncertainty

## 1 Introduction

Flood disasters have occurred frequently in China for several thousand years, which have badly influenced China's socio-economic development. Since 1990s, regional flood disasters have nearly occurred every year. Among them, the river basin flood respectively happened in 1991, 1998, 2003, 2007 and 2010 year. A Flood is a kind

G.-f. Liu (✉)
State Key Laboratory of Hydrology-Water Resources and Hydraulic Engineering, Hohai University, 210098 Nanjing, China
and
Business school, Hohai University, 213022 Changzhou, China
e-mail: gaofengliu@yahoo.cn

H.-m. Wang and J.-p. Tong
State Key Laboratory of Hydrology-Water Resources and Hydraulic Engineering, Hohai University, 210098 Nanjing, China
e-mail: huimingwang@hhu.edu.cn; tong19781978@126.com

Y. Zhou and D.D. Wu (eds.), *Modeling Risk Management for Resources and Environment in China*, Computational Risk Management,
DOI 10.1007/978-3-642-18387-4_14, 

of natural disasters, which is inevitable and can be relieved by disaster prevention/mitigation measures. In 2006, the State Council issued the overall emergency plan for public incidents, in which the flood and drought disaster is listed as the first place of the natural disaster class (The overall emergency plan for public incidents, http://www.gov.cn/yjgl/index.htm). Therefore, flood emergency management (FEM, for short) is especially important, and becomes the research hotspot.

Currently, Researches on FEM at home and abroad are fewer. Theoretical studies mainly concentrate in flood disaster emergency assessment(Sujit and Lee 1988; Srikantha and Dushmanta 1998; Feng et al. 2001; Li et al. 2003), monitoring and early warning (Steven 1986; Chen 1997), flood control (Windsor 1973; Uhver and Mays 1990; Yu et al. 1983; Xie and Yi 2002; Chen et al. 2008), emergency management system and psychological behaviors (Qiu 2007; Chen 2007; Hu et al. 2008; Josefa et al. 2001; Simon et al. 2007), etc., using methods such as Mathematics, Statistics, Operation Research, and integration methods of technologies of remote sensing (RS), geographic information systems (GIS), global position system(GPS), etc.. In practice, emergency management agencies of comprehensive disasters, related laws and regulations, have been established in developed countries. In China, the State Flood Control and Drought Relief Head-quarters, the command structures for flood control of seven big river basins and every province city have also established. Otherwise, comprehensive information platforms and decision support systems are researched continuously, such as 3D electronic river systems, country flood control command system, and so on.

It is clear that all researches about FEM are founded on "facts-coping" or "forecast -coping". However, along with aggravating climatic change, the characteristics of non-conventional, complexity, uncertainty, urgency are highlighted and the traditional negative emergency mode is difficult to adapt to the national needs of disaster prevention and mitigation. From the perspective of development, make sure that what are future scenarios of flood, and what are influencing factors, then theories and methods of FEM based on "scenarios-coping" can be established. "Scenario" is the basis for FEM, and this paper will present a preliminary discussion on scenario analysis of FEM in river basin.

## 2 Characteristics Analysis of FEM in River Basin

Flood disasters and FEM in river basin have typical characteristics:

1. Complexity. Flood is a complex natural phenomena, and flood disaster is formed by complicated interaction among inducing environment, hazard, disaster object (Wei and Jin 2002). Flood has two basic properties, one characterized by natural property of randomness, dynamics, non-uniformity, open, burst and regularity, the other characterized by social property of the interaction between complex human activities and flood. The process of FEM is often multi-agent, specialized, multi-domain, multi-layered and so complex.

2. Uncertainty. A flood is a jerky stochastic phenomena, and rainfall distribution, size, flood process are not deterministic. There are many uncertainties in the emergency response procedure, such as time, range, resources and rescuers, psychological cognition and decision-making behaviors.
3. Urgency. Due to the characteristics of great harm in a certain range and rapid spread in a hurry, it will result in a further deterioration of the flood disaster situation without timely effective measures. Timeliness and effectiveness of decision-making and action will affect the result of FEM.

The above mentioned characteristics determine that the decision makers must roundly identify key factors, accurately hold the developing trend of flood, and work out scientific emergency plan. Exactly, scene perception and scenario analysis method provides a realistic path to achieve this goal.

# 3 Scenarios Constructing of FEM in River Basin

## *3.1 Scene Perception and Scenario Analysis*

The aim of scene perception is to acquire various kinds of information about the spot of river basin flood. With the development of information technologies and spatial technologies, means of scene perception are progressive and diversity. For example, 3S technologies (RS, GIS, GPS, etc.) gradually substitute the ways of traditional manual work, widely applied in flood risk analysis, disaster damage assessment, etc..

The word "Scenario" first appeared in the book of the year 2000 in 1967 (Herman and Wiener 1967). In their opinion, the future is diverse, ways to various kinds of future are not unique, and the description of a possible future and ways to this future constitutes a scenario. Scenario is a series of factual description of future situations and developments from initial state to future state. Scenario analysis is conceiving various possible future scenarios by detailed reasoning and description of future, on the basis of critical assumptions in the major economical, technical, or industrial evolution. The value of Scenario analysis is to enable managers to discover certain trends of future change, to prepare for them and take positive actions to minimize the negative impact. This method has been widely used in enterprise management, transport planning, agricultural development, energy demand, climate change, etc., but seldom seen in FEM area.

## *3.2 Analysis of Key Factors*

There are many kinds of flood, here mainly taking the rainstorm flood as an illustration. Rainstorm flood occurs more frequently, more extensively, and is more uncertain. Flood which is coming will have many uncertainties in its process. The research objective of this paper is to analyze the future trend of flood, to definite its possible changes, and to consider that what ideal plans can be chosen by managers to deal with those uncertainty. The whole cycle of a flood is taken as the time length of study, and it is often unequal for different flood events.

The process of FEM is extremely complicated, having many impact factors. Overall, the effect of FEM depends on flood disaster situation and measures of flood emergency (Fig. 1). The former directly reflects the severity of flood and the latter reflects the ability of FEM. From perspective of disaster science, flood is the situation of hazards acting on disaster objects under a certain inducing environment. Hence, flood disaster contains three elements: hazard, inducing environment, hazard bearing body. These three elements interact on each other. Measures of flood emergency primarily involve emergency response system (the "one plan three systems"), emergency response technologies (information technologies, scheduling techniques) and emergency response resources (rescuers, materials and equipments, capital, etc.). Emergency response technologies and resources are the basic safeguards of emergency system running efficiently. It should be noted that, different from flood risk management, measures of flood emergency concerns non-project measures in scenario analysis of FEM, and flood control projects belong to inducing environment.

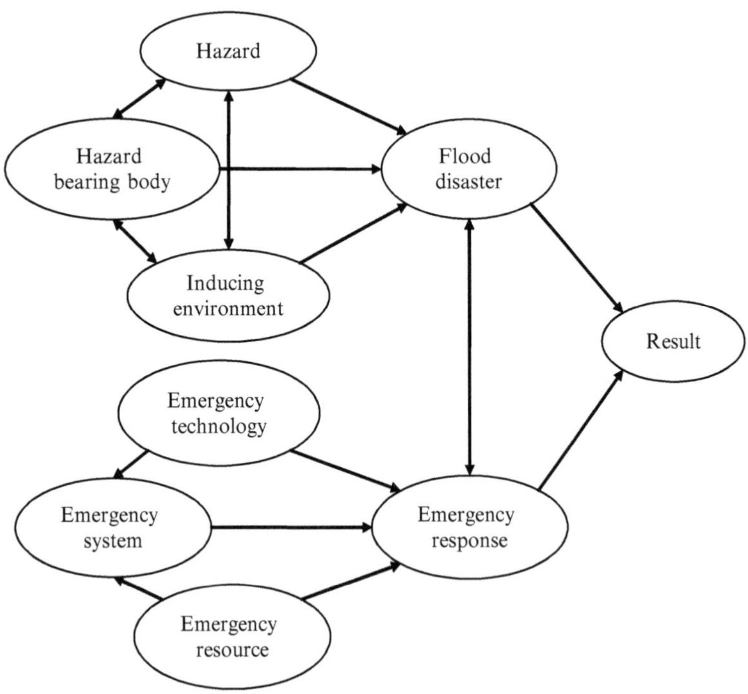


**Fig. 1** Factors diagram of FEM

## 3.3 Scenario Constructing Model of FEM in River Basin

FEM in river basin should be multi-stage and progressive because floods have the characteristics of randomness, gradually and fluidity. In this paper, a scenario constructing model is built (Fig. 2). First, related factors statistics of the typical historical flood scenes are made. After mergence and induction, historical scenario warehouse is formed, of which each scenario record includes scene characteristics and the feasible emergency measures corresponding. When a flood occurs, first the scene perception of flood at a certain time is made to obtain the key factors by using RS/GIS technologies, then this factor assemble is compared with those in the warehouse, thereby the possible scenarios of this flood are formed and added to the warehouse simultaneously.

Suppose that $M_i$ ($0 < i \leq m$ and $i \in N$) represents $m$ typical historical floods. $e_1$, $e_2$ and $e_3$, as the key factors of scenario design, respectively denote trend precipitation, flood intensity and flood locality, and

$$e_1 \in \{weakened,\ strengthened\},$$

$$e_2 \in \{under\,s\tan\,dard, over\,s\tan\,dard\},$$

$$e_3 \in \{plateau,\ mountainous\,region,\ hill\,area,\ plain\}$$

The values of $e_1$ and $e_3$ can be obtained from weather forecasting and RS/GIS monitoring. $e_2 = f(L, Q, W)$, in which $L$ is water level, $Q$ is peak flow, $W$ is flood volume, $f$ is functional relation among $L$,$Q$,$W$ and $e_2$. $\varphi_i$ denotes valid emergency

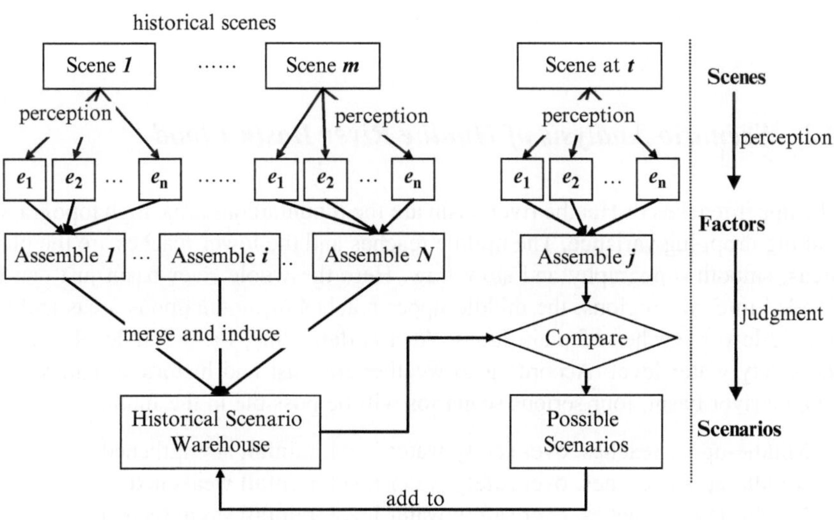


**Fig. 2** The scenario constructing model of FEM

plan under scene $i$. Therefore, scenario records in the historical scenario warehouse can be expressed as $S_k$ ($0<k \leq m$ and $k \in N$), $S_k = \{e_{1k}, e_{2k}, e_{3k}, \varphi_k\}$,.

At time $t$, the key factor assemble of flood scene is $S_t(t>0)$, expressed as $S_t = \{e_{1t}, e_{2t}, e_{3t}\}$,. Then, $S_t$ is compared with $S_k$, and the possible scenarios $S_{lt'}(l>0$ and $l \in N)$ at next decision time $t'$. If the scenario is the same as some scenario in the warehouse, the emergency plan existing can be used in emergency rescue; if not, then a new scenario is formed and new emergency plan should be raised based on historical scenarios, simultaneously recorded into the scenarios warehouse.

Although there are many commons among river basins, they are complex and vary. Because of the different characteristics of each river basin, the design and analysis of scenario depends on the specific river basin.

# 4 Scenario Analysis of FEM in Huaihe River Basin

## *4.1 Introduction of Huaihe River Basin*

Huaihe river basin is located in east of China, between the Yangtze river and the Yellow river, across 40 cities and 163 counties of 5 provinces-Henan, Hubei, Anhui, Jiangsu, Shandong. The total area of Huaihe river basin is $27 \times 10^4$ km$^2$. In this valley, the rainstorm weather system is diverse and the space and time distribution of rainfall is uneven. The average annual precipitation is 888 mm, while 50% ~ 80% precipitations concentrates in the flood season (June to September). Due to the particularity of climate and geographical conditions, flood disasters occur frequently in Huaihe river basin, especially in 1991, 2003, 2007 year. Therefore, FEM in Huaihe river basin is extremely urgent.

## *4.2 Scenario Analysis of Huaihe River Basin Flood*

The upper reaches of Huaihe river basin are the mountainous area, high topography and big dropping variance. The middle reaches and the lower reaches are the plain areas, smooth topography and slow flow. Here the whole river basin just can be divided into two regions, the middle-upper reaches of mountainous areas and the middle-lower reaches of plain areas. At a certain time, the water level exceeds the safety water level. According to weather forecast and historical materials of Huaihe river basin, four serious scenarios will be possible in the future.

- Middle-upper reaches, over safety water level, rainfall strengthened
- Middle-upper reaches, over safety water level, rainfall weakened
- Middle-lower reaches, over safety water level, rainfall strengthened
- Middle-lower reaches, over safety water level, rainfall weakened

**Scenario 1:** rainstorms concentrate in the middle and upper reaches, water level is over safety water level and the trend of rainfall continues to enlarge. The situation of this scenario is extremely stern, not only the upstream facing disaster, but the key embankments and areas of Huaibei dike, Hongzehu dike, Liyunhe dike, Bengbu and Huainan cities in downstream under threat due to flood flows assembling together from upstream or branches. At the moment, it is urgent and the key points are: (1) macroscopically, start emergency plan, organize relevant departments to fight floods in a cooperative manner; (2) microscopically, rationally operate flood by-pass, flood passage areas, flood storage and detention areas and reservoirs by advanced information technologies and scheduling technologies. The objective of operations is to leave more space for the subsequent strong rainfall, under guaranteeing the safety of flood control projects and minimizing flood losses as far as possible. The personal safety and the property security come first. It seems to be little probability of People in mountainous areas of upstream being inundated. They should pay attention to secondary disasters such as torrential flood and debris flow. People in storage-detention flood areas of mid-downstream are greatly affected, so the work of people migration and resettlement, relief supplies, is also an important aspect of emergency management.

**Scenario 2:** rainstorms concentrate in the middle and upper reaches, water level is over safety water level and the trend of rainfall becomes weak. Although less rainfall in the future, water level has exceeded the limit. So, it is urgent and the key points are: rationally operate flood by-pass, flood passage areas, flood storage and detention areas and reservoirs by advanced information technologies and scheduling technologies. The objective of operations is to minimize flood losses as far as possible under guaranteeing the safety of flood control projects.

**Scenario 3:** rainstorms concentrate in the middle and lower reaches, water level is over safety water level and the trend of rainfall continues to enlarge. Under this scenario, waterlog is the main disaster in middle-lower reaches. Emergency time is urgent and the situation is serious. The middle-lower reaches, fertile soil and pleasant climate, is known as the "land of fish and rice". There is a large and dense population, a high socio-economic development. Once floods happen, the loss will be huge. At the moment, it is urgent and the key points are: (1) macroscopically, start emergency plan, organize relevant departments to fight floods in a cooperative manner; (2) microscopically, make sure the personal safety and the property security and shift victims if necessary. Simultaneously, rationally operate flood by-pass, flood passage areas, sea/river channels by advanced information technologies and scheduling technologies, and pump water by advanced machinery. The objective of these operations is to minimize flood losses as far as possible, under guaranteeing the safety of key embankments and areas of Huaibei dike, Hongzehu dike, Liyunhe dike, Bengbu and Huainan cities in downstream. In addition, prevent from the secondary disasters, such as infectious diseases.

**Scenario 4:** rainstorms concentrate in the middle and lower reaches, water level is over safety water level and the trend of rainfall becomes weak. Under this scenario, it is urgent and the key points are: drain waterlogged fields and minimize flood losses as far as possible. Besides drainage by natural river channels and lakes, various irrigating stations, draining stations, drainage machines can be dispatched as soon as possible, with the assistance of drainage experts.

## 5 Conclusion

Flood is one of the biggest threats to the survival and development of human beings. Its prominent characteristics are complexity, uncertainty, inevitable and relievable. On the basis of these characteristics, this paper presents an analysis of FEM in river basin by applying the scenario analysis method, and the possible future scenarios of FEM are primarily built. Finally, a case study of Huaihe River Basin is researched, in which the future scenarios of FEM are described and corresponding recommendations of emergency measures are made. The application of scenario analysis in emergency management is an advanced subject though widespread applications in many other domains. Flood disaster has its particularity, differing from other emergencies. The research of this paper is just an attempt, and more thorough studies will be going to continue.

**Acknowledgment** This research was supported by National Natural Science Foundation of China (No.90924027), Specialized Research Fund for non-profit Public Industry of Ministry of Water Resources (No.200801027), National Social Science Foundation of China (No.10CGL069, 09CJY020, 10AJY005), China Postdoctoral Science Foundation (No.20100471372).

## References

Chen XW (1997) The application of remote sensing and geographical information system (GIS) in flood damage analysis (in Chinese). J Hydraul Eng 3:70–73

Chen XT (2007) New challenges of emergency response capability of flood and drought in China (in Chinese). Disaster Reduction China 1:26–27

Chen SY, Yuan JX, Guo Y (2008) Variable fuzzy decision-making theory and its application to decision-making supporting system for reservoir flood control operation (in Chinese). J Dalian Univ Technol 48(2):259–262

Feng P, Cui GT, Zhong Y (2001) On the evolution and prediction of urban flood economic loss (in Chinese). J Hydraul Eng 8:64–68

Herman K, Wiener AJ (1967) The year 2000: a framework for speculation on the next thirty-three years. Macmillan Publishing Company, London

Hu XH, Wang HM, Ma SJ (2008) A new flood risk management mode in Chinese city: government, market, public partnership (in Chinese). East China Econ Manage 22(9):121–125

Josefa Z, Hernandez J, Serrano M (2001) Knowledge-based models for emergency management systems. Expert Syst Appl 20:73–186

Li JR, Ding ZX, Huang ShF, Hu YL (2003) Research of flood and water logging loss assessment model based on spatial distribution social-economic database(in Chinese). J China Inst Water Resour Hydropower Res 1(2):104–110

Qiu RT (2007) Flood and drought emergencies and emergency management in China (in Chinese). China Emerg Rescue 4:4–8

Simon M, Sylvia T, Dennis P (2007) Risk communication in emergency response to a simulated extreme flood. Environ Hazards 7:179–192
Srikantha H, Dushmanta D (1998) Flood inundation modeling and loss estimation using distributed hydrologic model, GIS and RS. In: Proceeding of International Workshop on the Utilization of Remote Sensing Technology to Natural Disaster Reduction, pp. 239–250
Steven D (1986) Computer models in Lower Colorado river operation. Water Resour Plann Manage 112(3)
Sujit, Lee R (1988) A nontraditional methodology for flood stage damage calculation. Water Res Bull, 110–135
The overall emergency plan for public incidents, http://www.gov.cn/yjgl/index.htm
Uhver OI, Mays L (1990) Model of real-time optimal flood control operation of a reservoir system. Water Resour Manage 4(l):20–45
Wei YM, Jin JL (2002) Theory of risk management of flood disaster (in Chinese). Science Press, Beijing
Windsor JS (1973) Optimization model for reservoir flood control. Water Resour Res 9 (5):1219–1226
Xie LQ, Yi SZ (2002) Optimal flood dispatch model for multi-reservoir system (in Chinese). J Hydraul Eng 6:38–42
Yu JJ, Liang NS, Tang JR (1983) The flood optimal control of hydropower station reservoir (in Chinese). Water Resour Power 1(1):65–69

# Multi-objective Optimization and Integration of Port Supply Chain Model

**Jing Song, Fengping Wu, Dong Xu, and Jiawei Fu**

**Abstract** The optimization and integration of port supply chain is studied. According to the characteristics of the port supply chain, from the perspective of the whole supply chain management, the strategic level and operational level objectives of the supply chain will be integrated and optimized; from the customers' point of view, their satisfaction of the supply chain operating hours will be set as a goal of the service optimization to help to establish the port supply chain integrated optimization model, which targets at cost, time and flexibility. The view that the integrated optimization of collaborative enterprises is similar solutions to supply chain integration optimization is put forward and the collaborative enterprises are optimally to be integrated. Finally, the improved Particle Swarm Optimization (PSO) algorithm is adopted in the model solutions and the effectiveness and practicality of the model and algorithm are verified through a simulation case.

**Keywords** Integrated optimization · Multi-objective · Port supply chain · PSO

## 1 Introduction

Optimization and integration of the supply chain is an effective method of integrating supply chain resources. It has practical significance on the construction as well as design of supply chain. At present, the research on supply chain integrated optimization is mainly focusing on the manufacturing field (Sabri and Beamon 2000; Lee et al. 2007; Ma 2005): management of production system and logistics system integration, coordination of multi-layer supply chain optimization, integrated optimization problems on manufacturing supply chain and so on. However, the service supply chain optimization is less involved with the integration of optimization while the studies are merely focusing on issues including the

J. Song (✉), F. Wu, D. Xu, and J. Fu
Business school, Hohai University, 210098 Nanjing, China
e-mail: songjingsjj@yahoo.com.cn

Y. Zhou and D.D. Wu (eds.), *Modeling Risk Management for Resources and Environment in China*, Computational Risk Management,
DOI 10.1007/978-3-642-18387-4_15, © Springer-Verlag Berlin Heidelberg 2011

coordination of supply chain services, and distribution of services capacity (Cheng and Li 2008; Lu 2010).

This paper refers to the research results of optimization and integration of the manufacturing supply chain and does research on integrated optimization of port supply chains. Sabri and Beamon (2000) proposed that based on the strategic and operational level of integration of manufacturing supply chain and the characteristics of the port supply chain, the two-level integrated optimization of the port supply chain can be implemented. In addition, with the optimized performance of the port supply chain and the customer satisfaction on the supply chain service time, a multi-objective optimization model is established. At the same time, service sectors are optimized from the perspective of the whole supply chain which selective integration of optimization on related cooperative enterprises are fulfilled. Eventually, the integrated optimization of the port supply chain is accomplished by using the improved Particle Swarm Optimization algorithm.

## 2 Mathematical Model

In the port supply chain, the port enterprise is the core business, effectively integrating the transportation, material handling, processing, storage, customs clearance, distribution, finance, business service, and other kinds of logistics service providers (LSP) and customers into one. The port supply chain is a service-oriented supply chain with a primary function of providing logistics services (Li et al. 2009). Optimization and integration of port supply chains should not only optimize local links, but also optimize the whole supply chain.

The optimization goal of the strategic level of a port supply is to reduce the total operation cost by minimizing the cost of logistics services provided by collaborative enterprises in the port supply chain. The optimization target of the operational level is maximizing the flexibility of services provided by collaborating enterprises. From the analysis above, optimizing integrated collaborating enterprises of port supply chain can be regarded as a similar solution of the optimization and integration of the entire supply chain since it achieves the same optimization goal. Therefore, this paper uses a multi-objective optimization model, which optimizes for the integrated collaborative enterprises of supply chain, achieving the target of an optimized and integrated supply chain. The performance evaluation of the supply chain optimization is based on four aspects: cost, flexibility, customer response, and time. Among them, the customer response characterizes the customer satisfaction with the services of the port supply chain including service time and quality, etc. Taking into account the model parameters which facilitate accessibility, this paper combines the customer response and time into a single performance evaluation index of the supply chain optimization, that is, minimizing the total time of the services provided by collaborating enterprises.

In this paper, the multi-objective optimization model has three objectives: minimize the total operational cost of the supply chain; minimize the service time of the support chain; and maximize the flexibility of the supply chain. In addition, in the supply chain network, the distance between node enterprises, also influences the selection of enterprise collaborations. If the distance is closer, the transfer time is shorter and the cost is lowering which is more beneficial to improvement of the optimization function. The distance factor will be added to both the cost objective function and the time objective function in the model, improving the model's practical significance.

## 2.1 Model Assumptions

In the modelling, the port supply chain only considers its primary node enterprises: suppliers, customers, ports, shipping companies, processor service providers, storage service providers, and delivery service providers. Similarly, the service cost only considers the five main service areas: transport costs, port service costs, processing costs, warehouse operating costs, and distribution costs.

When the model mentions port services, it only refers to basic services such as loading and unloading services.

Within a certain period, the cost parameters are assumed to be stable.

The distance between node services in the supply chain is taken into account in both the transportation cost and the transport time of the cargo transfer. Transport costs are affected by the distance between two nodes and the amount of cargo being transported. Transit time, however, is only affected by distance. The per-unit transport cost coefficient and time coefficient are set to a constant value.

## 2.2 Model Parameters and Decision Variables

$k$: The index number of cargo types, $k \in \{1,2,\ldots,K\}$; $a$: The index number of origins of cargo (suppliers), $a \in \{1,2,\ldots,A\}$; $b$: The index number of sales locations (customers), $b \in \{1,2,\ldots,B\}$; $p$: The index number of ports, $p \in \{1,2,\ldots,P\}$; $e$: The index number of shipping companies, $e \in \{1,2,\ldots,E\}$; $i$: The index number of processing service providers, $i \in \{1,2,\ldots,I\}$; $j$: The index number of warehousing service providers, $j \in \{1,2,\ldots,J\}$; $g$: The index number of delivery service providers, $g \in \{1,2,\ldots,G\}$; $v$: The index number of logistics service providers, $v \in \{p,e,i,j,g\}$; $f$: The index number of the candidate members of port supply chain, $f \in \{a,b,p,e,i,j,g\}$; $m$, $n$: The index number of three types of service nodes of supply chain (ports, processors service providers, storage services providers). Cargo can be transported from three nodes to sales locations. $m,n \in \{p,i,j\}$, $m \neq n$; $L_{ap}$: The distance from the cargo type $a$ to the port $p$; $L_{mb}$: The distance from the supply chain service nodes $m$ to sales location $b$; $L_{mn}$: The distance between the supply chain service nodes $m$ and $n$; $CU_{vk}$: The service cost charged by $LSP_v$ for the cargo $k$; $CR$: The transport cost occurred from the cargo transfer between the supply chain service nodes $m$ and $n$; $\alpha$: The per-unit transport cost coefficient for the cargo transfer; $TU_{vk}$: The time needed by $LSP_v$ to offer service for the cargo $k$; $TR$: The time needed for the cargo transfer between the supply chain service nodes $m$ and $n$; $\beta$: The coefficient of the time for the cargo transfer; $T_{vk}$: The service time committed by $LSP_v$to customers; $T'_{vk}$: The actual time to complete services for the cargo $k$ by $LSP_v$; $M_v$: The maximum service capacity provided by $LSP_v$; $\omega_v$: The weighting coefficient of flexibility of different types of services on the supply chain; $X_k$: The quantity of the cargo $k$; $Q$: The total quantity of cargo in the supply chain; $C$: The total operation cost of the supply chain; $T$: The service time of the supply chain; $F$: The flexibility of supply chain; $Y_f$: It is a 0–1 variable. When it is 1, it means the candidate member $f$ is chosen as the collaboration enterprise of the port supply chain; otherwise, not selected.

## 2.3 The Objective Functions

### 2.3.1 Minimize the Total Operation Cost of the Supply Chain

$$\begin{aligned}\min C = &\sum_{e=1}^{E}\sum_{k=1}^{K} Y_e CU_{ek}\left(X_k, \sum_{a=1}^{A}\sum_{p=1}^{P} Y_a Y_p L_{ap}\right) + CR + \sum_{i=1}^{I}\sum_{k=1}^{K} Y_i CU_{ik}(X_k) \\ &+ \sum_{j=1}^{J}\sum_{k=1}^{K} Y_j CU_{jk}(X_k) + \sum_{g=1}^{G}\sum_{k=1}^{K} Y_g CU_{gk}(X_k, L) + \sum_{p=1}^{P}\sum_{k=1}^{K} Y_p CU_{pk}(X_k)\end{aligned} \tag{1}$$

$$L = \begin{cases} \sum_{p=1}^{P}\sum_{b=1}^{B} Y_p Y_b L_{pb}\ (\text{if } \sum_{i=1}^{I} Y_i + \sum_{j=1}^{J} Y_j = 0) \\ \sum_{i=1}^{I}\sum_{b=1}^{B} Y_i Y_b L_{ib}\ (\text{only if } \sum_{j=1}^{J} Y_j = 0) \\ \sum_{j=1}^{J}\sum_{b=1}^{B} Y_j Y_b L_{jb}\ (\text{elsewhere}) \end{cases} \tag{2}$$

$$CR = \sum_{k=1}^{K} \alpha X_k \left( \sum_{p=1}^{P}\sum_{i=1}^{I} Y_p Y_i L_{pi} + \sum_{p=1}^{P}\sum_{j=1}^{J} Y_p Y_j L_{pj} + \sum_{i=1}^{I}\sum_{j=1}^{J} Y_i Y_j L_{ij} \right) \tag{3}$$

The total operation cost C of the port supply chain includes: the service cost of at each links on the supply chain (shipping costs, port service costs, processing costs, warehouse operating costs, distribution costs); and the transport costs for the cargo transfer. Equation (1) is an objective function of the minimum total cost of supply chain. Equation (2) is a function to seek the distribution distance, if cargo do not need processing and storing, then deliver them directly to the sales sites after unloading form port; if the cargo need processing without storing, then the cargo will be shipped directly from the process sites to sell lands; if the cargo need storing, it can be regarded that the cargo shipped from the warehouses to sell sites. Equation (3) is the function of the cost for the cargo transfer.

### 2.3.2 Minimize the Service Time of the Supply Chain

$$\begin{aligned} \min T = & \sum_{e=1}^{E}\sum_{k=1}^{K} Y_e TU_{ek}\left(\sum_{a=1}^{A}\sum_{p=1}^{P} Y_a Y_p L_{ap}\right) + TR + \sum_{i=1}^{I}\sum_{k=1}^{K} Y_i TU_{ik}(X_k) \\ & + \sum_{j=1}^{J}\sum_{k=1}^{K} Y_j TU_{jk}(X_k) + \sum_{g=1}^{G}\sum_{k=1}^{K} Y_g TU_{gk}(L) + \sum_{p=1}^{P}\sum_{k=1}^{K} Y_p TU_{pk}(X_k) \end{aligned} \tag{4}$$

$$TR = \beta\left(\sum_{p=1}^{P}\sum_{i=1}^{I} Y_p Y_i L_{pi} + \sum_{p=1}^{P}\sum_{j=1}^{J} Y_p Y_j L_{pj} + \sum_{i=1}^{I}\sum_{j=1}^{J} Y_i Y_j L_{ij}\right) \tag{5}$$

The service time T includes: the time required by the links' services in supply chain (shipping time, port service time, processing time, storage operation time, delivery time); the time for the cargo transfer. Equation (4) is an objective function of the minimum service time of supply chain. Equation (5) is a function of the time for transfer cargo, which related to the distance between the two services nodes.

### 2.3.3 Maximize the Flexibility of the Supply Chain

$$\begin{aligned} \max F = & \omega_p \sum_{p=1}^{P} Y_p(M_p - Q) + \omega_e \sum_{e=1}^{E} Y_e(M_e - Q) + \omega_i \sum_{i=1}^{I} Y_i(M_i - Q) \\ & + \omega_j \sum_{j=1}^{J} Y_j(M_j - Q) + \omega_g \sum_{g=1}^{G} Y_g(M_g - Q) \end{aligned} \tag{6}$$

The port supply chain flexibility F consists of five parts: port flexibility, transport flexibility, processing flexibility, inventory flexibility, and delivery flexibility. Equation (6) is an objective function of the maximum flexibility of supply chain. The size of the supply chain flexibility, related to the gap between the service capacity limit of the links and the amount of services required. The weighting coefficient of flexibility $\omega$ can be obtained by comparing the relative importance of service sectors.

### 2.4 Constraints

$$0 \leq \sum_{k=1}^{K} X_k \leq M_v \tag{7}$$

$$0 \leq T'_{vk} \leq T_{vk} \tag{8}$$

$$\sum_{k=1}^{K} X_k = Q, \quad \forall k \in \{1, 2, \ldots, K\} \tag{9}$$

$$\omega_p + \omega_e + \omega_i + \omega_j + \omega_g = 1 \tag{10}$$

Equation (7) is the capacity constraints that the volume of services provided is less than the maximum services capacity. Equation (8) is the time constraints, which the actual service time is no more than the hours promised to customers. Equation (9) expresses the amount of kinds of cargo should be equal to the total cargo. Equation (10) indicates that the sum of the flexibility weighting coefficient should be equal to 1.

## 3 Particle Swarm Optimization

The integrated optimization model of port supply chain presented above is a multi objects optimization model. PSO will be applied to optimize the multi objects problem in this text. PSO (Kennedy and Eberhart 1995) is an optimization method based on iteration algorithm with a basic thinking that each potential solution is a "particle" in a D dimension search-space. The particle is moving in the search-space with a specific speed which is updated by its own moving experience and its fellow particles' experience. Each particle not only has a fitness value which is identified by the object function, but also knows its best found position by far which is called particle best (its position is indicated by pbest). This is the individual experience of the particle itself. In addition, each particle acknowledges the global best which is globally the best found position by all the particles in the swarm (its position is indicated by gbest) and this could be treated as the experience of the fellow particles. The movements of the particles are guided by its current position, the distance between its current position and its best found position, as well as the distance between its current positions and the globally best found position, and all current positions are continually updated until the closest Pareto solutions are found.

In order to better control the exploration and exploitation capabilities of the PSO, Shi and Eberart (1998) advocated the improvements on the PSO that is to add inertia weight $w$ into the speed updating formula. Inertia weight $w$ will affect the global and local search capabilities; the larger $w$ can strengthen the global search capability, while smaller $w$ can enhance local search ability.

Based on the improved PSO algorithm, this text is going to develop an algorithm suitable for multi-objective optimization model of port supply chain management. The candidate options on the port supply chain will be treated as particles in the D dimension space. To make sure the value of every dimension of each particle is one candidate enterprise, every dimension of the vector of the particle's speed will be an integer and the range of the integer is from 1 to the number of candidate enterprises at each point on the supply chain. For example, $Y_e$ is the particle to denote the shipping company and every dimension's value of its speed vector should be assigned an integer value within the interval $[1, E]$. In addition, the position of the particle will undergo a rounding up after each iteration process. Both the velocity and position of the particle have lower and upper bounds, such as the speed range of $Y_e$ is [-($E$-1), ($E$-1)] while its position range is [1, $E$]. The three objective functions of the port supply chain integrated optimization model will be considered as fitness functions.

The steps of improved PSO algorithm to implement the optimization of the multi-objects of the port supply chain are shown as below:

Step 1: Create a swarm and initialize particles' positions, velocities, and swarm size.

Step 2: Update the velocity and position of each particle.

Position updates:

$$v^{k+1} = wv_{id}^k + c_1r_1(pbest_{id}^k - x_{id}^k) + c_2r_2(gbest_{id}^k - x_{id}^k)$$

Velocity updates:

$$x_{id}^{k+1} = x_{id}^k + v_{id}^{k+1}$$

Step 3: Calculate each particle's fitness value which includes function values of the cost target, time target, and supply chain flexibility target.

Step 4: Compare each particle's current fitness with individual best and if it is better than pbest, then update the pbest.

Step 5: Compare each particle's fitness value with the global best and if it is better than gbest, then update the gbest.

Step 6: If the maximum number of iterations or the accuracy requirements is met, then stop, export the global best fitness value and its location; otherwise, return to step 3 to continue searching.

## 4 The Numerical Example of Simulation

Port $P$ is the core business of a port supply chain. There are three options for shipping companies, two processing and service providers, three storage service providers, two delivery service providers. $A$ volume of 100 t cargo $X$ needs to be

Table 1 The main parameters of the optimization model (1/3)

| | $CU$ | $TU$ | $M$ |
|---|---|---|---|
| $e_1$ | $1{,}200 + 200X$ | 3.2 | 110 |
| $e_2$ | $1{,}800 + 350X$ | 2.0 | 150 |
| $e_3$ | $1{,}000 + 280X$ | 2.7 | 120 |
| $P$ | $200 + 140X$ | $0.04 + 0.03X$ | 300 |
| $i_1$ | $500 + 120X$ | $0.10 + 0.025X$ | 110 |
| $i_2$ | $200 + 130X$ | $0.08 + 0.02X$ | 110 |
| $g_1$ | $160 + 0.6XL_{iB}$ | $0.02 + 0.001L_{iB}$ | 130 |
| $g_2$ | $200 + 0.3XL_{iB}$ | $0.01 + 0.0015L_{iB}$ | 120 |

Table 2 The main parameters of the optimization model (2/3)

| | $e$ | $p$ | $i$ | $g$ |
|---|---|---|---|---|
| $\omega_v$ | 0.3 | 0.3 | 0.2 | 0.2 |
| $T_v$ | 4 | 3.5 | 3.5 | 0.2 |

Table 3 The main parameters of the optimal model (3/3)

| $\alpha$ | $\beta$ | $L_{Pi_1}$ | $L_{Pi_2}$ | $L_{i_1B}$ | $L_{i_2B}$ |
|---|---|---|---|---|---|
| 0.4 | 0.001 | 8 | 10 | 68 | 60 |

shipped from $A$ to $B$ and land $B$ is near by the port $P$. Customers require that the cargo unloaded should be simply processed, and then directly sent to $B$ and sold there. According to customer's requirements, the best collaborative enterprise integration solution should be decided through considerations from shipping companies, processing service providers and distribution service providers .The parameters required in the example are in Table 1–3 below.

Using particle swarm optimization (PSO) through programming by MATLAB 6.5, two Pareto optimal solutions are obtained, respectively: [1 0 0; 1; 0 1; 0 1], [1 0 0; 1; 1 0; 0 1]. Value of 1 means that the cooperative enterprise with the corresponding parameter value of 1 is selected while the value 0 denoted it is not selected. For example, the first Pareto optimal solution indicates that if shipping company $e_1$, port $P$, processing service provider $i_2$, delivery service provider $g_2$ are selected as the collaborative enterprises in the supply chain, modestly satisfactory integrated optimization can be achieved. The optimal solution for the multi-objective optimization of port supply chain does not exist. It cannot meet the optimization of operating costs, service time and flexibility target simultaneously. PSO can obtain the sufficient Pareto optimal solutions which are widely and smoothly distributed. They are the closet optimal solutions which fulfil the scientific and rational selection of supply chain cooperative enterprises among a variety of goals and constraints. Two options of Pareto optimal solutions are available for the port manager or Cargo owner to choose.

## 5 Conclusion

This paper studied the optimization and integration of the port supply chain. As the port supply chain is different from the manufacturing supply chain, the characteristics of the port supply chain must be taken into account when the strategic and operational levels are optimized. From the whole supply chain perspective, with the consideration of the total cost, the service time and the flexibility of the port supply chain, a multi-objective optimization model is established. This paper employed an improved Particle Swarm Optimization algorithm to solve the multi-objective problem. The simulation example showed that the model and algorithm were reasonable. Parameters in the model were easy to obtain which made the practicality and feasibility of the model was approved.

**Acknowledgment** This research was supported by the Jiangsu Province foster the construction of national key disciplines of Technical Economics and Management point of "Technical Economics and Management of water resources" project.

## References

Cheng J, Li C (2008) Review of research on service supply chain. Modern Manage Sci

Kennedy J, Eberhart RC (1995) Particle swarm optimization. In: Proceedings of IEEE International Conference on Neural Networks, Perth, Australia

Lee CW, Kwon IG, Severance D (2007) Relationship between supply chain performance and degree of linkage among supplier, internal integration and customer. Supply Chain Manage: Int J

Li Z-k, Guo B-b, Yang Z ( 2009) Multi-logistics-task allocation in port logistics service supply chain. Port Waterway Eng

Lu Y (2010) Review of coordinated research on logistics service supply chain. Sci-Technol Manage

Ma X (2005) Multi-objective optimization and simulation of the supply chain. Chinese J Mech Eng

Sabri E, Beamon B (2000) A multi-objective approach to simultaneous strategic and operational planning in supply chain design. Int J Manage Sci

Shi Y, Eberart RC (1998) A modified particle swarm optimizer. In: IEEE World Congress on Computational Intelligence, Anchorage

# Research on Stadia Operation of Nanjing Summer Youth Olympic Games Based on Fuzzy Clustering

**Tang Peng, Pan Rong, and Jiayi Liu**

**Abstract** This paper made effective spatial agglomeration for 10 venues of the Nanjing 2014 Summer Youth Olympic Games based on Fuzzy Clustering. The conclusions of this paper will benefit to the classification of the Olympic stadia, to the integration of the resources, to the healthy development as well as sustainable utilization of the stadia. On the one hand, it will give reference for managers to adjust the present operational planning. On the other hand, it will also be evidence for mangers to select the operational pattern and it is also of exploratory significance for the future operational program of all kinds of venues.

**Keywords** Fuzzy clustering · Stadia · Stadia operation

## 1 Introduction

With 12 days duration, Nanjing 2014 Summer Youth Olympic Games will be held during 17–28 August, 2014. From obtaining the host right to hold Summer Youth Olympic Games, Nanjing is confident in its ability to hold an attractive Summer Youth Olympic Games in its 54-day preparation with its thorough arrangement,

T. Peng (✉)
School of Business, Hohai University, Nanjing 210098, China
and
Department of Sports, Hohai University, Nanjing 210098, China
e-mail: pansophie@yahoo.com.cn; tangpeng315@126.com

P. Rong
School of Business, Hohai University, Nanjing 210098, China

J. Liu
School of Public Administration, Hohai University, Nanjing 210098, China

Y. Zhou and D.D. Wu (eds.), *Modeling Risk Management for Resources and Environment in China*, Computational Risk Management,
DOI 10.1007/978-3-642-18387-4_16, © Springer-Verlag Berlin Heidelberg 2011

extensive volunteers' attendance, self-contained facilities as well as green, safe and convenient city environment.

The utilization of Olympic venues after contests is a worldwide problem. Notably, none of 15 stadia for hosting Nanjing 2014 Summer Youth Olympic Games will be established specially for it. There are two temporary venues (Xuanwu Lake Triathlon Venue and Sailing Venue of Nanjing Jinniu Lake Scenic Area) and one newly-built venue (Nanjing Hockey Stadium) in the 15 stadia. The only newly-built venue, however, is not built especially for this Summer Youth Olympic Games but will serve as a branch campus of Nanjing Sport School in the future. We wonder whether Nanjing 2014 Summer Youth Olympic Games which is featured by prudence will face the worldwide problem of utilization of Olympic venues after contests.

In the ear of Post-Olympic, how to make reasonable orientation for Post-Olympic venues with long construction duration and huge investment so as to bring noumenon function of venues into full play and realize value preservation and increment of state-owned assets is a great challenge that all managers are facing. To classify the venues according to their degree of sustainable utilization and conduct classified target location is beneficial to the integration of the resources, to the healthy operation and sustainable utilization of the venues. In the short term, it can provide reference for managers to adjust the current operational plan. In the long term, it can be evidence for mangers to choose the operational pattern and it is also of exploratory significance for the future operational planning of various types of venues.

Many scholars from both China and abroad have made much research work on the sustainable utilization of venues from different perspectives such as economic influence of venues (Baim; Preuss 2002; McKay et al.), development situation of hosting cities (Andranovich et al. 2002) and utilization of Olympic venues after contests (Zhao 2008). Currently, most domestic research (Dai 2007; Li et al. Dec. 2008) analyzes actuality of stadia by literature consultation, expert interview, data investigation and logic analysis first before finding out main problems existing and putting forward relevant countermeasures and suggestions. Li Chunliu and Ma Xiangqiang (Ma and Xu 2008; Li et al. 2009) mainly studied the methods of sustainable utilization of specific venues while Hu (1995) tried to scale sports stadiums and gymnasiums with a multi-factor comprehensive scaling model. Wang and Zhan (2008) established performance evaluation index system for stadia based on balanced scorecard and analyzed by AHP Method. Liu Bo (Liu and Zou 2006) analyzed goal orientation of benefit for domestic large-scale stadia and put forward the view to combine economic benefit and social benefit.

At present, systematic evaluation research on classified target location based on sustainable utilization of post-Olympic venues is rare, which is not in accordance with the concept of making sustainable utilization of sports infrastructure. Making evaluation research on sustainable utilization of Olympic venues before the hosting of Nanjing 2014 Summer Youth Olympic Games can help managers to know the operational situation and discover and solve problems in time. Putting forward classified goal orientation suggestions based on the results of sustainable utilization evaluation is of exploratory significance for the future operational planning of various types of venues.

In view of this, this paper establishes evaluation index system for sustainable utilization of venues and makes effective spatial agglomeration to 10 venues of Nanjing 2014 Summer Youth Olympic Games based on fuzzy clustering and puts forward classified goal orientation suggestions according to the clustering results with a view to provide evidence and reference for evaluating sustainable utilization of post-Olympic venues. The remainder of the paper is organized as follows. We elaborate in Sect. 2 the hypothesis of research, before introducing our established evaluation index system of sustainable utilization for Post-Olympic venues. We present research method (Sect. 4) before making empirical analysis with ten venues of Nanjing 2014 Summer Youth Olympic Games as examples (Sect. 5). We conclude in Sect. 6.

## 2 Hypothesis

The objects evaluated in this paper are the 10 Olympic venues which are the exclusion of two temporary venues, one newly-built venue, Nanjing International Expo Centre and Nanjing Laoshan National Forest Park. After the hosting of Nanjing 2014 Summer Youth Olympic Games, the temporary venues will be demolished while Nanjing International Expo Centre and Nanjing Laoshan National Forest Park do not fall into the category of stadium. Considering the availability of data, Nanjing Hockey Stadium is also exclusive from evaluation objects in this paper.

Therefore the 1st hypothesis of this paper is the assumed evaluation objects of ten Olympic venues.

As post-Olympic venues are just an epitome of other various venues, the research on classified goal orientation of them will be of exploratory significance for the future operational planning of various kinds of stadia.

Therefore the 2nd hypothesis of this paper is the assumed representativeness of venues.

Stadia with many same attributes, high degree of similarity and were classified as the same category by the evaluation model exist high degree of similarity and association in the orientation of goal and management method. The stadium which falls into the same category can operate with same kind of targets and management method.

Therefore the 3rd hypothesis of this paper is the assumed same kind of goal orientation matches the same kind of stadium.

The aim of evaluation and classification research on sustainable utilization is to classify the post-Olympic venues and find out features and rules of sustainable utilization of all kinds of stadia.

Therefore the 4th hypothesis of this paper is the assumed objective of sustainable utilization of all kinds of stadia.

## 3 Establishment of Index System

It is studied that there has not yet established a specific evaluation index system of sustainable utilization of venues, while there is a spot of research on the performance of venues and the indexes it values (Zhang Oct. 2005; Hui and Wang Dec. 2007). After repeated research and thrash, abiding by the scientific, systematic and operable principle of design evaluation indexes, 12 influencing factors were designed from four dimensions including customer dimension, financial dimension, internal business dimension and development dimension. The 12 influencing factors are respectively customer satisfaction, service quality, atmosphere construction of sports, cost control, profit creation, investment management, organizational mechanism, human resource management, informatization degree, market competitiveness, employee training, research situation. Reconstructed costs of venue, building area of venue, total capacity of venue and temporary capacity of venue were selected as evaluation indexes in the light of the availability of data.

## 4 Method

### *4.1 Fuzzy Clustering*

Traditional cluster analysis belongs to hard division which strictly divides each sample into a certain type. Fuzzy clustering, which fits elastic classification, introduces the concept of fuzzy mathematics to cluster analysis. Fuzzy clustering establishes uncertainty description for sample. Each sample has no longer only belonged to a certain kind but belongs to a certain type respectively according to definite membership degree. (Yang 2007)

Basic principle of system clustering method (Gui et al. 2005) is as follows. Firstly consider a certain amount of samples as a type ($n$ types) and regard each sample as dot in space with $m$ dimension ($m$ statistical index). That is, treat every statistical index as coordinate axis with $m$ dimension. Then, calculate $n(n-1)/2$ distance and category two samples with the shortest distance into one type. Thereafter, calculate distance between the abovementioned type and other $n-2$ samples in accordance with a specific clustering method and combine two kinds with the shortest distance. Repeat this process until all samples are combined into one type.

## 4.2 Algorithm Steps of Fuzzy Clustering

### 4.2.1 Establish Data Matrix

Suppose universe $U = \{X_1, X_{2,\ldots,} X_n\}$ is clustering object and each object has $m$ index to describe its figure. Thus, the original data matrix is:

$$X = \begin{bmatrix} x_{11} & x_{12} & \ldots & x_{1m} \\ x_{21} & x_{22} & \ldots & x_{2m} \\ \ldots & \ldots & \ldots & \ldots \\ x_{n1} & x_{n2} & \ldots & x_{nm} \end{bmatrix}$$

### 4.2.2 Range Normalization of Data

Usually, different data has different dimension in practice. In order to compare data with different dimension, pretreatment should be conducted to data. Two methods are popular to realize data condensation to $[0, 1]$.

(a) Translation transform

$$X'_{ij} = \frac{X_{ij} - \overline{X}_j}{S_j} \qquad (\mathrm{i} = 1, 2, \ldots, \mathrm{n};\ \mathrm{j} = 1, 2, \ldots, \mathrm{m}) \tag{1}$$

Therein, $\overline{X}_j = \frac{1}{n}\sum_{i=1}^{n} X_{ij}$; $\quad S_j = \sqrt{\frac{1}{n}\sum_{i=1}^{n}(X_{ij} - \overline{X}_j)^2}$

(b) Range transform

$$X^n_{ij} = \frac{X_{ij} - \min\limits_{1 \le i \le n}\{X'_{ij}\}}{\max\limits_{1 \le i \le n}\{X'_{ij}\} - \min\limits_{1 \le i \le n}\{X'_{ij}\}} \qquad (\mathrm{j} = 1, 2, \ldots, \mathrm{m}) \tag{2}$$

Obviously, $0 < X^n < 1$.

### 4.2.3 Establish Fuzzy Similar Matrix

Since universe is a finite set, classification issues are usually discussed based on fuzzy similar matrix in practice. Fuzzy similar relationship of $X$ is shown as a fuzzy similar matrix. That is, symmetric fuzzy square matrix $R$ whose element in diagonal is 1. Many methods are used to calculate similarity coefficient such as magnitude method, cosin method, correlation coefficient method.

### 4.2.4 Make Fuzzy Cluster Analysis

Transform $R$ into $R^*$. Then calculate transitive closure $t(R)_i$ of $R$ by square method.

### 4.2.5 Determine Optimal Threshold Value $\lambda$

Cluster can be done after the establishment of fuzzy equivalent matrix. The cluster results differ with different $\lambda$ matrix. Reasonable selection of threshold value $\lambda$ has a direct impact on the final clusters number.

## 5 Empirical Analysis

Step 1: Establish original data matrix $C_0$.
Step 2: Data normalization. Transform data into standard $Z$ fraction and obtain normalized matrix $C_1$.
Step 3: Establish similar matrix by correlation coefficient method.

$$r_{ij} = \frac{\sum_{k=1}^{m} \left|X_{ik} - \overline{X_i}\right| \left|X_{jk} - \overline{X_j}\right|}{\sqrt{\sum_{k=1}^{m} \left(X_{ik} - \overline{X}\right)^2} g \sqrt{\sum_{k=1}^{m} \left(X_{jk} - \overline{X_j}\right)^2}} \tag{3}$$

Therein, $\overline{X_i} = \frac{1}{m} \sum_{k=1}^{m} X_{ik}$. Obtain matrix $R_0$.
Step 5: Make cluster analysis by SPSS.
Step 6: Results are obtained in Fig. 1 after above-mentioned steps (Table 1).

**Table 1** Statistics of cluster analysis

| Cluster | Venues | Scale |
|---|---|---|
| 1 | Nanjing Rowing-Canoeing School | 5 |
| | Nanjing Sport Institute | |
| | Nanjing International Equestrian Field | |
| | Jiangning Sports Centre | |
| | Fangshan Sports Training Base of Jiangsu | |
| 2 | Nanjing Baima Slalom Course | 3 |
| | Nanjing Longjiang Gymnasium | |
| | Jiangning Football Training Base of Jiangsu | |
| 3 | Wutaishan Sports Centre of Jiangsu | 1 |
| 4 | Nanjing Olympic Sports Centre | 1 |

**Table 2** Center of gravity of each cluster

| Cluster | Reconstructed costs (million dollar) | Seat capacity | Temporary capacity | Building area (10,000 $m^2$) |
|---|---|---|---|---|
| 1 | 0.11 | 8,800 | 130 | 6.84 |
| 2 | 0.28 | 2,333 | 66.67 | 2.03 |
| 3 | 0.6 | 39,500 | 0 | 1.8 |
| 4 | 0.25 | 88,900 | 300 | 40.1 |

# 6 Conclusions and Perspective

This paper makes effective spatial agglomeration to ten stadia of Nanjing 2014 Summer Youth Olympic Games based on fuzzy cluster and classified them into four clusters. By extraction, comparison and analysis of index characteristics of various samples on the basis of the average value of various index data (Table 2), the conclusions are as following: most stadia belonged to the first cluster with obvious characteristics of the minimum investment cost, building area and total capacity. Notably, two venues in the universities all fell into this category. The scale of the second cluster is slightly inferior than the first cluster, which is featured by small building scale, limited capacity and moderate reconstructed costs. Scale of both the third and the fourth cluster is one, which means only one special case was classified to that kind. Venue in the third cluster has moderate building scale and capacity. Compared with other venues, it has very small building area and high reconstructed costs. The stadium in the fourth cluster is special too. It is similar in the characteristics of reconstructed costs with the third cluster but contrary with it in its building area.

Since stadium is characteristic by long construction duration, huge amount of investment cost and low benefit, seldom enterprise and person invest it. Currently, most domestic stadia are newly-built or reconstructed by funds from central or local government. Therefore, they belong to the category of state-owned assets. In operational goal orientation, quasi-public goods like stadium should not only meet the needs of the whole society, but also ensure the value preservation and increment of stated-owned assets and be guided by the market while regard making profits as object. When setting operational goal for specific stadium, cluster location should be conduct firstly in accordance with its sustainable utilization degree. Then overall guiding ideology and decision-making principle can be determined by drawing lessons from goals of other stadia in the same category and analyzing its resource elements as well as both internal and external environment.

Apply fuzzy cluster to sustainable utilization analysis on stadia can make accurate, dynamic forecast of data cluster and avoid man-made subjectivity and superficiality. It can better illustrate essential law lying in samples and of exploratory significance for the future operational planning of various kinds of stadia.

# References


Andranovich G et al (2002) Olympic cities: lessons learned from mega-event politics [J]. J Urban Aff 23(2)

Baim D The post-games utilization of Olympic venues and the economic impact of the Olympics after the games [C]. In: Proceeding of 1st Olympic Economy and City Development and Cooperation Forum.

Dai C (2007) Recycle economy and building and management of stadia [J]. Bus Situation 4:8

Gui X, Jin W, Hu Y (2005) Fuzzy cluster analysis and its application in transportation planning [J]. Transp Computer 2:80–83

Hu X (1995) Design of methods and model for multi-factor comprehensive scaling of sports stadiums and gymnasiums. J China Sports Sci Assoc 15(6):1–4

Hui Y, Wang Z (Dec. 2007) Establishment of performance evaluation index system of sports venues in universities. J Inner Mongolia Norm Univ 20(12):1–4

Li C, Wang H, Jiang M, Feng Y (2009) Research on the sustainable utilization of sport venues in Qinhuangdao after Beijing Olympic games. Constr Econ 2:1–4

Li C, Wang H, Cao Z, Jiang M (Dec. 2008) Research for the sustainable utilization of sport venues in Qinhuangdao after Beijing Olympic Games. J Hebei Norm Univ Sci Technol 22(4):1–4

Liu B, Zou Y (2006) Benefit target of large-scale stadium from the angle of the public economics [J]. J Shandong Inst Phys Educ Sports 10:24–26

Ma X, Xu G (2008) Research on the sustainable utilization of Beijing Olympic venues. China Market 5:1–2

McKay M et al. Reaching beyond the gold: the impact of the Olympic games on real estate markets [M]. www.joneslanglasalle.com.hk.

Pruess H (2002) The economics of Olympic games [M]. Walla Walla Press, Sydney

Wang Z, Zhan W (2008) Research on performance evaluation index system of Olympic venues [J]. Stat Decis 5:80–82

Yang W (2007) Research on tax decision support system based on fuzzy cluster analysis [J]. Bus Res 5:40–41

Zhang Y (Oct. 2005) Factors for assessing the performance of exhibition venues. J Guangzhou Univ 4(10):1–4

Zhao Y (2008) Research on management and sustainable development of large-scale stadia [J]. Knowl Econ 11:126–127

# Performance Evaluation of Scientific Research Program in Zhejiang Colleges Based on Uncertainty Analysis

**Lian-fen Yang and Yun Tang**

**Abstract** This paper concentrates on the performance evaluation of scientific research program in Zhejiang colleges based on uncertainty analysis. Firstly, it sets up the index system. It considers the characteristics of scientific research program in Zhejiang colleges and determines some indexes. It introduces the membership function to determine the membership degree of indexes. Secondly, the paper confirms the index weights. It uses the principle of Analytic Hierarchy Process (AHP) and the entropy weight method to confirm the index weights. This paper establishes an assessment model and gives a theoretical supportive to case study on performance evaluation of scientific research program in Zhejiang colleges.

**Keywords** AHP · Colleges S&T program · Entropy · The membership degree · Uncertainty analysis

## 1 Introduction

These years, the scientificresearch fund increased from ¥16.68 billion in 2000 to ¥73.27 billion in 2008. The R&D expenditure in colleges increased from ¥7.67 billion in 2000 to ¥39.02 billion in 2008. The number of R&D projects in colleges reached 429,096 in 2008, in Zhejiang province the number was 31,746, accounting for 7.4% of the total.(Data Source: Chinese S&T Statistic Yearbook 2009). The scientific research fund has been pumped, the R&D expenditure has been largely spent, and also, the number of the R&D projects is huge. While, there are little analysis on performance evaluation of colleges scientific research projects.

L.-f. Yang (✉)
Zhijiang College, Zhejiang University of Technology, Hangzhou 310014, China

Y. Tang
College of Business Administration, Zhejiang University of Technology, Hangzhou 310014, China

Y. Zhou and D.D. Wu (eds.), *Modeling Risk Management for Resources and Environment in China*, Computational Risk Management,
DOI 10.1007/978-3-642-18387-4_17, 

By literature searches, model cases interviews and experts consulting, this paper analyses the unique of colleges on undertaking scientific research projects, and sets up a performance evaluation system for colleges scientific research projects.

## 2 Definition and Principles

We define performance evaluation for S&T projects as follows: According to some certain standards and assessment procedures, scientific and technological departments use scientific and feasible ways to assess and examine the projects, which serves as the guidance and reference of making plans and project budget afterwards. The principle design of index system is: specific, measurable, data availability and time bound.

## 3 Index System

According to the four principles above, this index system contains 7 first-grade indexes and 21 second-grade index. "Performance evaluation of scientific research program in Zhejiang colleges" named as capital "A", these 7 first-grade indexes named as $A_1$–$A_7$, they are $A_1$ -implementation status, $A_2$-personnel, $A_3$-funds, $A_4$-output and achievement, $A_5$-direct economic benefit, $A_6$-indirect social influence, $A_7$-qualified personnel trained. The meanings of capitals present the second-grade indexes will be shown in Sects. 3.1–3.7. This index system is as Fig. 1.

### *3.1 Implementation Status*

Implementation status ($A_1$) includes 3 second-grade indexes. They are $A_{11}$- the progress of the project, $A_{12}$-the accomplishment of the essential technique and economic indexes, $A_{13}$-acceptance conclusion of the project. The final results are all adopted non-dimensional scores. The calculation of evaluating values is

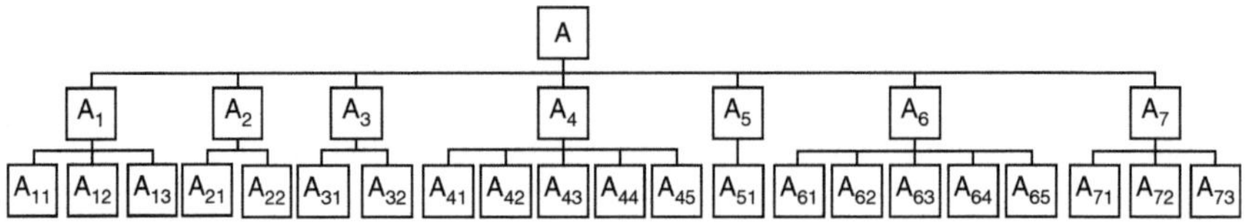


**Fig. 1** The index system

multiply the weight of each index by each score. (Method of determining the weight of index can be found in the Part 5, hereinafter the same).The details are as follows.

**$A_{11}$-The progress of the project:** adopt the way of choosing the different level to get the different score. There are five levels, the standards of counting as below. "*Proceeded according to plan*" graded for 10 points, "according to plan" means the actual proceeding is basically conformity with the planning proceeding; "*ahead of schedule*" graded for 9, it means the actual tempo is ahead of the plan; "*stalled*" graded for 6, it means the tempo of the project is fell behind the plan at least 6 months; "*pause*" graded for 4, it means the project is stopped, but it will be continued soon; "*canceled or ceased*" graded for 1, if the project cannot proceed or have no reason to proceed, it has been end.

**$A_{12}$-The accomplishment of the essential technique and economic indexes:** the way getting score is the same as above. There are three levels, the standards of counting as below: "*Completely done*" graded for 10 points, it refers to the essential technique and economic indexes have done above 80% of total; "*basically finished*" graded for 6, refers they have done 50%–80%; "*incomplete*" graded for 1, refers the indexes have done below 50%.

**$A_{13}$-Acceptance conclusion of the object:** adopt the way of choosing level score. There are three levels, the standards of counting as below: "*Qualified*" graded for 10 points; "*basically qualified*" graded for 6; "*unqualified*" graded for 1. The final conclusion should base on the acceptance conclusion from the administrative department.

## *3.2 Personnel*

Personnel ($A_2$) includes 2 second-grade indexes. They are $A_{21}$- professional title of the project head, $A_{22}$-educational background of participants. Calculation of evaluating values is multiply the weight of each index by each score. The details of indexes at all levels in this indicator are as follow.

**$A_{21}$-Professional title of the project head**: adopt the way of choosing level score, here are criterions: "*senior title*" graded for 10 points; "*vice-senior*" graded for 8; "*middle title*" graded for 6; "*junior title*" graded for 4; "*others*" graded for 1 points.

**$A_{22}$-Educational background of participants:** counted by numbers of persons. The calculation of the calculate value as formula (1).

$$\begin{aligned}
&\textit{the calculate value} = 10 \times \textit{the numbers of graduate students} + \\
&8 \times \textit{the numbers of university diplomas} + 6 \times \textit{the numbers of college diplomas} + \\
&4 \times \textit{the numbers of technical secondary school background} + \\
&1 \times \textit{the numbers of others}.
\end{aligned} \tag{1}$$

After get the calculate value, use the membership degree to get the score. The detail about membership is in Part 4.

## 3.3 Funds

Funds ($A_3$) contains 2 second-grade indexes. They are $A_{31}$-the budget of project and the actual capital utilized, $A_{32}$-the budget of project and the actual expenses. Calculation of evaluating values is multiply the weight of each index by each score. The details about these two indexes are as follow:

**$A_{31}$-The budget of project and the actual capital utilized:** choose level to get score, the standards of counting as below: "*Perfectly consistent*" graded for 10 points. It means the actual capital utilized amounting to above 90% of the budgets; "*Basically consistent*" graded for 6, 70–90% of the budgets; "*Discrepancy*" graded for 1, below 70% of the budgets.

**$A_{32}$-The budget of project and the actual expenses:** the way get scores is the same as above, the standards of counting as below: "*Perfectly consistent*" graded for 10 points. It means the actual expenses controlled within the pale of ±5% of the budgets; "*Basically consistent*" graded for 6, pale of ±10% of the budgets. "*Discrepancy*" graded for 1, almost out of 10% of the budgets.

## 3.4 Output and Achievement

Output & achievement ($A_4$) contains 5 second-grade indexes. They are $A_{41}$-papers and publications, $A_{42}$-the invention patent, $A_{43}$-the level of the achievement, $A_{44}$-the technical standard and $A_{45}$-the award of the project. Calculation of evaluating values is multiply the weight of each index by each score. The details about them as follow.

**$A_{41}$-Papers and publications:** counted by the number of the papers and publications. The calculation of the calculate value as follow:

$$\begin{aligned}&\textit{calculate value} = 10 \times \textit{the number of published books} + \\&10 \times \textit{the number of SCI(SSCI), EI, ISTP} + \\&8 \times \textit{the number of paper published in core journal} + \\&2 \times \textit{the numbers of paper in other journal.}\end{aligned} \tag{2}$$

Published book contains the central level and the local level. The list of core journal subjects to the core journal catalogue of Beijing university library. After got the calculate value, use the membership degree to get the score.

**$A_{42}$-The invention patent:** counted by the number of the invention patent, the calculation of the calculate value as follow:

$$\begin{aligned} \textit{calculate value} &= 10 \times \textit{the number of patents applies} + 2 \\ &\quad \times \textit{the number of patents authorized} \end{aligned} \tag{3}$$

The method of calculation is the same as "papers and publications".

**$A_{43}$-The level of the achievement:** choose the level score, the standards of counting as below: "*International leading*" graded for 10 points; "*international advancing*" graded for 8; "*national leading*" graded for 6; "*national advancing*" graded for 4; "*leading in the province*" graded for 2 points.

**$A_{44}$-The technical standard:** counted by the number of achievements. The calculation of the calculate value as follow:

$$\begin{aligned} &\textit{calculate value} = 10 \times \textit{ the number of international standards} + \\ &8 \times \textit{ the number of national standards } + 6 \times \textit{ the number of industry standards} + \\ &4 \times \textit{ the number of local standards } + 2 \times \textit{the number of enterprise standards.} \end{aligned} \tag{4}$$

The method of calculation is the same as "papers and publications".

**$A_{45}$-The award of project:** counted by the number of awards. The calculation of the calculate value as follow:

$$\begin{aligned} \textit{calculate value} &= 10 \times \textit{ the number of national grade } + 8 \\ &\quad \times \textit{the number of provincial} \\ \textit{or ministerial grade } &+ 4 \times \textit{the number of urban grade} \end{aligned} \tag{5}$$

Then the method of calculation is the same as "papers and publications".

## 3.5 *Direct Economic Benefit*

Direct economic benefit ($A_5$) includes only 1 second-grade index–the ratio of input and output ($A_{51}$). The calculation of the calculate value as follow:

$$\textit{calculate value} = \textit{project output value(yuan)/the actual expenses(yuan)} \tag{6}$$

After got the calculate value, use the membership degree to get the score.

## 3.6 *Indirect Social Influence*

Indirect social influence ($A_6$) contains 5 second-grade indexes. They are $A_{61}$-enegy conservation, emissions and cost reduction, $A_{62}$-ecological environment improvement,

$A_{63}$-public facility and utility, $A_{64}$-public security and ability to prevent and mitigate disasters, $A_{65}$-health conditions of the population.

In these five indicators, this paper chooses the answer yes or no to get score. If the answer is "yes", it is graded for 10 points, otherwise, graded for 0 point. The calculation of evaluating values is multiply the weight of each index by each score.

### 3.7 *Qualified Personnel Trained*

Qualified personnel trained ($A_7$) includes 3 second-grade indexes. They are $A_{71}$-job opportunity increasing, $A_{72}$-the number of taking degrees and $A_{73}$-an advance in professional title. Calculation of evaluating values is multiply the weight of each index by each score. The details are as follow:

**$A_{71}$-Job opportunity increasing:** counted by the numbers of the job opportunities increased. Then use the membership degree to get the score of "papers and publications" with the method of calculating difference in value.

**$A_{72}$-The number of taking degrees:** counted by the numbers. The numbers happen after the projects. The calculation of the calculate value as follow:

$$\begin{aligned} calculate\ value = {} & 10 \times the\ number\ of\ doctor\ degrees + \\ & 5 \times the\ number\ of\ master\ degrees + \\ & 2 \times the\ number\ of\ bachelor\ degrees \end{aligned} \tag{7}$$

After got the calculate value, use the membership degree to get the score.

**$A_{73}$-An advance in professional title:** counted by the numbers. The calculate value is:

$$\begin{aligned} & calculate\ value = 10 \times the\ number\ of\ getting\ high\ academic\ title \\ & + 5 \times the\ number\ of\ getting\ middle\ title \end{aligned} \tag{8}$$

Then use the membership degree to get the score.

## 4 The Membership Degree of Indexes-Curve Parameter Calibration Method

The indexes in this system have two ways of calculation: (1) the evaluating values of second-grades indexes whose scores can be can be used directly according to some standards. (2) the calculate value of second-grades indexes which the evaluating values must be unified first.

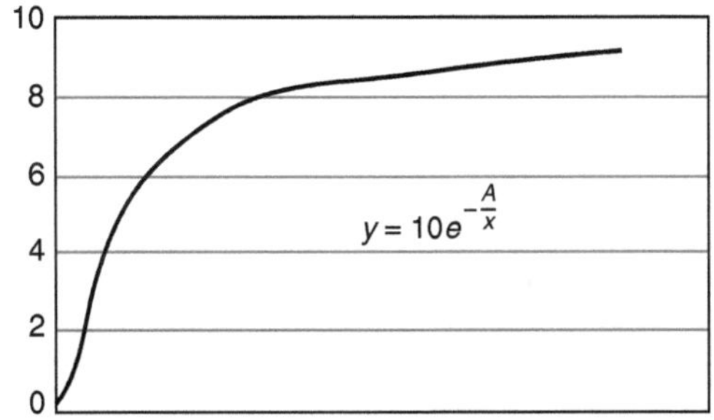


**Fig. 2** The shape of the curve

Adopt the calibration for curve parameters to determine the membership degree of indexes; the first step is to define a curve. This curve (function) should satisfy certain conditions as follow: (1) monotonicity. (2) Convergence (3) Parameters to be determined, the less the better. Based on reasons above, this paper chooses the function:

$$y = 10e^{-\frac{A}{x}} \tag{9}$$

The shape of the curve as follow (Fig. 2):

We define ($\bar{x}_i$, 5) the membership degree of the middle value equate to the average. Then we can count the value of the parameter A. Also, we insist $x_i = 0$, $y_i = 0$. Other $y_i$ rounds number to nearest integer.

## 5 Weight Determined

This model is based one the principle AHP and the entropy weight method to determine the weight. It stands for subjective and objective combined together, qualitative and quantitative analysis combined together. The steps of building the model as follow: (1) build the evaluation system as finger 1; (2) with several rounds of experts marking, get the scores with AHP. (3) At the stage of case study, use all the data we collect, determined weight with the entropy weight method. (4) To calculate the combination weight as final result.

### *5.1 Based on AHP to Determine the Weight*

**Hierarchical Structure.** AHP is a classical method on subjectively determining weight. This model has 7 first-grade indexes and 21 second-grade indexes.

**Construct Judgment Matrix.** After finish hierarchical structure, we invite experts to judge the importance between every factor in the same level. Use 1–9

Table 1 Judgment matrix

| $E_k$ | $F_1$ $F_2$ $F_3$ …… $F_n$ |
|---|---|
| $F_1$ | $f_{11}$ $f_{12}$ $f_{13}$ …… $f_{1n}$ |
| $F_2$ | $f_{21}$ $f_{22}$ $f_{23}$ …… $f_{2n}$ |
| $F_3$ | $f_{31}$ $f_{32}$ $f_{33}$ …… $f_{3n}$ |
| | ………………… |
| $F_n$ | $f_{n1}$ $f_{n2}$ $f_{n3}$ …… $f_{nn}$ |

Table 2 Meanings of numbers

| fij | Meanings |
|---|---|
| 1 | Fi and Fj equal important |
| 3 | Fi is a little more important than Fj |
| 5 | Fi is more important than Fj |
| 7 | Fi is much more important than Fj |
| 9 | Fi is extremely more important than Fj |
| 2, 4, 6, 8 | The important degree of Fi and Fj is between the levels above |
| 1/2, …, 1/9 | The meanings are opposite |

ratio scale method to construct the judgment matrix. We suppose there is importance relationship between $F_1$ $F_2$…$F_n$, the judgment matrix as follow (Table 1):

$f_{ij}$ means in $E_K$ the relative significance between $F_i$ and $F_j$. In normal, the value of $f_{ij}$ is 1–9 and their reciprocal. The meanings of numbers as follow (Table 2):

This model is built on theory aspect. W stands for the weight. The calculations as below:

$$M_i = f_{i1} \times f_{i2} \times \ldots f_{in} \tag{10}$$

$$V_i = n\sqrt[n]{M_i}\ ;\ V = (V_1, V_2, \ldots V_n)^T \tag{11}$$

$$W_i = V_i/(V_1 + V_2 + \ldots + V_n). \tag{12}$$

**Consistency Check.** $\lambda_{\max}$ is the maximum feature root of judgment matrix of order n. CI is the indicator of checking the consistency. Formula is

$$CI = (\lambda_{\max} - n)/(n - 1) \tag{13}$$

RI is average random consistency scale. The value of it as below (Table 3):

$$CR = \frac{CI}{RI} \tag{14}$$

When CR < 0.10, the matrix pass the consistency test, otherwise, it need to adjust till reach the satisfactory consistency.

**Table 3** The value of RI

| Order | 1 or 2 | 3 | 4 | 5 |
|---|---|---|---|---|
| RI | 0.00 | 0.58 | 0.90 | 1.12 |
| Order | 6 | 7 | 8 | 9 |
| RI | 1.24 | 1.32 | 1.41 | 1.45 |

## 5.2 Entropy Weight Method

Entropy weight theory said that to an index with different data, the distinction is the bigger the better. It contains more information, so it should be given a higher weight.

$e_j$ ($\mathbf{e_j} > \mathbf{0}$)stands for entropy of index j, $f_{ij}$ means the proportion of index j in system i. $w_j$ stands for the entropy weight. The formulas are

$$e_i = -\frac{1}{\ln n}\sum_{i=1}^{n} f_{ij}\ln(f_{ij}) \tag{15}$$

$$f_{ij} = \frac{x_{ij}}{\sum_{i=1}^{n} x_{ij}} \tag{16}$$

$$w_i = \frac{1 - e_j}{n - \sum_{i=1}^{n} e_i} \tag{17}$$

## 5.3 Combination Weight Determined

The capital $w^*$ stands for combination weight. $w^s$ is weight determined through AHP, $w^t$ stands is entropy weight. Construct these three weights to a linear combination. Suppose the subjective preference coefficient weight distributions $\beta$, the function as below:

$$w_j^* = \beta w_j^s + (1-\beta)w_j^t \tag{18}$$

Then construct a function through the minimum square of deviance, the function as follow:

$$\min z = \sum_{j=1}^{m} [(w_j^* - w_j^s)^2 + (w_j^* - w_j^t)^2] \tag{19}$$

Take the Eq. 18 into Eq. 19, it can be calculated that $\beta = 0.5$, and this result is without loss of generality. So the function of combination weight for all the indexes is below:

$$w_j^* = 0.5w_j^s + 0.5w_j^t \tag{20}$$

This paper focuses on the theoretical research. Data collection and case study will be shown in the following study.

## References

Chen Z-s, Yang L-f, Wang H-m (2008) On the comprehensive evaluation of modern service industry in CBD of Central City [J]. Sci Technol Progress Policy
Chi G-t, Zhu Z-c, Zhang Y-l (2008) The science and technology evaluation model based on entropy and G1 and empirical research of China [J]. Stud Sci Sci
Lu L-y (2009) The optimization of investment projects using evaluation method of multi-level entropy [J]. J Xihua Univ (Natural Science Edition).
Xie S-h (2009) Assessment methodology for R&D projects in transportation engineering base on comprehensive performance index [D]. Chang'an University, Xi'an, China
Yang L-f (2009) Cluster effect evaluation of modern service industry in CBD of Central City [J]. Stat Decis
Zhang X-a (2009) Construct and the performance evaluation indexes of science research team in university based on AHP [J]. Sci Technol Manage Res

# The Analysis and Empirical Test on Safe Headway

**Han Xue, Shan Jiang, and Zhi-xiang Li**

**Abstract** In order to solve the enormous difference about safe headway between the theoretical value from academic researches and the observational value from practical road conditions, this paper establishes a new vehicle-following model to accurately simulate the vehicle-following process. The model well simulates the actual situation, and provides an operable safety distance reference value for car drivers. Combined with the experiences of outstanding drivers, the model reengineers an integrated process as guidance including maintaining safe headway and changing traffic lane or braking. In addition, the paper discusses the limit of safe headway through sensitive analysis, and finds the reliable measures, so the safe headway could be shortened to improve the efficiency of road utilization. The broad applicability of the conclusion is verified by empirical data.

**Keywords** Car following model · Hazard · Process reengineering · Safe headway · Sensitive analysis

## 1 Introduction

In China, The Article 15 of the Highway Traffic Management (HTM), implemented in the March 3 of 1995, provided that, "The vehicles on the same lane must maintain adequate vehicle distance between them, while traveling along the highway. In regular circumstances, the vehicle distance is over 100 m, when the vehicle travels at the rate of 100 km/h; the vehicle distance is over 70 m, when the vehicle travels at the rate of 70 km/h". Obviously, the regulation does not have the practical operability because that the required safe headway is too long for drivers to comply

H. Xue (✉), S. Jiang, and Z.-x. Li
School of Management and Economics, Beijing Institute of Technology, 100081, People's Republic of China
e-mail: xuehanliang@163.com; jiangshanyou@163.com; zhxli@263.net

Y. Zhou and D.D. Wu (eds.), *Modeling Risk Management for Resources and Environment in China*, Computational Risk Management,
DOI 10.1007/978-3-642-18387-4_18, 

with. For instance, the Second Ring Road in Beijing is 32.7 km in length, one-way three lanes. Provided the vehicles travel at 60 km/h, according to the Article 15, the vehicle distance should be 60 m, as each lane is capable of 545 vehicles, so 1,635 for three lanes, which is unrealistic.

In this research, a sampling questionnaire survey was carried out among the automobile drivers in Beijing, 3,000 pieces of questionnaires were received and 2,175 were meaningful. According to this survey, most of drivers maintain a shorter distance compared with the regulated distance by the HTM measures. There have been no amendments or explanations to the Article 15 for 15 years since its enactment. Then, it is still necessary to discuss the vehicle distance.

Ben Gurion (Cahlon et al. 1979) believes that a confidence interval is determined for the response of the follower to a sufficiently small perturbation of the leader velocity to be a safe one according to a natural safety criterion. Groeger (1998) evaluate the effectiveness of a radar-based warning system on the headway drivers chose to adopt while driving in real traffic. Goto et al. (1999) propose a safe traffic speed control based on safe headway distances in AHS: advanced cruise-assist highway systems. Through introducing a generalized optimal speed function to consider spatial position, slope grade and variable safe headway, Li et al. (2008) investigated the effect of slope in a single-lane highway on the traffic flow with the extended optimal speed model. Yet the enormous difference about safe headway between the theoretical value and the observational value still exists.

This paper is trying to start from the microscopic individual vehicle by model process reengineering, sensitive analysis and the empirical testing to provide a practical operational procedures and standard values of safe headway for drivers.

## 2 The Formula Derivation of the Safe Headway and Sensitive Analysis

### 2.1 *Basic Assumption*

To get the proper safe headway, some factors need to be simplified for theoretical analysis.

1. Suppose that the two cars are on the same lane, the following one fails to take the steering wheel to avoid.
2. Dynamic fluctuation and air resistance are neglected.
3. Suppose that the two vehicles have same brake performance, mass $m$ and the drivers share the same response.
4. Suppose that the initial velocity of the two vehicles are $v_0$.
5. Suppose that the braking mechanical process is a uniformly retarded motion.

## 2.2 Model Building and Safe Headway Derivation

It is supposed that $t = 0$ is when the lead vehicle begins decelerating. At the moment of $t_1$, the velocity reduces from $v_0$ to $v_t$. The movement of the vehicle is a uniformly retarded motion and the distance is $S_1$. Then, the lead vehicle is in uniform motion at the velocity of $v_t$ $(0 \leq v_t \leq v_0)$.

The trailing car takes active measures when it discovers that the lead car slow down. It includes the following four stages: When the emergency caught by the eyes of the following driver (such as brake light up ahead or shortened vehicle space, etc.), the scene is flashed to the driver's visual cortex (discovery period); valuating the emergency, the driver decides whether emergency brake is in need or not (judgment period);the driver jams the brakes instead of steps on the gas to cancel free pedal travel (action period); because of the gap existed in the brake system, a certain period of time is required from the brake pedal operation to the brake responding (response period) (Evans 1991). It takes $t_2$ to finish the whole process, which is brake reaction time for the trailing car. In the period of $0 - t_2$, the travelling distance of the trailing car is $S_2$. The process of braking is as follows (see Fig. 1).

According to Table 1, the brake reaction time is approximately 0.6 s.

From the above mentioned, the trailing car will maintain constant velocity in $t_2$. The braking deceleration $a$ is

$$a = k\frac{mg}{m} = kg \quad (k \text{ is constant quantity}). \tag{1}$$

As for the lead car,

$$v_t = v_0 - kgt_1. \tag{2}$$

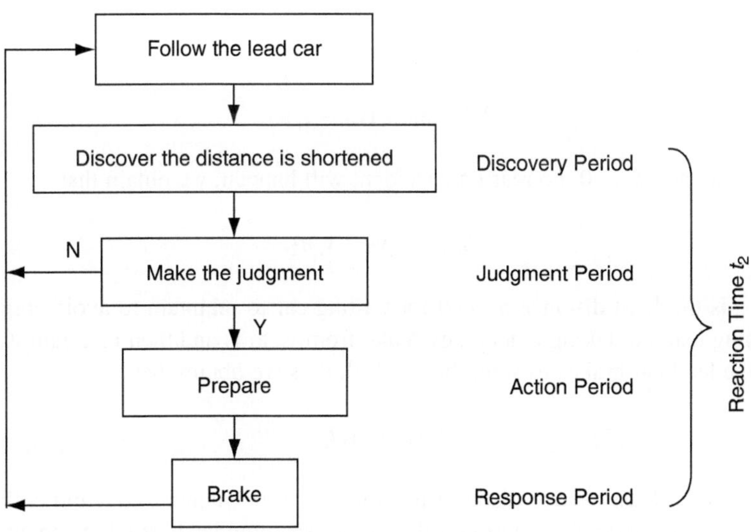


**Fig. 1** Process of braking in general

**Table 1** The brake reaction time for the drivers (Yu et al. 2009)

| Age bracket | Elements of the brake reaction time (ms) | | | | Reaction time (ms) |
|---|---|---|---|---|---|
| | Light-retina-visual center | Judgment period | Action period | Response period | |
| 20 ~ 29 | 40 | 376 | 110 ± 4.5 | 45 | [566.5, 575.5] |
| 30 ~ 39 | 40 | 371 | 121 ± 3.7 | 45 | [573.3, 580.7] |
| 40 ~ 49 | 40 | 374 | 135 ± 5.3 | 45 | [588.7, 599.3] |
| ≥50 | 40 | 383 | 147 ± 9.3 | 45 | [605.7, 624.3] |

$$S_1 = \int_0^{t_1} (v_0 - kgt)dt = \int_0^{t_1} v_0 dt - \int_0^{t_1} kgtdt = v_0 t_1 - \frac{1}{2} kgt_1^2. \tag{3}$$

During the same time, for the trailing car,

$$S_2 = v_0 t_2. \tag{4}$$

When $t > t_2$, the trailing car begins to decelerate and duplicate the same process of lead car (from $t = 0$ to $t = t_1$) until it reaches the same velocity $v_t$, and stop deceleration at time $t_1 + t_2$. During the same time (from $t = t_1$ to $t = t_2 + t_1$), the lead car is in uniform motion with the velocity $v_t$, and with the distance $S_3$, thus

$$S_3 = v_t t_2. \tag{5}$$

So the distance between two cars $\Delta S$ is

$$\Delta S = S_0 - [(S_2 + S_1) - (S_1 + S_3)]. \tag{6}$$

From (4) and (5), we obtain that

$$\Delta S = S_0 - (v_0 - v_t) t_2. \tag{7}$$

As long as $\Delta S \geq 0$, no rear-end accident will happen, we obtain that

$$S_{\min} = (v_0 - v_t) t_2. \tag{8}$$

$S_{\min}$ is the least distance needed for trailing car to maintain to avoid collision, assuming lead car taking emergency brake from $v_0$ to $v_t$ and then maintain $v_t$.

If the lead car brakes to stop, then $v_t = 0$, thus we obtain that

$$S^*{}_{\min} = v_0 t_2. \tag{9}$$

$S^*_{\min}$ is the least needed distance for trailing car to maintain to avoid collision, when two vehicles run at the same velocity $v_0$, assuming lead car taking emergency brake to $v_t = 0$.

**Table 2** The relationship between the safe headway and the initial velocity of trailing car ($v_t = 0$, $t_2 = 0.6$ s)[a]

| $v_0$(km/h) | 10 | 20 | 30 | 40 | 50 | 60 | 70 | 80 |
|---|---|---|---|---|---|---|---|---|
| $S^*_{min}$(m) | 1.68 | 3.36 | 4.98 | 6.66 | 8.34 | 10.02 | 11.64 | 13.32 |

[a]In order to facilitate discussion, list the situations when velocity is under 80 km/h, but this conclusion is also useful in situations when velocity is above 80 km/h, same below

**Table 3** The relationship between the safe headway and the final velocity of the lead car ($v_0 = 80$ km/h, $t_2 = 0.6$ s)[a]

| $v_t$ (km/h) | 0 | 10 | 20 | 30 | 40 | 50 | 60 | 70 |
|---|---|---|---|---|---|---|---|---|
| $S_{min}$ (m) | 13.32 | 11.64 | 10.02 | 8.34 | 6.66 | 4.98 | 3.36 | 1.68 |

[a]For different velocities, we could get different outcomes of relationships

Let reaction time $t_2 = 0.6$ s, $v_t = 0$, we obtain Table 2 from (9).

In general, suppose that the lead car take emergency brake to $v_t(v_t > 0)$, then maintain the velocity of $v_t$, we obtain Table 3 from (8).

If the driver of trailing car judge that the velocity of lead car would decrease a lot, according to Table 3, a longer distance would be needed. In addition, if the velocity of lead car reduces to 0, there would be the conclusion that safe headway is 13.32 m, the same as column 9, Table 2.

## 2.3 *Discussion of Shortening the Safe Headway*

To avoid the hazard of accidents, but also improve the efficiency of road utilization, we need to minimize the critical safe headway on the basis of (8). Clearly, the initial velocities of the two vehicles are under the highest road velocity limit control. Therefore, to reduce the safe headway, we should consider $t_2$ and $v_t$.

Improvement 1: rediscussion on the reaction time $t_2$.

It is no room for shortening discovery time, action time and response time, so to shorten the reaction time $t_2$, we have only two options.

One is to shorten the judgment time. Judgment time is the time for drivers in the emergency situation, to judge whether immediate action should be taken and when to taken. This period of time is related to the levels and experiences of drivers, more importantly, to the content needed to be determined. The more complex situation, the more time should be consuming. Therefore, it can greatly shorten the reaction time if we transfer the whole judge process into a conditioned reflex (Lisa 2003).

To anticipate the dangerous condition is another choice. That means to complete some work before the emergency brake, so as to shorten the whole reaction time, which is useful just for skilled drivers, and the whole brake reaction process should be reforged.

Thus, combined the experience of some excellent drivers, we suggest that when the distance between the vehicles is close to the critical safe headway, the driver of trailing car move foot from the accelerator pedal to the brake pedal and depress it to eliminate the free path, and judge for lane-changing. If it is feasible, change the lane and reentry into car-following process. If not, brake immediately to reduce the velocity to ensure safety when the trailing car reaches the critical safe headway. The flow chart of process reengineering we construct is Fig. 2.

As this process move the judgment time and action time before the trailing car reaches the critical safe headway, the brake motion of trailing car driver is a conditioned reflex which shortens the Reaction Time $t_2$ into 0.3 s (Han 1997).

Let $t_2 = 0.3$ s, from (9), we obtain Table 4.

Suppose that the movement of vehicles is divided into two types, forward or lateral. For lateral, the vehicle is just move to the adjacent lanes; its velocity does not change. We modify Nagatani symmetric lane-changing rules (Kurata and Nagatani 2003) to yield two cases and propose the motivation condition and security condition which should be satisfied. Here are specific rules for each case (Chen and Gao 2007).

Case 1: Along the driving direction, the distance between the current car and the lead car of the target lane is far greater than the distance between the Current car and the lead car of the current lane, at the same time, if the distance between the

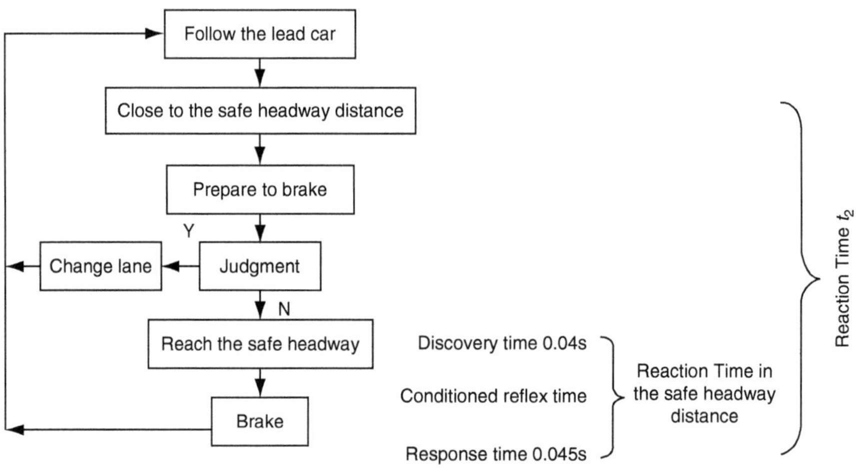


**Fig. 2** Brake judgment process for excellent drivers

**Table 4** The relationship between the safe headway and the initial velocity of the trailing car ($v_t = 0$, $t_2 = 0.3$ s)

| $v_0$ (km/h) | 10 | 20 | 30 | 40 | 50 | 60 | 70 | 80 |
|---|---|---|---|---|---|---|---|---|
| $S^*_{\min}$ (m) | 0.84 | 1.68 | 2.49 | 3.33 | 4.17 | 5.01 | 5.82 | 6.61 |

current car and the trailing car of the target lane is longer than the critical safe headway, it is feasible to change lanes.

Motivation Condition: $\Delta f_i(t) \gg \Delta x_i(t)$.

Security Condition: $\Delta b_i(t) > x_c$.

Case 2: Along the driving direction, the distance between the current car and the lead car of the target lane is longer than (not far greater than) the distance between the Current car and the lead car of the current lane, while the velocity of the lead car of the target lane is faster than the lead car of the current lane, at the same time, if the distance between the current car and the trailing car of the target lane is longer than the critical safe headway, it is feasible to change lanes.

Motivation Condition: $\Delta f_i(t) > \Delta x_i(t)$, $v_{other,f}(t) \geq v_i(t)$.

Security Condition: $\Delta b_i(t) > x_c$.

Then we get the flow chart of judgment for lane-changing. (See Fig. 3)

In the above rules, $\Delta f_i(t)$ is the distance between the vehicle $i$ and the lead vehicle of the target lane, $\Delta b_i(t)$ is the distance between the vehicle $i$ and the trailing vehicle of the target lane, $v_{other,f}(t)$ is the velocity of lead vehicle of the target lane, $v_i(t)$ is the velocity of the vehicle $i$ at time $t$.

Improvement 2: to improve $v_t$.

Suppose that two cars are running on the urban expressway with 80 km/h (the greatest velocity limit), from (8), we obtain Table 5.

If the driver of the trailing car can predict the velocity which the lead car will reduce to, he can maintain a relatively shorter safe headway according to Table 5. That is, if the driver can fully observe the whole road conditions, makes the correct

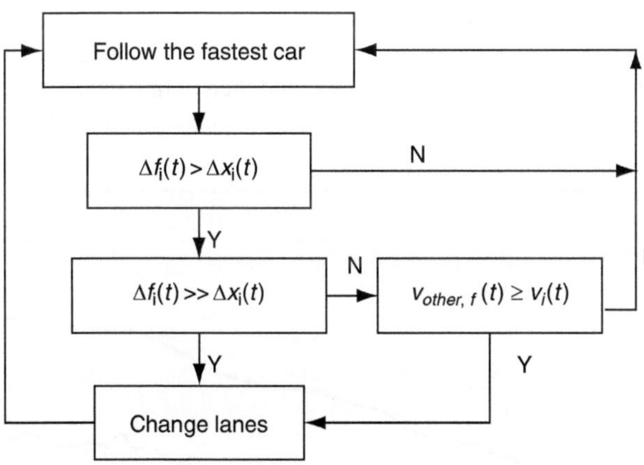


**Fig. 3** Flow chart of judgment for lane-changing

**Table 5** Relationship between the safe headway and the final velocity of the lead car ($v_0 = 80$ km/h, $t_2 = 0.3$ s)

| $v_t$ (km/h) | 0 | 10 | 20 | 30 | 40 | 50 | 60 | 70 |
|---|---|---|---|---|---|---|---|---|
| $S_{min}$(m) | 13.32 | 11.64 | 10.02 | 8.34 | 6.66 | 4.98 | 3.36 | 1.68 |

prejudgment, then, even maintain a relatively very short distance would not cause an accident. Of course, when the vehicle travels very fast, we do not advocate doing so for unexpected events. But it does can be used to explain that, in some special cases, a very short distance between vehicles did not lead to any accident.

# 3 Empirical Studies

## 3.1 Empirical Study on Safe Headway

We recruit 100 volunteers through the sample survey we mentioned before, and continuous record each driver's driving velocity and the distances from the vehicles in front by GPS velometers and laser distance meters. The entire testing process takes 2 months to ensure each one was collected steady-state data for 10 h. Among them, we intercept velocity data from 20 km/h to 80 km/h as valid data, excluding the impact of the installation location and bad weather.

We obtained a series of data sequence (*S*, *v*) for each driver by curve-fitting techniques. The vehicle's velocity *S* is from data recorded by GPS velometers according to each corresponding moment of the minimum distance *S* recorded by laser distance meters. As the driver does not always have to maintain critical safe headway, we divide velocity from 21 km/h to 80 km/h into 60 integral points, take the driver's minimum corresponding distance *S* for each integral point velocity, if there is no corresponding record, pad it by moving average method, then we obtain a new series of data sequence (*S*, *v*) for each driver.

According to these 100 series of sequence (*S*, *v*), we calculate the arithmetic mean of *S* at each integral points of velocity, and then obtain the minimum distances these drivers maintain (curve *b* in Fig. 4). For comparison, we select nine skilled

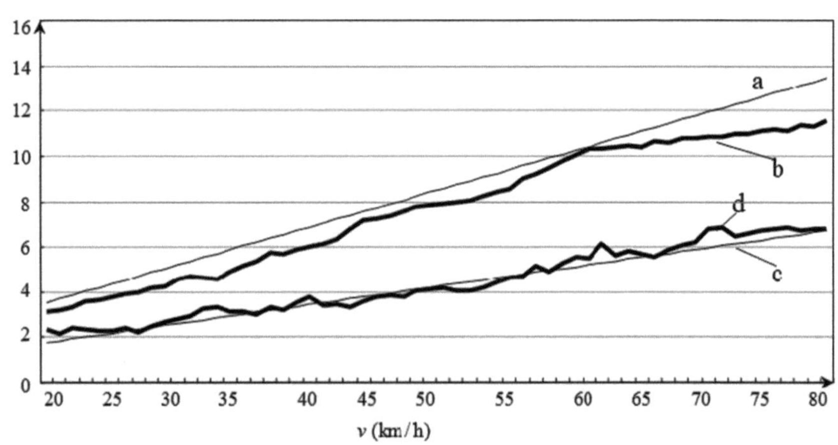


**Fig. 4** The empirical test on the safe headway

drivers who claim that they have excellent driving skills to predict dangerous situations. We calculate the arithmetic mean of $S$ at each integral points of velocity from these nine series of sequence ($S$, $v$), and then obtain the minimum distances these skilled drivers maintain (curve $d$ in Fig. 4).

The quantitative relationship between the safe headway and the initial velocity of the trailing car under ordinary braking process (See Table 2) is shown as Line $a$ in Fig. 4. The quantitative relationship between the safe headway and the initial velocity of the trailing car under reengineering braking process (See Table 4) is shown as Line $c$ in Fig. 4.

Figure 4 shows that the theoretical guidance values of the critical safe headway under our ordinary braking process dovetail nicely with the actual observations of ordinary drivers. It also shows that the theoretical guidance values of the critical safe headway under reengineering braking process which describes the skilled drivers who can predict dangerous situations dovetail nicely with the actual observations of them. The theoretical guidance values of the safe headway are realistic and entirely feasible.

## 3.2 Description of Some Special Traffic States

Data records show that the distance of vehicles could jump bigger because the trailing car may change lanes. The adjacent car insert into current lane lead to the distance of vehicles jump smaller. Taking these minimum values into account, we take all minimum $S$ from the samples, and obtain the following curve (see Fig. 5).

As long as the driver can predict the whole situation, distances between vehicles within 2 m do not induce particular dangerous accidents within 80 km/h velocity, which fully shows that sufficient sense of safety is more important than maintaining a longer distance mechanically.

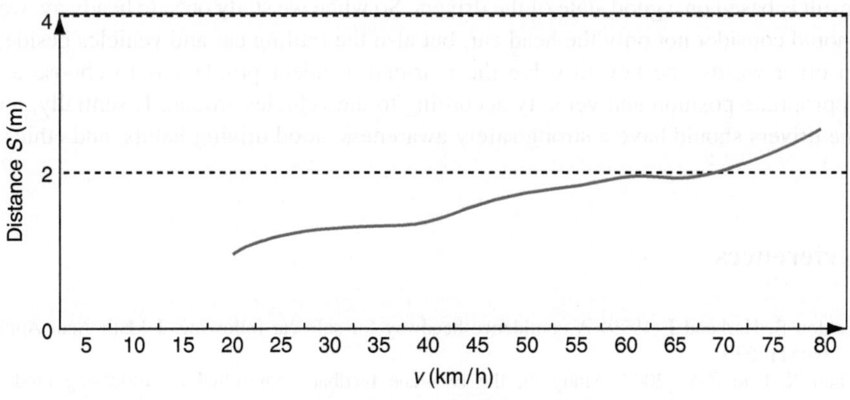


**Fig. 5** The empirical test in exceptional circumstances

**Table 6** The most likely distance between vehicles which was inserted by adjacent cars[a]

| | Velocity (km/h) | Distance of Vehicles (m) |
|---|---|---|
| A | 20–40 | >7 |
| B | 40–60 | >8 |
| C | 60–80 | 9.5–20 |

[a]The most likely distance between vehicles to be inserted by adjacent cars

### *3.3 The Problem Caused by Maintaining a Long Distance between Vehicles*

According to the size of velocities of trailing cars, we divide the data records into three groups as follows. (See Table 6)

As long as the distance between vehicles is more than 7 m, the possibility of being inserted by adjacent cars is increasing, which further illustrates the importance to maintain a reasonable relative distance. If the distance is too large, the adjacent cars may insert into current lane at any time and dramatically shorten the distance between vehicles, which more easily leads accident.

## 4 Conclusion

According to this paper, the traditional safe headway is not absolutely safe. It only works in the extreme assumptive cases. It is meaningless to talk about absolute safety because even maintaining an infinite distance, rear-end accident will still happen inevitably in many situations, such as deviation or tire bursting of adjacent lane vehicles.

This paper deduces a result which is relatively reasonable and practicable, and much smaller than many other research results. The bigger the distance is, the more danger it is because of the queue jumper. It should also be emphasized that the result is based on a good state of the drivers. So when we study on safe headway, we should consider not only the head car, but also the trailing car and vehicles beside. In other words, the key to solve the rear-end accident problem is to choose an appropriate position and velocity according to the vehicles around. Essentially, all the drivers should have a strong safety awareness, good driving habits, and ethics.

## References

Cahlon B, Harband J (1979) A confidence headway for safe car-following. J Math Anal Appl 69:511–530

Chen X, Gao Z-Y (2007) Study on the two-lane feedback controlled car-following model (in Chinese). Acta Phys Sin 56:2024–2029

Dorn L (2003) Driver behaviour and training. Ashgate Publishing Limited, Hampshire

Evans L (1991) Traffic safety and the driver. Van Nostrand Reinhold, New York
Goto Y, Furusawa H, Araki M, Fukuda T (1999) A safe traffic speed control in AHS. In: Proceedings 1999 IEEE/IEEJ/JSAI international conference on intelligent transportation systems, pp 459–464. IEEE, Tokyo, 1999
Groeger JA (1998) Close, but no cigar: assessment of a headway warning device. In: Proceedings of the 1998 IEE colloquium on automotive radar and navigation techniques, pp 51–54. IEE, London, 1998
Han Y-c (1997) Driver reaction time test and its mathematical treatment (in Chinese). Psychol Sci 20:436–440
Kurata S, Nagatani T (2003) Spatio-temporal dynamics of jams in two-lane traffic flow with a blockage. Phys A Stat Mech Appl 318:537–550, Elsevier, Netherlands
Li X-L, Song T, Kuang H, Dai S-Q (2008) Phase transition on speed limit traffic with slope. Chin Phys B UK 17:3014–3020
Yu Z-p, Wang Y, Gao F (2009) Interval analysis method for safety distance of car-following (in Chinese). Trans Chin Soc Agric Machinery 40:31–35

# A Sensitive Analysis on China's Managing Float Regime

**Shan Jiang, Han Xue, and Zhi-xiang Li**

**Abstract** In this paper, according to the PBoC's official statement, based on the data from 2005 to 2010, a basket of currencies is constructed with the goal of stabilizing the trade volume. The time series of this basket of currencies is compared with the ones of RMB by means of Eview 6.0. As a result, it is concluded that the trend of RMB exchange rate is propelled by supply and demand in the market under the current managing float regime, which is an inevitable outcome of the "the dilemma of PBoC" and surely, not manipulated by the Chinese government.

**Keywords** A basket of currencies · RMB · Sensitive analysis · The exchange rate elasticity of trade · Uncertainty

## 1 Introduction

On July 21, 2005, China introduced a new currency regime that ended the decade-long fixed nominal exchange rate of the renminbi vis-à-vis the US dollar (People's Bank of China 2005). The authorities not only immediately revalued the official bilateral rate by 2.1%, but also announced that the renminbi (RMB)henceforth would be managed "with reference to a basket of currencies" rather than being pegged to the dollar. Most importantly, the central bank said that the exchange rate was to become "more flexible" with its value based more on "market supply and demand."

Despite the policy change, China's currency strengthened very little. With the current account surplus up to 8% of GDP and increasing, China is now blamed to manipulate the rate of exchange between RMB and the U.S. dollar for purposes of

S. Jiang (✉), H. Xue, and Z.-x. Li
School of Management and Economics, Beijing Institute of Technology, 100081, People's Republic of China
e-mail: jiangshanyou@163.com; xuehanliang@163.com; zhxli@263.net

Y. Zhou and D.D. Wu (eds.), *Modeling Risk Management for Resources and Environment in China*, Computational Risk Management,
DOI 10.1007/978-3-642-18387-4_19, © Springer-Verlag Berlin Heidelberg 2011

preventing effective balance of payments adjustments or gaining unfair competitive advantage in international trade.

Frankel (1992) used purchasing power over a consumer basket of domestic goods as numeraire to define the "value" of each of the currencies in a basket of currencies; Frankel and Wei (1993) used the SDR; Bénassy-Quéré (1999) used the U.S. dollar; Ohno (1999) and Eichengreen (2006), the Swiss franc.

Merrill Lynch, Royal Bank of Canada (RBC) and many other institutes predict the types and weights of the basket of currencies of RMB. However, there is no trend of any basket of currencies be consistent with the trend of the RMB, which makes the relationship between the RMB and a basket of currencies in doubt, the RMB exchange rate may be manipulated.

Peter G. Peterson Institute for International Economics develops a new symmetric matrix inversion method for estimating consistent fundamental equilibrium exchange rates (FEERs) for leading advanced and emerging-market economies. Earlier this year, C. Fred Bergsten, director of the Peterson Institute for International Economics, told the Congress that his calculation showed RMB was undervalued by as much as 41% against the dollar (Cline and Williamson 2010). This statement has been repeated on many occasions by American officials as well as by Paul Krugman, the Nobel laureate economist.

Taking account of the likelihood that the regime has evolved over the 5 years, this paper conducts an updated evaluation of what exchange rate regime China has actually been following.

## 2 Phases of the RMB Exchange Rate Trend

When China in fact follows a perfect basket peg, the technique is an exceptionally apt application of OLS regression. It should be easy to recover precise estimates of the weights. The sum of the coefficients of regression of currencies should be one. The true regime is more variable than a rigid basket peg, and then the choice of numeraire does make some difference to the estimation. This paper used a remote currency, the Swiss franc, for its value will not be influenced by these currencies directly.

To show a basket of currencies, we assume that there are currencies $R_1$, $R_2$, $\cdots$, $R_n$, with weights equal to $w_1$, $w_2$, $\cdots$, $w_n$ respectively, then

$$E^{R/W} = c + \sum_{i=1}^{n} w_i E^{R/R_i} + \varepsilon. \quad (1)$$

$E^{R/W}$ is the exchange rate of the whole basket of currencies (valued by currency $R$). Here, $R$ denotes the Swiss franc.

According to the People's Bank of China (PBoC) Governor Zhou Xiao-chuan's speech about the principles of composing a basket of currencies, we select exchange

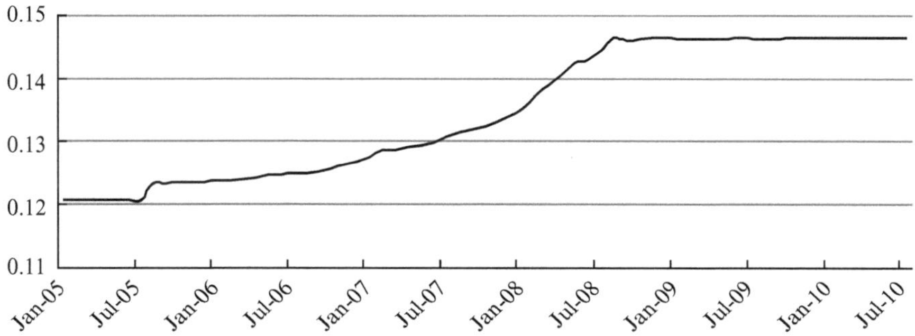


**Fig. 1** The exchange rate of RMB against U.S. dollar (January 2005–July 2010) (State Administration of Foreign Exchange. http://www.safe.gov.cn)

rate of 17 nations,[1] whose values of import and export with China are over ten billion U.S. dollars. Those are the exchange rates of U.S. dollar, Hong Kong dollar, Indian rupee, Japanese yen, Korean Won, Taiwan dollars, Indonesian rupiah, Malaysian ringgit, Philippine peso, Singapore dollar, Thai Baht, Euro, Pound, Russian ruble, Brazilian Real, Canadian dollar and Australian dollar against U.S. dollars. Then, we calculate the exchange rate of each currency against Swiss francs according to the exchange rate of Swiss francs against U.S. dollars.

According to the features of the curve in Fig. 1, we separate the monthly data into 3 phases with the turning point of August 2005 and July 2008 and analyze them by Eview 6.0 as follows:

Select monthly observations of data that ran from January 2005 to July 2005, the calculation result shows that the RMB was tightly pegged to the dollar, and no other currencies. The standard correlation coefficient is 1.

Select monthly observations of data that ran from August 2008 to June 2010, the calculation result shows that the RMB was still tightly pegged to the dollar, and no other currencies. The standard correlation coefficient is 0.995.

Select monthly observations of data that ran from August 2005 to July 2008, the calculation result of Eview 6.0 shows the RMB is influenced by many currencies, yet Korean Won and Indonesian Rupiah play a decisive role while the share of Euro, Sterling and U.S. Dollar is very low. Apparently, the multicollinearity makes Eview 6.0 can not accurately estimate the weights of a variety of currencies.

As Fig. 2 shows, according to the result above, we calculate the $E^{R/W}$and fit the exchange rates curve of the basket of currencies against the Swiss francs and the RMB exchange rate against the Swiss francs in time series, despite the good fit, the weight clearly not reflect the reality of the situation, thus there is not much reference value in it, yet one thing is certain that apparently, during this period, the PBoC intervenes the exchange rate of RMB with reference to a basket of currencies.

[1]Since the exchange rate of VND (Viet-Namese Dong) and the SAR (South African Rand) did not disclose by PBoC, we can only remove them from a basket of currencies.

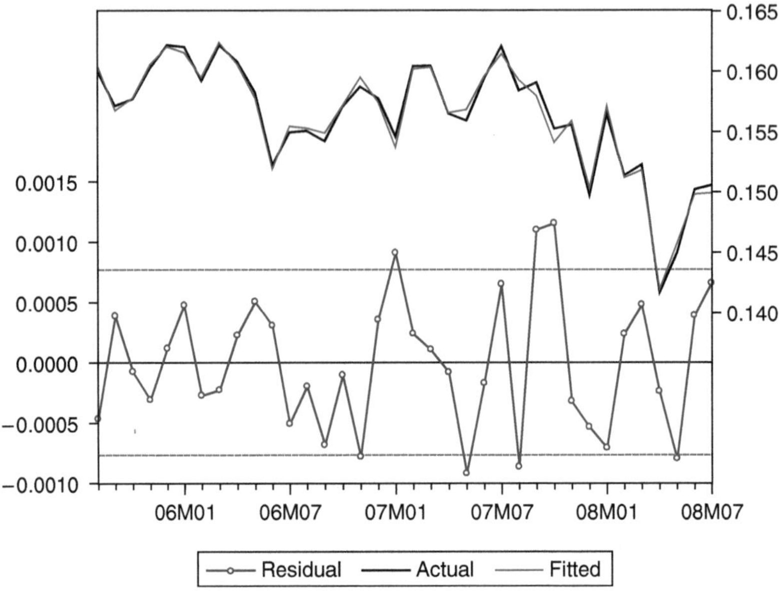


**Fig. 2** The exchange rates curve of the basket of currencies and the RMB against Swiss francs

## 3 The Analysis and Empirical Test on Exchange Rate of RMB

How does one ascertain what is the true exchange rate regime, if a country announces the adoption of a basket peg, and reveals a list of currencies that may be included in the basket, but does not reveal the exact weighting of the component currencies? Bénassy-Quéréet al. (2004) have used a particular technique to estimate the implicit weights. The weight-inference technique is very simple: one regresses changes in the value of the local currency, in this case the RMB, against changes in the values of the dollar, euro, yen, and other currencies that are candidate constituents of the basket. In the special case where China in fact follows a perfect basket peg, the technique is an exceptionally apt application of OLS regression. It should be easy to recover precise estimates of the weights. The fit should be perfect, an extreme rarity in econometrics: the standard error of the regression should be zero, and $R^2 = 100\%$.

### *3.1 The Weights of a Basket of Currencies*

As over 80% import and export commodities of China are dollar-denominated, taking the capital flows into account, if the stability of the balance of international payments is the goal of the new currency regime, the weight of the US dollar must be pretty close to 0.9 in that basket, and therefore it is not nearly as effective as it should be. Thus, we only consider the flow of goods and ignore capital flows, and take the stability of trade as a primary objective. Obviously, in the design of the currency weight, exchange rate elasticity of trade plays a key role in the stability of trade volume.

Suppose that the local nation has $n$ trade partners with currency $R_1, R_2, \cdots, R_n$ respectively. The bilateral trade volume between the local nation and its partners is $T_1, T_2, \cdots, T_n$. $T$ denotes the trade volume of the local nation. Thus,

$$T = \sum_{i=1}^{n} T_i. \tag{2}$$

Which is equivalent to

$$\hat{T} = \sum_{i=1}^{n} v_i \hat{T}_i. \tag{3}$$

$v_i$ denotes the proportion of each nation's bilateral trade volume in the trade volume of the local nation. So,

$$\sum_{i=1}^{n} v_i = 1. \tag{4}$$

Let $\varepsilon_i$ denote the exchange rate elasticity of the bilateral trade of each nation, rewrite (3) as

$$\hat{T} = \sum_{i=1}^{n} v_i \varepsilon_i \hat{E}^{R/R_i}. \tag{5}$$

Assume that the goal of pegging to a basket of currencies regime is to make the trade stable by $w_i$. That is to minimize the change of the volume of trade $\hat{T}^2$, so $w_i$ should be subject to

$$\sum_{i=1}^{n} w_i \hat{E}^{R/R_i} = 0. \tag{6}$$

Therefore, let $R_1$ denote \$, from (6) and the Equilibrium Equation of Arbitrage, based on \$, we obtain that

$$\hat{T} = \sum_{j=2}^{n} \left[ w_j \sum_{i \neq j, i=1}^{n} v_i \varepsilon_i + w_j \varepsilon_j (1 - w_j) \right] \hat{E}^{R_j/\$}. \tag{7}$$

To minimize $\hat{T}^2$, we write the first-order condition as follows:

$$\frac{\partial \hat{T}^2}{\partial w_j} = 0 \quad j = 2, 3, \cdots, n. \tag{8}$$

As the change of the exchange rate $\hat{E}^{Rj/\$}$ is unknown, we substitute (9) for (8):

$$w_j \sum_{i \neq j, i=1}^{n} v_i \varepsilon_i + w_j \varepsilon_j (1 - w_j) = 0 \quad j = 2, 3, \cdots, n. \tag{9}$$

And so we obtain the optimal currency weights

$$w_i = \frac{v_i \varepsilon_i}{\sum_{i=1}^{n} v_i \varepsilon_i}. \tag{10}$$

Obviously, the optimal weights depend on the trade shares of the partners and the exchange rate elasticity of bilateral trade. From (7) to (9), we obtain $\hat{T} = 0$, which shows that the optimal currency weights could make the change of trade volume to 0, or make the trade volume stable.

## 3.2 The Exchange Rate Elasticity of Bilateral Trade

Algebraically, suppose that the RMB is pegged to currencies $R_i$ with weights $w_i$, then, rewrite (1) to Logarithm (11):

$$\Delta \ln E^{CHF/RMB} = w_0 + \sum_{i=1}^{n} w_i \Delta \ln E^{CHF/R_i} + \varepsilon. \tag{11}$$

We should include a constant term $w_0$ to allow for the likelihood of a trend appreciation in the RMB, whether against the dollar alone or a broader basket (Frankel and Wei 1993; Frankel 2009).

According to (11), we select the monthly data of the 17 currencies mentioned above from August 2005 to July 2008. By the method of the test for multicollinearity of Eview 6.0, we remove Hong Kong dollar, Indian rupee, Taiwan dollars, Indonesian rupiah, Malaysian ringgit, Philippine peso, Singapore dollar, Thai Baht, Russian ruble, Brazilian Real and Canadian dollar one by one.

And then we get the logarithmic regression equation of U.S. dollar, Japanese yen, Korean Won, Euro, Pound, Australian dollar. The next step is to calculate the exchange rate elasticity of bilateral trade of these currencies.

$Y^*$ denotes the income level of the outside world as a whole; $Y$ denotes the income level of native nation; $E$ denotes the exchange rate. Suppose that both the import and export demand function has a zero-order homogeneous nature with regard to the exchange rate elasticity of bilateral trade, accordingly, the import and export demand function can be expressed as

$$Xd = AxY^{*\alpha_x} E^{\beta_x} u_x. \tag{12}$$

$$Md = AmY^{\alpha_x} E^{\beta_\mu} u_m. \tag{13}$$

$\alpha_x$ and $\alpha_m$ is the coefficient of export and import demand respectively; $\beta_x$ and $\beta_m$ is the coefficient of export and import prices respectively; $u_x$ and $u_m$ are random variables in the natural logarithm form.

The net trade volume $T$ is denoted by the ratio of exports and imports:

$$T = \frac{Xd}{E \times Md}. \tag{14}$$

Substitute the trade partners for the world, and then we rewrite (14) to Logarithm (15) (Liu-fu and Xue-feng 2007):

$$\ln T_{i,t} = \alpha + \beta \ln Y_t + \gamma \ln Y_{i,t} + \lambda \ln E_{i,t} + \varepsilon_t \quad \varepsilon_t \sim N(0, \sigma^2). \tag{15}$$

Where $T_{i,t}$ denotes the net trade volume of Sino-Partner nation $i$; $Y_{i,t}$ denotes the income level of Partner nation $i$ in period $t$; $Y_t$ denotes the income level of China; $E_{i,t}$ denotes the bilateral exchange rate against the currency of partner nation $i$ in period $t$. In general, $\beta > 0$ or $\beta < 0$; $\gamma > 0$ or $\gamma < 0$; $\lambda > 0$.

Considering the hysteresis effect of the exchange rate elasticity of trade, we select the monthly data of the bilateral trade volume of China between United States, Japan, Euro-zone countries, Britain, Australia and Korea from 2000 to 2008 (General Administration of Customs of the People's Republic of China. http://www.customs.gov.cn/) .

As the import and export trade volume shows stronger seasonal characteristics, they were seasonally adjusted by X12 before using. Thus, we calculate the six currencies exchange rate elasticities of bilateral trade against the RMB by Johansen Cointegration Test with Confidence level of 99% (Table 1).

Thus, from (10), we obtain the weights of these six currencies in the basket.

**Table 1** The exchange rate elasticities of bilateral trade against the RMB

| Currency | JPY | KRW | EUR | GBP | AUD | USD |
|---|---|---|---|---|---|---|
| Elasticity | −0.46 | −0.51 | 0.30 | 0.55 | 0.42 | 0.20 |

### 3.3 A Basket of Currencies Based on the Exchange Rate Elasticity of Bilateral Trade

From (1), we obtain a time series of the exchange rates of the basket of currencies against the U.S. dollar, and then extrapolate the exchange rate of RMB against the U.S. dollar by Eview 6.0

The following is the simulation result from August 2005 to July 2008 (Fig. 3).

### 3.4 A Basket of Currencies Based on the Bilateral Trade

The simulation result shows that the time series of RMB exchange rates and the basket of currencies exchange rates are very different. The huge gap is mainly generated by the impact of the exchange rates elasticity of bilateral trade for it varies very large. For example, only the weight of GBP is more than 50%.

Therefore, we might step back, assuming that differences in the exchange rate elasticities of bilateral trade can be ignored, then, $\varepsilon_1 = \varepsilon_2 = \cdots = \varepsilon_n$. So, according to (10),

$$w_i = v_i \quad i = 1, 2, \cdots, n. \tag{16}$$

That is, trade shares could be used as weights of the basket of currencies for trade stability, suppose that the exchange rate elasticities of trade are equal.

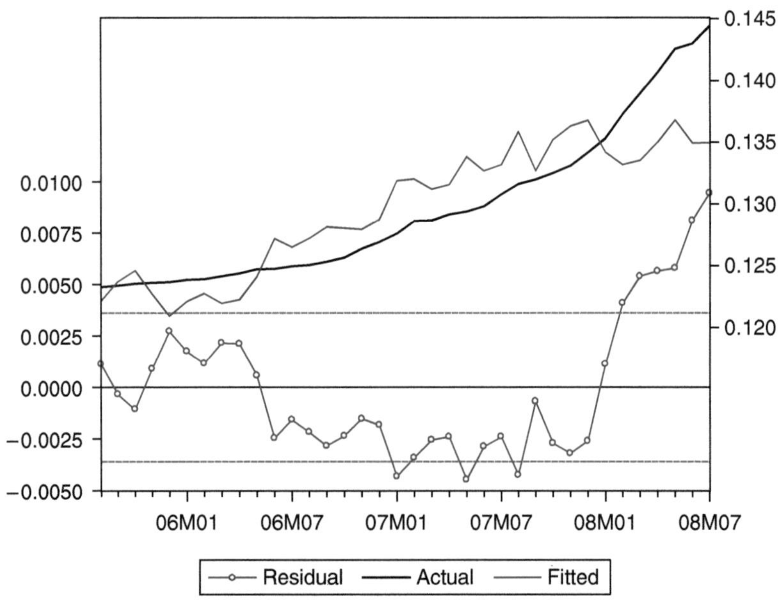


**Fig. 3** The exchange rates curve of a basket of currencies and the RMB against U.S.D

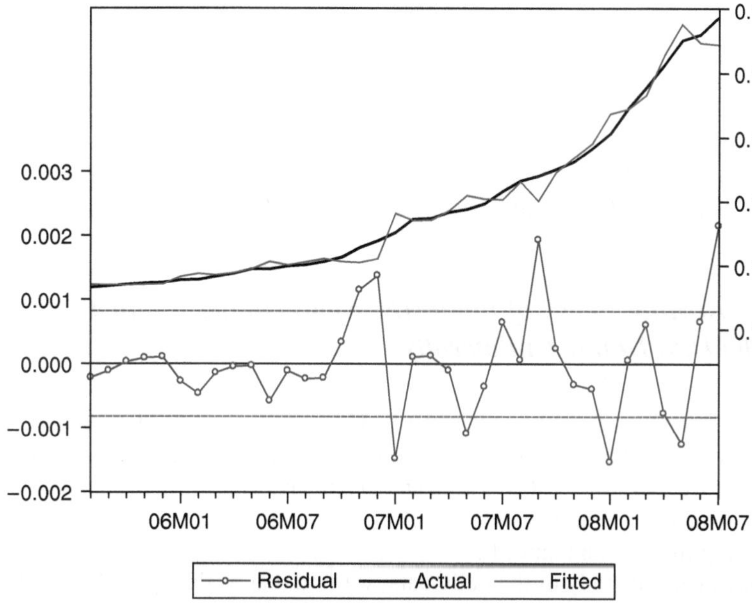


**Fig. 4** The exchange rates curve of another basket of currencies and the RMB against U.S.D

We take the trade shares of the above 16 nations (except U.S.)in last year as the weights of their currencies in current year to construct a basket of currencies(valued by U.S. dollar) and obtain a time series of the exchange rates of the basket of currencies against the U.S. dollar, and then extrapolate a time series of the exchange rates of RMB against the U.S. dollar by Eview 6.0.

The following is the simulation result from August 2005 to July 2008 (Fig. 4).

The above result shows that, the exchange rate of a basket of currencies decided directly by trade shares, better explains the fluctuation of RMB exchange rate against the U.S. dollar, than that considering the elasticity of bilateral trade. This sensitive analysis shows that the PBoC acknowledges the role of exchange rates on trade, but do not believe that exchange rate adjustments need the exchange rate elasticity of trade as reference.

Another reason is that in our exchange rate determination mechanism based on stabilizing trade volume, the effect of exchange rate elasticity is assumed as short-term or instant factor, but, we calculate that the impact of the exchange rates on trade volume has a significant hysteresis, and varies between countries.

Therefore, it is unreasonable to consider the exchange rate elasticity in determining the weights of currencies in a basket of currencies, without a perfect model to take into account the country factors and hysteresis factors.

In addition, the exchange rates of RMB against the U.S. dollar, from August 2005 to July 2008, can only be approximated by a quadratic faction of the exchange rates of a basket of currencies against the U.S. dollar, which is unreasonable. If the

PBoC really let the RMB pegged to a basket of currencies, the RMB and the basket of currencies should be linear, rather than quadratic.

Furthermore, the gaps between the simulated curve and the actual dispersion are still relatively large, so the exchange rates of RMB against the U.S.D can not be extrapolated by a basket of currencies against the U.S.D accurately.

Therefore, the RMB exchange rate is not pegged to a basket of currencies, but be managed with the reference to a basket of currencies, as the official said.

## *3.5 "Crawling Pegs" of the RMB, Appreciating Against the U.S. Dollar Unilaterally*

So, the question remains – how does the RMB exchange rate be determined?

Since August 2005, the nominal exchange rate of RMB against U.S. dollar only move from +0.44% to −0.86% within any given day, with average appreciation of 0.015%, (except the appreciation of 2% on July 21, 2005, the day the newly currency regime was announced).

Through careful study of the daily data, we find that as time goes by, the RMB exchange rates against the U.S. dollar maintain a steady upward trend. So, we extrapolate the exchange rate of RMB against the U.S. dollar by Eview 6.0 with reference to a uniform time series.

Figure 5 is the simulation results from August 2005 to July 2008 which is more accurate than Fig. 4, Considering the Sum squared resid 1.54E-5 is less than 2.22E-5.

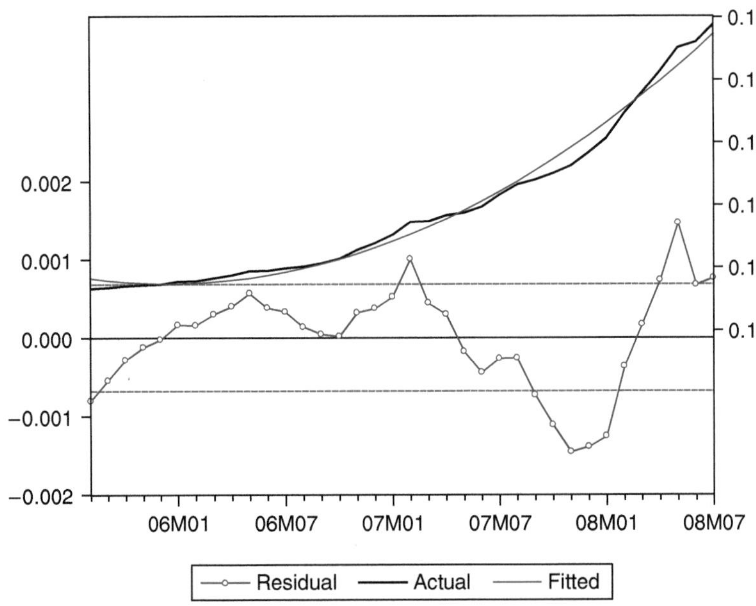


**Fig. 5** A time series curve of the exchange rates of RMB against U.S.D

Obviously, with the time pass by, the exchange rate of RMB increases along with a conic. This Crawling Pegs, which shows a typical unilateral appreciation against the U.S. dollar, is the real determination mechanism of the RMB exchange rate.

## 4 Crawling Pegs: the Dilemma of PBoC

As in earlier studies, the RMB was tightly pegged to the dollar before July, 2005. It followed the appreciation of the RMB against the dollar from August 2005 to July 2008, which was attributable to the appreciation of the currency of major countries against the dollar. Thereafter, the RMB was tightly pegged to the U.S. dollar again until June, 2010. This process is not surprising, because it was caused by the long-standing and deep-seated problems of China's economy.

Historically, developing countries have faced the same problems like China. Of course, if a country grows to reach a per capita income level of $2,000–$3,000 and continues to grow, it will have to appreciate its currency, which is determined by two factors: the major reason is the rising productivity for such growth; the other reason is that any large country can not ignore the tremendous wealth effect by the appreciation of its currency. So RMB appreciation is inevitable.

As Fig. 6 shows, after three decades of high growth, China's annual average growth of GDP is 9.7%, which is far higher than America's 4.1%. As a developing economic power, China will certainly not let RMB be pegged to a currency in a long term because the sustained increase of labor productivity and GDP would make the appreciation of RMB against the U.S. dollar inevitable.

On the other hand, due to China's foreign trade volume accounted for 44.7% of GDP, the external dependence is still very high, while a relatively stable exchange

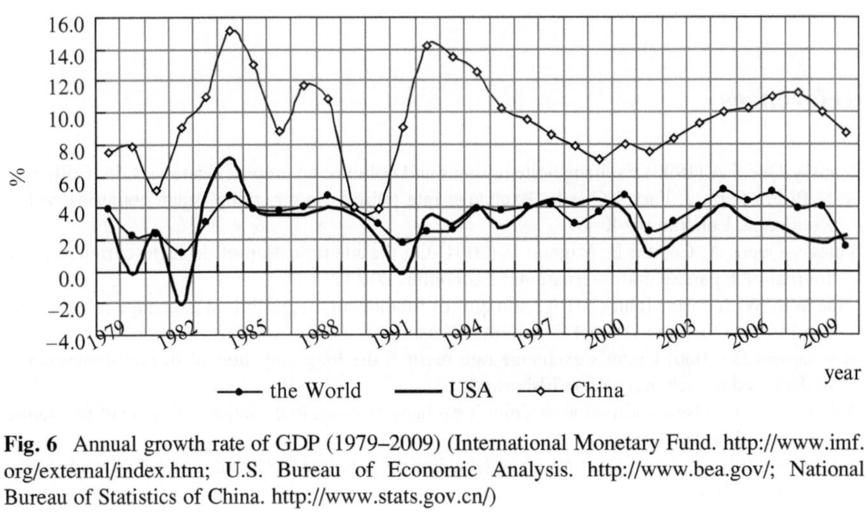


**Fig. 6** Annual growth rate of GDP (1979–2009) (International Monetary Fund. http://www.imf.org/external/index.htm; U.S. Bureau of Economic Analysis. http://www.bea.gov/; National Bureau of Statistics of China. http://www.stats.gov.cn/)

rate is very important for export. In addition, if we take the international payments of China as a whole, then more than 78% of it is valued by U.S. dollar. So, to maintain the high growth of GDP, it is necessary to maintain the exchange rate of RMB stable against the U.S. dollar and avoid fluctuations.

The PBoC should leverage the RMB's upward pressure while keeping the exchange rate of RMB stable. This difficulty can be called as "the dilemma of PBoC".

Thus, since PBoC announced that the daily exchange rate of RMB is based on medial rate of RMB-U.S. dollar transactions, according to market supply and demand fluctuations in a certain range, the only choice left for PBoC is to control the medial rate and allow a movement of up to +/−0.3% in bilateral exchange rates within any given day (actually narrowed to +/−0.15% at average). However, the result of the market reaction made by the great pressure of RMB appreciation is that RMB rose steadily as the curve shown in Fig. 1.

## 5 Conclusion

The result suggests that the regime is more an outcome propelled by supply and demand in the market under the existing managing float regime with reference to the U.S. dollar, than a managing float regime "with reference to a basket of currencies" proclaimed by PBoC, or a result of government manipulation proposed by U.S. Congressmen.

However, if the PBoC open its exchange rate target and the floating range, there is no excuse for U.S. to ask for the revaluation of the RMB. Actually, in the near future, China should implement a number of measures to increase exchange rate flexibility, reform the foreign exchange regime, and relax some capital control measures.

## References

Bénassy-Quéré A (1999) Exchange rate regimes and policies: an empirical analysis. In: Collignon S, Pisani-Ferry J, Park YC (eds) Exchange rate policies in emerging Asian countries, vol 3. Routledge, London, pp 40–64

Bénassy-Quéré A, Coeuré B, Mignon V (2004) On the identification of de facto currency pegs. Journal of Japanese and International Economies 20:112–127

Cline WR, Williamson John (2010) Estimates of fundamental equilibrium exchange rates. Peterson Institute for International Economics, Washington, DC

Eichengreen B (2006) China's exchange rate regime: the long and short of it, http://www.econ.berkeley.edu/~eichengr/research/short.pdf

Frankel JA (2009) New estimation of China's exchange rate regime. National Bureau of Economic Research, Massachusetts

Frankel JA (1992) Is Japan creating a Yen Bloc in East Asia and the Pacific? http://www.nber.org/papers/w4050.pdf

Frankel JA, Shang-Jin Wei (1993) Trade blocs and currency blocs. http://www.nber.org/papers/w4335.pdf
General Administration of Customs of the People's Republic of China. http://www.customs.gov.cn/
International Monetary Fund. http://www.imf.org/external/index.htm
Liu-fu Chen, Xue-feng Qian (2007) Research on asymmetric elasticity of RMB effective exchange rate: an empirical analysis based on Chinese trade data with G-7 members (in Chinese). Nankai Economic Studies 1:3–18
National Bureau of Statistics of China. http://www.stats.gov.cn/
Ohno K (1999) Exchange rate management in developing Asia. http://www.grips.ac.jp/teacher/oono/hp/docu02/read4.pdf
State Administration of Foreign Exchange http://www.safe.gov.cn
U.S. Bureau of Economic Analysis. http://www.bea.gov/

# Part III
# Risk Management in Sustainable Enterprise

# Financial Risk Assessment Model of Listed Companies Based on LOGISTIC Model

**Wang Fei and Cheng Jixin**

**Abstract** According to the logistic model, this thesis uses the A share listed companies in China as the research object, and selects 50 ST and non-ST companies as a sample in 2009. Facing the actual situation of listed companies' financial risk assessment, we develop 12 indicators of financial forecasts, use SPSS13.0 software to make factor analysis, and then make further analysis by using Logistic Regression Model to form a financial risk assessment model. The results show that this model is effective and it may serve as a basis for policy research.

**Keywords** Factor analysis · Financial indicators · Financial risk · Logistic regression analysis

With the rapid development of China's economy and capital market, the number of listed companies is increasing while the listed company's financial risk is gaining more and more attention from investors, listed companies and regulatory authorities. Therefore, an effective financial risk evaluation model for forecast will help stakeholders make accurate judgments, and facilitate capital markets' healthy development.

Some scholars have made good views of the study from indicators of choice and methods of application aspects on the financial risk assessment model. This paper selects 12 financial indicators which can reflect the performance of the listed company better based on the Logistic model. First, we use spss13.0 software to make factor analysis to reduce the levy and replace all the indicator variables with a few factors. Then we make Logistic regression analysis to form a simple evaluation model to reduce the huge amount of work. Finally, we incorporate test samples into the evaluation model to discriminate the validity of the model.

W. Fei (✉) and C. Jixin
Business School of Hohai University, ADD:1 Xikang Road, Nanjing 210098, China
e-mail: hhwf0889@sina.com

Y. Zhou and D.D. Wu (eds.), *Modeling Risk Management for Resources and Environment in China*, Computational Risk Management,
DOI 10.1007/978-3-642-18387-4_20, 

## 1 Sample Selection

In this study, to facilitate analysis, we define ST companies as enterprises suffered in financial crisis, and take enterprises which were specially treated due to unusual financial situation as samples from Shanghai and Shenzhen A share market in 2009. There are two reasons to account for this. Firstly, a share listed companies carry out the domestic accounting standards and accounting system; the external financial information can be collected easily and more comprehensively. Secondly, listed companies are more obvious with special treatment. According to the disclosure system of listed companies, the deadline of publishing their annual reports is on April 30 of the following year. Listed companies' Annual Report in (t-1) year and whether or not to have special treatment in t year are almost simultaneous, so there is no practical significance to predicate whether it has special treatment in t year with (t-1) year data. This article uses the company's financial data of (t-2) year (2007) to establish a model to forecast whether the company was special treated due to financial crisis in t year (2009). Among the 27 A-share listed companies in 2009, eliminating two companies with abnormal movements, we can get 25 ST companies. In addition, we also select 25 comparable non-ST companies under the same conditions.

The data is chosen from the wind information databases and Chinese listed company information network http://www.cnlist.com/.

## 2 Variable Selection

Financial indicators should be well selected to reflect the principle of corporate performance. In the selection of financial indicators, there are some important research results in this area whatever in China and other countries, such as the predictor variables used by Altman model and Index of Standard & Poor's. China also has business comprehensive performance Evaluation System. Considering the actual situation of the financial risk assessment in listed companies, this paper develops 12 indicators of financial forecasts, which mainly reflect five aspects of financial situation including profitability, operating capacity, solvency, development capacity and cash flow capacity. They are earnings per share($X_1$), ROE($X_2$), return on total assets($X_3$), velocity of liquid assets($X_4$), Accounts receivable turnover ratio($X_5$), current ratio($X_6$), quick ratio($X_7$), Asset-liability ratio($X_8$), OIG ($X_9$), Net profit growth rate($X_{10}$), Net asset growth($X_{11}$), and Ratio of net operating cash($X_{12}$).

## 3 Factor Analysis

The basic principle of factor analysis is: by studying the correlation coefficients between multiple variables matrix (or covariance matrix) of internal dependencies, to identify a few random variables named the principal component factors, which could represent all the variables. Then according to the size of pertinence, we divide the variables into groups. Within the same group, there is a higher correlation between the variables; among different sets of variables, there is a lower one. And the various factors are unrelated to each other, so all variables can be expressed as a linear combination of common factors. The purpose of factor analysis is to reduce the number of variables, and use a small number of factors to analyze the economic issues instead of all the variables.

## 3.1 *Statistical Test*

KMO and Spherical Bartlett's Test are used to test whether the data used is appropriate for factor analysis. KMO tests whether the partial correlation between variables is relatively small, and Spherical Bartlett test is used to determine whether the correlation matrix is a unit matrix. When the KMO statistic is below 0.5, it is not suitable for factor analysis. From the test results, it can be seen that KMO statistic is 0.712, greater than 0.5, and Spherical test chi-square statistic equals to 385.386, accompanied by a probability of 0.000 which is less than 0.01. Therefore, it is suitable to do factor analysis. It can be seen from Table 1.

## 3.2 *Calculating Factor*

Factor analysis requires the common factors extracted have practical meaning. In this paper, average orthogonal rotation is applied. The combination between Varimax method that simplifies factor interpretation and Quartmax method that simplifies variables interpretation, results in that a factor could represent a high load of variables and factors need the least number of explanatory variables. Rotated output is in Table 2.

The Total Variance Explained table gives all the loading cases of rotated factors. This paper uses principal component analysis to extract the common factors, and in the event that eigenvalue is greater than 1, we extract four common factors. From the table we can see that the cumulative contribution rate of the four principal component factors is 75.375%, which includes 75.375% of the original index information.

**Table 1** KMO and Bartlett's test

| | | |
|---|---|---|
| Kaiser-Meyer-Olkin measure of sampling adequacy | | .712 |
| Bartlett's Test of Sphericity | Approx. Chi-Square | 385.386 |
| | df | 66 |
| | Sig. | .000 |

**Table 2** Total variance explained

| | Initial eigenvalues | | | Rotation sums of squared loadings | | |
|---|---|---|---|---|---|---|
| Component | Total | % of variance | Cumulative % | Total | % of variance | Cumulative % |
| 1 | 3.897 | 32.478 | 32.478 | 3.413 | 28.440 | 28.440 |
| 2 | 2.783 | 23.189 | 55.667 | 2.485 | 20.704 | 49.145 |
| 3 | 1.319 | 10.990 | 66.657 | 1.697 | 14.142 | 63.287 |
| 4 | 1.046 | 8.718 | 75.375 | 1.451 | 12.088 | 75.375 |
| 5 | .759 | 6.327 | 81.702 | | | |
| 6 | .615 | 5.123 | 86.825 | | | |
| 7 | .573 | 4.778 | 91.603 | | | |
| 8 | .451 | 3.758 | 95.361 | | | |
| 9 | .293 | 2.445 | 97.806 | | | |
| 10 | .151 | 1.260 | 99.066 | | | |
| 11 | .089 | .745 | 99.811 | | | |
| 12 | .023 | .189 | 100.000 | | | |

Extraction method: principal component analysis

**Table 3** Component score coefficient matrix

| | Component | | | |
|---|---|---|---|---|
| | 1 | 2 | 3 | 4 |
| Earnings per share£″yuan£© | .273 | .021 | −.022 | −.038 |
| ROE£″%£© | .260 | −.004 | −.024 | .007 |
| Return on total assets£″%£© | .248 | −.076 | .058 | −.025 |
| Velocity of liquid assets £″times£© | −.068 | −.007 | .000 | .527 |
| Accounts receivable turnover ratio£″times£© | −.180 | −.054 | .383 | .441 |
| Current ratio£″multiple£© | −.008 | .426 | −.116 | .082 |
| Quick ratio£″multiple£© | −.030 | .411 | −.050 | .090 |
| Asset-liability ratio£″%£© | −.003 | −.167 | −.201 | .180 |
| OIG£″%£© | 0.28 | −.077 | .390 | −.049 |
| Net profit growth rate£″%£© | .140 | .274 | −.367 | .491 |
| Net asset growth£″%£© | .279 | −.010 | −.079 | −.092 |
| Ratio of net operating cash £″multiple£© | −.086 | .030 | .490 | −.012 |

Extraction method: principal component analysis
Rotation method: equamax with Kaiser normalization
Component scores

## 3.3 Establish Factor Score Coefficient Matrix

Factor score coefficient matrix, which shows the linear relationship between various financial indicators and each factor, is the factor score of each main factor. Using the factor analysis method, the disorderly complex index will be processed structurally, which can both reduce and simplify the observation dimension. It also guarantees the information of original data and makes preparations for logistic regression analysis.

According to the factor scores coefficient matrix (component score coefficient matrix, see Table 3), the expression of various factors can be listed.

$$\begin{aligned}F_1 &= 0.273x_1 + 0.260x_2 + 0.248x_3 - 0.068x_4 - 0.180x_5 - 0.008x_6 \\ &\quad - 0.030x_7 - 0.003x_8 + 0.028x_9 + 0.140x_{10} + 0.279x_{11} - 0.086x_{12}\end{aligned} \tag{1}$$

$$\begin{aligned}F_2 &= 0.021x_1 - 0.004x_2 - 0.076x_3 - 0.007x_4 - 0.054x_5 + 0.426x_6 \\ &\quad + 0.411x_7 - 0.167x_8 - 0.077x_9 + 0.274x_{10} - 0.010x_{11} + 0.030x_{12}\end{aligned} \tag{2}$$

$$\begin{aligned}F_3 &= -0.022x_1 - 0.024x_2 + 0.058x_3 - 0.000x_4 + 0.383x_5 - 0.116x_6 \\ &\quad - 0.050x_7 - 0.201x_8 + 0.390x_9 - 0.367x_{10} - 0.079x_{11} + 0.490x_{12}\end{aligned} \tag{3}$$

$$\begin{aligned}F_4 &= -0.038x_1 + 0.007x_2 - 0.025x_3 + 0.527x_4 + 0.441x_5 + 0.082x_6 \\ &\quad + 0.090x_7 + 0.18x_8 - 0.049x_9 + 0.491x_{10} - 0.092x_{11} - 0.012x_{12}\end{aligned} \tag{4}$$

## 4 Logistic Regression Analysis

In accordance with the results of factor scores of companies, we use Binary Logistic regression analysis of SPSS to make further analysis. Logistic model is a probabilistic decision model, and its dependent variable can only take two values: 1 and 0 (Virtual dependent variable). We set y as an incident that measures whether a listed company is in financial crisis, recording ST Company as $Y = 1$ and non-ST companies as $Y = 0$. Assuming P shows the probability of the occurrence of event Y, which is $P = P(Y = 1)$, then 1-P expresses the probability of no financial risk. Order $F_i (i = 1, 2, ..., N)$ for the model inputs, namely, the principal component factor extracted from financial index by factor analysis, $\beta_i (i = 1, 2, ..., N)$ for the weight of the principal component factor, and $\alpha$ for model constants. In the multiple regression in terms of P (probability) for the dependent variable, generally the equation is $P = \alpha + \beta_1 F_1 + \beta_2 F_2 + \beta_3 F_3 + \beta_4 F_4$. However, by using the equation to calculate, unreasonable situations often appear such as $P > 1$ and $P < 0$. Therefore, we make p as a logarithmic unit conversion, and then

$$\begin{aligned}\log\mathrm{it}(p) = \ln[p/(1-p)] &= \alpha + \sum_{i-1}^{n} \beta_i F_i \\ &= \alpha + \beta_1 F_1 + \beta_2 F_2 + \beta_3 F_3 + \beta_4 F_4\end{aligned}$$

So the general formula can be obtained from this evaluation model,

$$p = \frac{e^{\alpha + \beta_1 F_1 + \beta_2 F_2 + \beta_3 F_3 + \beta_4 F_4}}{1 + e^{\alpha + \beta_1 F_1 + \beta_2 F_2 + \beta_3 F_3 + \beta_4 F_4}}$$

Obviously, the evaluation model uses logit $(0.5) = 0$ as the symmetric center, and the value of logit (p) at $p = 0$ or $p = 1$ in the vicinity is from $-\infty$ to $+\infty$.

This model is applied to the 0–1 decision. On the one hand, when $P \geq 0.5$, it can determine the occurrence of the event Y which means that the company's financial risk of a crisis will occur. On the other hand, when $P < 0.5$, it can determine that the financial risk of the crisis will not occur. The most essential advantages of the evaluation model are that it does not require strict assumptions, overcomes the constraints of linear equations subjected to the limitations of statistical assumptions, and it has a wider range of applications. Here we use the four main factors (F1, F2, F3, F4) obtained from the factor analysis as independent variables to make Logistic regression analysis to further examine the financial risks of the assessment model.

## 4.1 Hosmer and Lemeshow Goodness of Fit Test

Logistic regression goodness-of-fit testing generally uses the Hosmer and Lemeshow goodness-of-fit to test statistics. In this paper, as we could see from the Hosmer and Lemeshow goodness of fit test statistics in Table 4, chi-square statistic is 6.605, and concomitant probability is 0.580, much larger than the given significance level 0.05. Therefore, in an acceptable level, this model is better fitting the data.

In addition, Table 5 shows that the model's accuracy rate is 88.0%. In details, both non-ST and ST companies have a rate of 88.0%.

## 4.2 Wald Statistics and Model Results

In the level of $\alpha = 0.05$, Table 6 shows the status of each factor.

From Table 6 we can get the following model,

$$\log it(p) = \ln \tfrac{p}{1-p} = -0.351 - 6.407F_1 - 0.690F_2 - 2.977F_3 + 0.016F_4$$

**Table 4** Hosmer and Lemeshow test

| Step | Chi-square | df | Sig. |
|---|---|---|---|
| 1 | 6.605 | 8 | .580 |

**Table 5** Classification Table[a]

| Observed | | | Predicted | | |
|---|---|---|---|---|---|
| | | | Serial number | | |
| | | | Non-ST company | ST company | Percentage correct |
| Step 1 | Serial number | Non-ST company | 22 | 3 | 88.0 |
| | | ST company | 3 | 22 | 88.0 |
| | Overall percentage | | | | 88.0 |

[a]The cut value is .500

**Table 6** Variables in the equation

| | | B | S.E. | Wald | df | Sig. | Exp(B) |
|---|---|---|---|---|---|---|---|
| Step 1[a] | FAC1_1 | −6.407 | 2.079 | 9.492 | 1 | .002 | .002 |
| | FAC2_1 | −.690 | .880 | .614 | 1 | .433 | .502 |
| | FAC3_1 | −2.977 | 1.288 | 5.345 | 1 | .021 | .051 |
| | FAC4_1 | .016 | .436 | .001 | 1 | .971 | 1.016 |
| | Constant | −.351 | .610 | .331 | 1 | .565 | .704 |

[a]Variable(s) entered on step 1: FAC1_1, FAC2_1, FAC3_1, FAC4_1

**Table 7**

| Code | $X_1$ | $X_2$ | $X_3$ | $X_4$ | $X_5$ | $X_6$ | $X_7$ | $X_8$ | $X_9$ | $X_{10}$ | $X_{11}$ | $X_{12}$ |
|---|---|---|---|---|---|---|---|---|---|---|---|---|
| 600868. SH | −0.80 | −0.78 | −0.87 | −0.96 | −0.57 | −0.50 | −0.36 | 0.87 | −0.21 | −6.70 | −0.80 | −0.04 |
| 000935. SZ | −0.71 | −0.77 | −0.99 | 0.14 | −0.44 | −0.47 | −0.31 | 0.73 | 0.84 | 0.10 | −1.08 | −0.14 |
| 600401. SH | 0.02 | 0.25 | 0.12 | −0.35 | −0.45 | −0.67 | −0.52 | 0.36 | −0.40 | 0.26 | −0.17 | −0.65 |
| 000657. SZ | −0.87 | −0.27 | −0.79 | 0.08 | 0.46 | −0.26 | −0.46 | −0.58 | 0.44 | −0.09 | −0.56 | 1.00 |
| 000722. SZ | −1.05 | −0.28 | −0.49 | −0.60 | −0.41 | −0.39 | −0.38 | 0.05 | −0.97 | −0.11 | −0.67 | −0.36 |

**Table 8**

| Code | 600868. SH | 000935. SZ | 600401. SH | 000657. SZ | 000722. SZ |
|---|---|---|---|---|---|
| P | 0.9950 | 0.9930 | 0.5920 | 0.8650 | 0.9940 |

Then

$$P = \frac{e^{-0.351-6.407F_1-0.690F_2-2.977F_3+0.016F_4}}{1+e^{-0.351-6.407F_1-0.690F_2-2.977F_3+0.016F_4}} \tag{5}$$

So (5) constitutes a financial risk assessment model.

## 5 Sample Test

Using the factor logistic regression models, we chose five listed companies at random that were specially treated in 2008 as an application sample to test. According to the financial statements in 2006, we select 12 financial indicators, and then calculate the probability of each company's risk.

First of all, the original data were standardized: see Table 7:

Taking the Variable Xi into the formula X(1) to formula X(5), and then we calculate the risk probability of each company, The results shown in Table 8:

From the results of test, all the P values of ST companies have more than 0.5 except that only one of the P value is close to 0.5. In addition, 80% are much higher than 0.5. It indicates that their probability of financial risks is much bigger, and it is consistent with the result that these companies were implemented special treatment of delisting warning in 2008.

# 6 Conclusions and Limitations

Admittedly, there are some limitations in financial risk assessment model of listed companies based on LOGISTIC model. Firstly, the sample selection does not consider the differences between different industries; different industries may not have strong comparability of financial indicators; so Index selection may lack the representative. Secondly, the span of time selected may not be broad enough because we only use two years in advance of the sample data to predict. However, advantages of the model are overwhelming.

Through the empirical analysis of domestic listed companies, we can see that the financial risk assessment model of listed companies based on LOGISTIC model has certain validity on the measures of financial risk in Chinese capital market. The importance of this evaluation model is that by using multiple variables, we can estimate the financial crisis of listed companies and give warnings two years in advance. From the application process, it can be seen that this evaluation model has certain operability and is simple to apply. Because there is no specific requirement on the data form and most of the financial data of listed companies is neither multivariate normal distribution nor the covariance, the model has broader development prospects.

# Sensitive Analysis of Intellectual Capital on Corporate Performance in Selected Industry Sectors in China

**Xuerong Wang, Li Liu, and Cuihu Meng**

**Abstract** Intellectual capital is increasingly being recognized as a driving force for the prosperity of economy and corporations. This paper applies the value-added intellectual coefficient (VAIC) model to investigate the link between components of intellectual capital and corporate performance in three industry sectors in China. We find that (1) the material capital employed efficiency has a positive effect on performance in both manufacturing and real estate sectors; (2) the human capital efficiency has a positive effect on performance in the manufacturing sector but not in the real estate or the IT sectors; (3) the structure capital efficiency has non-significant effect on performance in any of the three industry sectors.

**Keywords** Capital employed efficiency · Human capital · Intellectual capital · Sensitive analysis · Structural capital · Value-added intellectual coefficient (VAIC)

## 1 Introduction

Over the last few decades, the world's economy has shifted from being primarily driven by the use of tangible assets such as plant, equipment and real-estate to an economy that increasingly dependent upon the use of intangible resources such as knowledge, technology, core competencies and innovation (Meritum Project 2002). It is important to understand how intellectual capital contributes to organization performance and how can an organization effectively assess the value of IC.

X. Wang (✉) and C. Meng
School of Accounting, Nanjing University of Finance and Economics, Nanjing, Jianshu Province, People's Republic of China
e-mail: wxr960@sohu.com

L. Liu
School of Civil Engineering, The University of Sydney, Sydney, Australia
e-mail: l.liu@usyd.edu.au

Y. Zhou and D.D. Wu (eds.), *Modeling Risk Management for Resources and Environment in China*, Computational Risk Management,
DOI 10.1007/978-3-642-18387-4_21, 

Prior research suggests that the development of IC resources creates value for organisations, especially since the majority of an organisation's assets are intangibles that cannot be reflected in the balance sheet (Stewart 1997). The identification and measurement of an organisation's IC is important because these provide insights into the impact that the measurement of IC may have on management action (John 2009).

This paper investigates the effects of IC on corporate performance using a measurement model developed by Pulic (2000) using audited accounting information from selected Chinese companies. During empirical analysis of 486 samples that were collected from main three industries of China in the year of 2006 to 2008, generally, the findings show that intellectual capital is significantly correlates with corporate performance and there is diverse effects of IC on value creation in different types of industries. Meanwhile, it is indicated that the corporate should to make best of structure capital in China markets.

The main contribution of this paper is the identification of the link between the corporate performance and IC in selected Chinese industry sectors. Further, we try to explain the differences in the strength of links in the three industry sectors. We differ from other empirical research on IC by using both ROA and ROE as performance indicators.

Below, literature is reviewed and hypotheses presented. Then, research method is elaborated and results reported. Finally, conclusions are drawn and relevant issues discussed.

## 2 Literature Review

The concept of intellectual capital (IC) is based on the recognition that organisational knowledge needs to be managed and that technology has allowed for greater dissemination of this knowledge (Meritum Project 2002; Unerman et al. 2007). IC is first used by the economist Galbraith in 1969, who considered that intellectual capital can lead to competitive advantage. Over the last decade, the change in global economy has created renewed interest in intellectual capital and increased demand for measuring and reporting its affect on business and profitability (Juniad 2004). Intellectual capital covers a multitude of area and usually be viewed as invisible capital or intangibles. Typically, IC is defined to include intellectual material such as knowledge, information, intellectual property and experience that can be used to create wealth.

### *2.1 Components of IC for Measuring*

Sveiby (2007) identifies 34 different frameworks for the measurement and reporting of IC, the majority of which attempt to identify the components of IC. One problem with the plethora of approaches about measuring the components of IC is that no dominant approaches has yet emerged (John 2009). Nevertheless, all the models, frameworks, discussions and literature appear to be saying that IC is interesting (Chatzkel 2004), complex (Cuganesan 2005) and needs to be understood better (Mouritsen 2006).

Although there is no generally accepted framework for measuring IC or the concept of intellectual capital, most of studies share the view that intellectual capital can be classified into dichotomy or trichotomy – i.e. human capital and structural capital or human capital, structural capital and relational capital or customer capital, respectively. In this paper, we adopt the former view by dividing IC into human capital and structural capital. Relational capital or customer capital – the essential part in the intellectual capital, is assumed to be embedded in human capital.

## 2.2 Methods for Measuring the Valuation of IC

IC generally have been considered intangible assets, so it is difficult to measure with conventional financial tools objectively. The increasing interest in accounting for intangibles on the knowledge-based economies has led to increased research on IC. Hong Pew believes that IC measuring methods can be grouped broadly under two categories: those that do not use a monetary valuation of IC, and those that put a monetary value on IC. Juniad (2004) consider that it can be simply divides into two parts – internal measures and external measures. The most common internal measures of intellectual capital focus on budgeting, training and human resources. The four most popular internal measures of intellectual capital are: Human Resource Accounting; The Intangible Assets Monitor; The Skandia Navigator™; and The Balanced Scorecard. The main external methods exist for facilitating the valuation of intellectual capital, that is:

- Market-to-book ratio (M/B) (Stewart 1997): The assumption is that the portion of the market value of a company in excess of its book value is the market value of its intellectual capital. That is, the difference between the book value and market value of a company is taken as equalling the level of intellectual capital of the business. But Brennan and Connell argue that IC dose not comprise the entire difference between MV and BV.
- Tobin's Q: This ratio is the ratio of market value to firm asset replacement cost (Tobin and Brainard 1968) and can be used for making comparisons among firms. The replacement cost concept was designed to circumvent the differing depreciation policies used by accountants around the world (Joia 2000). If Tobin's Q exceeds one, the company is likely to seek to acquire more intellectual capital. I think the calculation of the replacement cost is difficult in new economy with a lot of value-added from IC.

- Calculated Intangible Value (CIV): This value uses industry norms to establish rates of return for tangible assets, and calculates the level of intellectual capital by attributing to it any return exceeding the industry norm.
- Return of Management (ROM): This is a measure of management efficiency in using total capital, including both physical and intellectual capital (Strassmann 1999). ROM is obtained by dividing management value by the sum of sales and administrative expenses. The weakness of this measure is that it assumes management to be the only value-adding layer and neglects the contribution of other employees to corporate success.
- Value-Added Intellectual Coefficient (VAIC): This measure is the total sum of the value creation efficiency of the physical capital of a company and two components of intellectual capital (namely human capital and structural capital) (Pulic 2000). This measure is designed to indicate the intellectual capital efficiency of a company, and high VAIC value is associated with good management utilization of the potential value creation from physical and intellectual capital (Williams 2001).
- Real Options Analysis (ROA): Real options is a recent approach which uses the methodology and theory of financial options to value intangible assets. A financial option is the right, but not the obligation, to buy or sell an underlying asset at a fixed price for a predetermined period of time. A real option is an option that is based on non-financial assets. Real options can be applied to determine the value to proceed, defer, expand or abandon investment.

Despite more than 30 different approaches to measuring IC, these can be classified into two types of approaches. The first type measures IC through questionnaire surveys (Bontis 1998, 2000). This approach is limited by the resources and time it takes to obtain reasonable sample size. In contrast, the second approach measures IC based on standard account information. For example, the Value-added intellectual coefficient (VAIC) method, applies an indicator system to evaluate the corporate value of tangible and intangible resources (Pulic 2000).

In this paper, we adopt the VAIC approach to measure the corporate IC from 486 samples from three industry sectors in China. The main advantages of using the VAIC approach include, firstly, all data used in the VDIC calculation are based on the audited information, so calculation can be considered objective and verifiable (Pulic 2000); Secondly, VAIC provides a standardized and consistent basis of measurement, thereby enabling the comparative analysis across various industrial sectors; Thirdly, VAIC is an output oriented, process method that can be applied across different business forms and at various levels of operations.

VAIC is an evaluation system to quantitatively measure intellectual capital from accounting information. VAIC comprise of three components: the coefficient of material capital (capital employed efficiency or CEE) and the coefficient of intellectual capital (measured by human capital efficiency and structure capital efficiency, respectively.

1. CEE (capital employed efficiency)
   Pulic constructed the intellectual coefficient model from the perspective of value-add to the firm from the utilization of intellectual capital. CEE indicates that the value-add created by per unit of material capital. The higher the ratio, the more efficient the enterprise is in employing material capital in creating value. CEE is defined as the ratio of VA/CE, where CE (capital employed) can be calculated as CE=Total assets-Current liabilities; VA=W+I+T+NP (W:Wages, I: Interest, T: corporate taxes, NP: After-tax profit).
2. HCE (human capital efficiency)
   Human capital refers to the capacity, attitude and creativity of staff, and with the relations with internal and external interest groups. Pulic believes that human capital should be able to reflect its contribution to the value-added, thus HCE can be used to response the relationship between human capital and the value-add. Therefore, the total wages and costs of staff for a firm have often been used to measure the human capital investment by the firm (HC). The ratio of the value added to human capital (total wages and costs of staff), HCE= VA/HC, indicates the value-adding brought by per unit of human capital, thus the quality of human capital.
3. SCE (structural capital efficiency).
   Structural capital refers to the systems and structure of a firm that facilitates business intellect. Examples include the organizational routines, processes, strategies and knowledge. Pulic proposes that intellectual capital is made up of human capital and structural capital. The coefficient of structural capital is calculated as SCE = (VA-HC)/VA.

## 2.3 Hypotheses

The operation of any industries is based on the material resources and financial resources. With the rapid development of economic, intellectual capital is playing an increasingly important role as a strategic resource with the characteristics as scarcity value, difficult to imitate, non-alternative, to maintain a sustained competitive advantage. Material capital is essential to the survival and competitive, meanwhile human capital is a carrier of employees in the companies possessing creativity, enable an increasing value to the organization. Structure capital of the enterprise guarantees the operation safely and orderly, high - quality and provides the work environment for exchange, couple with human capital to achieve the most profit.

H1: The capital employed efficiency has a positive effect on firm performance;
H2: The human capital efficiency has a positive effect on firm performance;
H3: The structure capital efficiency has a positive effect on firm performance.

# 3 Research Method

## 3.1 Data Collection

Data were collected from a Chinese database called WIND which contains the annual reports of Chinese companies. Only A-share listed companies on the Shenzhen and Shanghai Stock Exchanges are included in the study sample. To make sure that the information collected reflect the true performance of the firms and to remove market disturbances, especially companies with short history on the stock exchanges, companies listed after December 31, 2003 are not included in the sample. Further, companies received special treatment (ST) for a substantial period of time (e.g. delisted or suspended) has been excluded from the sample. In total, 486 companies, including 338 manufacturing companies, 77 information technology companies, and 71 real estate developers are included in the sample. Further, to remove the year-to-year volatilities of the data, the data from the sample are taken as the averages of the corresponding data from 2006, 2007 and 2008.

We choose the three sectors, manufacturing, real state, and IT because these are the main contributors to the Chinese national economy and present variations in the likely dependency on the three components of intellectual capital. Manufacturing industry in China is primarily a traditional, labor-intensive industry, where size (economy of scale) and HCE are likely to matter most for performance. In contrast, the IT industry is technology-intensive, innovation driven, has short life cycle and high degree of market volatility. Real estate industry is a capital-intensive sector where HCE may not as important as the other two sectors but CEE is important due to its capital intensive nature (Table 1).

Table 2 reports the means and standard deviations of the independent variables. It shows that, relative to the means of each sub sample, the volatility of HCE in each

**Table 1** Types of sample

| Industry | Sample size | Total assets (RMB Yuan, bil) | | Number of employees | | Annual sales (RMB Yuan,bil) | |
|---|---|---|---|---|---|---|---|
| | | Aver. | Range | Aver. | Range | Aver. | Range |
| Manu. | 338 | 1.896 | 0.174–10.786 | 1845 | 14 –31629 | 1.095 | 0.085–7.633 |
| Real Estate | 71 | 5.377 | 0.237–34.703 | 334 | 17–15460 | 1.054 | 0.031–3.148 |
| IT | 77 | 1.602 | 0.264–11.923 | 974 | 21–100201 | 8.945 | 0.016–1.166 |

**Table 2** Descriptive statistics

| | Manufacturing | | Real estate industry | | IT industry | |
|---|---|---|---|---|---|---|
| | Mean | Std.Dv | Mean | Std.Dv | Mean | Std.Dv |
| CEE | 0.1984 | 0.1483 | 0.0985 | 0.0879 | 0.1885 | 0.1659 |
| HCE | 1.6177 | 2.2323 | 5.5947 | 5.4242 | 1.8157 | 2.8358 |
| SCE | 0.4075 | 2.6241 | 0.8231 | 0.2174 | 0.4748 | 0.8072 |
| Size | 21.3630 | 1.0238 | 22.4054 | 1.2639 | 21.1943 | 1.1804 |

of the sub sample is higher than that of the CEE and SCE in the corresponding sub samples, suggesting big variances in human capital utilization practices across all the three sub samples.

## *3.2 Analysis*

### 3.2.1 Dependent Variables

Firm performance is measured using return on total assets (ROA) (Aboody et al. 1999) and return on equity (ROE). The use of two performance measures mitigates the potential inaccuracy in any of the two measures.

ROA is the ratio of the net income (less preference dividends) divided by book value of total assets as reported in the 2006~2008 annual report. ROE is defined as the ratio of the net income (less preference dividends) divided by book value of total shareholders' equity as reported in the 2006~2008 annual report.

### 3.2.2 Independent Variables

Following Pulic (2000) and Pulic and Bornemann (1999), CEE, HCE and SCE are used as independent variables. As discussed above, the formulae for deriving the three independent variables are as below:

CEEi = VAi/CEi; capital employed value-added coefficient for firm i;
HCEi = VAi/HCi; human capital value-added coefficient for firm i;
SCEi = SCi/VAi; structural capital value-added coefficient for firm i;

Where: VAi=Wi + Ii + Ti + Npi (Wi: Wages, Ii: Interest, Ti: Corporate Taxes, NPi: After–tax profit); CEi = book value of net assets for firm i; HCi = total investment in salary and wages for firm i; SCi = VAi – HCi; structural capital for firm i.

### 3.2.3 Control Variables

Firm size and financial leverage ratios are used as control variable to rule out the plausible explanations that these two rather than the three independent variables that explain the hypothesized causal relationships.

In this study, the natural logarithm of the total assets of a firm is used as a surrogate measure of firm size. Where there exists economy of scale, such as in the manufacturing sector, firm size is likely to be associated with profitability.

The leverage ratio (LEV) is defined as total debt divided by its book value of total assets. Higher LEV is typically associated with higher risks as well as the potential for higher profitability.

### 3.2.4 Regression Model

The regression model used is as follows:

$$Perf = \alpha_0 + \alpha_1 CEE + \alpha_2 HCE + \alpha_3 SCE + \alpha_4 Size + \alpha_5 Lev + \varepsilon$$

A significant regression coefficient (e.g. $p \leq 0.05$) indicates the significant effect of the corresponding variable on performance. For example, a significant $\alpha 1$ indicates that CEE has a significant effect on performance.

The above equation will be tested in the three samples (manufacturing, IT, real estate), respectively.

## 4 Results

The multiple regression results in Table 3 show that the IC components as well as the control variables have no significant effects on either the ROA or ROE in the IT industry (negligible adjusted R2s and non-significant F). Further, there is no evidence of SCE on either the ROA or ROE in any of the three industry sectors – Hypothesis 3 has been rejected.

In contrast, both CEE and HCE have some effects on performance in the manufacturing and the real estate sectors. Specifically, CEE has a significant effect on ROE in the manufacturing sector ($\beta = 0.21$, $p = 0.00$) and on ROA in the real estate sector ($\beta = 0.32$, $p = 0.00$). HCE has a significant effect on ROE in the manufacturing sector ($\beta = 0.20$, $p = 0.00$) but has non-significant effects on either ROE or ROA in the real estate sector.

In addition, Table 3 reports that firm size has a significant, negative effect on ROA in the manufacturing sector ($\beta = -0.30$, $p = 0.02$). Interestingly, Table 3 also shows that the financial leverage has a negative, significant effect on ROA ($\beta = -0.27$, $p = 0.00$) but a positive, significant effect on ROE ($\beta = 0.30$, $p = 0.05$) in the real estate sector.

**Table 3** Multiple regression results

| Industry | Manufacturing | | Real estate industry | | IT industry | |
|---|---|---|---|---|---|---|
| Variable | ROA | ROE | ROA | ROE | ROA | ROE |
| CEE | 0.08(0.52) | 0.21(0.00) | 0.32(0.00) | 0.07 (0.66) | −0.18(0.24) | −0.03(0.84) |
| HCE | 0.09(0.43) | 0.20(0.00) | 0.08(0.10) | 0.24(0.12) | −0.11(0.51) | −0.09(0.48) |
| SCE | −0.10(0.42) | 0.01(0.89) | 0.03(0.47) | −0.07(0.62) | −0.01(0.94) | 0.03(0.83) |
| Size | −0.30 (0.02) | −0.04(0.52) | −0.04(0.39) | 0.07(0.60) | 0.14(0.28) | 0.04(0.78) |
| Lev | −0.15(0.23) | −0.02(0.77) | −0.27(0.00) | 0.30 (0.05) | −0.03(0.84) | −0.03(0.83) |
| $R^2$ | 0.15 | 0.11 | 0.186 | 0.15 | 0.09 | 0.01 |
| Adj-$R^2$ | 0.10 | 0.09 | 0.124 | 0.08 | 0.02 | −0.06 |
| F- Sta | 2.59(0.03) | 7.97(0.00) | 19.82(0.00) | 2.205(0.06) | 1.21(0.32) | 0.16(0.98) |

The number reported inside brackets are p-values for the corresponding standardized regression coefficients or F-statistics

## 5 Discussion

The positive effect of CEE in the manufacturing and real estate sectors, respectively, is not surprising because both are capital intensive in nature. Therefore, efficient utilization of material capital should have a positive effect on performance. What is surprising is the negative effect of size on ROA in manufacturing. The explanation could be that as manufacturing operation reaches certain size, any further growth in size will result in reduced ROA as asset growth outstrips efficiency gains. Further study should investigate this phenomenon.

Out of the three industry sectors, HCE only has a positive effect on ROE in the manufacturing sector suggesting that the effect of HCE depends upon the labour intensiveness of the sectors studies. HCE as measured by wages and labour costs reflects the efficiency of blue collar workers but not that of knowledge workers in the IT industry. The non significant effect of HCE in the IT and real estate sectors is likely to be explained by the way HCE is measured. Future studies should find ways to measure the efficiency of organizations to harness the energy and creativeness from knowledge workers.

The lack of effect of SCE on performance across all the three sectors perhaps relates more to the way SCE has been measured in the VAIC model. Future studies should develop better conceptual as well as measurement models for structural capital.

The findings need to be interpreted with caution as the findings should be tested in other sectors or in other countries.

## 6 Conclusion

This paper investigates the relationship between the elements of the intellectual capital and financial performance of listed companies in selected industry sectors in China. We find that (1) the material capital employed efficiency has a positive effect on performance in both manufacturing and real estate sectors; (2) the human capital efficiency has a positive effect on performance in the manufacturing sector but not in the real estate or the IT sectors; (3) the structure capital efficiency has non-significant effect on performance in any of the three industry sectors.

**Acknowledgements** This paper is supported by the National Nature Science Foundation of China (No.71071072)

## References

Aboody D, Barth ME, Kasznik RR (1999) Evaluations of fixed assets and future firm performance: evidence from the UK. J Acc Econ 26(1–3):149–178

Bontis N (1998) Intellectual capital: an exploratory study that develops measures and models. Manage Decis 36(2):63–76

Bontis N, Dragonetti NC, Jacobsen K, Roos G (1999) Knowledge toolbox: a review of the tools available to measure and manage intangible resources. Eur Manage J 17(4):391–402

Bontis N, Keow WCC, Richardson S (2000) Intellectual capital and business performance in Malaysian industries. J Intellect Cap 1(1):85–100

Bornemann M (1999) Potential of value systems according to the VAIC™ method. Int J Technol Manage 18(5–7):463–475

Caddy I (2000) Intellectual capital: recognizing both assets and liabilities. J Intellect Cap 1(2):129–146

Chatzkel J (2004) Moving through the crossroads. J Intellect Cap 5(2):337–339

Cuganesan S (2005) Intellectual capital-in-action and value creation: a case study of knowledge transformations in an innovation project. J Intellect Cap 6(3):357–373

Firer S, Williams SM (2003) Intellectual capital and traditional measures of corporate performance. J Intellect Cap 4(3):348–360

John CD (2009) Intellectual capital measurement: a critical approach. J Intellect Cap 10(2):190–210

Joia LA (2000) Measuring intangible corporate assets-linking business strategy with intellectual capital. J Intellect Cap 1(1):68–84

Juniad MS (2004) Managing and reporting intellectual capital performance analysis. J Am Acad Bus Camb 3:439–448

Marr B, Chatzkel J (2004) Intellectual capital at the crossroads: managing, measuring, and reporting of IC. J Intellect Cap 5(2):224–239

Meritum Project (2002) Guidelines for managing and reporting on intangibles (Intellectual Capital Report). European Commission, Madrid

Mouritsen J (2006) Problematising intellectual capital research: ostensive versus per formative IC. Acc Auditing Account J 19(6):820–841

Pulic A (2004) Intellectual capital–does it create or destroy value? Measuring Bus Excell 8(1):62–68

Pulic A (2000) VAIC™ – an accounting tool for IC management. Int J Technol Manage 20(5–8):702–714

Shaikh JM (2004) Measuring and reporting of intellectual capital performance analysis. J Am Acad Bus 4(1–2):439–448

Stewart TA (1997a) Intellectual capital: the new wealth of organizations. Bantam Doubleday Dell Publishing Group, Inc., New York

Stewart TA (1997b) Intellectual capital. The new wealth of organizations. Doubleday – Currency, London

Strassmann PA (1999) The value of knowledge capital. Available online: http://www. strassmann.-com

Sveiby KE (1997a) The new organizational wealth: managing and measuring knowledge-based assets. Berrett-Koehler, San Francisco, CA

Sveiby KE (1997b) The new organizational wealth: managing and measuring knowledge-based assets. Berrett-Koehler, San Francisco

Sveiby KE (2007) Methods for measuring intangible assets. Available at: www.sveiby.com/portals/0/articles/Intangible Methods.htm. Accessed 15 May 2007

Tobin J, Brainard W (1968) Pitfalls in financial model building. Am Econ Rev 58:99–122

Unerman J, Guthrie J, Striukova L (2007) UK reporting of intellectual capital. ICAEW, University of London, London

Williams SM (2001) Corporate governance diversity and its impact on intellectual capital performance in an emerging economy. Working Paper, Haskayne School of Business, The University of Calgary, Canada.

# Research on Influence Factors Sensitivity of Knowledge Transfer from Implementation Consultant to Key User in ERP

**Jie Yin, Shilun Ge, and Feng Li**

**Abstract** ERP implementation is a knowledge transfer process of actors. The research collects questionnaires from the key users who participate in manufacturing ERP project which has been completed. With the 155 effective questionnaires of 45 manufacturing ERP projects from 15 areas, it empirically researches the influence factors sensitivity of knowledge transfer from implementation consultant to key user from four aspects of the characteristic of knowledge to be transferred, the risk and uncertainty of transfer process, the characteristic of transfer context, the characteristic of knowledge source and the characteristic of knowledge recipient.

**Keywords** Implementation consultant · Influence factors · Key user · Knowledge transfer · Random · Sensitive analysis · Uncertainty

## 1 Introduction

Knowledge transfer is the knowledge transfer process of main part from the main part with high knowledge potential energy to low one (Zander and Kogut 1995). The quantity of knowledge, the quality of knowledge and the construction of knowledge determine the knowledge energy. Implementation consultant and key

J. Yin (✉) and S. Ge
School of Economics and Management, Jiangsu University of Science and Technology, Zhenjiang, Jiangsu 212003, People's Republic of China
e-mail: yinjie9633@sina.com; jzgsl@jzerp.com

F. Li
Department of Postgraduate Administration, Jiangsu University of Science and Technology, Zhenjiang, Jiangsu 212003, People's Republic of China
and
State Key Laboratory of Hydrology Water Resources and Hydraulic Engineering, HoHai University, Nanjing, Jiangsu 210098, People's Republic of China
e-mail: justxwb@gmail.com

Y. Zhou and D.D. Wu (eds.), *Modeling Risk Management for Resources and Environment in China*, Computational Risk Management,
DOI 10.1007/978-3-642-18387-4_22, © Springer-Verlag Berlin Heidelberg 2011

user are the participators of ERP project. Implementation consultant is the implementation service for ERP, which offers specialized service of system implementation to enterprises and solves the various problems in the process of ERP implementation. It's a common mechanism for explore of huge information system to introduce key user. Key user consists of pillar workers and managers with rich experience, which accomplishes the implementation of ERP project with implementation consultant.

This passage will research knowledge transfer proceeding during the implementation of ERP from implementation to key user. With the 155 effective questionnaires of 45 manufacturing ERP projects from 15 areas it empirically researches the influence factors of knowledge transfer from implementation consultant to key user.

# 2 Research Suppose

## *2.1 Dependent Variable*

Knowledge transfer effect of dependent variable can be measured subjectively or objectively (Mowery et al. 1996). This passage adapts the method of subjective ness to assess knowledge transfer effect from implementation consultant to key user through getting remark from key user directly. How much key user has mastered the knowledge that implementation consultant own reflect knowledge transfer effect directly. The terminal goal of knowledge transfer is to promoting the successful implementation of ERP project, the performance of ERP system reflect knowledge transfer effect in some degree. Therefore, this passage measures knowledge transfer effect of dependent variable from two aspects of key user understands degree of ERP knowledge and ERP effect efficiency.

## *2.2 Independent Variable*

The independent variable includes: characteristic of knowledge, characteristic of transfer context, characteristics of knowledge sources, characteristic of knowledge receiver.

**Characteristic of Knowledge.** The characteristics of knowledge include: Tacitness of ERP knowledge, Causal ambiguity of ERP knowledge.

*Tacitness of ERP knowledge*. Polanyi is the first one who brings about the recessive characteristic of knowledge and divides knowledge into dominant knowledge and recessive knowledge based on the degree of knowledge expression. Dominant knowledge is easy to be wrote down and expressed and it can be encoded with visualize symbols like text, chart, formula, etc. Tacitness of knowledge reflects recessive degree of knowledge. The higher the tacit degree is, the lower the constructive degree of knowledge is. Meanwhile, it's harder to encode and expressed by language, text, or other ways clearly.

*Causal ambiguity of ERP knowledge.* Due to lacking logic recognition of relationship between input and output of knowledge, reason and result, the sender of knowledge may have difficulty in connect knowledge with activities under specific environment and also the receiver of knowledge has difficulty in apply knowledge to new environment (Simonin 1999). Therefore, this passage lists such supposes:

$H_{1.}$ The negative correlation on ERP knowledge tacit and knowledge transfer effect
$H_{2.}$ The negative correlation on causal ambiguity of ERP knowledge and the effect of knowledge transfer

**Characteristic of transfer context.** The characteristics of transfer context include: The degree of leader recognition, Learning culture, Relationship.

*The degree of leader recognition.* Top manager's support is regard as determined factors of whether ERP project can succeed or not. Top managers are advocators and promoters of ERP project, offering strategic direction to the implementation of project and secure all the resource ERP implementation need and make decision on the key issue.

*Learning culture.* Building learning-oriented organizations and creating a learning culture is an important part of business strategy (Lane and Lubatkin 1998). Learning culture reflects that if the enterprise encourages member companies to learn, advocates experiences sharing supports innovation and tolerates innovative errors.

*Relationship.* Knowledge exchange is built on mutual trust based on voluntary cooperation; cooperation in an atmosphere of mutual trust is a prerequisite for knowledge transfer. Knowledge transfer process requires frequent interaction between the two sides, easy communication, and pleasant cooperation. Thus, this paper assumes that:

$H_{3.}$ The positive correlation on leadership emphasis effect and effect of knowledge transfer
$H_{4.}$ The positive correlation on learning culture and the effect of knowledge transfer
$H_{5.}$ The positive correlation on bilateral relations and the effect of knowledge transfer

**Characteristics of knowledge sources.** The characteristics of knowledge sources include: communication coding ability of knowledge sources, the transfer desire of knowledge sources.

*Communication coding ability of knowledge sources.* Consultants and key users to implement effective communication between the protections of knowledge transfer is an important condition for the smooth. The ability to communicate knowledge source code is sending a clear knowledge of the expression of ideas and the ability to respond quickly to questions. Good knowledge communication sender encoding help information, particularly tacit knowledge transfer.

*The transfer desire of knowledge sources.* Out of exclusive information, special skills monopoly, implementation consultants in order to maintain its own unique value, dominant position, the core of knowledge transfer is often lack of desire, or even just transfer distortion, vague, fragmentary knowledge. Thus, this paper assumes that:

$H_{6.}$ The positive correlation on implement consultant's communication capacity and the effect of knowledge transfer
$H_{7.}$ The positive correlation on the transfer will of Implementation consultant and the effect of knowledge transfer

**Characteristics of knowledge receiver.** The characteristics of knowledge receiver include: knowledge receiver communication decoding ability, the knowledge receiver's will to obtain, knowledge absorption capacity of the receiver.

*Knowledge receiver communication decoding ability.* Ability to communicate knowledge receiver decoding is to listen to information, rapid response capability. Knowledge of the receiver's decoding ability and knowledge to communicate the source code of communication and knowledge transfer capacity reflects the person's communication skills.

*The knowledge receiver's will to obtain.* If the knowledge receiver lacks of access to knowledge, then will result in the block of knowledge, directly affect the effectiveness of knowledge transfer. Key users' lack of acquisition will, often due to information asymmetry, lack of confidence in new technologies, lack of trust on the implementation of the consultant, worried about their pay will vary with the failure of ERP project down the drain.

*Knowledge absorption capacity of the receiver.* Absorptive capacity refers to the understanding of the knowledge recipient evaluation, digestion and absorption, to understand the application of new knowledge (Tsai 2001). Absorption ability reflects both knowledge receiver for external knowledge ability, also reflects the receiver conversion, use knowledge innovation, to serve the organization's objectives. Thus, this paper assumes that:

$H_{8.}$ The positive correlation on the key user's capacity of decoding and communication and the effect of knowledge transfer
$H_{9.}$ The positive correlation on the key user's access will and the effect of knowledge transfer
$H_{10.}$ The positive correlation on the key user's absorptive capacity and the effect of knowledge transfer

## 3 Questionnaire Reliability and Validity Testing

### *3.1 Questionnaire Reliability Test*

In this study, total scale alpha coefficient was 0.956. According to the result of reliability analysis (as shown in Table 1), the questionnaires achieved ideal reliability.

**Table 1** Reliability analysis result of questionnaire

| Subscale | Measuring dimension | Cronbach |
|---|---|---|
| Knowledge of their characteristics | Implicit | 0.900 |
| | Causal ambiguity | |
| Transfer of contextual characteristics | Leadership emphasis | 0.952 |
| | Learning culture | |
| | Relations | |
| Knowledge of source characteristics | Communication encoding | 0.910 |
| | Transfer will | |
| Knowledge of the receiver characteristics | Communication decoding | 0.948 |
| | Access will | |
| | Absorptive capacity | |
| The effect of knowledge transfer | ERP management concept | 0.956 |
| | Project methodology | |
| | ERP systems and technological knowledge | |
| | ERP Performance | |

**Table 2** Validity analysis result of questionnaire

| Subscale | Measurement dimension | KMO | Common factor to explain the total variance (%) | |
|---|---|---|---|---|
| | | | Single | Accumulative |
| Characteristic of knowledge to be transferred | Implicit | 0.878 | 11.690 | 70.940 |
| | Causal ambiguity | | 59.249 | |
| Characteristic of transfer context | Leadership emphasis | 0.922 | 8.823 | 78.299 |
| | Learning culture | | 5.949 | |
| | Relations | | 63.527 | |
| Characteristic of knowledge source | Communication encoding | 0.888 | 9.114 | 74.317 |
| | Transfer will | | 65.204 | |
| Characteristic of knowledge recipient | Communication decoding | 0.931 | 7.762 | 79.721 |
| | Access will | | 5.889 | |
| | Absorptive capacity | | 66.071 | |
| Knowledge transfer | ERP management concept | 0.918 | 6.390 | 84.075 |
| | Project methodology | | 7.183 | |
| | ERP systems and technological knowledge | | 62.315 | |
| | ERP Performance | | 8.187 | |

## 3.2 *The Questionnaire Validity Test*

As shown in Table 2, the questionnaire achieved good construct validity.

# 4 Test Hypothesis

We consulted the test results of the research hypothesis:

$H_1$. The negative correlation on ERP knowledge tacit and knowledge transfer effect: strong supported.

$H_2$. The negative correlation on causal ambiguity of ERP knowledge and the effect of knowledge transfer: supported.
$H_3$. The positive correlation on leadership emphasis effect and effect of knowledge transfer: strong supported.
$H_4$. The positive correlation on learning culture and the effect of knowledge transfer: supported.
$H_5$. The positive correlation on bilateral relations and the effect of knowledge transfer: not supported.
$H_6$. The positive correlation on implement consultant's communication capacity and the effect of knowledge transfer: supported.
$H_7$. The positive correlation on the transfer will of Implementation consultant and the effect of knowledge transfer: not supported.
$H_8$. The positive correlation on the key user's capacity of decoding and communication and the effect of knowledge transfer: strong supported.
$H_9$. The positive correlation on the key user's access will and the effect of knowledge transfer: not supported.
$H_{10}$. The positive correlation on the key user's absorptive capacity and the effect of knowledge transfer: not supported.

## 5 Conclusions

The management strategies for ERP business practice guidance are promoted:

1. Optimize the knowledge structure, use modern information technology to create tacit knowledge sharing environment, maximizing the encoding of tacit knowledge.
2. Senior managements pay full attention to project implementation, encouraging active participation in relevant meetings to protect the full supply of resource.
3. To promote knowledge sharing, to encourage freedom of speech, to establish channels of communication and training positions, reasonable distribution of pay, to create a good learning culture.
4. To establish a scientific system of key users, implementation consultant selection and training. Put the communication skills into the reference standard of selection.

## References

Lane PJ, Lubatkin M (1998) Relative absorptive capacity and inter-organizational learning. Strateg Manage J 19(5):461–477

Mowery DC, Oxley JE, Silverman BS (1996) Strategic alliances and interfirm knowledge transfer. Strateg Manage J 17:77–91

Simonin BL (1999) Ambiguity and process of knowledge transfer in strategic alliances. Strateg Manage J 20(2):595–623

Tsai W (2001) Knowledge Transfer in Intra-organizational Networks: Effects of Network Position and Absorptive Capacity on Business Unit Innovation and Performance. Acad Manage J 44(5):996–1004

Zander U, Kogut B (1995) Knowledge and the speed of the transfer and imitation of organizational capabilities: an empirical test. Organ Sci 6(1):76–92

# The Majority of Stockholders' Subscription Option in SEO and Escalation of Commitment

**Wei Li, Zerong Liu, and Yang Tang**

**Abstract** This paper empirically verifies the impact of majority of shareholders' SEO subscription option on escalation of commitment under the Agent Theory's framework based on the data of Chinese listed companies from 2005 to 2008. Results show that listed companies always encouraged to perform the escalation of commitment no matter major stockholders participate the subscription or not. However, compared with the participation of all major shareholders, giving up the SEO subscription option by the major shareholders will increase the possibility of escalation of commitment; furthermore, the most serious escalation will be happened if all of the major shareholders give up the subscription option, and the second serious escalation will be incurred if part of the major shareholders gives up their subscription. These conclusions contribute to comprehend the economic consequences of subscription behavior from the majority of shareholders in SEO, as well as provide a theoretical basis for China Securities Regulatory Commission to supervise and standardize the listed companies' investment and financing activities, and reducing their investment risks.

**Keywords** Capital investment · Escalation of commitment · Investment risk · Seasoned equity offering (SEO) · Subscription option

## 1 Introduction

Escalation of commitment refers to the phenomenon of unpromising projects that have already been invested by accepting a large amount of resources are permitted to be continued and extra investment is still chosen by the decision makers.

Meng (2007) pointed out that only when two necessary conditions are both possessed can a company have a tendency to implement the escalation of commitment:

---

W. Li, Z. Liu, and Y. Tang (✉)
Department of Accounting, Business School, Tianjin University of Commerce, 300134, China
e-mail: tjtan2003@yahoo.com.cn

Y. Zhou and D.D. Wu (eds.), *Modeling Risk Management for Resources and Environment in China*, Computational Risk Management,
DOI 10.1007/978-3-642-18387-4_23, © Springer-Verlag Berlin Heidelberg 2011

one, decision makers have the will to continue the project, and another is, decision makers have the ability to continue the investment. The former emphasizes the factors that encourage the decision makers to invest, and the latter stresses the factors that restrict the funds; namely, only when a company has sufficient funds to further invest, can the escalation of commitment really happens. Otherwise, even if decision makers have the intention to continue the investment, the escalation of commitment will not happen anyway. However, by observing the existing researches, it is not hard to see that nearly all the literatures specifically concern the first condition. For instance, Self-justification Theory and Perspective Theory assume the escalation of commitment results from irrational decision makers with the limited cognition and information dealing capability, which lead to the motivation of further investment to the decision makers so that the decision is neither personal efficiency maximized nor business efficiency maximized; while Agency Theory states that the escalation of commitment is from rational decision makers' self-benefit incentives and inspiration. So up to now, there is hardly any research on the second condition. Therefore, Meng (2007) studied the effect of financing constraint on the escalation of commitment in the situation of the investing funds are got only from bank loans.

It is well known that Chinese listed companies show extremely strong prepensity in seasoned equity offering (SEO); after IPO, SEO has become the major means for listed companies to raise equity funds. However, we can find that it is common for the majority of shareholders to give up their subscription option of SEO, especially after 1998. Does this behavior affect the escalation of commitment? If so, what effect it should be? Current researches did not give an answer to this question.

This paper is under the condition that limit companies' funds only used in the investment to SEO, based on the Agency Theory framework, and empirically verifies the impact of majority of shareholders' SEO subscription option on escalation of commitment. This study contributes to comprehend the economic consequences of subscription behavior from the majority of shareholders in SEO, as well as provide a theoretical basis for China Securities Regulatory Commission to supervise and standardize the listed companies' investment and financing activities.

## 2 Theoretical Analysis

According to Agent Theory, managers, despite being agent of shareholders, will, under the asymmetric information situation, make decisions that maximize their personal benefit rather than the company's shareholders', and then, the agent problems appeared. Ever since La Port et al. (1999) discovered that ownership structure tends to be concentrated all over the world, agent problems between majority and minority of shareholders have attracted the attention of the researchers. According to the Supervision Efficiency Theory, the existence of major shareholders can assuage the agent conflict between the shareholders and the managers. But the Tunnel Efficiency Theory argues that the emergence of major shareholders also brings about negative effect, namely, they may embezzle companies' resources as their own by their supervision power over the companies. In La Port et al. (1999)'s opinion, major shareholders can take various measures to occupy the interest of minorities, including to seize the company's investing opportunities or enforce the company to invest in the projects that will bring about no profit to the business but benefit to themselves.

Applying Agent Theory to escalation of commitment in Kanodia et al. 1989 studied the issue regarding the escalation of commitment. He stated that managers, as the agent of a company have the private information regarding the project in the capital investment decisions, and in the alternatives of abandoning the project and the escalation of commitment, the managers prefer to choose the decision that will maximize their personal benefits. Managers would escalate a project even if its abandonment would adversely affect their reputations as competent managers, or influence their potential promotion opportunities, as long as they can get extra economic benefit from the project because the stockholders, as the bailor cannot supervise the managers' activities without the complete information of the project. Since, Ghosh (1997), Salter and Sharp (2001) also confirmed the above conclusion.

It is well know that one's net income result from a project equals to its investment revenue subtract its investment cost. Escalation of commitment is the decision re-making process related to a company's capital investment. Compare with the initial investment, apart from the cost of the decision re-making process, it also involves the liquidation value of the project (i.e. the opportunity cost for continue the project).

The investment cost of major shareholders is the same as that of minorities; it consists of the re-investing value and the liquidation value when the majority of shareholders participates a subscription in SEO, but their revenues are different. Generally, it is mostly probable that the residual value will flow towards the major shareholders because the majorities have the privilege to control the revenue while the minorities can only earn the related residual revenue in terms of their share proportion.

Consequently, continuing the project is still beneficial to the majority of shareholders (i.e. the net investing value is positive) while it is unbeneficial to the minorities (i.e. the net investing value is negative). In such situation, the majorities' decision would tend to further the investment, and then the escalation of commitment would be happened thereafter.

The investment cost of major shareholders is much lower than that of minorities when major shareholders give up the subscription option of SEO because of all funds invested to continue a project comes from the minorities. Once the funds invested into the project, it will be shared by both the major and the minor stockholders of a company, and the majorities enjoy the shares according to their proportion. The more the investing amount, the more the value unconditionally possessed by the majorities. Therefore, the investment cost of major shareholders only consists of liquidation value, while the minor shareholders have to charge all the re-investing cost amount (both the re-investing value and the liquidation value); this greatly reduce the majorities' investing cost while increase the minorities', and further results in the effect that the margin investing cost of the majorities is much lower than that for minorities in a same project. Thus, when the investing revenue is certain, escalation of commitment would be most likely happened because the majorities' margin revenue keeps going up while the minorities' keeps going down or even gets into negative. Therefore, although the escalation of commitment is probable happened when the major shareholders participate the option of SEO, the possibility of the escalation would be even high when the majorities give up their subscription option in a SEO.

# 3 Research Design

In this section, we will empirically verify the impact of major shareholder giving up SEO subscription option on escalation of commitment by using the method of Logit Regression Method base on the controlled agent problems. Here, we take the Escalation of Commitment as the dependent variable, while the Subscription Option as the independent variable.

## *3.1 Sample Selection and Dependent Variable*

We define escalation of commitment has three essential characters. Firstly, it is a re-investing-decision making process rather than an initial one; secondly, continue project will result in negative NPV (i.e. this project is not a promising one), here we focus on the negative co-relationship between the variables, and thirdly, the investment of escalation of commitment is the over-investment that beyond normal level of investment to a project with a negative NPV.

We think that variable of escalation of commitment should fully give the expression to the above three characters towards the escalation of commitment, but the existing variable not yet do it so far (Meng 2007; Zhang 2009). Therefore, we build the Escalation of Commitment Model by developing the Richardson's Over-investment Model (2006) (See Tang and Liu 2010), based on our principle: to fully embody these three characteristics we stated above.

The model (1) is the Escalation of Commitment Model. The variables of model (1) are expressed in Table 1.

$$\begin{aligned} INV_t = \alpha_0 + \alpha_1 Growth_{t-1} + \alpha_2 Lev_{t-1} + \alpha_3 Cash_{t-1} + \alpha_4 Age_{t-1} \\ + \alpha_5 Size_{t-1} + \alpha_6 RET_{t-1} + \alpha_7 INV_{t-1} \end{aligned} \tag{1}$$

We start with 2284 companies which finance in SEO that are listed in Shanghai and Shenzhen Stock Exchanges during the period from 2005 to 2008; the samples involve the enterprises that escalate and did not escalate. The samples regarding the

**Table 1** Variables definition of escalation of commitment model

| Variable | Definition |
|---|---|
| $INV_t$ | Capital Investment = the net variation of fixed assets ,under-constructed project, intangible asset , long-term deferred and prepaid expenses divide average total assets in the year of $t$ |
| $Growth_{t-1}$ | Growth Opportunities = sales revenue growth rate$_{t-1}$ |
| $Lev_{t-1}$ | Asset-Liability Ratio$_{t-1}$ |
| $Cash_{t-1}$ | Cash Holding = [( cash + temporary investment)/total assets]$_{t-1}$ |
| $Age_{t-1}$ | The firm's age come into the market up to the year of $t-1$ |
| $Size_{t-1}$ | The firm's size = Ln(total assets$_{t-1}$) |
| $RET_{t-1}$ | The Rate of Stock Return |
| *Year* | Dummy of Year, the benchmark is 2005 |
| *Industry* | Dummy of industry, we class them by CSRC benchmark |

**Table 2** Regression result of escalation of commitment's model

| Variable | Coefficient | T statistics |
|---|---|---|
| Intercept | −0.428 | −4.446*** |
| $INV_{t-1}$ | −0.145 | −4.790*** |
| $Growth_{t-1}$ | 0.002 | 2.931*** |
| $Lev_{t-1}$ | −0.003 | −2.730*** |
| $Cash_{t-1}$ | 0.240 | 5.282*** |
| $Age_{t-1}$ | −0.005 | −3.718*** |
| $Size_{t-1}$ | 0.021 | 4.782*** |
| $RET_{t-1}$ | 0.033 | 3.548*** |
| Industry and Year | Controlled | |
| Adj-$R^2$ | 0.082 | |
| F | 8.284*** | |
| N | 1773 | |

Dependent variable is the $INV_t$; Industry and year have been controlled

*significant at 10%; ** significant at 5%; *** significant at 1%

escalation of commitment are selected by the following: (1) Selecting non-financial enterprises which profit from fundamental operating transactions decreasing year after year during the period of 2003–2008. We choose their data just from 2005 to 2008. (2) Using these data into the Escalation of Commitment Model for regression and selecting the companies with residuals as positive. (3) Eliminating the enterprises which have been changed the use of funds financed in SEO. The remaining companies were considered to escalate. We use dummy variable to express the variable of escalation of commitment which equals 1 if the company is escalation of commitment, and 0 if the firm is non-escalation of commitment. Table 2 is the regression result of escalation of commitment's model.

Table 2 is the regression effect from the Escalation of Commitment Model; Table 3 is the final selected samples. All data come from CCER database. The software using by this paper is SPSS16.0.

**Table 3** Sample distribution

| | Finance in SEO, N = 2284 | | | Finance in SEO and escalation of commitment, N = 523 | | |
|---|---|---|---|---|---|---|
| | All | Some | None | All | Some | None |
| N | 1060 | 908 | 316 | 247 | 190 | 86 |

(1)"All" refers to the companies which all major shareholders give up subscription in SEO. "Some" refers to the companies which some major shareholders give up subscription in SEO. "None" means the companies which all major shareholders participate subscription in SEO.; (2) We eliminate the companies, finance in SEO over the past 3 years, which the subscription selection of their major shareholders is unknown; N means the numbers of samples

**Table 4** Empirical variables and their definitions

| Variable | Definition |
|---|---|
| *PESC* | Dependent variable which equals 1 if the company escalate and 0 otherwise. |
| *TF, BF* | Independent variables. The *TF* equals 1if all shareholders give up subscription and 0 otherwise. The *BF* equals 1if some shareholders give up subscription and 0 otherwise. |
| *AJ* | A dummy variable which equals 1 if the firm dismiss board chairman from his post, and 0 otherwise. |
| *AY* | Ln(total annual salary of three managers whose salary were highest in firm) |
| *Ctrl* | A dummy variable which equals 1 if the firm is state-owned enterprise, and 0 otherwise. |
| $FCF_t$ | Free cash $flow_t$ = (operating cash $flow_t$-new $investment_t$)/ average total $assets_t$. The new $investment_t$ = Expected investment estimated from escalation of commitment's model. |
| $ADM_t$ | Ln(overhead $expenses_t$/main business $income_t$) |
| $ORE_t$ | other $receivables_t$/total $assets_t$ |

## 3.2 Independent Variable, Controlled Variable and Testing Model

The independent Variables are the subscription of major shareholders in SEO. They are dummy variables described in Table 4. We select a set of controlled variables from other literatures including overhead expenses divide main business income as the agency question variable of managers-shareholders and other receivables divide total assets as the agency question variable of major-minor shareholders. We use the Logit regression Model (2) to examine the impact of subscription on escalation of commitment.

$$Ln\left(\frac{PESC}{1-PESC}\right) = \beta_0 + \beta_1 TF + \beta_2 BF + ContorlVarables + \varepsilon \qquad (2)$$

# 4 Empirical Analysis Results and Its Explanation

Table 5 depicts the results from the estimation of (2). The samples of Model A are the companies which finance in SEO over the past 5 years. The samples of Model B include the companies which finance in SEO and all major shareholders give up

**Table 5** Logit regression result

| Variable | Model A | Model B | Model C |
|---|---|---|---|
| Intercept | −0.100 | −1.602 | 1.312 |
| | (0.009) | (0.988) | (0.890) |
| *TF* | 1.085*** | | 1.110*** |
| | (15.867) | | (16.136) |
| *BF* | 1.023*** | 1.097*** | |
| | (13.613) | (14.911) | |
| *AJ* | 0.021 | 0.151 | −0.159 |
| | (0.016) | (0.369) | (0.543) |
| *AY* | −0.216*** | −0.110 | −0.325*** |
| | (8.9881) | (0.995) | (11.702) |
| *Ctrl* | 0.377** | -0.089 | 0.620*** |
| | (6.009) | (0.157) | (9.382) |
| $FCF_t$ | 2.978*** | 2.386** | 3.918*** |
| | (14.349) | (4.321) | (13.128) |
| $ADM_t$ | 0.000 | 0.000 | 0.000 |
| | (0.079) | (0.001) | (0.456) |
| $ORE_t$ | 0.439 | 0.663 | 0.400 |
| | (0.307) | (0.318) | (0.718) |
| Industry and Year | Controlled | Controlled | Controlled |
| *N* | 2284 | 1224 | 1376 |
| $R^2$ | 0.127 | 0.141 | 0.182 |
| Testing of H-L | $\chi^2 = 8.546$, Sig. = 0.382 | $\chi^2 = 9.987$, Sig. = 0.266 | $\chi^2 = 10.956$, Sig. = 0.204 |

(1) The data in the brackets are Wald statistics; (2) $R_N^2$ is the Nagelkerke $R^2$; (3) Testing of H-Lis the Hosmer & Lemeshow testing
*significant at 10%; ** significant at 5%; *** significant at 1%

subscription plus all major shareholders participated subscription. The samples of Model C include the firms which finance in SEO and some major shareholders give up subscription plus all major shareholders participated subscription.

Logit regression results in Table 5, all of model A, model B, model C, show a significant positive association both between *TF* and dependent variable, as well as between *BF* and dependent variable under the controlling the agency question variables. These results show that giving up the SEO subscription option by major shareholders will increase the possibility of escalation of commitment; this provides the support for the theoretical analysis. The coefficient of *TF* is higher than the *BF*'s. This shows that the most serious escalation will be happened if all of the major shareholders abandon the subscription, and the second serious escalation will be happened if some of the major shareholders give up the subscription.

## 5 Conclusions

The major shareholders are likely to force a company to invest in projects that bring no profit to the company but benefit themselves when there are agent conflicts between the major and the minor shareholders, and then result in escalation of commitment. When major shareholders give up their SEO subscription option, the investment cost of major shareholders would be decreased greatly while the minorities' investing cost increased, and more serious escalation will happened thereafter.

This paper empirically verifies the impact of major shareholders giving up SEO subscription option on escalation of commitment under the Agent Theory framework. Results show that giving up the SEO subscription option by major shareholders will increase the possibility of escalation of commitment. Furthermore, the most serious escalation will be incurred if all of the major shareholders abandon their subscription, and the second serious escalation will happened if some of the major shareholders give up their subscription.

## References


Ghosh D (1997) De-Escalation strategies: some experimental evidence. Behav Res Accounting 9:88–112

Kanodia C, Bushman R, Dickhaut J (1989) Escalation errors and the sunk cost effect: an explanation based on reputation and information asymmetries. J Acc Res 27:59–77

La Port R, Lopez-de-Slianes F, Shleifer A, Vishny R (1999) Corporate ownership around the world. J Finance 54:471–517

Meng M (2007) The formation mechanism of escalation of commitment during the transition period in China. D. Nan Kai university, Tianjin

Richardson S (2006) Over Investment of Free Cash Flow. Rev Acc Studies 11:159–189

Salter S, Sharp D (2001) Agency theory and escalation of commitment: do small national culture difference Matter? Int J Accounting 36:33–45

Tang Y, Liu Z-y (2010) Measurement of Project Escalation in Listed Companies. Theory Pract Finance Econ 31:350–360

Zhang D-l (2009) Study on factor of the escalating commitment in capital budget behaviour. Securities Market Her 2:66–71

# Research on Talent Introduction Hazard and Training Strategy of University Based on Data Mining

**Feng Li, Shilun Ge, and Jie Yin**

**Abstract** Based on uncertainty personnel basic information, data of teaching and scientific research in university, data mining and customer classification method are used, university talent are classified to four types by sensitivity, and talent development influence factors are found that include: initial graduate school, first degree, professional title, etc. According to the research result, linked with university development strategy and subject characteristics, the hazard and strategy for university talent introduction, talent training and human resource management are provided.

**Keywords** Data mining · Hazard · Sensitivity · Talent introduction and training · Uncertainty · University human resource management

## 1 Introduction

Traditional personnel management can not meet the development of modern university, and passive, transactional mode of administration should be replaced by initiative, strategic human resource management. It is an important research topic to explore the law of university human resource management. With the construction of university information system, the information of university personnel, technology, education and other, which hide the knowledge of a large amount of laws, have been accumulated abundantly. With the use of data mining

F. Li (✉)
Department of Postgraduate Administration, Jiangsu University of Science and Technology, 212003 Zhenjiang, Jiangsu, People's Republic of China
and
State Key Laboratory of Hydrology Water Resources and Hydraulic Engineering, HoHai University, 210098 Nanjing, Jiangsu, People's Republic of China
e-mail: justxwb@gmail.com

S. Ge and J. Yin
School of Economics and Management, Jiangsu University of Science and Technology, 212003 Zhenjiang, Jiangsu, People's Republic of China
e-mail: jzgsl@jzerp.com; yinjie9633@sina.com

Y. Zhou and D.D. Wu (eds.), *Modeling Risk Management for Resources and Environment in China*, Computational Risk Management,
DOI 10.1007/978-3-642-18387-4_24, © Springer-Verlag Berlin Heidelberg 2011

technology, we will extract the implicit, useful knowledge to explore university human resources development and management laws, which will provide support to the formulation of university human resources management policy.

The introduction and training of talent is an important component of university human resource management, which is directly related to the development speed of university human resources level. At present, many scholars have started relevant research of introduction and training of university talent. Such as Bai and Meng (2005), by analyzing the relation between teaching staff and structure of the title of teachers, she had an isolated point analysis for the introduction of high-level talent to find the law of the introduction of talent. Xu and Tan (2001) compiled the Shanghai talent indices system which includes the talent quality index, the talent academic degree index, the talent professional title index, the talent age index, the talent industry index and the talent aggregate index, and used the models to make a quantitative study on the inter-relationship between Shanghai talent indices and Shanghai macro-economic variables.

The research focus on the concrete problems or one point of university human resource management, and the data of university human resource management almost only include the talent science research data. Using the complete data of university human resource management and applying data mining technology to the full process of university human resource management is the main research direction in the future study.

# 2 Theoretical Basis

## 2.1 *The Construction of Customer Classification Model*

Wendell Smith, a U.S. scientist, put forward the concept of customer classification in the mid 1950s of the 20th century (Wu and Lin 2005). The so-called customer classification, that is, under the situation of a clear business strategy, business model and dedicated market conditions, according to the customer's values, needs and preferences and other comprehensive factors to have an classification of customers, the consumers have some degree of similarity who are belonged to unified customer base, while significant differences exist between different segments of customer base. Customer classification model selects certain segments variable and according to a certain criteria to have a classification of customs. Customer classification aims to better understand customers and provide personalized service.

## 2.2 *WK-means Clustering Algorithm*

Because of the traditional K-means clustering algorithm limitations that treated all the variables equally, scholars (Joshua and Michael 2005) put forward variables automatically weighted clustering algorithm (WK-means) which based on K-means clustering algorithm. WK-means clustering algorithm could automatically distribute different weight according to the different importance of variables. WK-means clustering algorithm is better at the ability of finding clustering than K-means type algorithm. WK-means algorithm effectively identifies noise variables by final variable calculated by weight gotten by iterative process of solving, and the ability for high dimensional data clustering of variable selection is important. WK-means clustering algorithm has already been aggregative in data mining software Alpha miner, Alpha miner is the open data mining platform explored by Hong Kong University and Harbin University of Industry, and the platform has realized clustering, classification and prediction, data association rules and other data mining algorithm.

## 3 The Process of Modeling

Analyzing the talent index model construction method, the following parameters are selected from many university teaching and researching projects of talents as the basis for building talent index model, including: the number of high-level discourse, the number of high-level research projects, the number of high-level books, the number of high patent, the number of high-level award, the standard teaching hours, the average student evaluation scores, the number of teaching reform subjects, the number of teaching achievements, the number of fine courses. Classify the university talents though clustering and divide them into following four types: both teaching and researching outstanding talent, biased towards researching talent, biased towards teaching talent, neither teaching nor researching outstanding talent.

Expression using the formula as follows:

UniversityTalent Type H = f (the number of high-level discourse, the number of high-level research projects, the number of high-level books, the number of high patent, the number of high-level award, the standard teaching hours, the average student evaluation scores, the number of teaching reform subjects, the number of teaching achievements, the number of fine courses.).

H = (A, B, C, D), where A on behalf of both teaching and researching outstanding talent; B on behalf of biased towards researching talent; C on behalf of biased towards teaching talent; D neither teaching nor researching outstanding talent.

The parameters that may be related to the development of university talent are selected from basic information of university talent personnel, including: gender, highest education, highest level of education graduate schools, the initial education, initial education graduate school, title, age, length of service, discipline, degree of change, job level. The association rules are mined from the basic information of talent and the types of talents to look for the main factors to talent development by using association algorithm.

## 4 Rules Mining

The data are from an engineering-based university, supplemented by economy management, arts, science, and agriculture and so on. The personnel, research, academic and other departments of this university have built information management system. To meet the need of research, this study has intercepted the relevant data since the various systems implemented to December 30, 2009.

After data collection, data cleaning, data discrete and other steps, using WK-means clustering algorithm, talents are classified into four categories. And then combined the basic personnel information with personnel categories relevance, the rule mining data sets are established. Setting the support level to 5% and confidence level to 50%, the association rules are mined by using Apriori algorithm (Table 1).

## 5 Strategy Study

University selected for this study has its subject characteristics as follows: engineering-based (ship is the featured subject), supplemented by economy administration, and humanities and social sciences is the new subject; school type: teaching

**Table 1** Part of the association rules

| Association rules | Support level (%) | Confidence level (%) |
|---|---|---|
| People who has senior title and doctoral degrees, ages between 39 and 47 and has education changed experience is A Talents. | 5.523 | 50.238 |
| People who has doctoral degree, ages between 39 and 47, and has education changed experience is A Talents. | 5.783 | 53.667 |
| People whose initial education graduate school is "985" university, the title is associate professor, gender is male is B class talent | 6.216 | 55.183 |
| People whose title is deputy high, initial education graduate school is "211" university, and learned science and engineering disciplines is B class talent | 6.306 | 58.672 |
| People whose title is Deputy High, has education changed experience, title below section chief, and learned science and engineering disciplines is B class talent. | 6.585 | 64.863 |
| People whose initial qualification is Master, length of service is 0–5 years, title is the middle class is C class talent. | 7.502 | 80.727 |
| People whose initial education graduate school is "211" university, and learned humanities and social sciences disciplines, and the highest educational background is undergraduate course is C class talent. | 6.126 | 76.539 |
| People whose title is the middle class, initial graduate school is a general university, ages between 23 and 28 years old and no education changed experience is C class talent. | 10.635 | 78.732 |
| People whose title is the primary class, the initial qualification is Master, initial education graduate school is "211" university and no education changed experience is D class talent. | 7.435 | 78.771 |
| People whose title is the primary class, the initial qualification is Master, no education changed experience is D class talent. | 10.492 | 73.596 |

university; school development strategy: become a teaching university of high level and character. According to the university's academic characteristics, type and development strategy, the following measures on talents introduction and development are introduced.

## *5.1 Talents Introduction*

**According to the character of university subject development, establish scientific and rational scheme of talents introduction.** Around the school's educational goals and level, development strategy, combine with the need of the construction of university teachers and disciplines, give full consideration of rational allocation problem such as disciplines, ages, learning edge structure, educational structure, establish short, medium and long combined talents introduction scheme of science, rationality, operation.

To the main subjects, talents introduction scheme is based on "look for outstanding talent" as the guiding ideology. The A class talent should be introduced, who has high title, high educational background, medium age, prominent teaching and reach. This type of talents can quickly become a subject leader, expand academic research direction, optimize subject echelon and drive the rapid development of disciplines.

To the developing subjects, talents introduction scheme should be based on "enrich the research talent". The B class talent should be introduced, who has higher title, higher educational background, and younger, strong research capability. This type of talents can improve the existing talents structure of academic emphasis on teaching people to drive research and raise the level of the school.

**Advocate the introduction of first-rate school talent to improve the introduction proportion of talents of high education background.** According to the rules, talents who have first-rate school learning experience are easier to develop to research talents; the increase of research talents will directly contribute to the improvement of the school scientific research level and boost school teaching and research level. Therefore, the staff that is newly introduced in teaching and research should give priority to doctoral student of "schools of directing," which will not only improve the educational structure of school talent and learning edge structure, but also directly improve school scientific research level, and keep a balance between school teaching and research development.

## *5.2 Talents Training*

**Set different training policy for different types of talents.** For teaching and research more prominent "A class" talent, providing significant support by giving research funding to promote their growth of being domestic leading scholars who have significant impact. For teaching or research more prominent "B class", "C class" talent, we should set up special funds to carry out its assistance, encourage them to achieve domestic first-class level in their disciplines.

**Establish scientific and sound performance appraisal system.** Scientific and effective performance appraisal system should be established to really link with performance. The distribution system should promote the overall healthy development of human resources. For "C class" talent emphasis on teaching, its talent assessment should give priority to fruits of research, design evaluation system in respect of research projects, research awards, scientific papers, patents, books and so on; for teaching and research more prominent "A class" talent, we should weaken their own performance assessment, focusing on achievements obtained in the construction of teams and disciplines.

## 6 Conclusion

From actual human resource management data of university, by using Data mining technology, many human resource management issues such as talent type classification, talent introduction and cultivation could be put forward by quantitative analysis. Meanwhile, the parameters such as staff enthusiasm, academic atmosphere, incentives and other soft factors are difficult to quantify, how to synthetically consider these factors effect to the university talent development needs further study.

## References

Bai F, Meng C (2005) Application of data mining technology in university talent introduction. J Taiyuan Univ science technology 4:66–67

Joshua ZH, Michael K (2005) Automated variable weighting in k-means type clustering. IEEE Trans Pattern Anal Mach Intell 5:657–668

Wu J, Lin Z (2005) Customer classification model study based on cardholders' consuming behaviour. J Cent Univ Finance 6:67–71

Xu G-X, Tan X-q (2001) Shanghai talent indices system and its applicable study. Study Finance Econ 12:36–43

# Supply Chain Performance Comprehensive Evaluation Based on Support Vector Machine

**Weiling Cai, Xiang Chen, and Xin Zhao**

**Abstract** The competition among enterprises has evolved into the supply chains competition. The evaluations of cross-process, cross-function, cross organization have been brought into supply chain performance evaluation system. Therefore, the study and analysis on supply chain performance evaluation, which adapts globalization supply chain competition environment, has important significant. Firstly, the paper analyzed the impact factors of supply chain performance, constructed the supply chain performance evaluation index system. Secondly, the paper has used information entropy to reduce the indices, established comprehensive evaluation model based on support vector machine (SVM). Finally, the paper investigated 26 supply chains data and used model to run simulative evaluation. The results were more precise than traditional back propagation (BP) neural network's evaluation results, which proved the feasibility and validity of the method.

**Keywords** Comprehensive evaluation · Index system · Information entropy · Supply chain performance · SVM

## 1 Introduction

In the economic globalization and market demand flexible environment, the enterprises' business activities are distributed. The enterprises performance evaluation within supply chain has the decentralized management unified under the same standard. Meanwhile, the supply chain enterprises more emphasize on time-based competition, emphasize creating value for customers, and emphasize win-win management concept. Therefore, supply chain performance evaluation chooses

W. Cai (✉)
School of mechanics & civil engineering, China University of Mining & Technology Beijing, China
e-mail: cwl9898@yahoo.cn

X. Chen and X. Zhao
Institute of Economics and Management Hebei University of Engineering Handan, China

Y. Zhou and D.D. Wu (eds.), *Modeling Risk Management for Resources and Environment in China*, Computational Risk Management,
DOI 10.1007/978-3-642-18387-4_25, 

the indices emphasized the concept which based on customer response: customer order fulfillment rates, product extension time, lead time, etc. With the new management technologies development of supply chain, network, and logistics integrated, it requires clear definition of the whole supply chain performance, which is the only way to make rational judgment from strategic point for supply chain management and operation (Jiang et al. 2002; Zhao 2002). In order to further apply and promote the advanced supply chain management theory in practice, it's necessary to get the supply chain performance evaluation away from the traditional functional performance evaluation, and access to the evaluation system of cross-process, cross-function, and cross-organization. Therefore, it's important to study and analyze the supply chain performance evaluation that adapts globalization supply chain competition environment.

There are many supply chain performance evaluation methods. So as to achieve the best evaluation result, it should choose the most suitable method according to the objects characteristics. Present evaluation methods at home and abroad are: expert evaluation, economic analysis, data envelopment analysis, analytic hierarchy process, fuzzy comprehensive evaluation, and neural network comprehensive evaluation, etc. The training process of SVM follows the structural risk minimization principle. In the process, the structural parameters will automatically adjust according to samples data, without over-fitting. Through solution of a linear constrained quadratic programming problem, it obtains global optimal solution, without local minimum problem. Therefore, SVM method successfully overcomes shortcomings in neural network (Zhang 2000).

As there are non-linear relationships among impact factors of supply chain performance, the paper constructed the support chain performance evaluation model based on SVM. On the basis of analyzing the impact factors of supply chain performance, the paper investigated 26 supply chains in Hangzhou, Handan, Tangshan, Beijing, etc., applied actual data to run simulative training for the model, and get very good results. The methods has important guiding significance and reference value in supply chain performance evaluation.

## 2 SVM Supply Chain Performance Evaluation Model

### *2.1 SVM Evaluation Model*

Support given training data $\{(x_i, y_j), i = 1, 2, \ldots\ldots l\}$, where $x_i \in R^d$ is the ith learning sample input value, and also is a d-dimensions vector. $x_i = [x_i^1, x_i^2, \ldots\ldots x_i^d]^T, y_j \in R$ is the corresponding target value. We define linear ε insensitive loss function as follows:

$$|y - f(x)_\varepsilon| = \left\{ \begin{array}{cc} 0 & |y - f(x)| \le \varepsilon \\ |y - f(x) - \varepsilon| & |y - f(x)| > \varepsilon \end{array} \right\}$$

That is, if the difference between target value y and learning constructed regression estimation function's value f(x)less than $\varepsilon$, then the loss is equal to 0.

How to handle it, if the learning sample set is non-linear? As we all know that the basic idea of SVM is: transform input sample space non-linear into another feature space, in where construct regression estimate function. This kind of non-linear transformation is realized through definition of appropriate kernel function $k(x_i, x_j)$. Where $k(x_i, x_j) = \phi(x_i) \bullet \phi(x_j)$, $\phi(x)$ is a non-linear function. So we can assume the non-linear regression estimate function as:

$$y = f(x) = w^T \phi(x) + b \tag{1}$$

In the formula, $\phi(x)$ is the non-linear mapping from input space to high dimensional feature space. Parameter $w$ and $b$ are estimated by minimized cost functional:

$$R_{svm}(C) = C\frac{1}{n}\sum_{i=1}^{n} L_\varepsilon(Y_i, y_i) + \frac{1}{2}\|w\|^2 \tag{2}$$

$$L_\varepsilon(Y, y) = \begin{cases} 0 & |Y - y| \le \varepsilon \\ |Y - y| - \varepsilon & |Y - y| > \varepsilon \end{cases} \tag{3}$$

In formula (2), $y_i = w^T \phi(x_i) + b$. In the given regularized risk functional, the first part $C\frac{1}{n}\sum_{i=1}^{n} L_\varepsilon(Y_i, y_i)$ is experience risk, which be measured by non-sensitive loss function $\varepsilon$. The second part $\frac{1}{2}\|w\|^2$ is the regularized part. $C$ is a positive constant, called the penalty parameter, which determines the balance between experience risk and regularized part. The bigger the $C$ is, the heavier the penalty to mistake is. Generally speaking, with the increase of the $C$, the accuracy of the test will be getting higher. But when it get a certain value, the further increase will lead to misclassification rise. In order to get optimal $w$ and $b$, it need to introduce slack variables $\zeta_i$ and $\zeta_i^*$, to get constrained optimization problem:

$$\begin{aligned} &\min_{w,b,\xi_i,\xi_i^*} \|w\|^2/2 + C\sum_{i=1}^{n}(\xi_i + \xi_i^*) \\ &s.t. Y_i - (w^T \phi(x_i) + b) \le \varepsilon + \xi_i \\ &(w^T \phi(x_i) + b) - Y_i \le \varepsilon + \xi_i^* \\ &\xi_i, \xi_i^* \ge 0, \forall i \end{aligned} \tag{4}$$

Definite the Lagrange function:

$$\begin{aligned} J(w, \xi, \xi^*, a, a^*, \gamma, \gamma^*) = {} & C\sum_{i=1}^{n}(\xi_i + \xi_i^*) + \frac{1}{2}\|w\|^2 - \sum_{i=1}^{n} a_i[w^T \phi(x_i) - Y_i + \varepsilon + \xi_i] \\ & - \sum_{i=1}^{n} a_i^*[Y_i - w^T \phi(x_i) + \varepsilon + \xi_i] - \sum_{i=1}^{n}(\gamma_i \xi_i + \gamma_i^* \xi_i^*) \end{aligned} \tag{5}$$

Then the dual optimization problem of formula (4) wrote out the matrix form as:

$$\max_a E^T a - \frac{1}{2} a^T P a$$
$$s.t. \sum_{i=1}^{n} (a_i - a_j^*) = 0 \qquad 0 \le a_i, a_i^* \le C \tag{6}$$

Where $a^T = [a_1, \cdots, a_n, a_1^*, \cdots, a_n^*]$

$$E^T = [\varepsilon + y_1, \cdots, \varepsilon + y_n, \varepsilon - y_1, \cdots, \varepsilon - y_n], \quad p = \begin{bmatrix} Q & -Q \\ -Q & Q \end{bmatrix}$$

$a_i$ and $a_i^*$ is called Lagrange multipliers. The corresponding sample of $(a_i - a_i^*) \neq 0$ is called SVM. Through control the two parameters $C$ and $\varepsilon$ in quadratic optimization, it can control the generalization of SVM. In general, the greater the $\varepsilon$ value, the less the number of SVM is, thus the more sparse the solution expression is, meanwhile, the large value of $\varepsilon$ also can reduce the data point's approximation. From this point of view, the value of $\varepsilon$ is the balance factor between sparse degree and data points density. $Q$ is a $n \times n$ symmetric matrix, and $Q_{ij} = \phi(x_i)^T \phi(x_j)$. $K(x_i, x_j) = \phi(x_i)^T \phi(x_j)$ is called kernel function. Commonly used kernel functions are (Li et al. 2003): polynomial kernel function $K(x_i, x_j) = [(x_i \cdot x_j) + 1]^q$, radial basis kernel function (RBF)

$$K(x_i, x_j) = \exp\left\{ - \frac{\|x_i - x_j\|^2}{2\sigma^2} \right\},$$

Sigmoid function $K(x_i, x_j) = \tanh(v(x_i, x_j) + c)$. The paper adopted the most common RBF. Solving the above quadratic optimization problem, can get $w = \sum_{i=1}^{n} (a_i - a_i^*)\phi(x_i)$. According to the KKT theorem, derived equation:

$$\begin{cases} \varepsilon - y_i + f(x_i) = 0 & a_i \in (0, C) \\ \varepsilon + y_i - f(x_i) = 0 & a_i^* \in (0, C) \end{cases} \tag{7}$$

The offset value $b$ can be calculated. The prediction decision function is as follows:

$$f(x) = \sum (a_i - a_i^*) K(x_i, x) + b \tag{8}$$

## 2.2 Implementation Steps

The specific implementation steps of supply chain performance evaluation model based on SVM are: Collected information, selected indices, constructed impact factors index system; Organized data, filtered sample, constructed sample set; Sample normalized. In order to ensure the SVM sum is 0, pre-process the input vector X and output vector Y

$$X(k,i) = \frac{X(k,i) - mean\_x(i)}{std\_x(i)} \quad Y(k) = \frac{Y(k) - mean\_y}{std\_y} \tag{9}$$

Where $mean\_x(i), std\_x(i)$ respectively are the $i$th arithmetic mean and standard deviation of input vector $X$. $mean\_y, std\_y$ respectively are the arithmetic mean and standard deviation of output vector $Y$. Determine the model training error test criteria: in order to measure the accuracy of prediction model, the paper used relative error, $Error(n) = |x(n, true) - x(n, pred)|/|x(n, true)|$ as the mean square error to evaluate the model's effect.

Model solution: solving the model parameters $(w, b)^T$, using Cross-Validat-ion method choose kernel function parameter $\sigma$ and penalty parameter $C$. Through simulation to test the evaluation result whether reached the accuracy requirement, get the optimum fitting function, that is, the evaluation model in the paper. Put the to be tested supply chain performance indices into the trained model, output the evaluation value.

## 3 Empirical Analysis

Because of the complexity of the supply chain management, it's difficult to evaluate using single index. It must evaluate from multi-angle, multi-perspective, and establish hierarchical index system. Here, learned from supply chain performance evaluation research results, surrounding its targets, according to its index system design principles, from planning, strategy and operation three layers, we analyzed the supply chain decision contents, clarified the related performance evaluation indices, then sum up the overall supply chain coordination performance evaluation index system (Li et al. 2003) (Table 1).

The paper investigated 26 supply chains in Hebei, Shandong, Beijing, Tianjin. Started from the core enterprise in a supply chain, we invested the enterprise's first-level suppliers and first-level distributors as the whole supply chain to research. The investigation based on index system; took the supply chain performance evaluation as general goal; considered planning, strategy and operation three aspects which can be subdivided to timeliness, efficiency, satisfaction, cooperation, environment protection, production development, supply, distribution, cost, benefits, and utilization ability, etc. 29 indicators. Reduced the information entropy of original data sheet,it can get indicators 14,16,19,21,24; that is, pollution control degree, production development cycle, distribution reliability, manufacturing costs, the total inventory costs, whose distinguish capacity is very limit and can be deleted. The actual data of investigation were scored by relevant experts, which results will be

**Table 1** Supply chain performance evaluation index system

| Target | Layer | Index | |
|---|---|---|---|
| Supply chain performance evaluation index | Strategy performance | Timeliness | Order cycle |
| | | | Perfect order fulfillment level |
| | | Efficiency | Cash flow time |
| | | | Investment rate return |
| | | Satisfaction | Flexible |
| | | | Response speed |
| | | | On-time delivery rate |
| | | | Products qualified rate |
| | | | Cost-profit ratio |
| | | Cooperative | Data sharing rate |
| | | | Node enterprises cooperative ability to business standard |
| | | | The awareness and level of node enterprises participate in problem-solving |
| | | Environment friendly | Hazardous material generation amount |
| | | | Pollution control degree |
| | | | Waste regeneration utilization |
| | Planning performance | Product development | Product development cycle |
| | | Supply | Level of co-operation to improve the quality |
| | | Distribution | Emergency distribution response level |
| | | | Distribution reliability |
| | | | Distribution arrangement effectiveness |
| | Operation performance | Cost | Manufacture cost |
| | | | Communication cost |
| | | | Transportation cost |
| | | | Total inventory cost |
| | | Effectiveness | Profit |
| | | | Economic value-added |
| | | Capacity utilization | Productivity |
| | | | Fixed assets effectiveness ratio |
| | | | Inventory turns |

trained and tested as SVM evaluation results, as well as the results to compare with the traditional BP network's

All the simulation procedures are realized in Matlab7.0 environment (Lin and Zhou 2005). Took time delay $\tau = 1$, and $m = 24$, through cross-validation chose the optimal penalty parameters $C = 50$, $\varepsilon = 0.0001$. Radial basis kernel function parameter $\sigma = 2.1$, and when $\{C, \sigma\} = \{50, 2.1\}$ trained the model: the first 22 samples were as training set, the last 4 samples were as test set, got MSE=0.0060. The meaning of MSE is the average deviation between actual value and predication value of test samples. The paper used MSE value to measure the accuracy of the prediction model. Fixed $\{C, \sigma\} = \{50, 2.1\}$ (Fig. 1).

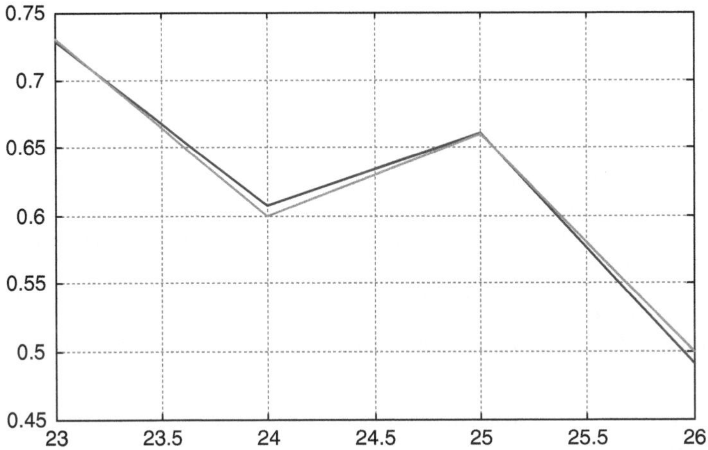


**Fig. 1** Forecast fitting chart

**Table 2** Comparison of SVM prediction, original and BP output values

| Sample | 23 | 24 | 25 | 26 |
|---|---|---|---|---|
| Original value | 0.73 | 0.60 | 0.66 | 0.50 |
| SVM prediction value | 0.72817 | 0.60702 | 0.66059 | 0.49052 |
| BP output value | 0.7261 | 0.5889 | 0.6001 | 0.4333 |

# 4 Conclusion

We can see from the comparison in Table 2, although traditional BP network and SVM all can get training results, but the result from SVM obviously better than the traditional BP network's, and errors is smaller than the traditional BP network's. The model is simple and convenient, which saved a great deal of tedious calculation and provided reliable help for improving efficiency. Therefore, the supply chain performance evaluation based on SVM is feasible, accurate and effective, which has a good promotion prospects. Nevertheless, in the model extension process, it should pay attention to select reasonable and effective indicators and samples according to the study objects' characteristics. The selection of model parameters also should accord with the actual situation.

To sum up, the SVM method can effectively deal with the complex supply chain performance evaluation problems, which has the network achieved global optimum and has the identification accuracy greatly improved. Therefore, the SVM-based supply chain performance evaluation method can be considered a feasible and effective method.

## References

Bian Z, Zhang X (2002) Pattern recognition. Tsinghua University Press, Beijing, pp 236–280

Erhun F, Keskinocak P, Tayur S (2008) Dynamic procurement in a capacitated supply chain facing uncertain demand. IIE Transactions. National Center for Biotechnology Information. http://www.ncbi.nlm.nih.gov

Green SD, Fernie S, Weller S (2005) Making sense of supply chain management: a comparative study of aerospace and construction. Constr Manage Econ 23:579–593

Jiang B, Wang L et al (2002) Supply chain management analysis from complex angle. Comput Eng Appl 15:52–54

Li Q, Song G, Zhang S (2003) Supply chain performance evaluation index system. China Mech Eng 14(10):881–884

Lin X, Qian Z (2005) Matlab 7.0 application collection. China Machine Press, Beijing

Premus R, Sanders NR (2008) Information sharing in global supply Chain alliances. J Asia Pac Bus 9(2):174–192

Zhang X (2000) Statistical learning theory & SVM. Acta Automatic Sin 1:21–22

Zhao L (2002) Supply chain management in knowledge economy. J Southeast Univ (Natural Science Edition) 32(3):514–522

# The Sensitivity Analysis of a Customer Value Measurement Model

**Liu Xiao-bin and Zhang Ling-ling**

**Abstract** Under the condition of increasing competition, it is more and more important to find customers' current needs, and tap the potential demand of customers. In the process of value creation, enterprises is not only concerned about the external physical variables which impacts customer value, but also should takes full account of intangible factors which affect consumer decisions. Based on the analysis of existing customer value measurement model, by importing the factor "lifestyle", we build a new model of customer value measurement. By means of sensitivity analysis of demand for customer value, it explains different valuations of the customer in different circumstances as well as different value judgments among different customers, and provides a dynamic guidance to marketing practice of enterprises.

**Keywords** Customer cost · Customer utility · Customer value · Lifestyle · Sensitivity analysis

## 1 Introduction

In an increasingly competitive market, the walls between enterprises and customers are blocking maximization of the company profits and maximization of customer satisfaction. It is imperative to dismantle these walls. However, with limited resource, enterprises are unable to break all the barriers, and their decisions can only depend on consumers' choices. Customers' resource is also limited. Their consumption process is the process of voting, while the value assessment is the most critical factor to customer choices. Because the nature of value is measurable, some scholars found that customer value is measurable. Only through measuring customer value, companies can apply it as a strategy to marketing practice. When

L. Xiao-bin (✉) and Z. Ling-ling
School of Management, Guangdong University of Business Studies, Guangzhou, China
e-mail: lxbcsjyy@163.com

Y. Zhou and D.D. Wu (eds.), *Modeling Risk Management for Resources and Environment in China*, Computational Risk Management,
DOI 10.1007/978-3-642-18387-4_26, 

the customers make their purchase decisions, they can compare the size of customer value more objectively that enterprises provide, and then they can have better choices.

Customer value measurement is a key element of customer value theory, and is premise and basis of the customer value management. It helps us to understand consumer behavior, interpret customers' needs and feelings. Accordingly, customer value measurement can explore the trends of customer demands. And in the range of the process of create value, customer value measurement makes companies do the right thing by the right way, and identify and develop a corporate strategy to attract and retain customers. Therefore, customer value measurement can create competitive advantages for enterprises.

## 2 Literature Review

In decades, both in academia and in business paid much attention to customer value. Zeithmal considered customer value as a "Premier abstraction", and inner judgments and overall evaluation of customers, which is related to individual experience, can be divided into several related factors. Different customer groups have different perceived values. Indrajit and Wayne insisted that perceived value is a multidimensional structure, and its scope should be defined according to the sort of product. Gronroos argued that driving factors of customer value should cover not only the core products and additional services, but also efforts to maintain the relationship. Similarly, based on epistemic logic of information processing, Woodruff (1997) proposed hierarchy model of customer value, and he thought the context of use play an important role in the production of customer value. Vantrappen, Slater and Jantrania changed with different contact stages between companies and customers for the dynamic of customer value. Flint and so on considered that some sudden events affecting customer value can change customer perception. Although there is no corresponding defines in academia field, many scholars considered customer value is impacted by situational factors as a multidimensional, dynamic concept. Therefore, in this paper, customer value is defined as: Based on situational factors personally, the utility a customer gains is divided by the cost he/she pays:

$$V = aU / C$$

"V" means customer value, "U" means the utility customer gains, "C" means the cost customer pays, and "a" means a coefficient generated by the customer's situational factors.

Similarly, there are many ways to measure customer value like one-dimensional, two-dimensional, multi-dimensional and so on. Gale proposed a two-dimensional measuring method contains two kinds of attributes, quality and price, which is given definite weight number to form a customer perceived value chart. However, this measuring method can only reflect the current situation, neglecting the potential of value creation. At the same time, customer value is divided into two dimensions, which is difficult to reflect the customer's real psychological process completely. Indrajit and Wayne (1998) proposed three-dimensional structure including brand, property, segmentation, and a value chart model to evaluate customer value. Flint considered in the consumer decision-making process, power of influence, the extent of expectation, temporary impulse, and environmental change can lead customers to tense and changing the expectations of customers to establish customer desired value. Holbrook developed a composition table of customer value. He divided customer value into three dimensions: (1) external/internal, (2) self-oriented/other-oriented, (3) active/passive. At the same time, Holbrook divided customer value into eight categories based on three dimensions: efficiency, excellence, status, respect, entertainment, aesthetics, ethics and spirituality. With customer value dimensions changed continuously in the academic study, research has become more in-depth in customer value measurement. In two-dimensional measurement, scholars tend to study the tangible product attributes and functional value. As scholars proposed multi-dimensional measurement, research on customer value has been gradually extended to the relational, emotional, situational and other intangible factors. For examples, Sweeney and Soutar divided functional value into the quality factor and price factor to developed a perval model including the emotional, social, quality and price factors based on different contributions on the overall value perception. In the multi-dimensional measurement, Shethetal proposed five relatively independent dimensions including functional, social, emotional, epistemic, situational value. In service industry, Petrick developed a SERV-PERVAL model, etc. However, these methods of measuring customer value also have some defects. Those theories are result-oriented rather than process-oriented, customer-oriented and competition-oriented, which focus on measuring current customer value, while ignoring dynamic change of customer value.

# 3 Main Customer Value Measurement Model

## *3.1 Gale's Customer Value Map*

Gale puts forward the way to measure customer value by means of quality and value. The weight of each attribute means and the score of each attribute is obtained through customer surveys. Namely: customer value = Σ (a property's relative score on quality dimensions × the weight of this dimension) + Σ (a property's relative score on price dimensions × the weight of this dimension)

$$\text{Property's relative score} = \frac{\text{the score of enterprises in the property}}{\text{the score of competitors in the property}}.$$

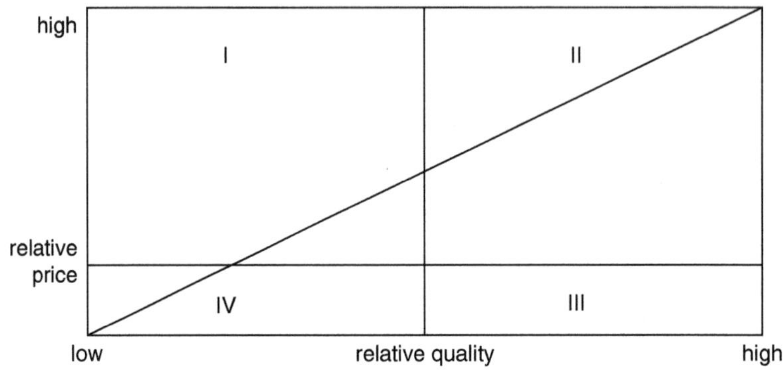


**Fig. 1** Gale's customer value map

Gale also drew customer value map and provided a more intuitive analysis (Fig. 1).

This method can directly compare customer value provided by company between provided by its competitors. It makes up for the deficiency of customer satisfaction measurement of the lack of competition-oriented in a certain extent. However, it has great limitations because customer value is measured only through two dimensions of quality and price. Therefore, the understanding of the value is narrow. In fact, consumers' behavior is also affected by situational factors. So this model cannot explain the heterogeneity when customers measure their values.

## *3.2 Kotler's Customer Delivered Value Model*

Kotler put forward the customer delivered value and customer satisfaction to measure customer value. He noted that customers pursue maximizing value with limited resources. Customers can determine which suppliers can provide a higher value. Customers delivered value is that total customer value minus total customer cost. The total customer value is the total income from products and services which include the product's value, functional value, personnel value and image value; Total customer cost is the total cost customers obtain products and services including the cost of money, time, cost, physical and energy. Therefore, the measurement of customer value can be expressed as:

$$\begin{cases} \mathrm{TCV} = \Sigma\,(\mathrm{Pd}, \mathrm{S}, \mathrm{P}, \mathrm{I}) \\ \mathrm{TCC} = \Sigma\ (\mathrm{M}, \mathrm{T}, \mathrm{C}) \\ \mathrm{TCDV} = \Sigma\,(\mathrm{Pd},\ \mathrm{S},\ \mathrm{P},\ \mathrm{I}) - \Sigma(\mathrm{M},\ \mathrm{T},\ \mathrm{C}) \end{cases}$$

TCV means total customer value, TCC means total customer cost, TCDV means customer delivered value; Pd means product value, S means service value, P means personnel value, I means image value; M means monetary costs, T means time costs, C means physical and energy costs.

Different from Gale's measurement way, Kotler's customer delivered value measurement model notices the impact of factors more than product which provides a theoretical framework for customer value measurement, and states that specific ways to increase customer value. And from the view of the enterprise as a provider the way of measuring customer value is little consider the impact factors of customer experience.

### *3.3 Woodruff's Customer Value Hierarchical Model*

Woodruff believed customer perceived value will change as time. Customers would pre-estimate value before purchase, and estimate value again after purchase. And this estimation will be pre- estimation of the next purchase in different stages of the buying process leading that customer perception of value is different. Woodruff not only raised the level of customer value measurement from the previous static to dynamic, but also began to study the customer's judgments for value through the buying process. Based on the principle of "means – end" chain, He construct customer value hierarchy model (Fig. 2)

Customer value hierarchy model takes into account the important role of customer use situation in the evaluation and expectation. To evaluate the product, customers use the expectation attributes, outcomes and objectives structure which was formed in the mind that can also well describe the perceived value. Woodruff introduced the qualitative data collection and analysis method of detect psychological perception of customer. But its disadvantages are that the customer will give a universal answer which can affect the accuracy of measurement if he found your intentions.

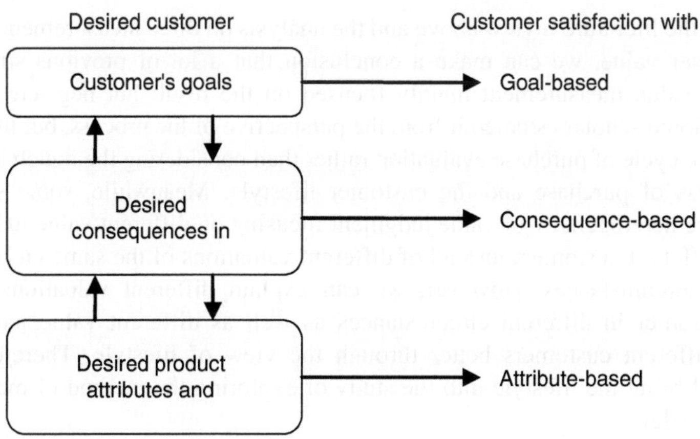


**Fig. 2** Customer value hierarchical model

To summarize analysis above, in the Gale's two-dimensional measurement model, customers have evaluation of product quality and customer value. So this method of measurement takes account of the important position of customers in the value evaluation and reflects the competition orientation in customer value measurement. It has its own advantages through comparing with the customer value provided by companies and competitors. But this method has a narrow understanding of customer value so that there will exist a wide gap in actual situation. Value evaluation of consumers is influenced by many factors more than quality and value. This defect makes the model unable to explain the diversity which generally exists when consumer measures the product's value. Comparing with Gale's two-dimensional measurement model, Kotler's customer delivered value model takes attention on impact factors but not in products to provide a theoretical frame for the measurement of customer value and indicate specific ways to increase customer value. But this model mainly stands on the viewpoint of company, takes less considerations of value judgment. In Woodruff's customer value hierarchical model, it takes account of important effect of customer use situation and introduces qualitative data collection and analytical method to detect customers' psychological perception. But there is a big defect that is when customer accepts investigations he probably makes a universal answer which will affect accuracy of measurement if he found your purpose. From these analyses above, the methods of customer value measurement currently are mostly static focusing on consequence or customer orientation. But it neglects competition orientation and the dynamic changes of customer value and lacks of understanding of customer situation which may affect value.

## 4 Building the Customer Value Measurement Model Based on Lifestyle

Based on the literature review above and the analysis on three measurement models of customer value, we can make a conclusion that a lot of previous studies of customer value measurement mainly focused on the result but neglected of the process. Some scholars studied it from the perspective of the process, but they only studied the cycle of purchase evaluation rather than considering the match between the process of purchase and the customer lifestyle. Meanwhile, some scholars studied the heterogeneity of value judgment focusing on different value judgments among different customers instead of different valuations of the same customer in different circumstances. However, we can explain different valuations of the same customer in different circumstances as well as different value judgments among different customers better through the view of lifestyle. Therefore, the paper will bring the lifestyle into the study of exploring the method of measuring customer value.

## 4.1 Lifestyle

In 1960s, marketing scholars introduced the concept “lifestyle” into the field of marketing in the consumers’ market segment particularly. A lot of scholars studied that the lifestyle affects consumer behavior. Thereinto, Lazer argued that the formation of lifestyle was related to culture, values, resources, beliefs, laws etc. It is a systemic, dynamic concept on behalf of different life characteristics in one group. So we believe different lifestyles will lead to different purchase decisions. Engel, Blaekwell and Kllat proposed E.K.B model which is account for the impact of lifestyle on consumer decision-making. Bermanand also pointed out that the lifestyle is an important variable affecting consumer decision-making. As can be seen from these studies that scholars combined the quantitative research with the qualitative research to prove that consumer lifestyle would affect consumer behavior, but they didn’t study how to influence consumer decision-making and what’s role in customer value evaluation through lifestyle. In this study, we bring the lifestyle into the study on customer value measurement, which can contribute enterprises to manage development direction to meet the needs of consumers in order to create the competitive advantage of enterprises.

## 4.2 The Customer Value Measurement Model Based on Lifestyle

Based on existing research about customer value measurement, in this study customer value is defined as: Based on situational factors personally, the utility a customer gains is divided by the cost he/she pays. Expressed as:

$$V = a\,U\,(F, Q, S)\,/\,C$$

“V” means customer value. “U” means the utility customer gains. “C” means the cost customer pays, and “a” means a coefficient generated by the customer’s situational factors. Therefore, “F” means the evaluation on product quality by customers. “Q” means the evaluation on service quality by customers. “S” means the acceptance level of customer on corporate brand image.

In this study, the potential impact factors are exposed, and the lifestyle is an important element of the model. “L” is a coefficient emerges from situational factors by consumers themselves, the differences of consumer lifestyles which will zoom in or out the customer value.

Based on the discussion above, the model is:

$$V = I[a_1F(L) + a_2Q(L) + a_3S(L)]/(\beta_1C_1 + \beta_2C_2 + \beta_3C_3)$$

"a" means the weight of factors in customer utility. "β" means the weight of factors in customer cost. "$C_1$" means monetary cost the customer takes for gathering product information. "$C_2$" means the price of products. "$C_3$" means the extra monetary cost the customer takes during the consumption or use. "L" is a coefficient which means the effect of lifestyle on value judgment. When "L" is greater than one, customer's consume events match their lifestyle, and the greater it is, the higher the matching is, which can zoom in customer value better; when "L" is smaller than one, customer's consume events don't match their lifestyle, and the smaller it is, the bigger the conflict is, which can reduce customer value greater. "I" is a coefficient which means the corporate image to customers. When "I" $>1$ means that corporate image is well, it can zoom in customer value. When "I" $<1$ means that the corporate image is poor, it will reduce the customer value.

## 4.3 Model Interpretation and Sensitivity Analysis

Based on the lifestyle of customer value zooming in or out, we study the important role of lifestyle in customer value measurement, and build a new model of customer value measurement. The model is the same as the value engineering in the form, which had already been proved to be an effective way to judge the value and a usual thinking model when customers make their purchase decisions. The division is adopted in the model that is more logical. When the customer holds value for money, the ratio is one, which means that customer utility is equal to customer cost; when the customer holds the best value, the ratio is greater than one, which means utility is greater to cost; when the customer holds value not for money, the ratio is smaller than one, which means utility is smaller to customer. Meanwhile, the format of the division is more accord with competition-oriented, which is favor of customers comparing and choosing among different values enterprises create. So on the whole, firstly, the division is adopted in this model, which is more logical, more compatible with the competition-oriented, and matching customers' thinking; Secondly, the concept "lifestyle" is infused into the model, which offers customers a new thread to make a matching-themselves purchase decision as well as expands the range that the business cognize consumer behavior; Finally, putting lifestyle into customer value measurement, we can explain different valuations of the same customer in different circumstances as well as different value judgments among different customers better to provide a dynamic guidance to marketing practice of enterprises.

Li Kouqing insisted that, by means of the general sensitivity analysis of the demand for customer value, the sensitivity value of demand for customer changing is the customer value elasticity of demand. And the level of commodity prices and customer demand for commodities inverse relationship between changes in different customer value and demand positive correlation between them. Moreover, that other things being equal, customers from the unit consumption of goods obtain the more interest, their willingness to purchase the greater the amount of goods.

Customers from the purchase and consumption of products or services of a company acquired interests in more then the more inclined to buy the company's products. The sensitivity of the demand for customer value can be measured by demanded elasticity indicators, expressed as:

$$E_{cv} = \frac{\Delta Q}{Q} / \frac{\Delta CV}{CV}$$

The basic meaning of this formula is that quantity demanded is affected by changes in the demand level when customer value has some degree of change. The different components of the total customer value such as changes in the interest of customer value resulting from changes of prices and the changes of then interest of perceived customer value resulting from changes in product quality, the sensitivity of customer demand is different. For example, $\Delta CV_{c_2}$ is changes in the interests of customer value resulting from changes in product prices; $\Delta CV_F$ is changes in perception of customer value generated by changes product quality. Corresponding calculation of the demand can also be sensitive to customer value. Especially because of their own reasons including the concept of changes in income and other lifestyle changes or external effects, customer value will completely change $\Delta CV_L$ and the demand for the product will also change. Therefore, people pursue different interests in the shopping and consumption of different goods and services. There is always the pursuit of different interests even though different consumers consuming the same goods and services. Understanding customers' needs on the sensitivity of different interests is to help companies design more targeted and improved products. To expand the market demand, companies must strive to increase the sensitivity to strong customer demand for benefits.

# 5 Conclusions

## *5.1 Theoretical Contributions*

Scholars have studied the customer value for a long time, but there isn't a united definition of customer value and there also exists a variety of ways and dimensions in the study of customer value measurement. But generally, most previous studies mainly focused on the result but neglected the effect of customers' value creation process, focused on the measure of current customer value but neglected the dynamic changes of customer value. Some studies focused on the customer orientation but neglected the competition orientation. By introducing customers' lifestyle into the customer value measurement and measuring customer value by ratios, we get the factors that influence customer value. These researches enrich the consumer decision-making theory and are important supplements of customer value theory.

## 5.2 Practical Significance

On one hand, this model which investigates the customer value measurement through lifestyle factors expands companies' scopes for knowledge of consumer behavior. It explains not only the diversity of value judgment between different consumers, but also the diversity of each consumer's purchase decisions in different situations. Therefore, it provides guidance for companies' marketing practice from the two-direction dynamic. On the other hand, from the customers' point of view, this model also provides customers some new ideas to make a better decision.

## 5.3 Limitations and Further Research Direction

This paper builds customer value measurement model by introducing lifestyle enriches the theory of customer value measurement and has an exploratory theoretical significance. However, this model needs to be tested and modified through further empirical study which will be another new research direction in future.

# References

Woodruff RB (1997) Customer value: the next source for competitive advantage. J Acad Mark Sci 25:139–153

# On the Relationship Between Capital Structure and Firm Value: Empirical Analysis Based on Listed Firms in Real Estate and Retail Trade

**Xiaohong Tai and Nan Chen**

**Abstract** With the improvement of business management and decision-making level of enterprises, corporate decision-makings on finance not only pay more attention to size of funding, but also to financing options and the financing structures in order to increase the market value of the enterprise and maximize investors' interests. It plays an important and significant role to improve financial decision-making level and optimize the capital structure of enterprises by studying the relationship between capital structure and firm value of listed firms. This paper reviews the history of capital structure theory, analyzes the relationship between capital structure and firm value, selects the real estate industry and the retail industry, and carries out an empirical analysis on the capital structure and the firm value by establishing comparative regression models.

**Keywords** Capital structure · Financial risk · Firm value · Listed companies · Uncertainty

## 1 Introduction

Capital structure refers to the structure and proportional of equity capital and debt capital of one enterprise. Capital structure theory is about the relationship among capital structure, consolidated cost of capital rate and firm value theory and is an important aspect of corporate finance theory, and is an important theoretical basis for decision-making on capital structure, too. Classical capital structure theories originated from 1958, until the late 1970s. In 1958, Modigliani and Miller published entitled 'The Cost of Capital, Corporation Finance and the Theory of Investment' in "American Economic Review." The article proposed a theorem known as MM theorem later, which placed the analysis of capital structure into

X. Tai (✉) and N. Chen
College of Business and Administration, Liaoning Technical University, 125105 Huludao, People's Republic of China
e-mail: tai-xiaohong@163.com

Y. Zhou and D.D. Wu (eds.), *Modeling Risk Management for Resources and Environment in China*, Computational Risk Management,
DOI 10.1007/978-3-642-18387-4_27, © Springer-Verlag Berlin Heidelberg 2011

the field of theory framework of general equilibrium, from only the field of applicative, descriptive traditional finance. The theorem explored the discipline of enterprise's capital structure so as to identify the intrinsic link between firm value and capital structure. In 1963, Modigliani and Miller published another article named 'Corporate income taxes and the cost of capital: a correction' which cancelled the assumption of T (tax) = 0, reached the conclusion that firm value will increase as the degree of financial leverage increases, i.e. the corrected MM theory. The theories later between capital structure and firm value developed, such as Agency Cost, Signaling and Pecking Order, etc.

Since the different national conditions, it has a great significance to analyze the relationship of capital structure and firm value on China's listed firms in different industries. This paper selects the real estate industry and the retail trade industry based on China's actual situation and carries out an empirical analysis on the capital structure and the firm value by establishing comparative regression model.

## 2 Our Approach

First, correlation analyzes the filtered parameters of firm value and capital structure, and then establishes regression models for further analysis. The regression models will include both the linear and quadratic models, because the values are not necessarily well fitted for the linear models, which also difficult to provide a theory threshold for capital structure. Second, obtain the regression coefficients for the models through parameters estimation and test significance of fitting goodness of regression model. At last reach the final conclusion through analysis.

## 3 Empirical Analysis

### *3.1 Indicator Selection*

Since our purpose is to analyze how the capital structure influences firm value, making indicators reflecting capital structure of listed firms independent variables, while indicators reflecting firm value the dependent variables. Capital structure indicators include the Liability/Asset-Ratio (L/A-R, debt ratio), Long-term Debt to Equity, Short-term Debt to Equity, the Proportion of Top Ten Shareholders, etc. Here we select debt ratio as the independent variable, which reflects how much one company's total assets were obtained through borrowing. Debt ratio reflects the company's comprehensive ability to repay debt, lower this ratio, more capable the company repays the debts; higher this ratio, less capable the company repays.

Usually indicators reflecting the firm value of listed companies include Earnings per Share (EPS), Return on Equity (ROE), ROA, Tobin's Q, Price-Earnings Ratio and so on. Here we select ROE as the dependent variable, because ROE reflects not only the performance, but also the capital structure of listed companies. ROE depends on the Total Return on Assets and Equity Multiplier, and increasing the equity multiplier is an important way of improving ROE and altering capital structure.

## 3.2 Sample Selection

In order to seize the industry-specific factors, the paper selects 2009 annual report of listed companies in Chinese real estate and retail trade as the sample data. Of course the ST companies should be removed for they cannot properly reflect the relationship between capital structure and firm value. The total sample consists of 40 listed companies through screening, number of which real estate and retail trade companies are 20 each. The type of data is cross-sectional data of December 31, 2009; data source is Tonghuashun stocks analysis software.

## 3.3 Correlation Analysis

According to analysis above, we set L/A-R (debt ratio) as independent variable, ROE dependent variable; cast Pearson correlation analysis, the results is as shown in Tables 1 and 2:

From Tables 1 and 2 above we know that under 95% and 99% confidence interval, the significance of t-test of the independent variable and the dependent variable during correlation analysis is 0.109 and 0.002 respectively, which indicates that real estate industry accepts null hypothesis: the firm value and capital structure

**Table 1** ROE and debt ratio of listed companies in real estate

| | | ROE | L/A-R |
|---|---|---|---|
| ROE | Pearson Correlation | 1 | .369 |
| | Sig. (2-tailed) | | .109 |
| | N | 20 | 20 |
| L/A-R (%) | Pearson Correlation | .369 | 1 |
| | Sig. (2-tailed) | .109 | |
| | N | 20 | 20 |

**Table 2** Correlation analysis result in retail trade

| | | ROE | L/A-R |
|---|---|---|---|
| ROE | Pearson Correlation | 1 | .641(**) |
| | Sig. (2-tailed) | | .002 |
| | N | 20 | 20 |
| L/A-R (%) | Pearson Correlation | .641(**) | 1 |
| | Sig. (2-tailed) | .002 | |
| | N | 20 | 20 |

** Correlation is significant at the 0.01 level (2-tailed)

has no significant correlation; the retail trade industry rejects the null hypothesis that there exist a significant correlation between the firm value and capital structure. This may be because the real estate is capital intensive industry, which has a huge amount of investment, complex operation and long payback period; the debt ratio tends to be large while comparing with other industries. So the cross-section debt ratio has no significant relationship with the firm value. The retail trade industry has shorter payback period and more even cash flow while comparing with real estate, so the debt ratio and the firm value is significantly correlated.

### 3.4 The Establishment of Regression Models

The models of real estate and retail trade industry will establish by following forms:

$$V = b_0 + b_1x + \varepsilon_1 \tag{1}$$

$$V = b_0 + b_1x + b_2x^2 + \varepsilon_2 \tag{2}$$

$b_i$ represent the constants and independent variable's coefficients; $\varepsilon_j$ represent random errors. Model i is the linear regression model used to determine whether the capital structure and firm value is positively or negatively correlated; Model j is the quadratic regression model that can be used to test whether there is a best debt ratio maximizing firm value.

### 3.5 Regression Analysis

As is shown in Table 3 above and Fig. 1 below, the regression equations should be:

$$V = -0.215 + 0.258x \tag{3}$$

$$V = 42.425 - 1.895x + 0.022x^2 \tag{4}$$

In Model 1, the determination coefficient of linear regression equation $R^2 = 0.136$, indicating that ROE and L/A-R has a vague linear relationship, but still we can see they are positively correlated.

In Model 2, the determination coefficient $R^2 = 0.562$, indicating that ROE and L/A-R has a clearer curve relationship than that of linear model, and the significance is higher than that of the linear regression equation too.

We can see from Fig. 1 that debt ratio of most listed real estate companies are high, between 60% and 70%. Curve reached its lowest point where the debt ratio is about 42%, i.e. the lowest point of the firm value. Beyond that point, ROE increases

**Table 3** Model summary and parameter estimates of the real estate industry

| Equation | Model summary | | | | | Parameter estimates | | |
|---|---|---|---|---|---|---|---|---|
| | R Square | F | df1 | df2 | Sig. | Constant | b1 | b2 |
| Linear | 0.136 | 2.844 | 1 | 18 | 0.109 | −0.215 | 0.258 | |
| Quadratic | 0.562 | 10.921 | 2 | 17 | 0.001 | 42.425 | −1.895 | 0.022 |

Dependent variable: ROE
The independent variable is LAR

**Fig. 1** ROE and debt ratio diagram of real estate

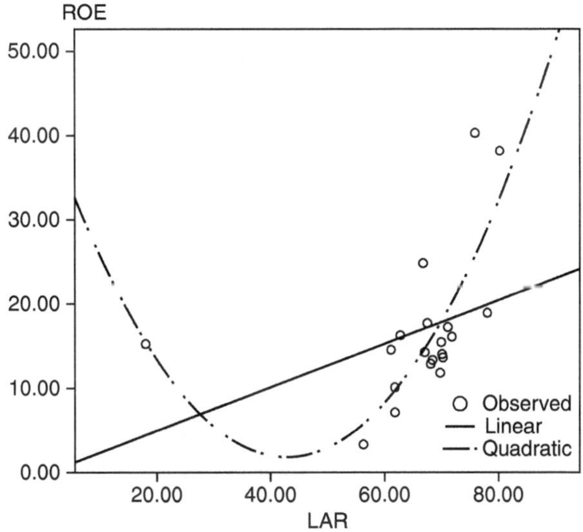


as the debt ratio increases, indicating that companies make good use of a high degree of leverage to bring benefits; but too high debt ratio could bring high financial risk.

As is shown in Table 4 and Fig. 2 below, the regression equations should be:

$$V = -3.686 + 0.37x \tag{5}$$

$$V = 6.584 - 0.042x + 0.004x^2 \tag{6}$$

In Model 3, the determination coefficient of linear regression equation $R^2 = 0.411$, indicating that ROE and L/A-R has a positive linear relationship, i.e. they are positively correlated.

In Model 4, the determination coefficient $R^2 = 0.426$, and the significance is higher than that of the linear regression equation. There is a curve relationship between ROE and total debt ratio, we can see from Fig. 2 that ROE increases as the debt ratio grows, and when the debt ratio reaches 60%, the ROE grows faster, indicating that 60% is the inflection point of listed retail trade companies.

**Table 4** Model summary and parameter estimates for the retail trade industry

| Equation | Model summary | | | | | Parameter estimates | | |
|---|---|---|---|---|---|---|---|---|
| | R Square | F | df1 | df2 | Sig. | Constant | b1 | b2 |
| Linear | 0.411 | 12.547 | 1 | 18 | 0.002 | −3.686 | 0.37 | |
| Quadratic | 0.426 | 6.307 | 2 | 17 | 0.009 | 6.584 | −0.042 | 0.004 |

Dependent variable: ROE
The independent variable is LAR

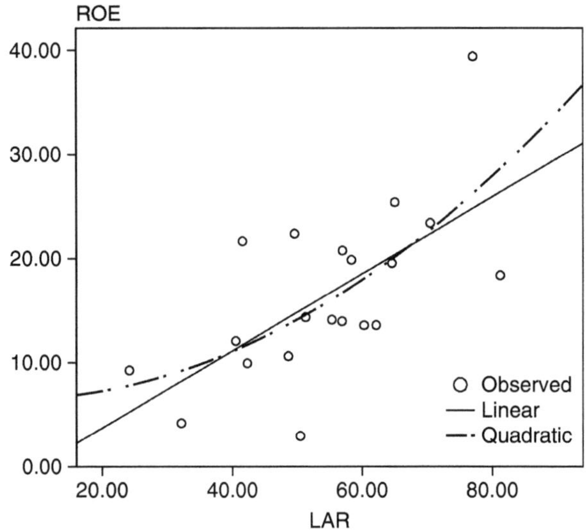


**Fig. 2** ROE and debt ratio diagram of retail trade industry

## 3.6 *Result Explanation*

Through comparison from regression results above, we can see that there exists an obvious industrial characteristic in relationship between capital structure and firm value in listed firms. In the retail trade industry, most of the debt ratios are between 40% and 60%, while most of which exceed 60% in real estate. In linear regression, the capital structure and firm value are positively related no matter in which industry; in quadratic regression, the firm value of retail trade industry increases as debt ratio increases gradually, and after it reaches 60%, companies become scale efficiency and firm value increases faster; while real estate industry shows a trend of decreasing first and then increasing in quadratic regression. Thus, different capital structures exist in different industries, whose firm values' characteristics vary greatly too. In addition, industry characteristics, industry barriers and degree of industry concentration of different industries vary. Retail trade industry has high concentration, intense competition and high technical risk, so companies should raise funds moderately, not borrow too much; the real estate is capital-intensive and mass-investment industry, and has long investment-repayment period, and has significant characteristics of having to depend on financial institutions such as banks. Therefore, different industries' capital structures vary greatly; whose firm value and financing methods are not the same either. Different firms should select the appropriate financing methods.

# 4 Conclusion

Through all the analysis above it can be concluded as follows:

1. Capital structure' impact on firm value does exist in listed companies, however, capital structure has industrial characteristics, different industries mean different conditions. No matter the retail trade or real estate industry, firms tend to have huge investments and long investment payback period, so these companies have higher total debt ratio than other industries; retail trade companies demand strong liquidity because of intense competition, so debt ratio is lower than that in real estate industry.
2. Whether in the retail trade industry or real estate, there exists linear or quadratic relationship between capital structure and firm value, but in real estate, the linear equation is not significant. But still the same is that both industries' firm value increases as scale of debt grows before achieving scale efficiency.
3. The real estate industry exist a lowest point because of poor capital structure. Beyond that point, ROE increases normally. But before reaching that point, companies do not make good use of financial leverage effectively. Yet this phenomenon does not exist in retail trade.

Listed companies' capital structure should reflect the ultimate goal of corporate finance: maximizing firm value. Capital structure is the basis of corporate governance, and sound and effective corporate governance is a strong guarantee of improving firm value. Therefore, Chinese listed companies should be combined with changes taking place in the business environment and own characteristics of operation and management and industry, optimize capital structure and arrange reasonable debt level; in the meantime, deal well with the relationship between capital market and product market management.

# The Influence of Securities Transaction Stamp Tax Adjustment on Shanghai Stock Market-Based on the adjustment on September 19, 2008

**Junhong Chu and Lei Zhang**

**Abstract** Securities transaction stamp tax is the main tax item in Chinese stock market. In the past few years, Chinese stock market fluctuated anomaly. The government adjusted the securities transaction stamp tax several times to stabilize the stock market. Using Event-Study analysis and GARCH model, this paper analyses the influence of securities transaction stamp tax adjustment on Shanghai stock market after the adjustment on September 19, 2008. It concludes that this adjustment had strong impact on Shanghai stock market in the short-term; but for the long term, the effect was weakened step by step. Meanwhile, it tries to explain the causes of the result and gives some suggestions about the securities transaction stamp tax.

**Keywords** Event-study analysis · GARCH model · Securities transaction stamp tax · Shanghai stock market

## 1 Introduction

Chinese stock market fluctuated abnormally in 2007 and 2008, and several measures were taken in the stock market. One of the measures was to adjust the securities transaction stamp tax. Ministry of Finance announced on September 19, 2008 that the securities transaction stamp tax collection was changed from bilateral to unilateral imposition, and the tax rate was to maintain 1‰.

Many scholars abroad studied that how much a stock market can be influenced by securities transaction stamp tax. Tobin (1974, 1978) and Stiglitz (1989) thought that the securities transaction tax could inhibit stock market volatility. Jackson and

This paper is sponsored by the Incubating Program for Outstanding and Creative Talented Youths in Guangdong Colleges and Universities (No.: WYM08050)

J. Chu (✉) and L. Zhang
Zhuhai Campus, Jinan University, Guangdong, China
e-mail: jhchu@126.com

Y. Zhou and D.D. Wu (eds.), *Modeling Risk Management for Resources and Environment in China*, Computational Risk Management,
DOI 10.1007/978-3-642-18387-4_28, © Springer-Verlag Berlin Heidelberg 2011

Dolmen showed that when stock trading stamp tax fell from 2% to 1%, market volatility would increase the rate of 70%. Lindgren and Westlund (1990) and Ericsson and Lindgren showed that the increase of securities transaction stamp tax by 1% would lead to market volatility to decrease by 50% and 70% in the long term. Opponents believed that levying securities transaction stamp tax would not necessarily reduce the volatility of the stock market. Roll (1989) researched 23 industrialized countries, and found that the relationship between the securities transaction tax and the stock market volatility were not significant. Saporta and Kan (1997) found that the securities prices and market volatility were not affected by the adjustment of the securities transaction stamp tax.

There are also many studies focused on Chinese stock market. Wang conducted analysis on the impact of the securities transaction stamp tax rate reduction on November 16, 2001, and showed that the adjustment had significant effect on the market index in short term, but these effects weakened gradually. Tong (2005) tested the impacts of the previous adjustments of the securities transaction stamp tax rate, and the results showed that the adjustments did not necessarily lead to significant changes of the market volatility. Yao and Yang showed that the adjustment of the securities transaction stamp tax rate had a stronger impact on stock price volatility in short term, but a limited impact in the long run.

All of the above empirical studies discussed the adjustment of the tax rate, while this paper focused on the impact of the adjustment of tax collection methods on September 19, 2008, from bilateral to unilateral imposition.

## 2 Short-Term Influence on Shanghai Stock Market

### *2.1 Analysis of Event Study Method*

Selecting September 19, 2008, the date of adjustment for the event day ($t = 0$), the event window is set to the 30 trading days before and after.

The sample includes 859 shares in the Shanghai Stock Exchange (including A shares and B shares), which are all the stocks listed except the stocks that are traded abnormally or affected by other events.

The normal rate of return $E[R_{it}/I_t]$ is estimated by the Shanghai Composite Stock Index.

Abnormal rate of return of each stock on time $t$:

$$AR_{it} = R_{it} – E[R_{it}|I_t] \tag{1}$$

The average abnormal rate of return of all the stocks on time $t$:

$$AAR_t = \frac{1}{N}\sum_{i=1}^{n} AR_{it} \tag{2}$$

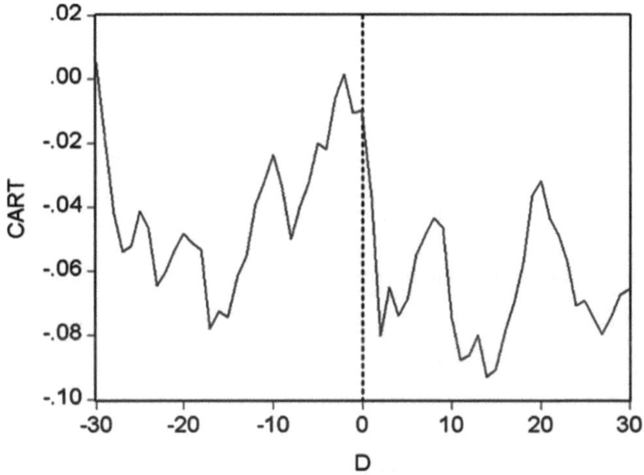


**Fig. 1** $CAR_t$ changes with time in the event window

The average cumulative abnormal rate of return of all stocks until time $t$:

$$CAR_t = \sum_{j=T1}^{t} AAR_j \tag{3}$$

## 2.2 *Positive Test Results*

Calculating the average cumulative abnormal rates of return of the sample in the research period. Figure 1 shows $CAR_t$ changes with time in the event window.

It can be seen from Fig. 1 that $CAR_t$ begins to rise about 17 days before the adjustment of securities trade stamp tax. It shows that there is an early response in the market. After the announcement of the adjustment, $CAR_t$ begins to rapidly decline, indicating that there is an over-reaction of the market.

# 3 Long-Term Influence on Shanghai Stock Market

## 3.1 *ARCH Test*

The sample is the daily return ($R_t$) of Shanghai Composite Index from September 19, 2008 to April 1, 2009.

First of all, examine the stationary. Using Eviews software, conduct unit root test. Table 1 shows the result.

**Table 1** The result of the unit root test on daily return of Shanghai Composite index

| | | t-Statistic | Prob.* |
|---|---|---|---|
| Augmented Dickey-Fuller test statistic | | −4.698287 | 0.0002 |
| Test critical values | 1% level | −3.484653 | |
| | 5% level | −2.885249 | |
| | 10% level | −2.579491 | |

*Mackinnon one-sided p-values

**Table 2** The result of Lagrange multiplier test on Shanghai Composite index daily return

| ARCH Test: | | | |
|---|---|---|---|
| F-ststistic | 2.774903 | Probability | 0.004421 |
| Obs*R-squared | 24.24792 | Probability | 0.006969 |

The *ADF* statistic is −4.698287, and less than the critical values under the significance level of 1%, 5%, 10%. So the sequence $R_t$ is stationary. Therefore, the mean equation of the $R_t$ is expressed by the formula:

$$R_t = \alpha + \beta * R_{t-1} + \varepsilon_t \tag{4}$$

Establish ARCH (q) model with the corresponding square of the residual sequence $e^2_t$ : $e^2_t = a_0 + a_1 e^2_{t-1} + a_2 e^2_{t-2} + \ldots\ldots + a_q e^2_{t-q} + u_t$, and obtain the coefficient of determination $R^2$. It can be proved that $N^*R^2$ obeys $\chi^2(q)$ distribution, where N is the number of observations. The lag order of $q$ is 10, and conduct the Lagrange multiplier test. Table 2 shows the result.

$N^*R^2$*(Obs*R-squared)* is 24.24792, and it is larger than the critical value (23.209) under the significance level of 1%. So the ARCH phenomenon exists, that is, the adjustment of securities transaction stamp tax increased the Shanghai stock market volatility.

### 3.2 Establish the GARCH Model

Establish the GARCH (1, 1) model: $var[\varepsilon_t/\Omega_{t-1}] = h_t = \alpha_0 + \alpha_1 \varepsilon^2_{t-1} + \beta_1 h_{t-1}$. The value of $(\alpha_1 + \beta_1)$ reflects the durability of the affect of external shocks on the volatility. The greater is the value, the slower is the decay. As the stock returns are affected by the level of risk, add $\sqrt{h_t}$ which represents the level of risk to the model. The results are shown in Table 3.

## 4 Analysis of the Reasons for Short and Long Term Impact

### 4.1 The Reason for Short Term Impact

Earlier reaction of stock market shows that the information trickled out, which is a vulnerability in the securities system. Some people got insider information, and

**Table 3** The results of GARCH (1, 1) model on Shanghai Composite index daily return

| | Coefficient | Std.Error | z-Statistic | Prob. |
|---|---|---|---|---|
| @SQRT(GARCH) | 38.41467 | 20.59411 | 1.865323 | 0.0621 |
| C | −0.928582 | 0.511007 | −1.817163 | 0.0692 |
| RT(−1) | −0.069680 | 0.086564 | −0.804953 | 0.4208 |
| | Variance Equation | | | |
| C | 0.000201 | 0.000134 | 1.49624 | 0.1346 |
| RESID(−1)^2 | −0.003540 | 0.002380 | −1.487429 | 0.1369 |
| GARCH(−1) | 0.661146 | 0.239746 | 2.757691 | 0.0058 |
| R^2 | 0.046900 | Mean dependent var | | 0.001513 |
| Adjusted R^2 | 0.007187 | S.D. dependent var | | 0.025749 |
| S.E. of regression | 0.025701 | Akaike info criterion | | −4.436643 |
| Sum squared resid | 0.079265 | Schwarz criterion | | −4.301582 |
| Log likelihood | 285.5085 | F-statistic | | 1.180982 |
| Durbin-Watson stat | 2.035523 | Prob(F-statistic) | | 0.322542 |

bought stocks in advance, then sold them at high prices to get profits after the announcement.

The over-reaction of the stock market might be attributed to the large quantity of small and medium investors in China stock market. Small investors respond to the transaction cost more sensitively than the institutional investors. Moreover, small investors are lack of rationality and often enlarge market signals. Although "policy market" seems to be more and more far away from us in recent years, the investors are still very sensitive to the policy factors.

## 4.2 *The Reason for Long Term Impact*

In the long term, market volatility weakened gradually and finally went to a relatively stable state.

The adjustment of the securities transaction stamp tax cannot reverse the long-term trend of the stock market, because the stock trend depends more on macro economy, listed companies fundamentals, market supply and demand, and other factors. The bear market for over three years before 2006 was not changed by the reduction of securities transaction stamp tax rate, but the real reversal in the stock market was due to the share reform and the revaluation of China's stock value. Market growth substantially in 2007 is because of constantly enlarging the market volume, expanding market funds, the high-growth of Chinese economy, the low inflation and the future expectations. The stock market declined in 2008: Firstly, the U.S. subprime mortgage crisis worsened, causing global stock markets fell sharply; Secondly, because of the overheating trend of China's economy, the government adopted a tight monetary policy and strengthened macro-control.

The adjustment is not affected the investors seriously who work on long-term operations. The real investors will not change their decisions which rely on the intrinsic value and the future returns of stocks.

## 5 Policy Recommendations

The above analyses show that the function of securities transaction stamp tax is limited. The government should make efforts from the following aspects. First, maintain the way of levying securities transaction stamp tax unilaterally to encourage investment, and reduce stock market volatility. Second, maintain low securities transaction stamp tax rate to reduce transaction cost for investors. Third, improve the law and the tax system. Fourth, get rid of the capital market over-reliance on the policy.

## References

Lindgren R, Westlund A (1990) How did transaction costs on the Stockholm stock exchange influence trade and price volatility? Sknadianviska Enskilda Banken Q Rev 2:30–35

Roll R (1989) Price volatility, international market links, their implications for regulatory policies. J Financ Serv Res 3(2–3):211–246

Saporta V, Kan K (1997) The effects of stamp duty on level and volatility of UK equity price. Working paper. Bank of England

Stiglitz JE (1989) Using tax policy to curb speculative short – term trading. J Financ Serv Res 3 (2–3):101–115

Tobin J (1974) The new economics one decade older, The Janeway lectures on historical economics. Princeton University Press, Princeton, pp 95–104

Tobin J (1978) A proposal for international monetary reform. East Econ J 4(7):153–159

Tong F (2005) Securities transaction tax and market volatility: evidence from Chinese stock market. Stat Decis 10:57–59

# Random Subspace Method for Improving Performance of Credit Cardholder Classification

**Meihong Zhu and Aihua Li**

**Abstract** The main task of credit card risk management for commercial banks and credit card retailers is analyzing credit cardholders' behavior and predicting bankruptcy. In this paper, we investigate dimension reduction techniques for improving performance of the high-dimensional credit cardholder data classification. We choose MCLP as base classifier. We theoretically analyze the characteristics of PCA, filter, wrapper, and RSM. Then, our experimental research focuses on RSM. Due to the specialities of dimension reduction and ensemble of RSM, we experimentally compare its performance with that of PCA, single classifier and Bagging on the same training set and test set. The results show that RSM can highly improve classification performance of the credit cardholder data set. From the idea of ensemble, RSM demonstrates its superiority over Bagging. From the angle of dimension reduction, RSM shows its predominance over single classifier, and the same advantage as PCA. Finally, we explain our results from the aspects of MCLP algorithm, RSM and data set.

**Keywords** Bagging · Credit card risk · Multiple criteria linear programming (MCLP) · PCA · Random subspace method (RSM)

## 1 Introduction

The main task of credit card risk management for commercial banks and credit card retailers is analyzing credit cardholders' behavior and predicting bankruptcy. Generally, according to historical consumption records of cardholders, they are divided

M. Zhu (✉)
School of Statistics, Capital University of Economics and Business, Beijing 100070, China
e-mail: zhumh070809@hotmail.com

A. Li
School of Management Science and Engineering, Central University of Finance and Economics, Beijing 100081, China
e-mail: neu_aihua@yahoo.com.cn

Y. Zhou and D.D. Wu (eds.), *Modeling Risk Management for Resources and Environment in China*, Computational Risk Management,
DOI 10.1007/978-3-642-18387-4_29, 

into two predefined classes: bankrupted customers and current ones. With these data, many credit card classification models have been proposed to construct the classification and prediction mechanism for new credit card applicants. However, due to high dimensionality and complicated structure of credit cardholder data set, few of the current classification models can present absolutely satisfying prediction performance on it. Thus, in improving classification and prediction performance of high-dimensional credit cardholder data set, dimension reduction method is worthy of considering.

Generally, dimension reduction techniques can be divided into two categories. One is feature transformation which transforms the original feature space into a low-dimensional feature space, such as principal component analysis (PCA) technique. Another is feature selection which consciously selects some features in the original feature space, such as filter and wrapper methods. Then analysis is executed in the new low-dimensional feature space.

PCA and filter are both data pre-process techniques and can be suitable for any classification algorithms. However, the new feature set is not necessarily optimal for all classification algorithms because a specific classification algorithm prefers to a specific feature subset. In addition, PCA has another two obvious drawbacks. One is that the meaning of each principal component is difficult to understand. Another shortcoming of PCA is that, in the process of feature transformation, some significant discriminative information in the high dimensional space will be lost. Comparatively, in wrapper, the classification algorithm and wrapper process are alternated and the selected feature subset is optimal for the classification algorithm employed in the alternation. In practice, we often use a specific classification algorithm to solve a problem, so wrapper should be suitable. But, wrapper is very time-consuming.

In classifying of high-dimensional credit cardholder data set, Li (Li 2007) first used PCA to reduce the data dimension and then performed Multiple Criteria Linear Programming (MCLP) (Shi et al. 2001; Shi et al. 2002) classification algorithm in the PCA feature space. Result of PCA+MCLP was better than that of single MCLP. However, due to the disadvantages of PCA, it is necessary to investigate other techniques for improving classification performance of the credit cardholder data set.

Actually, random subspace method (RSM) (Ho 1998) can also be used as an effective and efficient feature selection and dimension reduction technique. In addition, it also has the advantages of ensemble classifier. The characteristics and operation of RSM are described in the subsequent part of the paper.

This paper investigates the effect of RSM on high-dimensional credit cardholder classification. At the same time, MCLP is used as the base classifier. For the last 10 years, as a new burgeoning linear programming based linear classification model, MCLP has been successfully applied in credit cardholder classification (Shi et al. 2001; Shi et al. 2002; Peng et al. 2004; Kou et al. 2005), and there are many comparable results for credit cardholder classification. Compared with other models, MCLP doesn't need any assumption, and it is comprehensible and non-parametric. The detailed description of MCLP classification algorithm can be seen in (Shi et al. 2001; Shi et al. 2002; Peng et al. 2004; Kou et al. 2005).

The paper is organized as below: Section 2 gives the process of RSM ensemble. Section 3 introduces some preparations for subsequent analysis. The result of RSM based MCLP on credit cardholder data set is presented in Sect. 4. Finally, Sect. 5 contains conclusions and comments on future research directions for the credit cardholder classification.

## 2 The Random Subspace Method

The idea of RSM came from two aspects: the theory of stochastic discrimination (SD) (Kleinberg 1990; Kleinberg 1996) and the method of cross-validation. In the SD theory, classifiers are constructed by combing many individual sub-classifiers which have weak discriminative power but can generalize very well. In the cross-validation method, random subsamples are selected from the raw training set and individual classifiers are trained. According to some combination rule, these individual classifiers are combined into an ensemble classifier. This combination method can be used on all classification algorithms. Two typical combination methods are Bagging (Breiman 1996) and Boosting.

Different from Bagging and Boosting, in RSM, each subsample and sub-classifier comes from feature space, rather than from instance space. According to RSM, each feature subset is randomly extracted from the original feature space, and each individual classifier is created based on those features in the feature subset using all training cases. Then, all individual classifiers are made up to construct an ensemble classifier by simple majority vote.

Given a training set $TR_{n \times p}$, which has $p$ predictive variables and $n$ training cases, let $A_i = (A_{i1}, \ldots, A_{ip})$ be the data for the $p$ variables in $ith$ case, where $i = 1, \ldots, n$. In RSM, we want to independently and randomly pick $T$ feature subspaces with the dimension of each subspace is $k$ ($<p$) and sample size is still $n$. Then, we construct $T$ individual classifiers based on these $T$ subspaces. Finally, we combine these $T$ individual classifiers into an ensemble classifier. The organization of RSM is as below Fig. 1:

The key to the success of RSM is the high accuracy of each individual classifiers and the diversity among these individual classifiers. Based on a specific random subspace, an optimal individual classifier for this subspace is constructed, but it maybe can't show good classification performance on other subspaces. Fortunately, due to the diversity among subspaces, the drawback of each individual classifier is complemented and the combined ensemble is superior to a single classifier constructed from the original training set in the complete feature space.

The subspace dimension is much smaller than that of the original feature space, while the number of training instances remains the same. So, the relative training sample size increases. Thus it is very suitable for the situation of small training sample and high dimension, especially useful in high dimension data set with many redundant features. It is a parallel process, and adaptable for any classification

Input: original training set $TR_{(n\times p)}$, integer $T$, integer $k(<p)$

1. Repeat for $t = 1,2,\ldots,T$:

(a) Randomly sample an $k$-dimension subspace $A^t$ from $TR_{(n\times p)}$ and form a new training set $A^t = TR^t = TR^t_{(n\times k)}$, with each case $A_i^{\ t} = (A_{i1}^{\ t}, A_{i2}^{\ t}, \ldots, A_{ik}^{\ t},)$, $i = 1,2,\ldots,n$.

(b) Construct an individual classifier $C^t$ based on $TR^t$.

2. Combine classifiers $C^t, t = 1,2,\ldots,T$, by simple majority vote into a final ensemble classifier

$$C^*(A) = \underset{y\in\{-1,1\}}{argmax} \sum_{t=1}^{T} \delta_{\text{sgn}}(C^t(A)), y$$

Where $\delta_{i,j}$ is the Kronecker symbol, and $y\in\{-1,1\}$ is the original class label of case $A$.

Output: C*

**Fig. 1** The process of constructing RSM ensemble classifier

algorithm. Compared with PCA, filter and wrapper, it is convenient to operate, timesaving, and very effective.

# 3 Preparation for Analysis

## *3.1 Characteristics of the Credit Cardholder Data Set*

The dataset we use come from a major US bank (Peng et al. 2004). It includes 5000 credit cardholder records with each described by 38 original attributes. In order to present more clear data information and to facilitate data analysis, 66 new attributes are derived from them. The class labels of these cardholders are "bankrupted" or "current". Here we call them as "Bad" or "Good" respectively. There are 815 "Bad" records and 4185 "Good" records in this data set.

## *3.2 Experimental Design*

When executing RSM, firstly, according to a pre-defined proportion of training set vs. test set, we randomly partition the whole credit cardholder D into training set TR and test set TE. In order to decrease the effect of random partition, we independently execute the above process four times according to the same TR-TE proportion. The average result of the four times is used for comparison. In addition, we split the data set according to three different training set vs. test set proportions, so that we can find the effect of RSM at different level of training set size. In order to lessen the impact of the imbalance of the two classes in credit cardholder data set, in each training set, the case numbers of two classes are designed as equal. All performance comparisons are executed on the corresponding test set. According to

```
For i=1 to 3(3 different training set-test set proportions)
For j=1 to 4(4 times partition according to the same TR-TE proportion)
(a)Randomly split D into TRij and TEij.
(b)From TRij, 9 independent random subspaces with 33 predictive
attributes are drawn, and 9 individual classifiers are constructed. Then,
according to the simple majority voting rule, an ensemble C*ij is made
up based on TRij.
(c) Respectively compute "bad" accuracy rate, "good" accuracy rate and
overall accuracy of C*ij on TEij, denoted as a*ij.
j = j+1
End of j.
Compute the average accuracy of 4 ensembles on TEij, i.e., ā*i = Σ(j=1..4) a*ij
i = i+1
End of i.
```

**Fig. 2** The RSM process at different training set-test set partition

**Table 1** The partition of the credit cardholder data set in three runs

| i | Size of training set TR | | | Size of test set TE | | |
|---|---|---|---|---|---|---|
| | Bad | Good | All | Bad | Good | All |
| 1 | 200 | 200 | 400 | 615 | 3985 | 4600 |
| 2 | 400 | 400 | 800 | 415 | 3785 | 4200 |
| 3 | 600 | 600 | 1200 | 215 | 3585 | 3800 |

the method of (Ho 1998), the dimension of each random subspace is defined as half of the original feature space: 33. For simplicity, each ensemble includes nine individual classifiers.The step of our method is as follows:

In order to show the advantage of RSM, the single classifier $C_{ij}$ is directly constructed on the original training set $TR_{ij}$, and the performance of $C_{ij}$ is evaluated on $TE_{ij}$. In addition, here we also construct Bagging ensemble with the similar step as RSM ensemble described in Fig. 2. The partition in the three runs is following Table 1:

## 4 Result and Discussion

The experimental results with different methods on different training set sizes and different partitions are described in Table 2. In Table 2, "Ave" means the "average" result of four random samplings.

We can see that, at each i-j pair, the accuracy rate of RSM is markedly higher than that of single classifier, especially when training set is small. However, Bagging ensemble doesn't show consistent advantage over single classifier, especially when the training set is large enough. Specifically, when training set is small

**Table 2** The accuracy rate of different methods with different training sets

| i | j | Accuracy rate of RSM | | | Accuracy rate of bagging | | | Accuracy rate of single classifier | | |
|---|---|---|---|---|---|---|---|---|---|---|
| | | Bad | Good | Overall | Bad | Good | Overall | Bad | Good | Overall |
| 1 | 1 | 76.3 | 73.7 | 74.1 | 72.5 | 72.6 | 72.6 | 69.8 | 70.5 | 70.4 |
| | 2 | 74.3 | 78.6 | 78.0 | 68.1 | 74.9 | 74.0 | 65.4 | 73.8 | 72.6 |
| | 3 | 78.5 | 75.5 | 75.9 | 75.1 | 70.2 | 70.9 | 71.2 | 71.6 | 71.5 |
| | 4 | 73.8 | 73.3 | 73.4 | 64.7 | 70.3 | 69.5 | 61.0 | 74.1 | 72.3 |
| | Ave | 75.7 | 75.3 | 75.4 | 70.1 | 72.0 | 71.7 | 66.8 | 72.5 | 71.7 |
| 2 | 1 | 77.8 | 74.9 | 75.2 | 73.3 | 70.8 | 71.0 | 71.6 | 74.7 | 74.4 |
| | 2 | 76.3 | 75.2 | 75.3 | 72.4 | 71.1 | 71.2 | 70.8 | 72.9 | 72.7 |
| | 3 | 78.3 | 74.2 | 74.6 | 74.5 | 70.3 | 70.7 | 72 | 73.5 | 73.4 |
| | 4 | 75.4 | 76.2 | 76.1 | 74.8 | 73.6 | 73.7 | 72.4 | 74.2 | 74.0 |
| | Ave | 77.0 | 75.1 | 75.3 | 73.8 | 71.5 | 71.7 | 71.7 | 73.8 | 73.6 |
| 3 | 1 | 80.9 | 76.7 | 77.0 | 74.4 | 76.0 | 76.0 | 76.3 | 74.5 | 74.6 |
| | 2 | 80.5 | 74.9 | 75.2 | 74.4 | 74.6 | 74.6 | 74.4 | 75.3 | 75.2 |
| | 3 | 75.3 | 75.2 | 75.2 | 73.5 | 72.9 | 72.9 | 70.2 | 72.9 | 72.8 |
| | 4 | 80.2 | 76.9 | 77.1 | 73.0 | 73.5 | 73.5 | 74.4 | 74.4 | 74.4 |
| | Ave | 79.2 | 75.9 | 76.1 | 73.8 | 74.2 | 74.2 | 73.8 | 74.3 | 74.2 |

compared with test set, with each single classifier, accuracy rate of the "Bad" class is lower than that of the "Good" class, i.e., each single classifiers has a prejudice against the "Bad" class, which is the result we don't expect to see. Bagging can improve the result of single classifier to some extent, While RSM can remarkably ameliorate the situation of single classifier. When training set is large enough, due to the decrease of the "Bad" class in the test set, RSM can be more obviously partial to the "Bad" class.

Li (Li 2007) used PCA method with MCLP on this data set, the research result shows that when training set size is 1,400 (with 700 "Bad" records and 700 "Good" records), the test set accuracy of "Good" is 76%, the "Bad" is 78%, and the overall is 76.1%. This result is close to that of this paper.

In all, RSM can highly enhance the performance of MCLP on the credit cardholder set. In this research, RSM also markedly shows its consistent advantage over Bagging, while other researches don't always support this conclusion. Why does RSM work well on credit cardholder data set with MCLP classification method? That can be understood from different aspects.

Firstly, due to the characteristics of dimension reduction and ensemble, generally, especially under high-dimension conditions, RSM has better performance than single classifier. While Bagging, when the attributes appear greater, doesn't have the strongpoint of RSM. From the classification algorithm aspect, as a linear programming based classification algorithm, MCLP has its own characteristics. From the former research on MCLP (Zhu et al. 2009), comparatively, MCLP is a high-bias and low-variance classification algorithm. Especially when the sample size is small, it has low classification accuracy and bad stability. So, when training set is small, RSM behaves much better than single MCLP model. From the data set angle, the credit cardholder data set has high dimension. Maybe there are many redundant and irrelevant attributes which often deteriorate the performance of some classification methods. In addition, in different data locality, the behavior of the data often varies considerably, and different subspaces may show better discrimination in different localities. Thus, the combination of these individual classifiers can show better performance than single classifier.

## 5 Conclusion and Future Work

In this paper, we investigate the dimension reduction techniques for improving the performance of the high-dimensional credit cardholder data classification. Due to the advantage of MCLP, we choose it as the base classifier. We theoretically analysis the characteristics of PCA, filter, wrapper, and RSM. Then, our experimental research focuses on RSM. Due to the specialities of dimension reduction and ensemble of RSM, we experimentally compare its performance with that of single classifier and Bagging on the same training set and the same test set. The results show that RSM can distinctly improve the classification performance of the credit cardholder data set. From the idea of ensemble, it demonstrates the superior ity of RSM over Bagging. From the angle of dimension reduction, it shows the predominance of RSM over single classifier, and the same advantage as PCA. Finally, we explain the results again from the aspects of MCLP algorithm, RSM and the data set.

However, there is still much work to be studied further. In this paper, we only quantitatively investigate the performance of PCA and RSM with MCLP on the credit cardholder data set, while other attribute selection and reduction methods are only analyzed theoretically. We need actualize a further and systematic analysis and comparison of various dimension reduction methods suitable for credit cardholder classification problem. In addition, we should employ different base classifiers to evaluate the effect of dimension reduction methods for credit cardholder data.

**Acknowledgements** This work is partially supported by the project "The Building of Characteristic Specialty for Statistics" from the Beijing Government.

## References

Breiman L (1996a) Bagging predictors. Mach Learning 24:123–140

Breiman L (1996b) Bagging predictors. Mach Learning 24:123–140

Ho TK (1998) The random subspace method for constructing decision forests. IEEE Trans Pattern Anal Mach Intell 20(8):832–844

Kleinberg EM (1990) Stochastic discrimination. Ann Math Artif Intelligence 1:207–239

Kleinberg EM (1996) An overtraining-resistant stochastic modeling method for pattern recognition. Ann Stat 4(6):2319–2349

Kou G, Peng Y, Shi Y, Wise M, Xu W (2005) Discovering credit cardholders behavior by multiple criteria linear programming. Ann Oper Res 135(1):261–274

Li A (2007) Research on consumption debit problems based on data technology frame. PhD Dissertation. School of Management, Graduate University of Chinese Academy of Sciences

Peng Y, Kou G, Chen ZX, Shi Y (2004) Cross validation and ensemble analyses on multiple-criteria linear programming classification for credit cardholder behavior. In: Bubak M et al (eds) ICCS 2004, vol 3039, LNCS., pp 931–939

Shi Y, Wise M, Luo M, Lin Y (2001) Data mining in credit card portfolio management: A multiple criteria decision making approach. In: Koksalan M, Zionts S (eds) Multiple criteria decision making in the new millennium. Springer, Berlin, pp 427–436

Shi Y, Peng Y, Xu W, Tang X (2002) Data mining via multiple criteria linear programming: applications in credit card portfolio management. Int J Inf Technol Decis Making 1:131–151

Zhu MH, Shi Y, Li A, Zhang D (2009) Comparison of bias-variance structure of three classification algorithms: MCLP, LDA and C5.0. J Grad Sch Chin Acad Sci 26(4):443–450

# Credit Risk Model and Bayesian Improvement for Companies in China

**Lan Luo, Jian Xiong, and Qing Zhou**

**Abstract** The paper focuses on credit risk measuring methods and tries to find suitable model for China's listed companies' credit risk measurement. According to the default condition in Chinese listed companies, we apply the factor analysis to the correlative data, and give the default discrimination model by Logistic regression. As the precision of the credit risk default model affect the bank's risk status and profit directly, the paper uses the Bayesian estimate to improve the predictive power of credit risk default models. Comparing the precision of two models by AUC value and Brier Score, the result shows that the value of AUC of Standard estimator is 0.834 while the same value of Bayesian estimator is 0.870. It shows that the Bayesian estimate has a higher predictive power of precision and stability.

**Keywords** Credit Risk · Empirical Bayesian Estimate · Factor Analysis · Logistic Regression

## 1 Introduction

With the rapid growth of global economy, problems related to credit risk have been paid much more attention than before. Credit risk has become one of the important risks which financial institutions have to face. Thus, how to control and get an accurate estimation of credit risk plays a key role in the decision making for the

L. Luo (✉)
College of Mathematics and Information Science, Guangzhou University
e-mail: luolan.0809@163.com

J. Xiong
College of Mathematics and Information Science, Guangzhou University
and
The Institute of Mathematical Sciences, the Chinese University of Hong Kong

Q. Zhou
Teradata Information Systems Limited, Guangzhou Brand

Y. Zhou and D.D. Wu (eds.), *Modeling Risk Management for Resources and Environment in China*, Computational Risk Management,
DOI 10.1007/978-3-642-18387-4_30, 

financial institution, investors and government supervisor. Listed companies are an important part of our national economy, and the basis of the securities market. The normative acts of the company and their financial situation will affect the development of China's securities market and the interests of investors directly. Based on this background, this paper will discuss the credit risk of China's listed companies.

The paper focuses on credit risk measuring methods and tries to find suitable model for China's listed companies' credit risk measurement. According to the default condition in Chinese listed companies, We apply the factor analysis to the correlative data of china's listed companies, and give the default discrimination model by Logistic regression.

As the precision of the credit risk default discrimination model affect the bank's risk status and profit levels directly, the paper uses the Bayesian Estimator to improve the predictive power of credit risk default discrimination models. Comparing the precision of two models by AUC value and Brier Score, the result shows that the value of AUC of Standard estimator is 0.834 while the same value of Bayesian Estimators is 0.870, and so 3.6 percentage points higher than the standard estimator. It shows that the Bayesian Estimator has a higher predictive power of precision and stability.

# 2 Logistic Regression and Default Prediction Model

## *2.1 Data Description*

The financial data and industry data used in the paper are mainly from: RESSET financial research database.

Some normal companies and default companies are chosen for analysis in our paper. When Altman developed the ZETA model (Altman et al. 1977), he chose a group of bankrupt companies as default companies. According to China's specific situation, we define the companies those were special treated (ST) cause of abnormal financial situation as a firm in the credit crisis. The reasons are:

1. Credit risk is one kind of default risk in simple words. The companies' credit risk is usually manifested in the form of financial distress. The most serious financial distress is bankruptcy. In fact, the company get in the credit crisis often develop gradually from the normal financial to the financial crisis. Therefore, the company's financial difficulties not only have omen, but also can be predicted. The most direct way to predict credit risk is predict the probability of default. For companies, although the emergence of individual defaults is not means to a significant credit risk, but when a company is in financial difficulties, there must be a large number of default cases, such as unable to pay the principal or interest of debt, and then lead to default. So we prefer the probability of the company running into financial difficulties than the company's default probability.

**Table 1** Financial indicators

| Variables | Financial indicators | Computing method |
|---|---|---|
| $x_1$ | Current Ratio | Current Assets/Current Liabilities |
| $x_2$ | Quick Ratio | (Current assets-merchandise)/Current debt |
| $x_3$ | Effective Fund Ratio | Working Capital/Total assets |
| $x_4$ | Asset-liability Ratio | Total debt/ Total assets |
| $x_5$ | EBIT/total assets | EBIT/total assets |
| $x_6$ | ratio of sales to cost | Cost of sales/revenue of sales |
| $x_7$ | Assets net profit margin | Net income/total assets |
| $x_8$ | Total Profit Increasing Ratio | (Profit of the period-Profit of prior period)/Profit of prior period |
| $x_9$ | Growth rate of total assets | Total assets increasing during the year/Total assets at the beginning of the year |
| $x_{10}$ | The effective of financial leverage | Profit before tax EBIT/(Profit before tax EBIT – interest) |
| $x_{11}$ | Per-share earnings | End net profit/Total number of shares end of period |
| $x_{12}$ | Adjusted net assets per share | (Year-end equity – Net receivables more than three years – Prepaid expenses – Long-term prepaid expenses)/ Last year the total number of ordinary shares |

2. In the public capital market of China, "ST", as a government regulation of stock market, is a special treatment to some listed companies with unusual situation. In the March 16, 1998, the CSRC issued "the notification of special tread to the abnormal condition of shares of listed companies", and required the Shanghai Stock Exchange and Shenzhen Stock Exchange give the stock shares which are abnormal special treat (ST). In China's stock market, the "ST" stocks are mostly due to two consecutive years deficit or the net asset value per share is lower then the per value of shares. The ST listed companies are always means poor performance and have serious financial problems. It is just the signs of default companies.

Therefore, we take the special treated listed companies in unusual financial situation as the companies in credit crisis. Let the ST companies be treated as default sample and non-ST companies be treated as normal sample. We selected 171 listed companies in Shanghai Stock Exchange as the research sample, including 84 listed companies which ware special treated in 2006. It means the companies related to sample data were ST in 2004. The financial indicators must be different between the ST companies and non-ST companies.

Now we select 12 financial variables as following Table 1:

## 2.2 *Factor Analysis*

The core objectives in factor analysis are reducing variables and extracting the factors. Before factor analysis, it is necessary to examine whether there are a certain degree of linear correlation among the financial variables, and if it is suitable for factor analysis to extraction factors.

**Table 2** KMO and Bartlett's test

| Kaiser-meyer-olkin measure of sampling adequacy | | .694 |
|---|---|---|
| Bartlett's Test of Sphericity | Approx. Chi-Square | 4.617E3 |
| | df | 66 |
| | Sig. | .000 |

Using SPSS statistical software, relevant test results are in Table 2.

Table 2 is the results of the KMO (Kaiser-Meyer-Olkin) test and Bartlett test. KMO test measures the difference of the squared simple correlation coefficient and the square of the partial correlation coefficient of these variables. Among them, the correlation coefficient reflects the role of public factors; the partial correlation coefficients reflect the effect of special factors. In KMO test, if the KMO is more close to 1, means the data is more suitable for factor analysis. The test shows the KMO value for our data is 0.694, therefore, we believe that the data is suitable for factor analysis. Same result we get from the Bartlett ball test.

Then, we do the factor extraction and factor rotation, and get the factor scores:

$$\begin{aligned} FAC1 = {} & -0.058\mathrm{X}_1 - 0.069\mathrm{X}_2 - 0.153\mathrm{X}_3 - 0.165\mathrm{X}_4 - 0.232\mathrm{X}_5 - 0.129\mathrm{X}_6 \\ & - 0.201\mathrm{X}_7 - 0.079\mathrm{X}_8 - 0.305\mathrm{X}_9 - 0.139\mathrm{X}_{10} - 0.395\mathrm{X}_{11} - 0.218\mathrm{X}_{12} \end{aligned}$$

$$\begin{aligned} FAC2 = {} & -0.039\mathrm{X}_1 - 0.057\mathrm{X}_2 + 0.484\mathrm{X}_3 - 0.509\mathrm{X}_4 + 0.609\mathrm{X}_5 + 0.108\mathrm{X}_6 \\ & + 0.114\mathrm{X}_7 - 0.008\mathrm{X}_8 - 0.208\mathrm{X}_9 - 0.189\mathrm{X}_{10} - 0.189\mathrm{X}_{11} + 0.006\mathrm{X}_{12} \end{aligned}$$

Thus, we can do logistic regression and default prediction model.

In order to interpret the results of the regression conveniently, the two state variables should be changed into values 0 or 1. We set Default Company as value 0 and the normal company as value 1. It means that the probability of default company does not default is 0, and the probability of normal companies do not default is 100%. About the regression results, the more the company's default status value closes to 1, the less the company does default. So, if we get the regression result of no default probability, say F (x), then the probability of default formula is $p = 1 - F(x)$.

Let the status of that company is or not default be dependent variable and the five factors we choose according to their scores be the explanation variables. Input the relevant data of ST companies and non-ST companies of 2004 into SPSS software, using Logistic regression analysis then easy to get the default prediction model:

$$p = \frac{1}{1 + e^{-(0.437+5.193*FAC1+3.234*FAC2+0.904*FAC3-0.458*FAC4+0.510*FAC5)}} \tag{1}$$

Table 3 is the results of estimation value of Regression coefficient and significance test. From the view of the Wald statistics and the corresponding value of P, the five coefficients corresponding P values are less than 0.05, indicating the five factors are significant.

**Table 3** Variables in the equation

| | | B | S.E. | Wald | df | Sig. | Exp(B) |
|---|---|---|---|---|---|---|---|
| Step 1[a] | FAC1_1 | 5.193 | .836 | 38.603 | 1 | .000 | 179.922 |
| | FAC2_1 | 3.234 | 1.149 | 7.916 | 1 | .005 | 25.384 |
| | FAC3_1 | .904 | .318 | 8.104 | 1 | .004 | 2.469 |
| | FAC4_1 | −.458 | .202 | 5.114 | 1 | .024 | .633 |
| | FAC5_1 | .510 | .235 | 4.713 | 1 | .030 | 1.665 |
| | Constant | .437 | .287 | 2.316 | 1 | .128 | 1.547 |

## 2.3 *Empirical Bayes Estimate and Improved Bayesian Model*

Establish of credit risk model usually bases on some default data. As the scarcity of default events, only a small number of international banks or credit bureaus can collect enough data. Many small and medium sized banks have only a small sample of default data, so the credit risk model has unguaranteed prediction accuracy and stability. Therefore, we should pay attention to the impact of small sample of default data when establishing credit risk models. We introduce Bayesian methods, with it to establish an internal model basis on the small samples of default data, to improve the predictability of the model by integrating prior information.

Adkins and Hill proposed the prior information of coefficients which to be estimated on Probit model in 1996.We use this result on Logistic model.

Logistic model is similar to Probit model, and it bases on the following two points: Firstly, the estimation on Logistic model is much simple than on Probit; Secondly, Logistic model is not sensitive to changes in default rates of sample, and the change of the proportion between default samples and normal samples couldn't affect the consistency of the variable coefficient estimates. Thus, Logistic model is used widely to analyze the default prediction both in academic research and practical field.

Adkins and Hill (1996) made equal weight on prior coefficient information, and gave the Empirical Bayes statistics

$$\beta_{EB} = w \cdot \beta_{pr} + (1 - w) \cdot \hat{\beta} \tag{2}$$

*Note*: (1) $\beta_{\text{pr}}$ is a coefficient column vector for the prior information which does not depend on the data used in regression. We can get it from experience or the opinion of professionals from large banks and scoring agencies. In this paper, we generate prior coefficient by simulating an outside large sample data sets:

$$\beta_{pr} = (\,0.682 \quad 5.010 \quad 3.850 \quad 0.836 \quad -0.369 \quad 0.614\,)' \tag{3}$$

(2) $\hat{\beta}$ is Maximum Likelihood of Logistic standard estimates of the coefficient column vector. We estimate the parameters of Logistic regression model by MLE method, and give the maximum likelihood observed value of listed company:

$$\hat{\beta} = (\,0.4365 \quad 5.1925 \quad 3.2314 \quad 0.904 \quad -0.4578 \quad 0.5099\,)' \tag{4}$$

(3) $w \in [0, 1]$ is a given weight. The weight here of Empirical Bayes estimator based on prior coefficient information is:

$$w = \frac{J-2}{(\hat{\beta}-\hat{\beta}_{pr})' I(\hat{\beta})(\hat{\beta}-\hat{\beta}_{pr})} \tag{5}$$

$J$ is the number of constraints imposed by Logistic regression coefficients. In this paper, variable of default prediction model (1 are five factors extracted through factor analysis, then $J = 5$; (4) $I(\hat{\beta})$ is information matrix, that is the inverse matrix estimated coefficient of variance. If there is a linear regression model: $y_i = \beta' x_i + \varepsilon_i i = 1, 2, \cdots, n, \varepsilon_i \sim iid\, N(0, \sigma^2), x_i \in \mathbb{R}^p$, so $\theta = (\beta', \sigma)$ is a $(p+1)$ dimensional parameter vector $y = (y_1, \ldots, y_n)'$, $\chi = (x_1', \ldots x_n')'$, so

$$L(y, \chi; \theta) = \prod_{i=1}^{n} \frac{1}{\sqrt{2\pi}\sigma} \exp\left\{-\frac{1}{2\sigma^2}(y_i - \beta' x_i)^2\right\},$$

$$\begin{aligned}
\ell(y, \chi; \theta) &= \log\left(\frac{1}{(2\pi)^{\frac{n}{2}}\sigma^n}\right) - \frac{1}{2\sigma^2}\sum_{i=1}^{n}(y_i - \beta' x_i)^2 \\
&= -\frac{n}{2}\log(2\pi) - n\log\sigma - \frac{1}{2\sigma^2}(y - \chi\beta)'(y - \chi\beta) \\
&= -\frac{n}{2}\log(2\pi) - n\log\sigma - \frac{1}{2\sigma^2}(y'y + \beta'\chi'\chi\beta - 2\beta'\chi' y)
\end{aligned}$$

partial derivative of the parameters is

$$\frac{\partial}{\partial\beta}\ell = -\frac{1}{2\sigma^2}(2\chi'\chi\beta - 2\chi' y) \tag{6}$$

$$\frac{\partial}{\partial\sigma}\ell = -\frac{n}{\sigma} + \frac{1}{\sigma^3}\{(y - \chi\beta)'(y - \chi\beta)\} \tag{7}$$

then

$$\chi'\chi\hat{\beta} = \chi' y \Rightarrow \hat{\beta} = (\chi'\chi)'\chi' y$$

$$\frac{n}{\hat{\sigma}} = \frac{1}{\hat{\sigma}^3}(y - \chi\hat{\beta})'(y - \chi\hat{\beta}) \Rightarrow \hat{\sigma}^2 = \frac{1}{n}\| y - \chi\hat{\beta}\|^2$$

$$\hat{\sigma}^2 = \frac{1}{n}\sum_{i=1}^{n}(y_i - \hat{\beta}' x_i)^2$$

is a residual sum of squares. fix $x_i$, we get

$$E(y) = \chi\beta \ \text{ and } \ Var(y) = \sigma^2 \mathrm{I}_n, \ \text{So}$$

$$E(\hat{\beta}) = (\chi'\chi)^{-1}\chi' E(y) = \beta$$

$$Var(\hat{\beta}) = \sigma^2((\chi'\chi)^{-1}) \quad (8)$$

Use MatLab7.0, we get estimated coefficient covariance matrix:

$$Var(\hat{\beta}) = \begin{pmatrix} 0.0838 & 0.0096 & -0.1210 & 0.0317 & -0.0083 & 0.0020 \\ 0.0096 & 0.1903 & 0.2120 & -0.0024 & 0.0079 & 0.0055 \\ -0.1210 & 0.2120 & 1.1871 & -0.1261 & 0.0491 & 0.0104 \\ 0.0317 & -0.0024 & -0.1261 & 0.1343 & -0.0105 & -0.0043 \\ -0.0084 & 0.0079 & 0.0491 & -0.0105 & 0.0504 & -0.0010 \\ 0.0020 & 0.0055 & 0.0104 & -0.0043 & -0.0010 & 0.0326 \end{pmatrix} \quad (9)$$

Thus inverse of $Var(\hat{\beta})$, say $I(\hat{\beta})$, can be easily get.

For the estimated coefficient with smaller variance, $\hat{\beta}$ will be given a higher weight. In order to avoid over shrinkage, When $w$ is greater than 1, we restrict it equal to 1. In this case, Bayesian estimator is equivalent to prior coefficient.

From (3, (4, and (5, we have

$$w = \frac{J-2}{(\hat{\beta} - \hat{\beta}_{pr})' I(\hat{\beta})(\hat{\beta} - \hat{\beta}_{pr})} = 0.8875$$

(5) From above, we can get Empirical Bayes coefficient:

$$\begin{aligned} \beta_{EB} &= w \cdot \beta_{pr} + (1-w) \cdot \hat{\beta} \\ &= (\, 0.654 \quad 5.030 \quad 3.781 \quad 0.844 \quad -0.379 \quad 0.602 \,)' \end{aligned}$$

Therefore, the improved model for the Bayesian is:

$$p' = \frac{1}{1 + e^{-(0.654+5.031^{*}FAC1+3.781^{*}FAC2+0.844^{*}FAC3-0.379^{*}FAC4+0.602^{*}FAC5)}}$$

## 2.4 *Accuracy Assessment of Predictive Model*

We compare the accuracy of the predictive power between $\hat{\beta}$ and $\beta_{EB}$ according to AUC value (Stein and Jordao 2003; Stein 2005) and Brier score (1950). The former

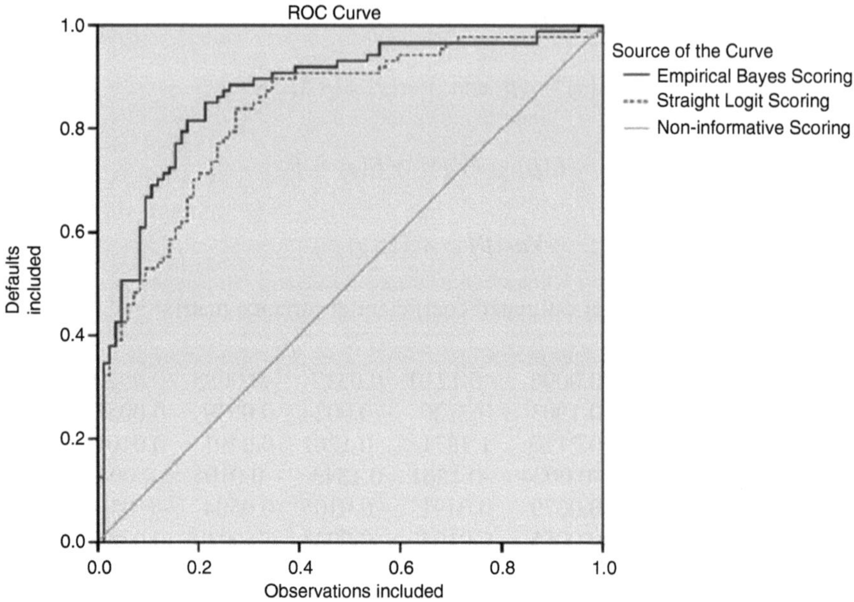


**Fig. 1** ROC curve of improved model and standard model

**Table 4** Area under the curve

| Test result variable(s) | Area | Std. error[a] | Asymptotic sig.[b] | Asymptotic 95% confidence interval | |
|---|---|---|---|---|---|
| | | | | Lower bound | Upper bound |
| Bayesian estimate | .870 | .028 | .000 | .815 | .925 |
| standard estimate | .834 | .031 | .000 | .773 | .895 |

based on the distinction between default samples and the normal samples, while the latter is used to measure the statistical prediction value and goodness of fit of sample values.

Figure 1 gives two ROC curves of Bayesian improved model and the standard logistic model. Dotted line represents the standard ROC curve of logistic model, solid line represents the ROC curve of Bayesian improved model. It shows that the improved model is better than the standard model. From Table 4, we can see AUC value of standard estimator is 0.834, AUC value of Bayesian improved estimator is 0.87, which is 3.6 percentage points higher than the standard estimates. It shows that Bayesian estimator has higher prediction accuracy and stability.

Calculating the Brier Score, we get it is $b=0.122$ for standard Logistic model, and $b_{EB}=0.117$ for Bayesian improved model. $b_{EB}$ is lower than $b$, and shows Bayesian estimator fitting better.

Stein and Jordao (2003) & Stein (2005) show that : It is significant that predictive accuracy appears 2–4% points of difference. In this paper, the prediction accuracy of Bayesian estimator is 3.6% points higher than the standard estimates, this means it's practical valuable.

# 3 Conclusion

From the view of above empirical analysis, the predictive accuracy of Bayesian estimator using prior information is better and increased 3.6% point than standard Logistic regression estimate. Based on our work, the customers may be accurately classified for different risk group by using their accounting and financial data, and the customers default probability can be counted. Therefore the probability of company financial crisis can be predicted. If the default data is not enough for regular statistical analysis, the prediction can be made by Bayesian estimator using prior information according to our work, and the model reliability can be ensured. Moreover, if banks plan to check up the default probability of new customer, Bayesian estimate model can be created by using prior information of general customers' data. This method we propose here is practical valuable.

# References


Adkins L, Hill RC (1996) Using prior information in the probit model: empirical risks of Bayes, empirical Bayes, and Stein estimators. In: Berry DA, Chaloner KM, Geweke JK (eds) Bayesian analysis in statistics and econometrics. John Wiley and Sons, New York, pp 79–90

Altman E, Haldeman R, Narayanan P (1977) ZETA analysis: a new model to identify bankruptcy risk of corporations. J Banking Finance (l):29–54

Brier GW (1950) Verification of forecasts expressed in terms of probability. Mon Weather Rev 78:1–3

Stein RM (2005) The relationship between default prediction and lending profits: integrating ROC analysis and loan pricing. J Bank Finance 29:1213–1236

Stein RM, Jordao F (2003) What is a more powerful model worth? Moody s KMV Inc, New York

# Research on Business Types Recognition Based on the Method of AHP-ELECTRE

**Hongbo Xu and Xiaoguo Jiang**

**Abstract** There are lots of risks and uncertainties in the process of enterprise operation. Enterprise business can be divided into core business, auxiliary business, peripheral business and market-oriented business. It is particularly important to identify a variety of business types. On the basis of improving the method of ELECTRE, this paper applies the method of AHP-ELECTRE to business types recognition based on the evaluation factors, including value, uniqueness, sustainability, competitiveness and concentration. So the research on the business types recognition may help enterprises to recognize different types of business and take their proper strategies.

**Keywords** AHP-ELECTRE · Assessment method · Business types · Risk · Uncertainty

## 1 Introduction

Every single enterprise may have different business about itself so that they will take different strategies for their own purpose. Therefore, it is particularly important to recognize and a variety of business types.

Currently, there is no uniform method to divide the business types. Lepak and Snell (1998) pointed out that Enterprise business was classified as peripheral businesses, core business, traditional business and unique business. However, from

H. Xu (✉)
Business school, Hohai University
and
Nanjing Institute of Industry Technology, 210098 Nanjing, China
e-mail: xhbo007@163.com

X. Jiang
School of Economics, Anhui University of Technology, 243032 Maanshan, China

Y. Zhou and D.D. Wu (eds.), *Modeling Risk Management for Resources and Environment in China*, Computational Risk Management,
DOI 10.1007/978-3-642-18387-4_31, 

the asset-specific dimension and the complexity of the production, Vining and Globeman (1999) divided the business types into low specificity low complexity, low specificity high complexity, high specificity low complexity and high specificity high complexity. Also enterprise business will be considered from the value chain point by scholars (Porter 2004; Xiao 2009; Li 2008), such as production, sales, research and development, information technology, financial management, human resources management, logistics management, customer relationship management, and so on. Xu (2008) and Dai (2010) divided the business types into the core business and non-core business based on the relationship between the business and its core competence. Nowadays, one of the most remarkable model is Arnold's (2000) classification. He proposed enterprise business would be classified as core business, auxiliary business, peripheral businesses and market-oriented business. Based on his classification, the author tries to employ the integration of AHP and ELECTRE method to recognize different business types of enterprises with the improvement of ELECTRE method.

## 2 Determination of the Principles of Core Business Evaluation Factors

Currently, the core business has still not been given a uniform definition, but on related activities of core business, they share the same views. It is the typical feature of the core operational activity that other operational activities tend to be adjusted accordingly. If an operational activity cannot reinforce the influence on the other activities, it might remain in an external position. In strategy, the core business-related activities will affect the long-term benefits, the existence and the development of enterprises (Zhao 2010). On the other words, an enterprise's core competency depends on its core business. To understand the core business activity of an enterprise is very important and essential. Economically, it seems be the source of value creation for an enterprise. In short, evaluation factors in core business can be summarized as follows:

**Value.** Core business activities are considered as the most critical part of value creation and the main source of value-added business. Hence, value is the most vital features of the core business.

**Sustainability.** Core business activities seem to be the area of expertise in the enterprise. Basically, core business activity is the tool and source of an enterprise's competition. So, the enterprise's core business activities are not only relatively stable but constantly unchanged. In other words, he should be provided with continuity of time.

**Uniqueness.** Core business activities embody core competence within the business, which makes core business activities have the same uniqueness as core competence. There are two main advantages here, one is not easily imitated by competitors; another is not easily replaced by other activities.

**Competitiveness.** Core business activities should achieve the advanced level in its field with respects of technology, cost, and efficiency. At the same time, it is allowed to remain a strong competitive advantage and market competitiveness.
**Concentration.** Changes in core business activities require the majority of enterprises have carried out the appropriate adjustments of other activities. Core activities can be understood as the process of diversity production management platform. It is also the most important part of various cross-businesses because of higher concentration.

## 3 The Construction of the Non-inferior Relationship of the Operational Activities Levels with AHP-ELECTRE Method

Based on the non-inferior relationship of the operational activities levels, suppose MADM problem $MA = (X, C, W, F, Y)$, alternative operational activities set is $X = (x_i | i = 1, 2, \cdots, n)$; evaluation factors (property) set is $C = (c_j | j = 1, 2, \cdots, m)$; evaluation factors weight set is $W = (w_j | j = 1, 2, \cdots, m)$; evaluation factors function set is $F = (f_j | j = 1, 2, \cdots, m)$; the score valuation of alternative operational activities in various evaluation factors is $Y = (y_{ij} | i = 1, 2, \cdots n; j = 1, 2, \cdots, m)$, $y_{ij}$ stands for the valuation of alternative operational activity xi in the evaluation factor $c_j$, that is, when the evaluation factor function is $f_j$, $y_{ij} = f_j(x_i), i = 1, 2, \cdots n; j = 1, 2, \cdots m$.

The construction of the non-inferior relationship of the operational activities levels is the core work of ELECTRE method on the basis of the decision matrix $Y = [f_j(x_i)]$. To judge if there exists the non-inferior relationship of the levels between $x_i$ and $x_k (x_i, x_k \in X)$, the distinction of harmony and non-harmony shall be made.

The specific process of construction of the non-inferior relationship of the operational activities levels is as follows,

Step 1: the definition of evaluation factors. With AHP method, the weight of evaluation factors $W = (w_j | j = 1, 2, \cdots, m)$ shall be defined through the relevant judgment matrix on the basis of multiple comparison and quantization.

Step 2: the classification of the evaluation factors of the business activities under the multiple comparison.

In the evaluation factors $c_j$, if $f_j(x_i) \geq f_j(x_k)$ $(\forall x_i, x_k \in X)$, then operational activity $x_i$ in the evaluation factors is not inferior to $x_k$. The set of all evaluation factors $c_j$ which meets $f_j(x_i) \geq f_j(x_k)$ is denoted as $I^+(x_i, x_k)$, that is,

$$I^+(x_j, x_k) = \{c_j | f_j(x_i) \geq f_j(X_k), 1 \leq j \leq m\} \tag{1}$$

In the evaluation factors $c_j$, if $f_j(x_i) \leq f_j(x_k)$ $(\forall x_i, x_k \in X)$, then operational activity $x_i$ in the evaluation factors is inferior to $x_k$. The set of all evaluation factors $c_j$ which meets $f_j(x_i) \leq f_j(x_k)$ is denoted as $I^-(x_i, x_k)$, that is,

$$I^{-}(x_j, x_k) = \{c_j | f_j(x_i) < f_j(X_k), 1 \le j \le m\} \tag{2}$$

Given that operational activity $x_i$ is as good as or better than $x_k$, $I^{+}(x_i, x_k)$is the set of all the evaluation factors. Then $W^{+}$ stands for their weight sum. Given that operational activity $x_i$ is worse than $x_k$ , $I^{-}(x_i, x_k)$is the set of all the evaluation factors. Accordingly, $W^{-}$stands for their weight sum.

$$W^{+} = \sum_{j=I^{+}} w_j \tag{3}$$

$$W^{-} = \sum_{j=I^{-}} w_j \tag{4}$$

Step 3: the definition of the harmony index.

As far as $\forall x_i, x_k \in X$, is concerned, the harmony index $a$ ($i$, $k$) defined as follows:

$$a(i,k) = \frac{W^{+}}{W^{+} + W^{-}} \tag{5}$$

Step 4: the definition of non-harmony index.

Suppose that the evaluation scale of evaluation factors $c_j$ lies the closed interval [$u_j$,$U_j$]. That is, in the evaluation factors $c_j$, the minimum score of the operational activity is $u_j$, and the maximum score is $U_j$. As far as $\forall x_i, x_k \in X$ is concerned, the non-harmony index $b$ ($i$, $k$) of multiple comparison is defined as follows,

$$b(i,k) = \begin{cases} 0, & f_j(x_i) \ge f_j(x_k) \\ \dfrac{\max\limits_j \left[ \left| f_j(x_i) - f_j(x_k) \right| \right]}{\max\limits_j (U_j)}, & f_j(x_i) < f_j(x_k) \end{cases} \tag{6}$$

Step 5: the establishment of harmony index matrix and non-harmony index matrix.

By multiple comparison of the alternative operational activities in operational activities set X, the harmony index can constitute the relevant index matrix A and the non-harmony index can constitute the relevant index matrix B.

$$A = \begin{bmatrix} - & a(1,2) & \cdots & a(1,n) \\ a(2,1) & - & \cdots & a(2,n) \\ \vdots & \vdots & \vdots & \vdots \\ a(n,1) & a(n,2) & \cdots & - \end{bmatrix} \quad B = \begin{bmatrix} - & b(1,2) & \cdots & b(1,n) \\ b(2,1) & - & \cdots & b(2,n) \\ \vdots & \vdots & \vdots & \vdots \\ b(n,1) & b(n,2) & \cdots & - \end{bmatrix}$$

The self comparison of every operational activity along the diagonal is omitted for they don't have any sense. Finally, n(n−1) harmony index and n(n−1) non-harmony index can be summed.

Step 6: the establishment of the harmony advantage matrix and non-harmony advantage matrix.

To achieve harmony advantage matrix and non-harmony advantage matrix, at first harmony threshold $p$ *and* non-harmony threshold $q$ shall be defined. If the harmony index $a(i, k)$ of alternative operational activities $x_i$ and $x_k$ is equal or greater than the harmony threshold $p$, decision makers conclude $x_i S x_{k.}$. If the harmony index $b(i, k)$ of alternative operational activities $x_i$ and $x_k$ is equal or less than the harmony threshold $q$, decision makers conclude $x_i S \bar{x}_{k.}$. $\bar{a}$ stands for the mean of all the harmony indices and $\bar{b}$ stands for the mean of all the non-harmony indices.

The definition of harmony threshold $p$ and non-harmony threshold $q$ usually refers to $\bar{a}$ and $\bar{b}$. Specifically, harmony threshold $p$ is equal or greater than the mean of harmony indices $\bar{a}$, Whereas non-harmony threshold $q$ is equal or less than the mean of non-harmony indices $\bar{b}$.

$$\bar{a} = \frac{\sum_{\substack{i=1 \\ i \neq k}}^{n} \sum_{\substack{k=1 \\ k \neq i}}^{n} a(i,k)}{n(n-1)} \tag{7}$$

$$\bar{b} = \frac{\sum_{\substack{i=1 \\ i \neq k}}^{n} \sum_{\substack{k=1 \\ k \neq i}}^{n} b(i,k)}{n(n-1)} \tag{8}$$

After the definition of harmony threshold $p$ and non-harmony threshold $q$, the comparison of harmony matrix $A$ and non-harmony matrix $B$ with $p$ and $q$ respectively shall be made. Then the harmony advantage matrix and non-harmony advantage matrix can be achieved with the representation of $R$ and $T$. If $a(i,k) \geq p$, $r(i,k) = 1$. If $a(i,k) < p$, $r(i,k) = 0$. If $b(i,k) \leq q$, $t(i,k) = 1$. If $b(i,k) > q$, $t(i,k) = 0$.

$$R = \begin{bmatrix} - & r(1,2) & \cdots & r(1,n) \\ r(2,1) & - & \cdots & r(2,n) \\ \vdots & \vdots & \vdots & \vdots \\ r(n,1) & r(n,2) & \cdots & - \end{bmatrix} \qquad T = \begin{bmatrix} - & t(1,2) & \cdots & t(1,n) \\ t(2,1) & - & \cdots & t(2,n) \\ \vdots & \vdots & \vdots & \vdots \\ t(n,1) & t(n,2) & \cdots & - \end{bmatrix}$$

Step 7: the definition of the non-inferior relationship of levels between the integration advantage matrix and operational activities.

According to the intersection of harmony matrix $R$ and non-harmony matrix $T$, the integration advantage matrix $E$ can be achieved. The elements of the corresponding positions of $R$ and $T$ are 1, the element of the position of integration advantage matrix $E$ is 1. Otherwise it is 0, that is, when $r(i,k) = 1$ and $t(i,k) = 1$, $e(i,k) = 1$. Otherwise, $e(i,k) = 0$.

$$E = \begin{bmatrix} - & e(1,2) & \cdots & e(1,n) \\ e(2,1) & - & \cdots & e(2,n) \\ \vdots & \vdots & \vdots & \vdots \\ e(n,1) & e(n,2) & \cdots & - \end{bmatrix}$$

When $e(i,k) = 1$, $x_i S x_k$, that is, $x_i$ of operational activities is not inferior to $x_k$.

# 4 The Improvement of ELECTRE Method and the Design of Criterion of Business Types

## 4.1 ELECTRE Improvement

In order to group the alternative scheme with ELECTRE method, this paper also makes further improvement of the method on the basis of scholars' study (Wang 2006). Some more definitions are shown in the following,

**Definition 1.** *In the alternative operational activities X, for* $\forall x_i \in X$, *after the construction of pairwise non-inferior relationship of the levels, if there exist* $\alpha_i (0 \le \alpha_i < n)$ *pieces of* $x_i S x_k (k = 1, 2, \cdots n; k \ne i)$, *the advantage number of the definition of* $x_i$ *is* $\alpha_i$; for $\forall x_i \in X$, if existing $\beta_i (0 \le \beta_i < n)$ pieces of $x_k S x_i (k = 1, 2, \cdots n; k \ne i)$, the disadvantage number is $\beta_i$.

**Definition 2.** *N pieces of alternative operational activities are grouped according to the advantage number and disadvantage number of alternative operational activities. The requirement of the first group is* $\beta_i = 0$ *and* $\alpha_i > \frac{n}{2}$. *The requirement of the second group is* $\alpha_i \ge \frac{n}{2} > \beta_i > 0$. *The requirement of the third group is* $\frac{n}{2} > \alpha_i > \beta_i > 0$ *or* $0 < \alpha_i \le \beta_i < n$. *The requirement of the fourth group is* $\alpha_i = 0$ *and* $\beta_i > 0$.

## 4.2 The Design of the Criterion of Business Types Recognition

According to the advantage and disadvantage numbers of an enterprise's operational activities, as well as the improvement of the ELECTRE method, some business types criterions are provided below,

**Guideline 1:** *Disadvantage number is 0 and the business on the condition that the advantage number is equal or greater than half the business total is core one.*

**Guideline 2:** *The advantage number and disadvantage number are not 0 and the business on the condition that the advantage number is greater than the disadvantage one and equal and greater than half the business total is auxiliary one.*

**Guideline 3:** *The advantage number and disadvantage number are not 0 and the business on the condition that the advantage number is greater than the disadvantage one but less than half the business total or the advantage number is less than the disadvantage one is peripheral one.*

**Guideline 4:** *The advantage number is 0 and the business on the condition that the disadvantage number is greater than 0 is market-oriented one.*

## 5 Example Analysis

Jiangsu Huajian Energe-saving Glass Co. Ltd. is situated in the Yu Huatai district of Nanjing, which mainly deals in the production of plating film raw glass sheet and energe-saving glass including various high quality and intensive processing combination products. His operational activities can be roughly divided into six parts: production and sales of project glass, production and sales of plating film glass, production and sales of locomotive glass, production and sales of insulating glass, glass package and glass transportation. The six parts are represented by $x_1, x_2, x_3, x_4$, $x_5$, $x_6$ in turn. The business types of the enterprise can be recognized with AHP method, ELECTRE method and its above mentioned relevant improvement as follows,

Firstly, with AHP method, the weight of evaluation factors ($c_1$ value, $c_2$ uniqueness, $c_3$ sustainability, $c_4$ competitiveness, $c_5$ concentration) is defined. The judgment matrix $A$ is constructed on the basis of experts' score and pairwise comparison.

$$A = \begin{bmatrix} & c_1 & c_2 & c_3 & c_4 & c_5 \\ c_1 & 1 & 3 & 1 & 2 & 3 \\ c_2 & 1/3 & 1 & 1/4 & 1/3 & 1 \\ c_3 & 1 & 4 & 1 & 2 & 3 \\ c_4 & 1/2 & 3 & 1/2 & 1 & 2 \\ c_5 & 1/3 & 1 & 1/3 & 1/2 & 1 \end{bmatrix}$$

The weight of evaluation factors ($c_1 \sim c_5$) is $W = (0.31, 0.08, 0.32, 0.19, 0.10)^{\mathrm{T}}$, through the calculation of judgment matrix $A$, $RI = 1.12$, $CR = 0.0103 < 0.1$, which conforms to the test of sustainability.

According to the experts' scores, the score (scale 1~5)of evaluation factor of every operational activity is shown in Table 1.

According to steps 2–4, the harmony index matrix $C$ and non-harmony index matrix $D$ can be achieved.

**Table 1** The score of evaluation factors of operational activities

| Scheme | $x_1$ | $x_2$ | $x_3$ | $x_4$ | $x_5$ | $x_6$ |
|---|---|---|---|---|---|---|
| Factors | | | | | | |
| $c_1$ | 4 | 5 | 3 | 4 | 2 | 2 |
| $c_2$ | 2 | 4 | 2 | 3 | 2 | 1 |
| $c_3$ | 1 | 4 | 1 | 4 | 4 | 1 |
| $c_4$ | 1 | 4 | 1 | 3 | 1 | 2 |
| $c_5$ | 3 | 3 | 1 | 3 | 1 | 1 |

$$C = \begin{bmatrix} & x_1 & x_2 & x_3 & x_4 & x_5 & x_6 \\ x_1 & - & 0.1 & 1 & 0.41 & 0.68 & 0.81 \\ x_2 & 1 & - & 1 & 1 & 1 & 1 \\ x_3 & 0.59 & 0 & - & 0 & 0.68 & 0.81 \\ x_4 & 1 & 0.42 & 1 & - & 1 & 1 \\ x_5 & 0.59 & 0.32 & 0.69 & 0.32 & - & 0.81 \\ x_6 & 0.51 & 0 & 0.61 & 0 & 0.6 & - \end{bmatrix}$$

$$D = \begin{bmatrix} & x_1 & x_2 & x_3 & x_4 & x_5 & x_6 \\ x_1 & - & 0.6 & 0 & 0.6 & 0.6 & 0.2 \\ x_2 & 0 & - & 0 & 0 & 0 & 0 \\ x_3 & 0.4 & 0.6 & - & 0.6 & 0.6 & 0.2 \\ x_4 & 0 & 0.2 & 0 & - & 0 & 0 \\ x_5 & 0.4 & 0.6 & 0.2 & 0.4 & - & 0.2 \\ x_6 & 0.4 & 0.6 & 0.2 & 0.6 & 0.6 & - \end{bmatrix}$$

According to the formula of $\bar{c}$ and $\bar{d}$ in step 5, by calculation of harmony index mean $\bar{c}$ and non-harmony index mean $\bar{d}$, the results are $\bar{c} = 0.63$, $\bar{d} = 0.29$. Furthermore, according to step 5, by comparison of the elements of Matrix $C$ and Matrix $D$ with $\bar{c}$ and $\bar{d}$ respectively, the harmony advantage matrix $U$ and non-harmony advantage matrix $G$ can be achieved.

$$U = \begin{bmatrix} & x_1 & x_2 & x_3 & x_4 & x_5 & x_6 \\ x_1 & - & 0 & 1 & 0 & 1 & 1 \\ x_2 & 1 & - & 1 & 1 & 1 & 1 \\ x_3 & 0 & 0 & - & 0 & 1 & 1 \\ x_4 & 1 & 0 & 1 & - & 1 & 1 \\ x_5 & 0 & 0 & 1 & 0 & - & 1 \\ x_6 & 0 & 0 & 0 & 0 & 0 & - \end{bmatrix} \quad G = \begin{bmatrix} & x_1 & x_2 & x_3 & x_4 & x_5 & x_6 \\ x_1 & - & 0 & 1 & 0 & 0 & 1 \\ x_2 & 1 & - & 1 & 1 & 1 & 1 \\ x_3 & 0 & 0 & - & 0 & 0 & 1 \\ x_4 & 1 & 1 & 1 & - & 1 & 1 \\ x_5 & 0 & 0 & 1 & 0 & - & 1 \\ x_6 & 0 & 0 & 1 & 0 & 0 & - \end{bmatrix}$$

According to step 6, there exists superior relationship of levels among every operational activity as follows

**Table 2** The advantage and disadvantage date in business activities

| | $x_1$ | $x_2$ | $x_3$ | $x_4$ | $x_5$ | $x_6$ |
|---|---|---|---|---|---|---|
| Advantage date | 2 | 5 | 1 | 4 | 2 | 0 |
| Disadvantage date | 2 | 0 | 4 | 1 | 2 | 5 |

$$x_1 S x_3, x_1 S x_6; \quad x_2 S x_1, x_2 S x_3, x_2 S x_4, x_2 S x_5, x_2 S x_6; \quad x_3 S x_6;$$
$$x_4 S x_1, x_4 S x_3, \; x_4 S x_5, x_4 S x_6; \quad x_5 S x_3, x_5 S x_6;$$

According to the superior relationship of levels among every operational activity, the advantage and disadvantage numbers can be collected as is shown in Table 2.

Based on business types Recognition criteria 1 to 4, from various types of business activities of enterprises, we can draw the following conclusions: operational activity $x_2$ is core one; $x_4$ is auxiliary; $x_1$, $x_3$, and $x_5$ are peripheral; $x_6$ is market-oriented.

# 6 Conclusion

Business type recognition problem is so complex that a large number of factors should be involved in. It is difficult to solve it only by quantitative analysis. AHP-ELECTRE method is a simple and effective way to solve the problem. Through the method, non-inferior relationship of levels can be constructed and the advantage and disadvantage numbers of every operational activity can be collected. At last, according to the judgment guidelines1 to 4, business types can be recognized. However, in practice, considering the complexity of involved factors and difficulty of considerations, the recognition becomes more complex. So, it deserves our further study.

# References


Arnold U (2000) New dimensions of outsourcing: a combination of transaction cost economics and the core competencies concept. Eur J Purch Supply Manage 6:23–29

Dai L (2010) Determination the order non-core business dissection. China Management Informationization 93–95

Lepak DP, Snell SA (1998) Virtual HR: strategic human resource management in the 21st century. Hum Resour Manage Rev 8:215–234

Li W (2008) Value chain reanalysis under the pressure of high costs. Commercial Times 44–47

Porter ME (2004) Competitive advantage. Simon & Schuster Ltd, United kingdom

Vining A, Globerman S (1999) A conceptual framework for understanding the outsourcing decision. Eur Manage J 17:645–654

Wang J (2006) Research on outsourcing decision model. Dalian University of Technology

Xiao W (2009) Value chain analysis and model construction based on steel and iron corporations. Accounting and Finance 74–77

Xu J (2008) Identification of core competence and outsourcing of. Market Modernization 70–71

Zhao H (2010) How to build core competence. Liaoning Economy 71–73

# Part IV
# Environmental Risk Management

# Research on Chaotic Characteristic and Risk Evaluation of Safety Monitoring Time Series for High Rock Slope

**Guilan Liang**

**Abstract** High rock slope engineering is typical nonlinear system, its evolution process is chaotic, dissipated and uncertain. Chaotic system can't be forecasted for long-term and needs discuss about maximum time scale of predictability. Nonlinear theory is proposed to research maximum time scale of predictability of safety monitoring chaotic time series and construct the model APSO-RBFNN to predict the chaotic time series in maximum time scale of predictability. The largest Lyapunov exponent and maximum time scale is calculate with small data sets method. In the maximum time scale of predictability, the essay applies APSO-RBFNN to chaotic time series for risk assessment. The engineering cases studies reveal that the forecasting values are in good agreement with the measured values and this model has high accuracy and a good prospect for risk assessment of nonlinear chaotic time series of geotechnical engineering.

**Keywords** Chaotic time series · high rock slope · nonlinear theory · risk assessment · uncertainty analysis

## 1 Introduction

Prediction and forecasting based on safety monitoring time series foretell the future of the slope system according to law of itself development. The usual approaches are to construct dynamic mathematic model which can describe the slope engineering system and by solving the mathematic model to predict and forecast (Liang et al. 2008; Liang et al. 2007). But the references about the prediction and forecasting were all one-step forecasting or a few steps and didn't mention the maximum time scale of predictability. In fact, high rock slope engineering are typical nonlinear and uncertain system and its evolution process is chaotic, dissipated, even more

G. Liang
College of Harbor, Coastal, and Offshore Engineering, Hohai University, Nanjing 210098, China
e-mail: wendylgl@163.com

Y. Zhou and D.D. Wu (eds.), *Modeling Risk Management for Resources and Environment in China*, Computational Risk Management,
DOI 10.1007/978-3-642-18387-4_32, © Springer-Verlag Berlin Heidelberg 2011

complicated. Chaotic system can't do long-term forecasting. Thus, the maximum time scale of predictability needs to be discussed when forecasting.

The paper proposes to solve above-mentioned problem using nonlinear theory, such as phase space reconstruction and chaotic theory and so on. The essay applies the nonlinear theory to the phase space reconstruction, time delay and embedding dimension when studying chaotic characteristic of the high rock slope. The aim is to calculate maximum time scale of predictability and forecast chaotic time series using APSO-RBFNN model in the maximum time scale of predictability. The essay applies the above-mentioned theory to safety monitoring time series of the left bank high rock slope of Jinping first stage hydropower station. The engineering cases studies reveal that the forecasting values are in good agreement with the measured values and this model has high accuracy.

## 2 Maximum Time Scale of Predictability of Chaotic Time Series

Mass information of safety monitoring time series for high rock slope is the reflection of dynamic evolution for rock mass system in the outer environment and under the action of all kinds of loads. The single time series contain rich chaotic message.

The property, being very sensitive to initial condition, is one of important characteristics of chaotic system. In order to describe and quantitatively analyze the property, the index, Lyapunov exponent, is introduced. If largest Lyapunov exponent $\lambda_1$ is greater than zero, the system is regarded as chaotic. Largest Lyapunov exponent $\lambda_1$ is an important index of predicting the time series and stands for the longest time of doubling state error of the chaotic system. When predicting the time series, we should, in the first place, judge if the system is chaotic. The maximum time scale of predictability need to be calculated when the system is chaotic. In general, the maximum time scale of predictability is defined as the inverse of the largest Lyapunov exponent, that is $T = 1/\lambda_1$ and the largest Lyapunov exponent is computed by phase space reconstruction.

### *2.1 Phase Space Reconstruction*

Since the 1980s, As Takens furthering study of topology on the basis of predecessor's result, the dynamics mechanism research of time series is promoted. The phase space reconstruction that is widely used in time series is its specific reflection.

Phase space reconstruction of safety monitoring time series for high rock slope is attractor reconstruction on basis of limited data. One-dimensional time series which represent information of m-dimensional independent system, as time goes on, can draw a track of reflecting dynamic change in m-dimensional state space. This is so-called phase space reconstructing which can judge if a system is chaotic.

# 3 C-C Method

Common methods to calculate the time delay $\tau$and embedding dimension $m$ are auto correlation, multiple autocorrelation and mutual information, but these methods are large and intricate calculation or can not fully embody nonlinear characteristic of the time series. In order to overcome these shortcomings, C-C method is introduced in this essay.

Here, a correlation integral on basis of G-P algorithm is adopted to describe correlation of nonlinear time series.

$$C(m,N,r,t) = (2/M(M-1))\sum \theta(R-d_{ij}), r>0 \tag{1}$$

$$S(m,N,r,t) = C(m,N,r,t) - C^m(1,N,r,t), r>0 \tag{2}$$

Time delay and embedding dimension are decided by statistic $S(m,N,r,t)$. In (1) and (2), $d_{ij} = \|x_i - x_j\|$, and when $x<0, \theta(x) = 0$; $x \geqslant 0, \theta(x) = 1$, $\theta(x)$ is Heaviside step function. Formula $M = N - (m-1)\tau$ expresses the number of embedding points in m-dimensional phase space. Because equation $S(m,N,r,t)$ contains two correlation integral functions, this method is called as C-C method.

## 3.1 *Small Data Sets Arithmetic*

After calculating the time delay and embedding dimension using C-C method, another goal of researching chaos system is to compute largest Lyapunov exponent of time series. If largest Lyapunov exponent is greater than zero, the researched system is considered as chaotic. And the maximum time scale of predictability is relevant to largest Lyapunov exponent. The small data sets arithmetic is applied to calculating the largest Lyapunov exponent in this essay.

The algorithm of small data sets method is as follows:

(1) Using the Fast Fourier Transform (FFT) algorithm to calculate average period $T$;
(2) Calculating time delay $\tau$ and embedding dimension $m$ to reconstruct phase space;
(3) Seeking the closest point and restricting short separation: seeking the nearest neighborhood for every point $X_i$ in reconstructed orbit $X$. The concrete calculation as follows:

$$\lambda_1(i,k) = \frac{1}{k\Delta t}\frac{1}{(M-k)}\sum_{j=1}^{M-k}\ln\frac{d_j(i+k)}{d_j(i)} \tag{3}$$

Where: $\Delta t$ is sample period, $d_j(i) = \|Y_{j+i} - Y_{\hat{j}+i}\|$ is the distance that the $j$ th nearest neighborhood dot pair experiences after $i$ discrete time steps, that is to say:

$$d_j(i) = C_j e^{\lambda_1(i\Delta t)}, C_j = d_j(0) \tag{4}$$

If the both sides of equation (4) take a logarithm, the result is as follows:

$$\ln d_j(i) = \ln C_j + \lambda_1(i\Delta t),\ j = 1, 2, \cdots, M \quad (5)$$

The largest Lyapunov exponent is almost the slope coefficient of above line which can get by approaching this line with least squares techniques. That is:

$$y(i) = (1/\Delta t)\langle \ln d_j(i) \rangle \quad (6)$$

(1) Fitting a straight line with least squares techniques and the slope coefficient of the line is the largest Lyapunov exponent $\lambda_1$.

The relation between maximum time scale of predictability $T$ and $\lambda_1$ is $T = 1/\lambda_1$.

## 4 APSO-RBFNN Algorithm

After obtaining the maximum time scale of predictability, the essay applies APSO-RBFNN algorithm to predicting of chaotic time series. About the detailed algorithm step are introduced in auctorial paper ≪ Study and application of PSO-RBFNN model to nonlinear time series forecasting for geotechnical engineering ≫ and ≪ wavelet neural network based on adaptive particle swarm optimization and its application to displacement back analysis ≫.

## 5 Application of the Engineering Example

The left bank high rock slope of JinPing first stage hydropower station has the characteristics of valley steep slope, exposed bedrock, the strong development of deep fissures and thousand meters of the relative height difference. The structural planes in the slope mainly contain the f5, f8, f42-9 as the representative of faults, and SL44 ~ 1 fissures as the representative of deep cracks. In general, slope excavation of the high slope is 60–100 m, but the highest slope excavation is almost 160 m. Therefore, the intensive monitoring equipments are laid out in these slopes in order to provide valuable information for the slope design, construction and information feedback on basis of a vast amount of detailed monitoring data.

### *5.1 Selecting of Monitoring Points*

The essay mainly researches the monitored deformation data of observation points TP12-2, TP13-1, TP14-1. The maximum time scale of predictability is studied by using nonlinear theory. Then, the time series are predicted with APSO-RBFNN algorithm in maximum time scale of predictability.

## 5.2 *Analysis of Chaotic Characteristic and Phase Space Reconstruction*

The time series of TP12-2 is transformed with the Fast Fourier Transform (FFT) algorithm by setting the time-interval $\Delta t = 1d$ during reconstructing.

The values of the time delay $\tau_d$ and time window $\tau_w$ which are calculated with C-C method are 9 and 18. The embedding dimension is 3 on basis of the formula $\tau_w = (m-1)\tau_d$ and the calculating curve is shown to be Fig. 1.

The correlation dimension equals 2.7398 on basis of G-P algorithm and the largest Lyapunov exponent is equal to 0.0185 with small data sets arithmetic. Because the correlation dimension is decimal and largest Lyapunov exponent is greater than zero, the system is considered as chaotic. Since the system is chaotic, the time series is short-term predictability, not long-term predictable. The maximum time scale of predictability is equal to 54 according to the formula $T = 1/\lambda_1 = 1/0.0185 = 54$.

Similarly, the time series of monitoring points TP13-1 and TP14-1 are analyzed. The calculated largest Lyapunov exponents are 0.0159 and 0.0204 and the maximum time scale of predictability respectively are 63 time-step and 69 time-step.

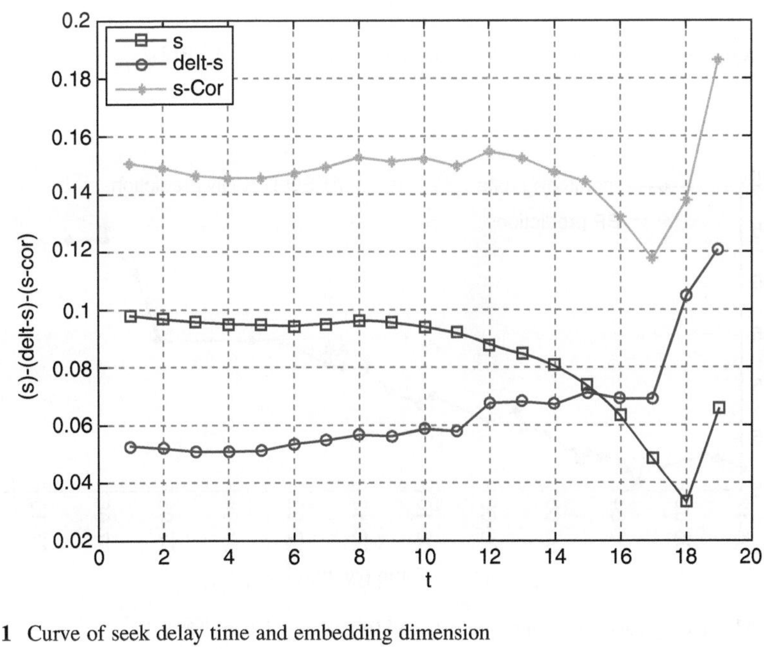


**Fig. 1** Curve of seek delay time and embedding dimension

### *5.3 Prediction of the Time Series and Comparison of Prediction Methods*

After calculating the maximum time scale of predictability, the APSO-RBFNN model is applied to predicting the time series. In order to verify the accuracy of the APSO-RBFNN model, the APSO-RBFNN model and typical BP model are simultaneously used to predict and extrapolate. The prediction curves are shown in Figs. 2–4.

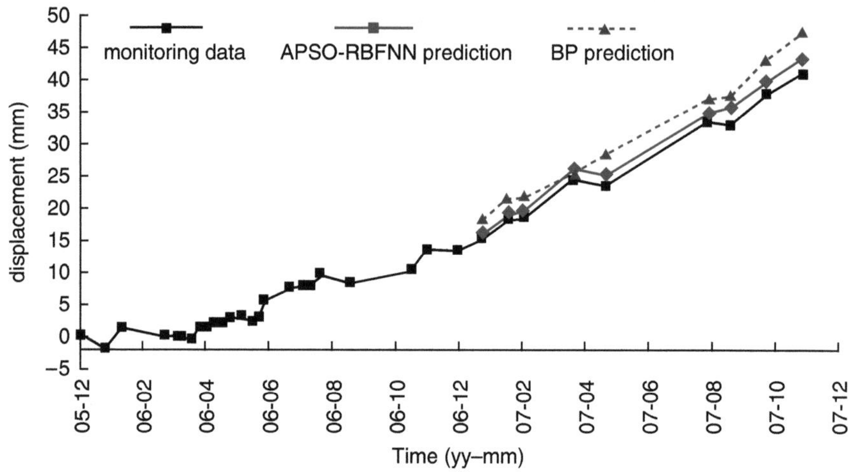


**Fig. 2** Comparing curve of monitoring data and forecasting data of TP12-2

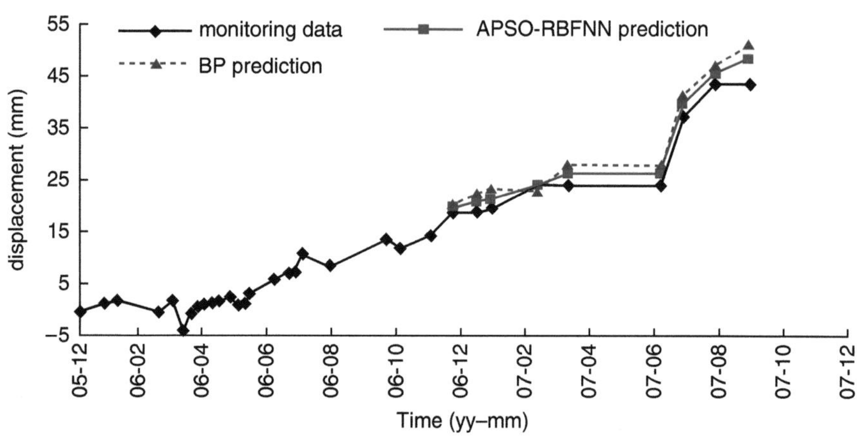


**Fig. 3** Comparing curve of monitoring data and forecasting data of TP13-1

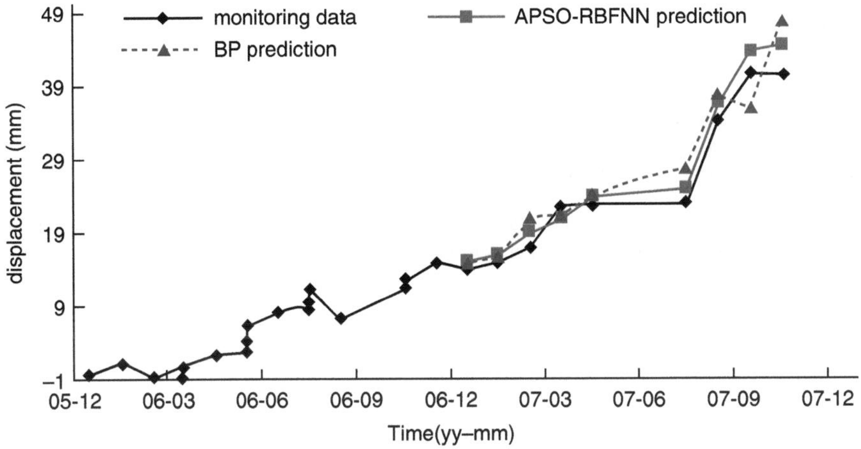


**Fig. 4** Comparing curve of monitoring data and forecasting data of TP14-1

The Figs. 2–4 show that the displacement gradually increases because of the effect of dynamic unloading during slope excavation and the APSO-RBFNN algorithm has higher predicting precision and quicker convergence speed comparing with BP model with the same level of the errors. The maximal absolute value of relative error of APSO-RBFNN and BP for points TP12-2,TP13-1 and TP14-1 are 9.009%, 21.941%; 11.239%, 19.289%; 11.976%, 23.556%, respectively.

# 6 Conclusions

Nonlinear theory is proposed to research the chaotic characteristic and calculate maximum time scale of predictability of safety monitoring time series for high rock slope (Chatterjee and Siarry 2006). In the maximum time scale of predictability, the essay applies APSO-RBFNN to chaotic time series for prediction.

Chaotic characteristic of safety monitoring time series of the left bank high rock slope of Jinping first stage hydropower station is studied and the APSO-RBFNN and BP model are applied to predicting the chaotic time series. Comparing with BP, in the maximum time scale of predictability, the forecasting values of the APSO-RBFNN are in better agreement with the measured values and this model has higher accuracy and a good prospect for nonlinear chaotic time series forecasting of geotechnical engineering.

**Acknowledgments** We are grateful for the monitoring data provided by CHIDI. We also acknowledge the financial support from the National Natural Science Foundation of China Project 50909038, Doctoral Fund of Ministry of Education of China Project 20090094120006, the Fundamental Research Funds for the Central Universities.

## References

Chatterjee A, Siarry P (2006) Nonlinear inertia weight variation for dynamic adaptation in particle swarm optimization. Comput Oper Res 859–871

Liang GL, Xu WY, Wei J (2007) Wavelet neural network based on adaptive particle swarm optimization and its application to displacement back analysis. Chinese J Rock Mech Eng 1251–1257

Liang GL, Xu WY, He YZ (2008) Study and application of PSO-RBFNN model to nonlinear time series forecasting for geotechnical engineering. Rock Soil Mech 995–1000

# Regional Eco-efficiency and Environmental Risk Analysis in China Based on NUO-DEA Model

**Xiufeng Zhu, Ning Zhang, and Yongrok Choi**

**Abstract** Eco-efficiency is an important approach for economic sustainability analysis indicating how efficient the economic activity is, considering environmental risk analysis simultaneously. Traditional DEA framework neglect the undesirable output during the process of production, in industrial society lots of harmful by-products are made at the same time, which lead to serious environmental pollution and risk. In this paper we adopt a non-radial undesirable output DEA model to measures the eco-efficiency of each province in China. The conclusion is that economically developed area eastern part has the highest eco-efficiency, while less developed area western part has the worst eco-efficiency that may cause great environment risk. The western area is in a dilemma situation facing the poor economic and bad environmental condition. We suggest that the western area can purchase waste discharge right from more developed area to settle this problem provisionally.

**Keywords** China · eco-efficiency · environmental risk · non-radial undesirable-outputs DEA (NUO-DEA)

X. Zhu
School of Management, Shandong Women's University, Jinan 250002, China

N. Zhang (✉)
School of Management, Shandong Women's University, Jinan 250002, China
and
Department of International Trade, Inha University, Incheon 402-751, South Korea
e-mail: zn928zn@hotmail.com

Y. Choi
Department of International Trade, Inha University, Incheon 402-751, South Korea

Y. Zhou and D.D. Wu (eds.), *Modeling Risk Management for Resources and Environment in China*, Computational Risk Management,
DOI 10.1007/978-3-642-18387-4_33, 

# 1 Introduction

Recent years, China has made many remarkable goals in economic construction and social development. For a long time, China's scale-orientation economic development led to inefficient natural resource utilization and energy use in the production process, as well as high consumption and high pollution that causing the environmental risk. Since UN Conference on Environment and Development (UNCED) in 1992, sustainable development has been used as a fundamental development strategy by many countries, including China. While sustainable development has been adopted as a goal, it dose not, in itself provide the means by which an unsustainable development could be transformed into a sustainable one. Strategies for optimizing the use of resources in more efficient way play an important role. Eco-efficiency is a good tool for environmental risk analysis, indicating an empirical relation in economic activities between environmental cost or value and environmental impact, has been proposed as a route to promote such transformation.

The concept of eco-efficiency can be found in 1970s as the concept of "environment efficiency" (Freeman et al. 1973). Schaltegger and Synnestvedt (2002) named eco-efficiency as a "business link to sustainable development". In 1990s eco-efficiency has received significant attention in the sustainable development literature Eco-efficiency plays an important role in expressing how efficient the economy is with regard to nature's goods and services. According to the definition, eco-efficiency is measured as the ratio between the value of production including income, high quality goods and services, GDP, jobs etc) and the environmental impacts of the product.

Data envelopment analysis (DEA) is a well-established linear programming approach for measuring the relative efficiency of each decision-making units (DMUs) that have multiple inputs and outputs, proposed by Charnes et al. (1978) and extended by Banker et al. (1984). DEA has recently been widely applied to evaluate the eco-efficiency.

Unfortunately, the recent research neglect the effects of undesirable output, or they consider the undesirable but they did not deal with them accord with the real production activities. In this paper we introduce a new Non-radial DEA model for considering undesirable output conditions (NUO-DEA) to measure Chinese regional economy eco-efficiency

In the current circumstances that it's difficult to account environmental risk, the most prominent feature of current research is providing a simple and feasible approach take into account the actual cost of environmental pollution to measuring the economic efficiency in China. The structure of the paper is organized as follows; Sect. 2 reviews previous studies on indicators and measurement of eco-efficiency analysis. Based on previous works, a set of regional eco-efficiency indicators and undesirable output DEA model are developed for regional economy eco-efficiency analysis in Sect. 3. Sect. 4 illustrates undesirable output DEA models with real data set of 30 regions including provinces, municipality and autonomous region in China. Section 5 provides discussions on the results of our research. Finally, overall conclusions and implications are presented.

# 2 Methodology

There are diverse studies analyzing the efficiency of environment and energy. In recent years, DEA has generally been used to analyze the efficiency of the eco-efficiency. For instance, Ramanathan (2000) adopted DEA to compare the efficiency of alternative transportation modes,

## 2.1 Literature Review

Hu and Wang (2006), Hu and Kao (2007) introduced a TFP energy efficiency index employing DEA. Fare et al. (1989) firstly developed a nonlinear programming approach to deal with pollutants. However, non-linear programming is very inconvenient to achieve, the application has been largely restricted. In his paper a directional function based on DEA model to measure undesirable outputs in environment performances evaluation problem this research has settled undesirable output well but the DEA model was radial and output oriented measurement which ignore the slack variable that lead to biased estimate.

Zhou et al. (2007) developed several DEA models to measure eco-efficiency considering energy inputs, non-energy inputs such as capital and labor, desirable outputs and undesirable outputs. Scheel (2001) presented some radial tools that assume any change of output level will involve both desirable and undesirable outputs. Considering undesirable outputs $Y^b$ as $1/Y^b$ to take it as desirable outputs, so this method can be solved in a traditional CCR model, but this approach runs counter to the actual production process, the efficiency result is a biased evaluation.

Seiford and Zhu (2002) developed a radial DEA model, in the presence of undesirable outputs, the method is All undesirable output is multiplied by -1, then find a suitable transition vector to transform all the negative into a positive undesirable output, this method can solve undesirable outputs well, but the shortcomings of this method is the solution can be solved only in the CRS (Constant returns to scale) condition.

In this paper we employ a Non-radial DEA Model which suggested by Tone (2003).This DEA is non-radial and non-oriented, and utilizes input and output slacks directly in producing an efficiency measure, in our paper, the Non-radial DEA is modified so as to take undesirable outputs into account.

## 2.2 Our DEA Framework

Suppose that there are n DMUs each having three factors : inputs, good outputs and bad outputs, as represented by three vectors $x \in R^m$, $y^g \in R^{s1}$, $y^b \in R^{s2}$ respectively,. We define the matrices $Y^g$, $Y^b X$ as follows. $Y^g = [y_1^g, ..., y_n^g] \in$

$R^{s1\times n}$ $Y^b = [y_1^b, ..., y_n^b] \in R^{s2\times n}$ $X = [x_1, ..., x_n] \in R^{m\times n}$, we assume $X > 0$, $Y^g > 0 Y^b > 0$

The production possibility set(p) is as follows

$$P = \left\{(x, y^g, y^b) | x \geq X\lambda, y^g \leq Y^g\lambda, y^b \geq Y^b\lambda, \lambda \geq 0\right\} \tag{1}$$

As Tone (2003)'s method the Non-radial DEA can be measured as follows

$$\rho^* = \min \frac{1 - \frac{1}{m}\sum_{i=1}^{m} \frac{s_i^-}{x_{i0}}}{1 + \frac{1}{s_1 + s_2}\left(\sum_{r=1}^{s_1} \frac{s_r^g}{y_{r0}^g} + \sum_{r=1}^{s_2} \frac{s_r^b}{y_{r0}^b}\right)} \tag{2}$$

Subject to

$$\begin{aligned} x_0 &= X\lambda + s^- \\ y_0^g &= Y^g\lambda - s^g \\ y_0^b &= Y^b\lambda + s^b \\ s^- &\geq 0, s^g \geq 0, s^b \geq 0, \lambda \geq 0 \end{aligned} \tag{3}$$

The vectors $s^-$ and $s^b$ correspond to excesses in inputs and bad outputs, respectively, while $s^g$ express shortages in good outputs. The DMU is efficient in the presence of undesirable outputs if $\rho^* = 1$, but the object function (1.1) is not a linear programming, Using the transformation by Charnes and Cooper 1962), we have an equivalent linear program in $t$, $\varphi$, $s^{-*} s^b$ and $s^g$ as displayed below (Charnes and Cooper 1978)

$$\begin{aligned} r^* &= \min t - \frac{1}{m}\sum_{i=1}^{m} \frac{s_i^-}{x_{i0}} \\ 1 &= t + 1 + \frac{1}{s_1 + s_2}\left(\sum_{r=1}^{s_1} \frac{s_r^g}{y_{r0}^g} + \sum_{r=1}^{s_2} \frac{s_r^b}{y_{r0}^b}\right) \end{aligned} \tag{4}$$

$$\begin{aligned} x_0 t &= X\varphi + s^- \\ y_0^g t &= Y^g\varphi - s^g \\ y_0^b t &= Y^b\varphi + s^b \\ s^- &\geq 0, s^g \geq 0, s^b \geq 0, \varphi \geq 0, t > 0. \end{aligned} \tag{5}$$

Let an optimal solution of [LP] be $(t^*, \varphi^*, s^{-*}, s^{g*}, s^{b*})$. Then we have an optimal solution as defined by $\rho^* = t^*$, $\lambda^* = \frac{\varphi^*}{t^*}$, $s^{-*} = \frac{s^{-*}}{t^*}$, $s^{g*} = \frac{s^{g*}}{t^*}$, $s^{b*} = \frac{s^{b*}}{t^*}$

The existence of $(t^*, \varphi^*, s^{-*}, s^{g*}, s^{b*})$.with $t^* > 0$ is guaranteed by (Charnes and Cooper 1978)

# 3 Statistical Results and Implications

In our paper, according to principal of economics, we adopted labor (10,000 persons) and investment of fix asset named capital (100 million RMB) as two non-resource inputs, energy consumption as resource input, GDP (100 million RMB) as a desirable output, Sulphur Dioxide Emission (10,000 t), Volume of Waste Water (10,000 t), Solid Wastes Utilized (10,000 t) as three undesirable outputs which can be called "three wastes" in China. The number of labor is an input, which include employee in all organizations and individual workers. The indicator of investment of fixed assets is usually used as an input in the literature, such as Ahujia and majumdar (1998), Ng and Chang (2003), Hu and Wang (2006). Since the energy input data of Tibet can not be found in our study, we collect the data for 30 provinces, municipalities, and autonomous regions from Statistical Year Book of China from year of 2005 to 2008.

## 3.1 *Results of Our DEA Model*

We utilize our NUO-DEA Model to measure eco-efficiency of 30 regions in China from year 2005 until 2008, because from the year of 2005 energy consumption is available in the Statistical Year Book. In the 11 "Five-Year Plan" it is clearly stated that GDP energy consumption should be reduced by 20% in the next five years. So energy consumption was be emphasized in National Bureau of Statistics of China from 2005. DEA-Solver Pro5.0 was employed to run the model. You can get the detailed results by connecting the author.

Beijing, shanghai, Jiangsu and Guangdong showed the highest eco-efficiency in our research period. All of them are located in the eastern part of China. Qinghai and Ningxia showed the worst eco-efficiency, both of them are in the western part of China

In order to analyze the difference of eco-efficiency among different regions, we divide the 31 regions into three parts as many researchers suggested (Hu and Wang 2006). The eastern area is constituted by 11 provinces including the eight coastal provinces such as Shandong, Jiangsu, Zhejiang, Guangdong the 3 municipalities of Beijing, Tianjin and Shanghai. This area has made great progress in economy in recent years, and its GDP is around half of total GDP in China. Most of light industries, foreign trades are located in this area, and most FDI and technologies are also attracted by this area. The central area consists of ten provinces which are all inland provinces such as Heilongjiang, Jilin, Inner Mongolia. This area has a large population and is a basement of farming industries. The western area covers more than half of the territory of China. It includes one municipality of Chongqing and nine provinces, including Gansu, Qinghai, Xinjiang, Sichuan. Compared to the other two areas, this area has low population density. This area is the least developed area in China.

According to our results, the three regional parts showed different eco-efficiency, the average eco-efficiency of eastern area is 0.767, which is the highest area than the other two areas. The central area's average eco-efficiency is 0.435 and the western area showed the lowest average score 0.333. According to Lindmark and Vikstrm (2003)'s research, the less developed area has fewer industry where the pollution should not be serious so the environment efficiency may be higher than more developed area. But in our paper, we are drawing a completely different conclusion. Not only the GDP of east area is higher than other areas, but also the eco-efficiency is the best.

The eastern area is a more developed area which has a high economic level, so the government can utilize great capital that benefits from rapid economic growth into environmental governance and pollution treatment to achieve a sustainable development. The economic growth and environment governance is in a harmonious condition. The central area is developing area in economy, which has abundant natural resources and strong industrial base, but industrial structure of this area is very low, the increase of economy is at the cost of high energy consumption and severe environment contamination. Ecological environment can not endure the negative externality. The western area is particularly rich in resources but extremely underdeveloped in economy. Ecological environment of western area is very vulnerable. Worse still, because of western area's Characteristics of public goods, eastern and central area may use western area's pollution discharge facility without any cost that causing an Eco-aggression to the western area.

It is different for government to make a policy in order to improve the eco-efficiency of western area. If the government emphasize on the environment protection, the poor economic conditions of western area can hardly be improved on, but if the government don not limit the pollution level it is not accord with the spirit of Chinese sustainable development plan. The western area is in a dilemma now.

## 4 Conclusion

In this paper, in order to measure the economic efficiency with environmental risk analysis in China, we used a new Data envelopment analysis model named non-radial undesirable output DEA to estimates the eco-efficiency of 30 regions from year 2005 to 2008 in China. The results that we have achieved is as follows, first, the eastern part showed the highest green efficiency score 0.767 with the highest economic level at the same time. The central area's average eco-efficiency is 0.435 which ranked the second. The western area showed the worst green economic efficiency 0.333, the economic level of this area is the poorest in china too. Our conclusion is opposite to literature. In Lindmark and Vikstrm (2003)'s study the less developed area showed higher eco-efficiency than the more developed industrial area. According to our results, The western area in China is in a dilemma situation now, If the environment protection is emphasized, the poor economic conditions of western area can not be improved , but if the local government set no limitation to the pollution level, it is not accord with the spirit of Chinese sustainable development plan. We suggest that the western area can purchase waste discharge right from more developed area to settle this problem provisionally. In the future research we will adopt stochastic DEA and sensitivity analysis which created by Wu (2010) to compare the results.

## References

Ahujia G, Majumdar SK (1998) An assessment of the performance of Indian state-owned enterprises. J Prod Anal 9:113–132

Banker RD, Charnes A, Cooper WW (1984) Some models for estimating technical and scale inefficiencies in data envelopment analysis. J Manage Sci 30(9):1078–1092

Charnes A, Cooper WW (1962) Programming with linear fractional functions. Nav Res Logistics Q 15:330–334

Charnes A, Cooper WW, Rhodes E (1978) Measuring the efficiency of decision making units. Eur J Oper Res 2:429–444

Fare R, Grosskopf S, Lovell CAK, Pasurka C (1989) Multilateral productivity comparisons when some outputs are undesirable: a nonparametric approach. Rev Econ Stat 71:90–98

Freeman MA, Haveman RH, Kneese AV (1973) The economics of environmental policy. John Wiley & Sons, New York

Hu JL, Kao CH (2007) Effcient energy-savings targets for APEC economics. J Energ Policy 35:373–382

Hu JL, Wang SC (2006) Total-factor energy efficiency of regions in China. J Energ Policy 34(17):3206–3217

Lindmark M, Vikstrm P (2003) Lobar convergence in Productivity – a distance function approach to technical change and efficiency improvements. Paper for the conference catching-up growth and technology transfers in Asia and Western Europe, Groningen, pp 17–20

Ng YC, Chang MK (2003) Impact of computerization on firm performance: a case of Shanghai manufacturing enterprises. J Oper Res Soc 54:1029–1037

Ramanathan R (2000) A holistic approach to compare energy efficiencies of different transport modes. J Energ Policy 28:743–747

Schaltegger S, Synnestvedt T (2002) The link between "green" and economic success. J Environ Manage 65:339–346

Scheel H (2001) Undesirable outputs in efficiency valuations. Eur J Oper Res 132:400–410

Seiford LM, Zhu J (2002) Modeling undesirable factors in efficiency evaluation. Eur J Oper Res 142:16–20

Tone K (2003) A slacks-based measure of efficiency in data envelopment analysis. Eur J Oper Res 130:498–509

Wu DD (2010) A systematic stochastic efficiency analysis model and application to international supplier performance evaluation. Expert Syst Appl 37:6257–6264

Zhou P, Poh KL, Ang BW (2007) A non-radial DEA approach to measuring environmental performance. Eur J Oper Res 178:1–9

# Environmental Hazard by Population Urbanization: A Provincial Clustering Analysis Based on IRF

**Yamin Wang**

**Abstract** This paper investigates responses of nine environmental indicators to the urban population increase. We apply a impulse response function model that based on population and environmental data of 31 provinces, municipalities and autonomous provincial regions (expect Hong Kong, Macao and Taiwan) for the period 1998–2008. According to the Cumulate Impulse Response Ratio of these provincial regions, clustering analysis is done. Evidences show that: ten central and western provincial regions, in which a shock in urban population leads to significant positive effect in the most of pollution indicators, present good convergence. But in the other 21 provincial regions, when urban population increases in a short term, different environmental pollution indicators in different provincial regions have rather different presentation.

**Keywords** Clustering analysis · environmental hazard · impulse response · provincial regions · urban population

## 1 Introduction

Urbanization is a gradually process of agricultural population changing into non-agricultural population, the rural population into urban population increasingly, and rural areas into cities (Lam 2003). China's long-standing urban-rural dual structure makes a deep gap between urban and rural. Large numbers of people are migrating from rural to urban in recent years. Changes in the distribution of population in urban and rural will impact the environment as the difference in the domestic life style. Additionally, population moving into the cities makes large concentrations of human resources and industries in the cities. More pollution is caused along with the rapid consumption of resource and energy, which generates a great environmental pressure. Also, the concentration of the population will increase the

Y. Wang
Department of Finance, Nanjing University of Finance & Economics, Nanjing City 210046, P.R, China
e-mail: yaminwang1990@gmail.com

Y. Zhou and D.D. Wu (eds.), *Modeling Risk Management for Resources and Environment in China*, Computational Risk Management,
DOI 10.1007/978-3-642-18387-4_34, 

possible loss during environmental crisis (Jiang and Yu 2008). Hence, when regional governments intervene the process of urbanization through the registered permanent residence policy, they should consider not only scale economic effect, but also the environmental problems which are caused by over-concentration of population.

It is commonly accepted that the population urbanization pressure in different regions is unbalanced. Overall, the population pressure in China's eastern coastal cities is greater than the central and western provincial regions. And their abilities to deal with environmental issues are also different. Therefore, it's necessary to investigate the environmental effect by population urbanization in different provincial regions before to carry out coordination of population urbanization in macro-policy.

This article attempts to answer two questions: (1) for a specific period of time, how indicators of environmental pollution in different provincial regions will respond to the process of population urbanization. (2) what is similar and different among the response characteristics from one region to another.

## 2 Literature Review

Firstly, of content, there are few studies directly related to population urbanization and the environment. Most of the studies involved many aspects of urbanization performance including urban economic development, urban population growth, urban regional scale, the increasing of cities' number, and so on.

Secondly, in term of the research methods, studies on the relativity between urbanization and environmental pollution are mostly in connection with Environment Kuznets Curve (referred to as EKC curve). This theory was first proposed in 1992 by the American economist Grossman and Krueger, the meaning of which is that: in the early stages of modern economic growth, resource-intensive industry is dominant which generally produce more serious pollution. Because of the lack of clean technologies and environmental consciousness, environmental pollution will become more serious with the economic development. To a certain level of economic development, environmental pollution gradually reduce after knowledge-intensive industries and clean technologies developed and promoted.

Yuping Wu found that Beijing's economic growth and environmental pollution showed significant features of the inverted U-shaped curve (Wu et al. 2003). Xiaosi Tian discovered Nanjing industrial wastewater emission and GDP per capita represented N-shaped relationship, and that industrial waste gas emission and solid waste output line with inverted U-shaped relationship. These two indicators are ahead of schedule the turning point (Tian et al. 2007). Wangyi Hu also found the environmental indicators and GDP per capita in Nanjing showed a certain succession of EKC trajectory characteristics. Since the 1990s of the twentieth century, environmental deterioration has been checked in Nanjing, gradually into the economic and environmental development of the advanced stage (Wang et al. 2006). Xiumig Hu's research indicated that industrial "three wastes" of pollutants and emission in Wuhan is in the left outside of EKC curve. The rest of the curves have passed or are in the EKC turning point. And industrial "three wastes" pollution began to develop in the direction of benign (Hu et al. 2005). Furthermore, some scholars find environmental indicators and urbanization between the U-shaped, inverted U-shaped law or N-shaped. Jiang Du, in the relationship between urbanization and the environment for research, verify the level of urbanization of China and various environmental indicators whether characteristics of EKC curve.

However, researches in these areas mostly focus on some single particular city. To the extent of our knowledge, only a limited number of studies examine comparison of various areas. What's more, the studies are still in the blank which aimed at environmental consequences for the sudden start of the urban population growth. But in populous countries such as China, the population policy will likely significantly change in the urban population of a region in the short term. So it's valuable to expend more targeted research for short-term environmental consequences of urban population change.

# 3 Data and Time Series Properties

We utilize annual data on Urban Population (P) , Volume of Industrial Dust Emission (Ind_D), Volume of Industrial Fume Emission (Ind_F), Volume of Industrial Sulphur Dioxide Emission (Ind_SD), Volume of Industrial Solid Waste Emission (Ind_SW), Volume of Industrial Waste Water Emission (Ind_WW), Volume of Domestic Fume Emission (Dom_F), Volume of Domestic Fume Emission (Dom_ SD), Volume of Domestic Fume Emission (Dom_SW),Volume of Domestic Fume Emission (Dom_ WW) for the period 1998–2008. All data are from "CHINA POPULATION STATISTICS YEARBOOK" (1999–2009), and the China Economic Information Network (http://db.cei.gov.cn/).

Unit root test results are needed to properly specify and estimate VARs in generalized variance decompositions, we utilize Augmented Dickey and Fuller (ADF). Taking Beijing as an example, the results of the unit root test are reported in Table 1. According to the unit root test results, first-order difference should be

**Table 1** Unit root test results

| Variable | ADF | Result | Variable | ADF | Result |
|---|---|---|---|---|---|
| Levels | | | First differences | | |
| P | −0.917457 | Refused c | P | −3.162963 | Received b |
| Ind_D | −0.858465 | Refused c | Ind_D | −2.947677 | Received c |
| Ind_F | −4.305004 | Received a | Ind_F | −4.164721 | Received b |
| Ind_SD | −1.182070 | Refused c | Ind_SD | −2.910368 | Received c |
| Ind_SW | −7.030937 | Received a | Ind_SW | −3.481668 | Received b |
| Ind_WW | −4.028240 | Received b | Ind_WW | −3.250248 | Received c |
| Dom_F | −2.227573 | Refused c | Dom_F | −3.028764 | Received c |
| Dom _SD | −5.777285 | Received a | Dom _SD | −5.643965 | Received a |
| Dom _SW | −0.461084 | Refused c | Dom _SW | −2.771221 | Received b |
| Dom _WW | −2.381752 | Refused c | Dom _WW | −7.977759 | Received a |

Superscripts a, b, and c indicate significance at 1%, 5%, and 10% respectively
Levels: 1%: −4.297073; 5%: −3.212696; 10%: −2.747676; First differences, 1%: −4.420595; 5%: −3.259808; 10%: −2.771129

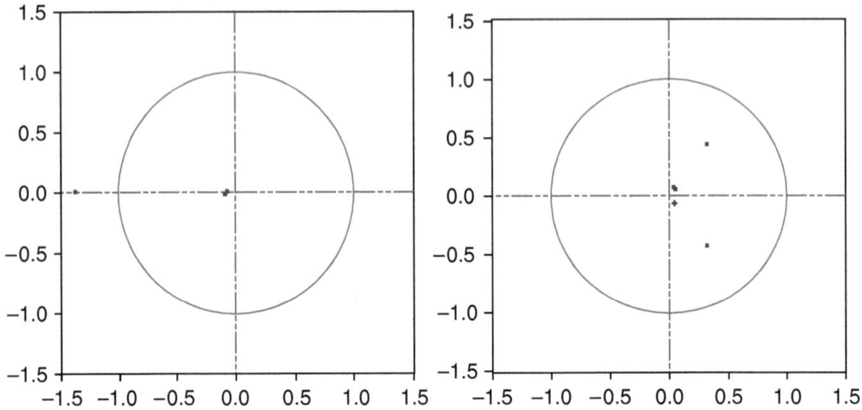


**Fig. 1** AR characteristic polynomial when p = 1 (*left*) and when p = 2 (*right*)

done in the series of P, Ind_D, Ind_SD, Dom_F, Dom _SW, and Dom _WW. The estimated VAR system is as below:

$$y_t = A_1 y_{t-1} + \cdots + A_p y_{t-p} + \varepsilon_t \quad t = 1, 2, \cdots, T \tag{1}$$

For example: $y_t = (P_t, Dom\_\ SD_t)'$, $P$ is lag lengths, $A_1 \cdots A_p$ are (2×2) coefficient matrices, and $\varepsilon_t$ denotes white noise residuals.

In order to find the optimal lag lengths (p), we need to employ an AR-Root test. We can receive the lag lengths (p) with which the inverse roots of AR Characteristic Polynomial are all in a unit circle. The graph of the inverse roots of AR Characteristic Polynomial when p = 1 (*Right*) and p = 2 (*Left*) is as below Fig. 1:

So, we can let p = 2.The estimated VAR(2) system is as below:

$$y_t = A_1 y_{t-1} + A_2 y_{t-2} + B x_t + \varepsilon_t \quad t = 1, 2, \cdots, T \tag{2}$$

## 4 Impulse Response Model (IRF)

Considering the impacts of Urban Population (P) on all variables in the VAR(2) system may provide useful insights about the short run. To that respect we employ impulse response decompositions of Koop and Pearson and Shin. Impulse responses show how a variable responds to a shock in the other variable initially and whether the effect of the shock persists or dies out quickly.

In this case of Beijing, it is clear from an examination of Fig. 2 that Volume of Industrial Fume Emission (Ind_F), Volume of Industrial Waste Water Emission (Ind_WW), Volume of Domestic Fume Emission (Dom_SW) and Volume of Domestic Fume Emission (Dom_ WW) do not respond at all to changes in Urban Population (P). Whereas the response of Volume of Industrial Dust Emission

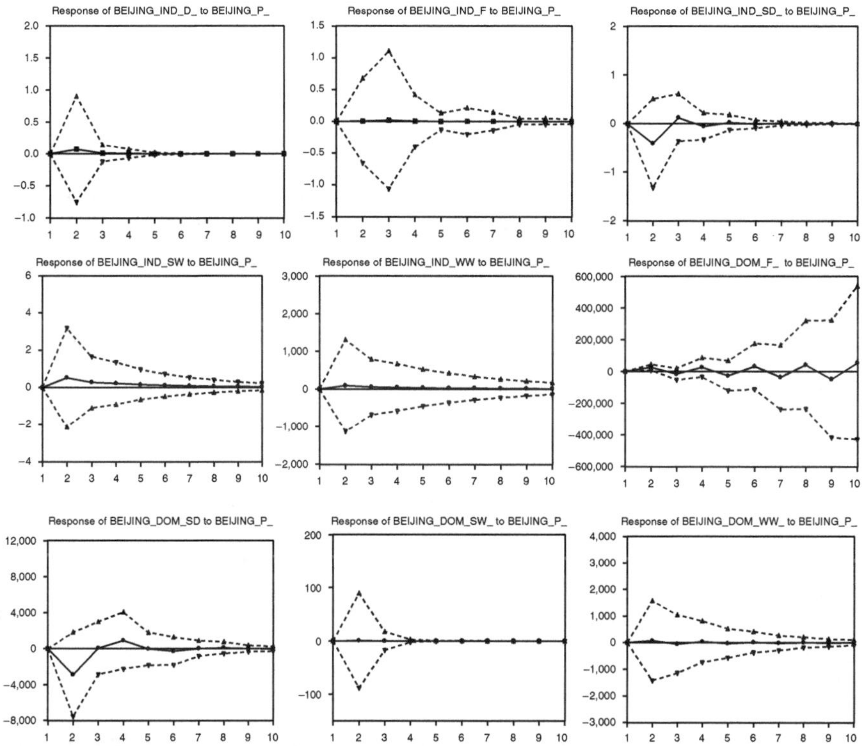


**Fig. 2** Impulse response results in the example of Beijing

(Ind_D), Volume of Industrial Sulphur Dioxide Emission (Ind_SD) and Volume of Domestic Fume Emission (Dom_F) to Urban Population (P) have negative and significant initial representation.

## 5 A Provincial Clustering Analysis

We define the Cumulate Impulse Response Ratio to show the relative pressure of pollution in various provincial regions. The definition of the Cumulate Impulse Response Ratio of r indicator in k region for the period of n is as below:

$$\Delta_{rk}(n) = \sum_{i=1}^{n} \frac{\varphi_{rki}}{x_{rk2008}} \tag{3}$$

$\varphi$ is the value of Impulse Response, $x$ is the value of r indicator in k region in the year 2008.Cumulate Impulse Response Ratio of 31 provincial regions for the period of 5 years are in the Table 2.

**Table. 2** Cumulate impulse response ratio of 31 provincial regions for the period of 5 years

| | | Ind_D | Ind_F | Ind_SD | Ind_SW | Ind_WW | Dom_F | Dom_SD | Dom_SW | Dom_WW |
|---|---|---|---|---|---|---|---|---|---|---|
| 1 | Anhui | 0.20 | 0.10 | 0.56 | −10.23 | 0.02 | 0.23 | 0.15 | 0.22 | 0.13 |
| 2 | Beijing | −2.39 | −0.05 | 0.24 | −0.45 | −0.27 | −0.04 | −0.12 | 1.54 | −0.01 |
| 3 | Chongqing | −4.95 | 0.04 | 0.24 | 6.64 | 0.04 | 0.00 | 0.01 | 0.02 | 0.12 |
| 4 | Fujian | 0.06 | −0.03 | 0.59 | −0.06 | 0.25 | 0.96 | 1.81 | 0.05 | 0.18 |
| 5 | Gansu | −0.82 | 0.20 | 0.11 | 0.66 | 0.00 | −0.02 | −0.29 | 0.29 | 0.28 |
| 6 | Guangdong | 0.07 | 0.08 | −0.04 | −0.35 | 0.43 | 0.12 | 2.36 | −0.01 | 0.00 |
| 7 | Guangxi | 0.12 | 0.09 | 0.22 | −0.33 | 0.63 | 0.35 | 0.08 | −0.01 | 0.13 |
| 8 | Guizhou | 0.33 | −0.23 | 0.25 | −0.20 | 0.04 | 0.33 | 0.08 | 0.02 | 0.08 |
| 9 | Hainan | −0.06 | −0.28 | 0.20 | 5.14 | −0.22 | 0.65 | −0.64 | 0.03 | 1.00 |
| 10 | Hebei | 0.07 | 0.14 | 0.10 | −0.05 | 0.12 | −0.03 | −0.01 | 0.00 | 0.17 |
| 11 | Heilongjiang | 0.50 | −0.05 | 0.54 | −0.92 | 0.06 | −2.20 | −0.01 | 0.03 | −0.01 |
| 12 | Henan | 5.78 | 0.06 | 0.41 | −0.31 | −0.02 | 8.47 | 0.06 | 0.10 | 0.04 |
| 13 | Hubei | −5.64 | 0.47 | 0.30 | 0.13 | 0.01 | 0.44 | 0.37 | 0.24 | −0.27 |
| 14 | Hunan | −0.01 | 0.07 | 0.21 | −1.31 | 0.04 | 0.03 | −0.02 | 0.07 | 0.20 |
| 15 | In_Mongolia | 0.72 | 0.58 | 0.42 | 0.19 | 0.39 | −0.08 | 0.00 | 0.06 | 0.18 |
| 16 | Jiangsu | 0.52 | −0.04 | 0.02 | −5.37 | −0.03 | −0.29 | 0.01 | −0.01 | 0.09 |
| 17 | Jiangxi | 0.14 | 0.06 | 0.60 | 0.05 | 0.01 | 0.13 | 0.30 | 0.12 | 0.09 |
| 18 | Jilin | −0.12 | 1.53 | 1.01 | 0.12 | 0.94 | 0.06 | −0.09 | −0.01 | 1.13 |
| 19 | Liaoning | 0.13 | 0.17 | 1.77 | −2.30 | 0.07 | 1.99 | 0.10 | 0.01 | 0.29 |
| 20 | Ningxia | −4.44 | −0.08 | 0.21 | −0.15 | 0.30 | −0.23 | 0.00 | 0.69 | −0.07 |
| 21 | Qinghai | 0.63 | 0.06 | 0.73 | −1.36 | 0.29 | 0.12 | −0.10 | −0.22 | 0.46 |
| 22 | Shaanxi | −0.19 | 0.19 | 0.26 | −0.39 | 1.99 | 1.63 | 0.10 | 0.21 | −0.14 |
| 23 | Shandong | −1.47 | 0.60 | 0.05 | −2.62 | 0.77 | 0.01 | 0.28 | −0.36 | 0.01 |
| 24 | Shanghai | 1.41 | −0.01 | −0.09 | 0.20 | 0.07 | 2.23 | −0.38 | 5.61 | −0.70 |
| 25 | Shanxi | 0.15 | −7.46 | 0.05 | 8.93 | 0.11 | 0.02 | 0.00 | −0.28 | 0.02 |
| 26 | Sichuan | 4.38 | −0.97 | 0.01 | −2.93 | 0.01 | −0.05 | 0.00 | 0.07 | 0.05 |
| 27 | Tianjin | −9.00 | −0.19 | 0.00 | −5.12 | 0.08 | −0.73 | −0.02 | −0.02 | −0.14 |
| 28 | Tibet | −0.06 | 0.04 | 0.10 | −0.20 | −0.06 | 0.10 | 0.06 | 0.19 | 0.09 |
| 29 | Xinjiang | 0.45 | 0.34 | 0.53 | 0.08 | 0.36 | −0.35 | −0.02 | 0.18 | 0.31 |
| 30 | Yunnan | 0.40 | 0.26 | 0.21 | −0.83 | 0.00 | 0.07 | 0.18 | −0.40 | 0.21 |
| 31 | Zhejiang | −0.01 | 0.00 | −0.27 | −1.18 | 0.00 | 0.02 | −0.07 | −0.18 | 0.01 |

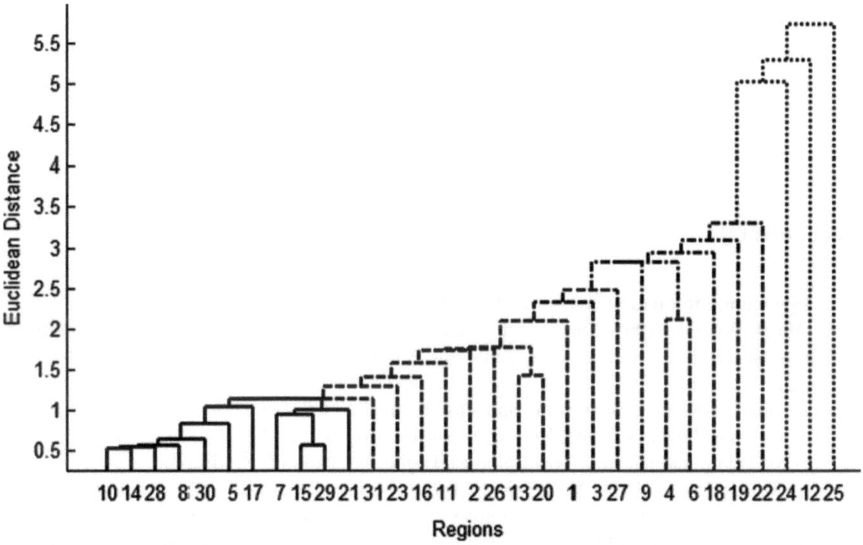


**Fig. 3** Clustering analysis graph

Based on the Cumulate Impulse Response Ratio for the period of 5 years in 31 provincial regions, we do the Clustering Analysis with the principle of shortest euclidean distance. Clustering Analysis Graph is as Fig. 3. It is clearly that in the nearest distance Hebei (10), Hunan (14), Tibet (28), Guizhou (8), Yunnan (30), Gansu (5), Jiangxi (17), Guangxi (7), Linner Mongolia (15), Xinjiang (29), Qinghai (21) adding up to ten provincial regions make up a community. And through the Table 2 we can know that a shock in urban population leads to positive effect in the most of pollution indicators in these provincial regions. But when it comes to the presentations of the other provincial regions, things are become various. The Cumulate Impulse Response Ratio of the other 21 provincial regions shows a relatively great difference. Shanxi, Henan and Shanghai show the greatest difference from others.

# 6 Conclusion

By employing the impulse response model(IRF) and the clustering analysis model, we were able to find that China's ten central and western provincial regions show good convergence. In these provincial regions, a shock in urban population leads to significant positive effect in the most of pollution indicators which might because of the dominant position resource-intensive industry having and the lack of clean technologies and environmental consciousness in these provincial regions. It seems to give some evidence to prove the EKC curve theory. But in the other 21 provincial regions, when urban population increases in a short term, different environmental pollution indicators in different provincial regions have rather different presentation. Thus, the population policy should be made under the considering of the similarity and differences of different environmental pollution indicators in these provincial regions.

## References

China Economic Information Network, http://db.cei.gov.cn/

Hu M, Hu H, Wang L (2005) Study on the quadratic model ofindustrial "three wastes" in Wuhan City – based on environmental kuzents curve (EKC). Resources and Environment in the Yangtze Basin 14:470–474

Jiang D, Yu L (2008) Urbanization and environmental pollution: empirical study sased on provincial panel data of China. Resources and Environment in the Yangtze Basin 6:825–826

Lam H (2003) Population science (in Chinese). Higher Education Press, Beijing

Tian X, Chen J, Zhu C (2007) Modeling the level of economic growth and the industrial "three waste" pollution in Nanjing City. Resources and Environment in the Yangtze Basin 4:410–413

Wang Y, Cui X, Chen W (2006) Empirical research on the relationship between economic growth and environmental degradation in Nanjing City. Resources and Environment in the Yangtze Basin 2:142–146

Wu Y, Ke S, Sung J (2003) Modeling economic growth and environmental degradation of Beijing. Geographical Research 2:239–245

# Study on Sustainable Utilization of Water Resources in Tieling City Based on System Dynamics Approach

**Yan Li, Cheng Hu, Yuanhui Zhao, and Xiaoqiang Tan**

**Abstract** The urban water supply and demand model plays an important role in the simulation and sustainable utilization of urban water resources. The system dynamics (SD) approach is applied to construct the water supply and demand model of Tieling, which is used to simulate the development tendency of water resources and forecast the water demand in planning years. The practical verification on historical data shows that the relative error was small and the model is reliable. Furthermore, we present four modes to manage water resources in the paper. Through the comparison and analysis of the simulation results simulated by the proposed model in the paper under the four different modes, we can find that water supply is greater than water demand from 2005 to 2020 according to the second and fourth modes. That is, the water resources in Tieling are sustainable utilization if we take saving measures and pollution control measures on water resources.

**Keywords** Sensitivity · supply-demand model · sustainable utilization · system dynamics · water resources

## 1 Introduction

Tieling locates in the north of Liaoning Province and the middle of Songliao Plain. It will become one of the Shenyang economic regions. Therefore, the water resources will affect its economic development. Currently, the supply of water resources in Tieling presents a downward trend. For example, the total amount of

Y. Li (✉), Y. Zhao, and X. Tan
College of Environmental and Chemical engineering, Shenyang Ligong University, Shenyang 110159, China
e-mail: liyan6865@tom.com

C. Hu
Liaoning Academy of Environmental Sciences, Shenyang 1100031, China
e-mail: hch8812@163.com

Y. Zhou and D.D. Wu (eds.), *Modeling Risk Management for Resources and Environment in China*, Computational Risk Management,
DOI 10.1007/978-3-642-18387-4_35, 

water resources is 1.877 billion $m^3$ in the year 2007, which is less than average 2.559 billion $m^3$ of several years, and the capacity of supply is 1.045 billion $m^3$ lower than 1.217 billion $m^3$ of the year 2006 (Tieling Water Conservancy Bureau 2008). So there will be a bad effect on the ability of water supply and even its sustainable utilization without proper management on water resources. Nowadays, there are many methods to assess water resources, such as analytic hierarchy process (Yi et al. 2007), multi-objective analysis (Cheng 2004), neural network (Lou and Liu 2004), optimization (Zuo 2005), principal component analysis (Yi et al. 2008), system dynamics method (Tian et al. 2009; Wang et al. 2005; Chen 2005; Zhao 2006). SD method has been used to simulate the relationship among factors of a complex system (Xu and Zou 2005). It is based on system theory, and integrates feedback theory, information theory, decision support theory, and computer technology, together. SD method mainly reflects the casual feedback relationship among variables of a module in the system through the first-order differential equations (Wang et al. 2009). It can forecast water supply and demand under different management programs, and predict the relevant variables so as to obtain the best solution of water management.

# 2 Construction of Water Supply and Demand Model in Tieling

During the water supply and demand modeling using SD method, we should determine the system boundary first, then analyze the system structure, reveal the contradictions and problems in the system, define evaluation objectives, and identify relevant variables and their characteristics (Ford 2009).

## *2.1 Determination of the System Boundary*

SD boundary not only can distinguish the internal and external part of the system, but also is the important connection between the system and its external part. In this paper, we determine the system boundary as follows: (a) the planning period ranges from year 2005 to 2020 by one year step, (b) the baseline year is 2005, and (c) Tieling administrative region is taken as the modeling region boundary.

## *2.2 Dividing Subsystem*

Water supply-demand system (WSDS) is a complex system, which contains population, society, environment, resources, and other factors. According to the real situation of water resources in Tieling and the modeling requirements, we divide the WSDS into four major subsystems, i.e., population subsystem, economic subsystem, water resources subsystem and water environment subsystem. Each subsystem contains a number of secondary subsystems. The four subsystems together influence the interaction of supply and demand of water resources in Tieling.

### 2.2.1 Population Subsystem

Population is the most active factor in water supply-demand balance system. The relationship between population and water resources manifests in two aspects. On one hand, excessive population growth will lead to environmental deterioration and water scarcity, on the other hand, the shortage or surplus of water resources will affect population growth. Therefore, population subsystem could reflect the relationship between population and water resources. The variables in population subsystem include total amount of population, population growth speed, population growth rate, the amount of urban population, the amount of rural population, urbanization level, and so on.

### 2.2.2 Economy Subsystem

Socio-economic development is closely related to water supply-demand. Economic growth will increase the demand and consumption of water resources, at the same time, the shortage of water will affect the supply of water and hinder economic development. Economy as a subsystem will play a critical important role in WSDS. The variables in economy subsystem include industrial added value (IAV), the growth rate of industrial added value, and so on.

### 2.2.3 Water Resources Subsystem

Water resources, which are the core of the system, are an important material basis for human survival and development, and can directly impact on water supply and demand. The variables in water resources subsystem include the total supply amount, total water demand amount, the difference between water supply and demand, the shortage degree of water, water desalination capacity, the amount of water in this region, the amount of available groundwater resources, the amount of surface water supply, the amount of sewage reuse, industrial water demand, agricultural water demand, agricultural irrigation area, the growth speed of agricultural irrigation area, domestic water demand, ecological water demand, and etc.

### 2.2.4 Water Environment Subsystem

The quality of water directly affects the use of water resources, i.e., the good water quality may improve the utilization of water resources, otherwise, the poor quality may reduce the utilization, even have a bad effect on water supply. So, it is a very significant part in the WSDS. This subsystem contains the following variables: total amount of sewage, industrial waste water discharged from sewage treatment plant, the amount of living sewage emissions, the coefficient of wastewater discharge, sewage treatment capacity, the rate of sewage treatment, the amount of wastewater reclamation, the wastewater reclamation rate, COD emissions (consistency with national water environment control indicator), living COD emissions, industrial COD emissions, and etc.

### 2.2.5 State Equations

Apart from the above four subsystems, there are three state equations and 29 variables which are introduced in the above paragraphs, in the model (see Fig. 1). The three state equations are as follows.

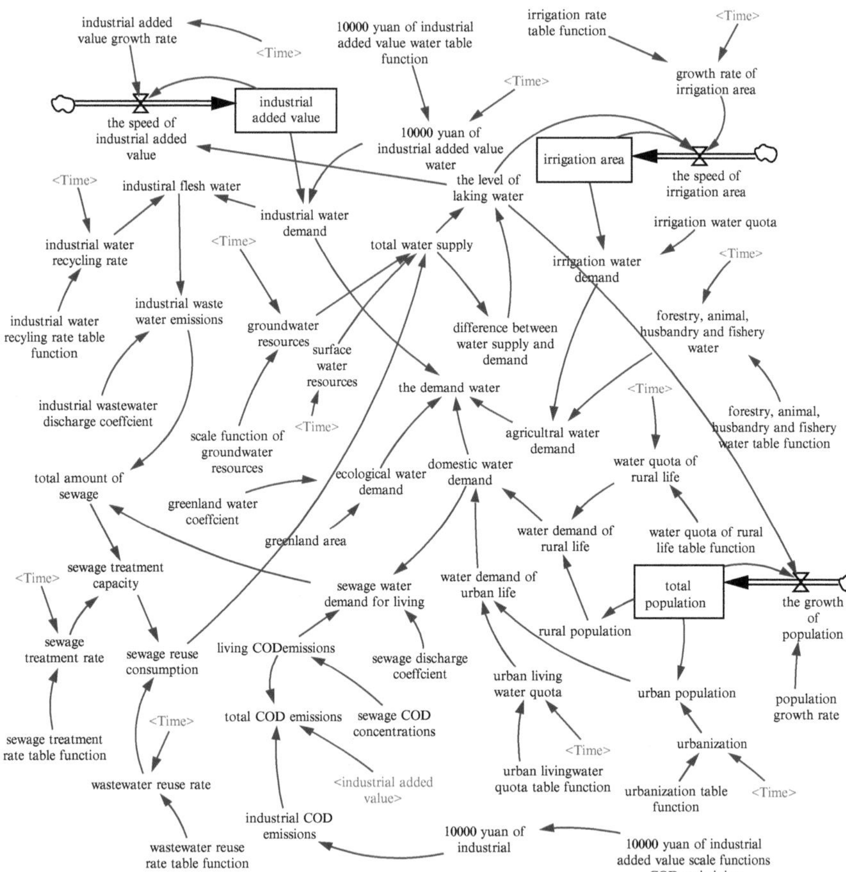


**Fig. 1** Water supply balance system flow diagram in Tieling

*Industrial added value* = INTEG(+the growth speed of IAV, the initial IAV);
*Total population* = INTEG(+the growth speed of population, the initial population);
*Irrigation area* = INTEG(+the growth speed of irrigation area, the initial irrigation area).

## 3 Model Examination Using Historical Data

Because there are many variables in the model, we can not list the whole examination results of all variables. So, we just list the examination results of industrial added value, irrigation area, and total population to illustrate the effectivity of the model constructed in the paper. The statistics data from 2001 to 2008 used in the examination come from statistic Yearbook (Tieling Statistics Bureau 2001–2008) and some Bulletins(Department of Environmental Protection of Liaoning Province 2002; Tieling Water Conservancy Bureau 2002–2009). Table 1 shows the examination results. The examination results show that all the relative errors on the variables are lower than 3%.

## 4 Results and Analysis

In this section, we present four modes for managing water resources, and give the parameter values of decision variables in Sect. 4.1. Then, in Sect. 4.2, we present the simulation results computed by the WSDS model proposed in the paper.

**Table 1** Comparison between historical value and the related simulated value by WSDS model

| Year | Industrial added value (10 thousand yuan) | | | Irrigation area (10 billion $m^3$) | | | Total population (10 thousand) | | |
|---|---|---|---|---|---|---|---|---|---|
| | Hist. data | Simu. value | Relative error (%) | Hist. data | Simu. value | Relative error (%) | Hist. data | Simu. value | Relative error (%) |
| 2001 | 30.3 | 30.3 | 0 | 15.3 | 15.3 | 0 | 298.9 | 298.9 | 0 |
| 2002 | 34.8 | 35.0 | 0.57 | 15.7 | 15.4 | 1.91 | 299.3 | 299.5 | 0.07 |
| 2003 | 52.2 | 51.9 | 0.57 | 15.7 | 15.8 | 0.64 | 299.4 | 300.1 | 0.23 |
| 2004 | 73.3 | 74.4 | 1.50 | 15.9 | 15.8 | 0.63 | 300.4 | 300.7 | 0.10 |
| 2005 | 95.6 | 93.1 | 2.62 | 15.9 | 16.0 | 0.63 | 302.6 | 301.3 | 0.43 |
| 2006 | 126.4 | 129.0 | 2.06 | 15.9 | 16.0 | 0.63 | 304.6 | 303.2 | 0.46 |
| 2007 | 171.6 | 174.1 | 1.46 | 15.9 | 16.0 | 0.63 | 305.4 | 303.8 | 0.52 |
| 2008 | 256.4 | 249.1 | 2.85 | 16.0 | 16.1 | 0.63 | 305.9 | 304.4 | 0.49 |

## 4.1 Water Management Mode Design

According to socio-economic conditions, water resources management schemes, economic development planning, water conservation planning, environmental planning, and etc in Tieling, we give four water resources management modes (see Table 2), and use the WSDS model proposed in the paper to predict the trend of water supply and demand from year 2005 to 2020.

Mode 1 is the current developing mode, which simulates the natural evolution process of the system in the future several years without adding any direct human intervention according to the historical development level of the system. The values of the variables in this mode are entirely based on the actual development level of the year 2005 as a reference, and are used to simulate and forecast the future of the system. Mode 2 is the only saving measures mode, which adopts some measures to improve the reuse rate of industrial water and wastewater, to lower the water consumption quotas of farmland irrigation, urban and rural water, and to reduce industrial water demand. Mode 3 is only pollution control measures mode, which applies economic and technical ways to improve sewage treatment rate and to reduce COD emissions of IAV. Mode 4 combines saving measures mode and pollution control measures mode, which uses saving measures and pollution control measures at the same time. The values of decision variables of each mode are listed in Table 3.

## 4.2 Analysis of Simulation Results

We use the proposed WSDS model to simulate the supply and demand of water resources under the four water management modes, respectively, in order to find effective water management modes to achieve water resources sustainable utilization. Simulated results are shown in Fig. 2.

Figure 2 shows the trend of water demand and supply. From Fig. 2a and c, we find that the amount of water supply is lager than the demand before the year 2014, after that year, the water demand is lager than water supply under Mode 1 and Mode 3.

That is, water resources are not sustainable utilization under these two modes. For Mode 2 and Mode 4 (see Fig. 2b and d), we find that the total water demands are relatively small due to cost-saving measures, and the total water requirements of the

**Table 2** Four modes to simulate the WSDS in Tieling

| Mode no. | Mode |
|---|---|
| Mode 1 | Current developing mode |
| Mode 2 | Only saving measures mode |
| Mode 3 | Only pollution control measures mode |
| Mode 4 | Saving measures and pollution control measures mode |

**Table 3** The value of each decision variable in each mode

| Decision variable | Mode 1 | Mode 2 |
|---|---|---|
| Industrial water reuse change table function(%) | [(2005,0)–(2020,1)], {0.91,0.91,0.91,0.91} | [(2005,0)–(2020,1)], {0.91,0.92,0.93,0.94} |
| Industrial wastewater discharge coefficient(%) | 0.80 | 0.80 |
| Disposal rate of sewage table function(%) | [(2005,0)–(2020,1)], {0.23,0.23,0.23,0.23} | [(2005,0)–(2020,1)], {0.23,0.23,0.23,0.23} |
| Rate of water reuse table function(%) | [(2005,0)–(2020,1)], {0.06,0.06,0.06,0.06} | [(2005,0)–(2020,1)], {0.06,0.12,0.18,0.24} |
| Domestic sewage coefficient(%) | 0.90 | 0.90 |
| COD emissions of IAV table function (kg/TTY) | [(2005,0)–(2020,20)], {13.5,13.5,13.5,13.5} | [(2005,0)–(2020,20)], {13.5,13.5,13.5,13.5} |
| area of greenland ($hm^2$) | 1530 | 1530 |
| Urban living water quota (L/(each *D)) | [(2005,0)–(2020,300)], {153,153,153,153} | [(2005,0)–(2020,300)], {153,130,110,100} |
| Rural water quota(L/(each*D)) | [(2005,0)–(2020,300)], {70,70,70,70} | [(2005,0)–(2020,300)], {70 ,65,60,55} |
| Farmland irrigation quota (Ten thousand $m^3$/ $hm^2$) | 0.41 | 0.35 |
| Water consumption of ten thousand Yuan IAV ($m^3$/TTY) | [(2005,0)–(2020,200)], {120.8,120.8,120.8,120.8} | [(2005,0)–(2020,200)], {120.8,100.0,80.0,60.0} |
| **Decision variable** | **Mode 3** | **Mode 4** |
| Industrial water reuse change table function(%) | [(2005,0)–(2020,1)], {0.91,0.91,0.91,0.91} | [(2005,0)–(2020,1)], {0.91,0.92,0.93,0.94} |
| Industrial wastewater discharge coefficient(%) | 0.60 | 0.60 |
| Disposal rate of sewage table function(%) | [(2005,0)–(2020,1)], {0.23,0.3,0.4,0.5} | [(2005,0)–(2020,1)], {0.23,0.3,0.4,0.5} |
| Rate of water reuse table function(%) | [(2005,0)–(2020,1)], {0.06,0.06,0.06,0.06} | [(2005,0)–(2020,0.5)], {0.06,0.12,0.18,0.24} |
| Domestic sewage coefficient(%) | 0.70 | 0.70 |
| COD emissions of IAV table function (kg/TTY) | [(2005,0)–(2020,20)], {13.5,9.0,5.0,1.0} | [(2005,0)–(2020,20)], {13.5,9.0,5.0,1.0} |
| Area of greenland (hm2) | 1530 | 1530 |
| Urban living water quota (L/(each *D)) | [(2005,0)–(2020,300)], {153,153,153,153} | [(2005,100)–(2020,300)], {153,130,110,100} |
| Rural water quota (L/(each*D)) | [(2005,20)–(2020,100)], {70,70,70,70} | [(2005,20)–(2020,100)], {70 ,65,60,55} |
| Farmland irrigation quota (Ten thousand m3/ hm2) | 0.41 | 0.35 |
| Water consumption of ten thousand Yuan IAV (m3/TTY) | [(2005,0)–(2020,200)], {120.8,120.8,120.8,120.8} | [(2005,20)–(2020,200)], {120.8,100.0,80.0,60.0} |

Remark: TTY–ten thousand Yuan; Sequence (2005,X), (2010,Y), (2015,Z), (2020,W) is shortened as {X,Y,Z,W}

two modes are basically same which is 907 million $m^3$. According to the water supply-demand model, we can draw conclusion that Mode 2 and Mode 4 can achieve water resources sustainable utilization until 2020, even more after. That is, the supply of water outweighs the demand and the system is in a surplus state.

From Fig. 2, we can also find that there is an upward trend of water supply in the four modes, but the second and the fourth ones have a greater increase in total water supply, while others are small. The descending order of water supply will be 4,2,3,1 till 2020. This observation is consistent with the commonsense.

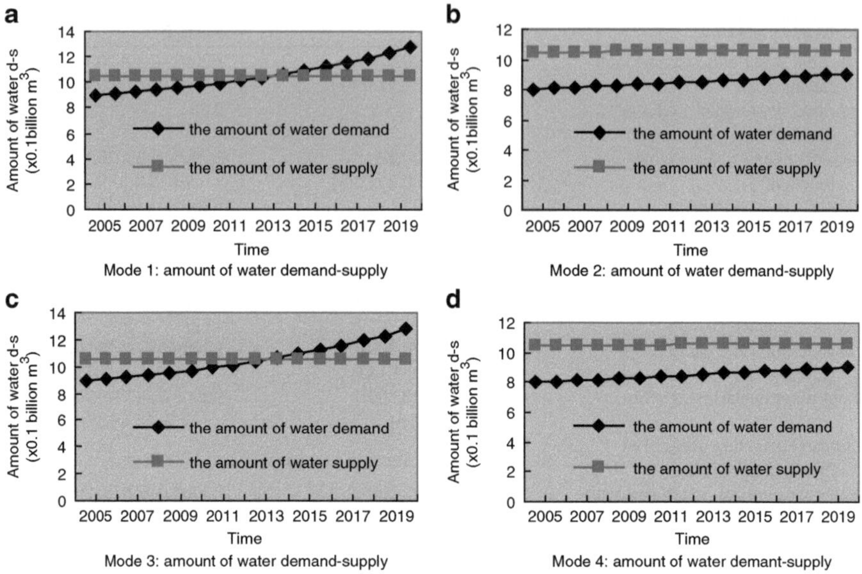


**Fig. 2** Simulation of the amount of water supply and demand

# 5 Conclusion

The system dynamics method is applied into the dynamic supply-demand balance of water resources in Tieling. The model is reliable through history data examination. The predicted results can be used in water resources management in Tieling, and the related management modes can achieve the sustainable use of water resources in Tieling. We can find that water supply-demand balance will be broken if we keep the current development schema or are only on the view of environment protection since 2014. We must take saving measures and pollution control measures together to achieve water resources sustainable utilization in Tieling.

**Acknowledgments** This research was supported by the National Prominent Special Project Foundation of China under Grant No. 2009ZX07526-006.

# References

Chen M (2005) The theory and methods for water resources carrying capacity assessment. Hehai University (in Chinese)

Cheng G (2004) Multi-objective analysis based on the sustainable use of regional water resources. Kunming Institute of Technology. (in Chinese)

Department of Environmental Protection of Liaoning Province (2002–2009) Bulletin of the state of environment 2001–2008 (in Chinese)

Ford J (2009) Environmental simulation-environmental systems introduction to system dynamics. Science Press, Beijing (in Chinese)
Lou W, Liu S (2004) On assessment of sustainable development level of regional water resource using artificial neural networks. J Agric Syst Sci Integr Res 20(2):113–119 (in Chinese)
Tian L, Zhang H, Zhang X (2009) A system dynamics approach for economic developing zone water demand forecasting: a case study of Tianjin Linkong area. J Tianjin Polytech Univ, 8 (3) (in Chinese)
Tieling Statistics Bureau (2001–2008) Tieling city statistical yearbook. Tieling Municipal Statistics Bureau (in Chinese)
Tieling Water Conservancy Bureau (2008) Water resources bulletin Tieling 2007. Tieling Daily. 25 Mar 2008 (in Chinese)
Tieling Water Conservancy Bureau (2002–2009) Water Resources Bulletin Tieling, 2001–2008 (in Chinese)
Wang W, Lei X, Yu X (2005) Study on the region carrying capacity of water resources based on system dynamics(SD) model. J Water Resour Water Eng 3(16):11–15 (in Chinese)
Wang J, Li X, Li F, Bao H (2009) Simulation and prediction of water environmental carrying capacity in Liaoning Province based on system dynamics model. J Appl Ecol 20(9):233–224 (in Chinese)
Xu G, Zou J (2005) The method of system dynamics: Principle, characteristics and new development. J Harbin Inst Technol Soc Sci 8(4):72–77 (in Chinese)
Yi L, Li J, Fan W (2007) Evaluation of sustainable exploitation and utilization of water resources based on analytic hierarchy process (AHP) method in Aksu Region. J Water Resour Water Eng 18(1):44–52 (in Chinese)
Yi Y, Haimiti Y, Wang T et al (2008) Application of principal component analysis in analyzing water quality of urban rivers. Arid Zone Res 25(4):498–501 (in Chinese)
Zhao C (2006) System dynamics to the regional water resources carrying capacity in applied research. Xi'an University of Architecture and Technology (in Chinese)
Zuo Q (2005) Urban water resources carrying capacity theory, methods and application. Chemical Industry Press, Beijing (in Chinese)

# Research on Ecosystem Service Value of Forests in the Upper Qiupu River

**Zhang Leqin, Fang Yuyuan, Xu Xingwang, Cao Xianhe, and Rong Huifang**

**Abstract** Taking LY/T1721—2008 as assessment standard, this study chosen these methods such as demonstration, literature study, expert interview and comparison. Results showed that the value of forest ecosystem services was about $48{,}556.77 \times 10^4$ Yuan, the production value $5{,}762.23 \times 10^4$ Yuan and the social services value $7{,}057.02 \times 10^4$ Yuan, respectively. Obviously, the ecosystem service value reached 8.42 times of the production value and 6.88 times of the social services value, respectively. So, the paper insists that the ecological services value was the upper limit of ecological compensation for the upper Qiupu River basin, and the values including the storage and retention of water and soil and water conservation were the lower limit.

**Keywords** Forest ecosystem service value · Qiupu river · Risk · Sensitivity · Sustainable development

## 1 Introduction

The Qiupu River is located in the Yangtze River branch in chizhou city anhui province. The upstream and downstream of Qiupu River are located in Shitai county and Guichi area respectively. However, the residents in Shitai country sacrificed the development opportunity to benefit the residents in Guichi area, which was called environmental external economical behavior. The private cost of environmental external economical behavior was more than social cost while the private income was less than social benefit.

Z. Leqin (✉), F. Yuyuan, X. Xingwang, and R. Huifang
Resource Environment and Tourism Department, Chizhou College, Chizhou, Anhui, China
e-mail: zhangleqing@sohu.com

C. Xianhe
State Forestry of Shitai, Shitai, Anhui, China


DOI 10.1007/978-3-642-18387-4_36, 

These theories such as externality theory, public goods theory, law of value theory and function theory inisit that the upstream area should compensate the downstream area for ecological environment protection to internalize the external behavior. So we should figure out the compensation standards and the forests ecosystem service value in the upper Qiupu River must be evaluated as a precondition. Besides, the comparison of the ecosystem service value and the economic and social value could arouse the environmental awareness of the residents in upstream area, which can benefit the sustainable development of the regional resources, environment and society.

## 2 The General Situation of the Research District

As the research object, the upstream area in Shitai country is a humid subtropical monsoon climate with low mountain terrain, high hill. In the research district, the area of forest was 111,000 $hm^2$ (Chizhou Bureau of Statistics 2008) with a forest coverage 81.7%, including the area of broad leaved forest 58,410 $hm^2$, coniferous forest 41,098 $hm^2$ and shrub forest 5,344 $hm^2$ (Cao 2008).

## 3 Assessment Method and Index Selection

The research method was that mentioned in The Specifications for Assessment of Forest Ecosystem Services in China (No: LY/T1721—2008) (The Research Institute of Forest Ecology, Environment and Protection, The Chinese Academy of Forestry 2008). The research indices were showed in Table 1. The date and

**Table 1** The indices system of ecosystem service value used in this study

| Value type | Index classification | Index |
|---|---|---|
| Production value | Production of forest products | Food production |
| Ecosystem service value | Storage and retention of water | Water volume regulating |
| | | Water purification |
| | Soil and water conservation | Soil conservation |
| | | Maintaining soil fertility |
| | Fixing carbon and releasing oxygen | Fixing carbon |
| | | Releasing oxygen |
| | Atmosphere environmental purification | Supply of negative ion |
| | | Absorption of pollutants |
| | | Adsorbing dust |
| | Nutrient accumulation | Nutrients accumulation of trees |
| | Biodiversity conservation | Biodiversity conservation |
| Social services value | Forest recreation, scientific and culture research | Forest recreation |
| | | Scientific and culture research |

parameter were obtained by three ways: social public data from LY/T1721—2008; literature data from published research results; and field date from Chizhou Statistical Yearbook 2008 and investigation data from Shitai Forestry Administration and Shitai Soil and Water Conservation Station.

# 4 Valuation Methods and Data Sources

## *4.1 Ecosystem Service Value*

### 4.1.1 Value Accounting of Water Resources Conservation

The value of water volume regulating can be calculated from (The Research Institute of Forest Ecology, Environment and Protection, The Chinese Academy of Forestry 2008):

$$U_{\text{regulating}} = 10 \times C_{\text{reservoin}} \times A \times (P - E - C). \quad (1)$$

$C_{\text{reservoin}}$ is the investment of unit capacity of reservoir construction with the value of 6.1107 Yuan·t$^{-1}$ (The Research Institute of Forest Ecology, Environment and Protection, The Chinese Academy of Forestry 2008). $A$ is the area of the forest in research with the value of 111,000 hm$^2$ (Chizhou Bureau of Statistics 2008), $P$ is the yearly precipitation with the value of 1,369.5 mm·a$^{-1}$ (Chizhou Bureau of Statistics 2008), $E$ is annual evaporation (742.2 mm·a$^{-1}$) (Tian 2006), $C$ is the surface runoff (396 mm·a$^{-1}$) (Chizhou Bureau of Statistics 2008).

The value of water purification can be calculated from (The Research Institute of Forest Ecology, Environment and Protection, The Chinese Academy of Forestry 2008):

$$U_{\text{purification}} = 10 \times K \times A \times (P - E - C). \quad (2)$$

Where $K$ is the expense of water purification with the value of 2.09 (Yuan·t$^{-1}$) (The Research Institute of Forest Ecology, Environment and Protection, The Chinese Academy of Forestry 2008).

### 4.1.2 Value Accounting of Soil and Water Conservation

The value of soil conservation can be calculated from (The Research Institute of Forest Ecology, Environment and Protection, The Chinese Academy of Forestry 2008):

$$U_{\text{soil conservation}} = \frac{A \times C_{\text{excavation}} \times (X_2 - X_1)}{\rho}. \quad (3)$$

$C_{excavation}$ is the expense of excavation and transportation (12.6 Yuan m$^{-3}$) (The Research Institute of Forest Ecology, Environment and Protection, The Chinese Academy of Forestry 2008). $X_1$ is the soil erosion modulus in the forest and $X_2$ is the soil erosion modulus without forest. So $X_2 - X_1$ equals 316.86 t hm$^{-2}$ a$^{-1}$ (Qiu and Li 2009). $\rho$ is soil bulk density (1.3 t m$^{-3}$) (Xu and Zhu 2004).

The value of maintaining soil fertility can be calculated from (The Research Institute of Forest Ecology, Environment and Protection, The Chinese Academy of Forestry 2008):

$$U_{fertility} = A \times (X_2 - X_1) \times \left( \frac{N \cdot C_1}{R_1} + \frac{P \cdot C_1}{R_2} + \frac{K \cdot C_2}{R_3} + M \cdot C_3 \right). \quad (4)$$

Where $R_1$, $R_2$, $R_3$ refer to nitrogen content of diammonium phosphate (14.0%), phosphorus content of diammonium phosphate (15.01%) and kalium content of potassium chloride (50.0%), respectively (The Research Institute of Forest Ecology, Environment and Protection, The Chinese Academy of Forestry 2008). $C_1$, $C_2$, $C_3$ refer to the price of diammonium phosphate (2,400 Yuan a$^{-1}$), potassium chloride (2,200 Yuan a$^{-1}$) and organic matter (320 Yuan a$^{-1}$), respectively (The Research Institute of Forest Ecology, Environment and Protection, The Chinese Academy of Forestry 2008). $N$, $P$, $K$, $M$ refer to soil nitrogen content (0.12%), soil phosphorus content (0.059%), soil kalium content (1.68%) and soil OM content (0.68%), respectively, which were offered by Chen Pengwei of Shitai soil and water conservation station.

### 4.1.3 Value Accounting of Fixing Carbon and Releasing Oxygen

The value of fixing carbon can be calculated from (The Research Institute of Forest Ecology, Environment and Protection, The Chinese Academy of Forestry 2008):

$$U_{fixing} = A \times C_{carbon} \times (1.63 R_{carbon} \cdot B_{productivity} + F_{carbon}). \quad (5)$$

$R_{carbon}$ is carbon content of carbon dioxide (27.27%) (Wang et al. 2007), $C_{carbon}$ is the cost of fixing carbon (1,200 Yuan·t$^{-1}$) (The Research Institute of Forest Ecology, Environment and Protection, The Chinese Academy of Forestry 2008), $B_{productivity}$ is net primary productivity of forest (22.7424 t·hm$^{-2}$ a$^{-1}$) (Qiu and Li 2009; Wu 2009), $F_{carbon}$ is the net amount of soil carbon fixing per unit area (3.297 t·hm$^{-2}$ a$^{-1}$) (Yu et al. 2007).

The value of releasing oxygen can be calculated from (The Research Institute of Forest Ecology, Environment and Protection, The Chinese Academy of Forestry 2008):

$$U_{releasing} = 1.19 \times C_{oxygen} \times A \times B_{oxygen}. \quad (6)$$

Where $C_{oxygen}$ is the price of oxygen (1,000 Yuan·t$^{-1}$) (The Research Institute of Forest Ecology, Environment and Protection, The Chinese Academy of Forestry 2008), $B_{oxygen}$ is the amount of releasing oxygen per unit forese area (2.589 t hm$^{-2}$·a$^{-1}$) (Yu et al. 2007).

### 4.1.4 Value Accounting of Atmosphere Environmental Purification

The value of supply of negative ion can be calculated from (The Research Institute of Forest Ecology, Environment and Protection, The Chinese Academy of Forestry 2008):

$$U_{\text{negative ion}} = \frac{5.265 \times 10^5 \times A \times H \times K_{\text{negative ion}} \times (Q_{\text{negative ion}} - 600)}{L}. \quad (7)$$

$H$ is the average height of forest (6 m), $K_{\text{negative ion}}$is the cost of producing negative ion ($5.8185 \times 10^{-18}$ Yuan·each$^{-1}$) (The Research Institute of Forest Ecology, Environment and Protection, The Chinese Academy of Forestry 2008), $Q_{\text{negative ion}}$ is the concentration of negative ion (5,500 number· cm$^{-3}$) (Xu 2004), $L$ is the life of negative ion (20 min).

The value of absorption of pollutants can be calculated from (The Research Institute of Forest Ecology, Environment and Protection, The Chinese Academy of Forestry 2008):

$$U_{\text{pollutants}} = K_{\text{sulfur dioxide}} \cdot Q_{\text{sulfur dioxide}} \cdot A + K_{\text{nitrogen oxides}} \cdot Q_{\text{nitrogen oxides}} \cdot A + K_{\text{fluoride}} \cdot Q_{\text{fluoride}} \cdot A. \quad (8)$$

$K_{\text{sulfur dioxide}}$, $K_{\text{nitrogen oxides}}$ and $K_{\text{fluoride}}$ are the cost of controlling sulfur dioxide (1.2 Yuan·kg$^{-2}$), nitrogen oxides (0.63 Yuan·kg$^{-2}$), fluoride (0.69 Yuan·kg$^{-2}$) pollution, respectively (The Research Institute of Forest Ecology, Environment and Protection, The Chinese Academy of Forestry 2008). $Q_{\text{sulfur dioxide}}$, $Q_{\text{nitrogen oxides}}$ and $Q_{\text{fluoride}}$ were the amount of absorbing sulfur dioxide, nitrogen oxides, fluoride per unit forese area, respectively. In our research, for broad-leaf forest, $Q_{\text{sulfur dioxide}}$, $Q_{\text{nitrogen oxides}}$ and $Q_{\text{fluoride}}$ are 88.65 kg·hm$^{-2}$·a$^{-1}$, 6.0 kg·hm$^{-2}$·a$^{-1}$ and 4.65 kg·hm$^{-2}$·a$^{-1}$, respectively (Wang et al. 2007). For coniferous forest, $Q_{\text{sulfur dioxide}}$, $Q_{\text{nitrogen oxides}}$ and $Q_{\text{fluoride}}$ are 215.60 kg·hm$^{-2}$·a$^{-1}$, 6.0 kg·hm$^{-2}$·a$^{-1}$ and 0.5 kg·hm$^{-2}$·a$^{-1}$, respectively (Wang et al. 2007). The area of broad-leaf forest and coniferous forest are 58,410 hm$^2$, 41,098 hm$^2$, respectively (Cao 2008).

The value of adsorbing dust can be calculated from (The Research Institute of Forest Ecology, Environment and Protection, The Chinese Academy of Forestry 2008):

$$U_{\text{dust}} = K_{\text{dust}} \cdot Q_{\text{dust}} \cdot A. \quad (9)$$

$K_{\text{dust}}$ is the cost of controlling dust (0.15 Yuan·kg$^{-2}$) (The Research Institute of Forest Ecology, Environment and Protection, The Chinese Academy of Forestry 2008), $Q_{\text{dust}}$ is the amount of absorbing dust per unit forese area. For broad-leaf forest $Q_{\text{dust}}$ is 10,110 kg·hm$^{-2}$·a$^{-1}$ (Wang et al. 2007), and coniferous forest 33,200 kg·hm$^{-2}$·a$^{-1}$ (Wang et al. 2007).

### 4.1.5 Value Accounting of Nutrient Accumulation

The value of nutrient accumulation can be calculated from (The Research Institute of Forest Ecology, Environment and Protection, The Chinese Academy of Forestry 2008):

$$U_{\text{nutrient}} = A \cdot B_{\text{productivity}} \times \left( \frac{N_{\text{nutrient}} \cdot C_1}{R_1} + \frac{P_{\text{nutrient}} \cdot C_1}{R_2} + \frac{K_{\text{nutrient}} \cdot C_2}{R_3} \right). \tag{10}$$

$B_{\text{productivity}}$ is net primary productivity of forest (22.7424 t hm$^{-2}$ a$^{-1}$) (Qiu and Li 2009; Wu 2009), $R_1$, $R_2$, $R_3$ refer to nitrogen content of diammonium phosphate (14.0%), phosphorus content of diammonium phosphate (15.01%) and kalium content of potassium chloride (50.0%), respectively (The Research Institute of Forest Ecology, Environment and Protection, The Chinese Academy of Forestry 2008). $C_1$, $C_2$ refer to the price of diammonium phosphate (2,400 Yuan·a$^{-1}$), potassium chloride (2,200 Yuan·a$^{-1}$), respectively (The Research Institute of Forest Ecology, Environment and Protection, The Chinese Academy of Forestry 2008). $N_{\text{nutrient}}$, $P_{\text{nutrient}}$ and $K_{\text{nutrient}}$ are the content of nitrogen (0.0067%), phosphate (0.0445%) and potassium (0.8904%) in trees, respectively (Song et al. 1999).

### 4.1.6 Value Accounting of Biodiversity Conservation

The value of biodiversity conservation can be calculated from (The Research Institute of Forest Ecology, Environment and Protection, The Chinese Academy of Forestry 2008):

$$U_{\text{biodiversity}} = A \cdot S_{\text{biodiversity}}. \tag{11}$$

Where $S_{\text{biodiversity}}$ is the opportunity cost of species disappearance per unit forest area (1,780.74 Yuan·hm$^{-2}$·a$^{-1}$) (Qiu and Li 2009).

## 4.2 *Value of Production*

Production value can be calculated as follow (Qiu and Li 2009):

$$U_{\text{production}} = A \cdot K_{timber} \cdot m \cdot n \cdot v. \tag{12}$$

Where $K_{timber}$ is the average timber price (600 Yuan·m$^{-3}$) (Qiu and Li 2009), $m$ is the strength of falling trees (35%) (Qiu and Li 2009), $n$ is the synthetical timber production ratio (50%) (Qiu and Li 2009), $v$ is the amoumt of timber storage per unit area (49.44 m$^3$·hm$^{-2}$·a$^{-1}$) (Qiu and Li 2009).

## 4.3 *Social Services Value*

Social services value can be calculated as follow (Qiu and Li 2009):

$$U_{\text{social}} = U_{\text{recreation}} + U_{\text{research}}. \tag{13}$$

$U_{\text{recreation}}$ is the value of forest recreation (6.633 × 10$^7$ Yuan·a$^{-1}$) (Chizhou Bureau of Statistics 2008; Qiu and Li 2009) and $U_{\text{research}}$ is the value of scientific and culture research (4.240 × 10$^6$ Yuan·a$^{-1}$) (Chizhou Bureau of Statistics 2008; Qiu and Li 2009).

# 5 Results and Analysis

## 5.1 *Results*

Table 2 indicated that in the upper Qiupu River the total value of ecosystem service of forests, production value and social services value is 4.856 × 10$^8$ Yuan·a$^{-1}$, 5.762 × 10$^7$ Yuan·a$^{-1}$ and 7.057 × 10$^7$ Yuan·a$^{-1}$, respectively.

## 5.2 *Analysis of the Ecosystem Service Value*

### 5.2.1 Comparison Among the Values of Different Ecological Service Types

The values of different ecological service are showed in Table 3.

From Table 3, the value of different ecological service types can be arranged as follow: fixing carbon and releasing oxygen > storage and retention of water > atmosphere environmental purification > biodiversity conservation > soil and water conservation > nutrient accumulation, which is similar to the results presented by Wang et al. (2007). The ratio of the values of different ecological service types in 2007 can be presented in pie chart as Fig. 1. Figure 1 shows that the sum including value of fixing carbon and releasing oxygen and value of storage and retention of water reached 87.119% of the total ecosystem service value.

**Table 2** The values of ecological service in the upper Qiupu River in 2007

| Value type | Index classification | Index | Value($\times 10^4$ Yuan·a$^{-1}$) | Sum($\times 10^4$ Yuan·a$^{-1}$) |
|---|---|---|---|---|
| Production Value | Production of forest products | Food production | 5762.23 | 5762.23 |
| Ecosystem service value | Storage and retention of water | Water volume regulating | 15668.44 | 48556.77 |
| | | Water purification | 5358.96 | |
| | Soil and water conservation | Soil conservation | 34.08 | |
| | | Maintaining soil fertility | 1130.83 | |
| | Fixing carbon and releasing oxygen | Fixing carbon | 17855.39 | |
| | | Releasing oxygen | 3419.8 | |
| | Atmosphere environmental purification | Supply of negative ion | 49.90 | |
| | | Absorption of pollutants | 118.32 | |
| | | Adsorbing dust | 2932.46 | |
| | Nutrient accumulation | Nutrients accumulation of trees | 11.97 | |
| | Biodiversity conservation | Biodiversity conservation | 1976.62 | |
| Social services value | Forest recreation, scientific and culture research | Forest recreation | 6,633 | 7057.02 |
| | | Scientific and culture research | 424.02 | |

**Table 3** The values of different ecological service types in 2007

| Ecological service types | Value($\times 10^4$ Yuan·a$^{-1}$) | Proportion (%) | Ecological service types | Value($\times 10^4$ Yuan·a$^{-1}$) | Proportion (%) |
|---|---|---|---|---|---|
| Storage and retention of water | 21027.4 | 43.304 | Atmosphere environmental purification | 3100.68 | 6.385 |
| Soil and water conservation | 1164.91 | 2.399 | Nutrient accumulation | 11.97 | 0.024 |
| Fixing carbon and releasing oxygen | 21275.19 | 43.815 | Biodiversity conservation | 1976.62 | 4.073 |
| Sum | 48556.77 $\times$ 10$^4$ Yuan·a$^{-1}$ and 100% | | | | |

### 5.2.2 Comparison Among Ecosystem Service Value, Social Services Value and Production Value

From the results, we can know that in 2007, the ecosystem service value reached 8.42 times of the production value and 6.88 times of the social services value, respectively, which is showed in Fig. 2.

### 5.2.3 Comparison Between Ecosystem Service Value and GDP of the Corresponding Period

In 2007, the total GDP of Shitai country reaches 6.840 $\times$ 10$^8$ Yuan (Chizhou Bureau of Statistics 2008), and per capita GDP was 6,267 Yuan (Chizhou Bureau

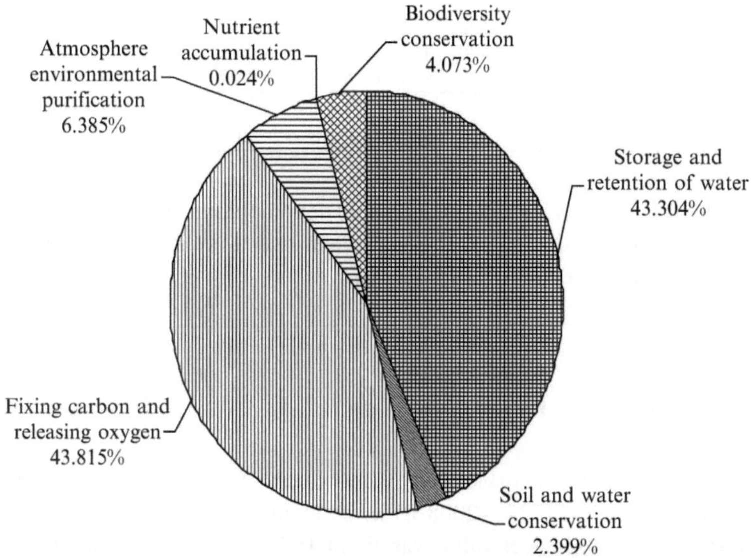


**Fig. 1** This Pie chart shos the ratio of the values of different ecological service types in 2007

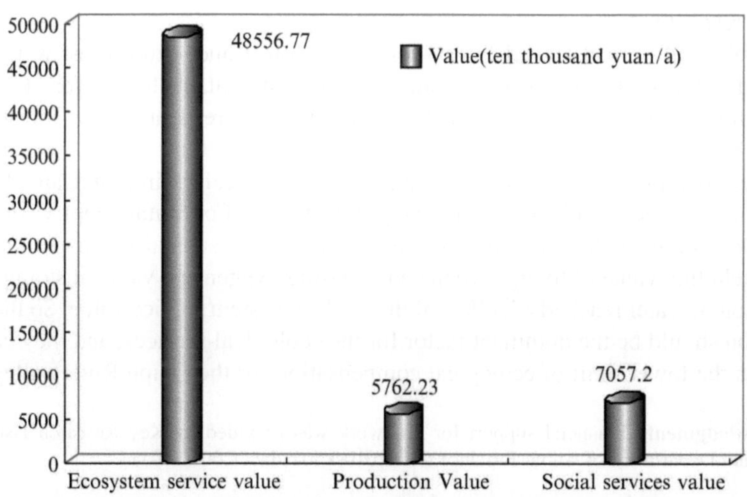


**Fig. 2** This bar graph shows ecosystem service function values of forest in the upper Qiupu River

of Statistics 2008). On the other hand, in 2007 the ecosystem service value reached $4.856 \times 10^8$ Yuan and per capita ecosystem service value was 4,448 Yuan. Obviously, the ecosystem service value had the proportion of 70.99% of GDP, which can be showed in Table 4.

From Table 4, we can conclude that the ecosystem service value is more than two-thirds of the GDP. It means that the forests in Shitai country have enormous

Table 4 The comparison between ecosystem service value and GDP in Shitai county in 2007

| Type | Total GDP | Total ecosystem service value | Per capita GDP | Per capita ecosystem service value |
|---|---|---|---|---|
| Number ($\times 10^4$ Yuan·a$^{-1}$) | 68,398 | 48556.77 | 0.6267 | 0.44948 |
| Proportion(%) | 100 | 70.99 | 100 | 70.98 |

ecological values, which insist the sustainable development and the development of ecological tour in Shitai country. It is thus clear that it is very important for us to protect effectively the ecological resources.

## 6 Conclusion and Discussion

In this paper, the forests ecosystem service value in the upper Qiupu River were researched taking LY/T1721—2008 as assessment standard. The results showed that forest ecosystem services value was about 48,556.77 $\times$ $10^4$ Yuan, the production value 5,762.23 $\times$ $10^4$ Yuan and social services value 7,057.02 $\times$ $10^4$ Yuan. The forests ecosystem service value had the proportion of 70.99% of GDP, and reached 8.42 times of the production value and 6.88 times of the social services value, respectively.

Surely, only broad leaved forest, coniferous forest and shrub forest were considered in the study. Actually, for Anhui province it contains leaved forest, coniferous forest, shrub forest, bamboo forest, economic forests and open forest, etc. (Xu 2005).

Therefore, protection and construction of the ecological environment are playing extremely important role and should be given priority. For Shitai country, sustainable Development should be the favorite choice. The results also showed that the sum including value of fixing carbon and releasing oxygen and value of storage and retention of water reached 87.119% of the total ecosystem service value. So the two function should be the dominant factor for the ecological services, and they can be used as the lower limit of ecological compensation for the Qiupu River valley.

**Acknowledgments** Financial support for this work was provided by Key Research Issues of Education Department of Anhui Province (NO:2010sk502zd;ZD2008009-1).

## References

Cao X (2008) Forest resources report of shitai 2008. Shitai Forestry Administration

Chizhou Bureau of Statistics (2008) Chizhou statistical yearbook 2008. pp 20–246

Qiu W, Li J (2009) The studies on Huangshan city's eco-compensation value of Xinanjiang upstream. J Biol 39–42

Song J, Wang B, Peng S, Wang M (1999) The storage and cycling nutrient of Ixonanthes chinensis in south subtropic broad leaf forests. Acta Ecol Sin 224

The Research Institute of Forest Ecology, Environment and Protection, The Chinese Academy of Forestry (2008) The specifications for assessment of forest ecosystem services in China (LY/T1721—2008). State Forestry Administration. pp 4–12
Tian X (2006) Analysis of spatiotemporal distribution and tendency of amount of evaporation in Anhui province. J Anhui Tech Coll Water Resour Hydroelectr Power 52
Wang B, Li S, Guo H (2007) The assessment of forest ecosystem services evaluation in Jiangxi province. Jiangxi Sci 554–559
Wu G (2009) Measurement on monetary value of forestry multifunction. J Anhui Agric Sci 17159–17161
Xu Z (2004) The anion resources research in tourism area of Anhui province. AnHui Agric Univ 4
Xu X (2005) Resource development & market. value of forest eco-system services in Anhui province. Resour Dev Market 96
Xu X, Zhu C (2004) Estimation methods of the economical loss of ecological destruction in mountainous regions of South Anhui. J Mt Res 735–741
Yu X, Wu L, Rao L, Li J, Yang R (2007) Assessment methods of ecological functions of soil and water conservation measures. Sci Soil Water Conserv 110–113

# Research on Environmental Financial Risk Management and Construction of Environmental Management System

**Zhao Yajing, Xiao Xu, and Zhang Caiping**

**Abstract** Environmental finance is developing as a field in response to an acceptance of the idea that sound environmental management is positively correlated with sound economic management. Thus, there is growing confidence that environmental quality is justified by the bottom line. However, because environmental quality cannot be packaged like a physical commodity and sold in a traditional marketplace, innovation has been required to develop new financial instruments that recognize and reward environmental virtue in the private sector. The paper studies such related problems as the environment management and shareholder value creation, environmental management system, tools of risk management accelerate the development of environmental finance.

**Keywords** Environmental finance · environmental management system · tools of risk management · shareholder value creation

## 1 Introduction

The rate of societal change has been accelerating since the inception of the industrial revolution. We are now increasingly aware that the negative environmental side effects of that revolution are not trivial. Nor are they insuperable or too costly to contemplate. However, until recently many of these side effects were largely ignored. People may have observed some impacts but they were not

---

Z. Yajing (✉) and X. Xu
Central South University Hu-nan, Changsha 410083, China
e-mail: antoni_zhao@live.cn; xiaoxucs@163.com

Z. Caiping
Central South University Hu-nan, Changsha 410083, China
and
University of South China, Hu-nan, Hengyang 421001, China
e-mail: zcp2002117@sina.com

Y. Zhou and D.D. Wu (eds.), *Modeling Risk Management for Resources and Environment in China*, Computational Risk Management,
DOI 10.1007/978-3-642-18387-4_37,  

systematically managed. Some were the responsibility of the public sector (especially nuclear power, water supply and treatment, and solid waste management) and hence rarely a concern of the private sector. Some of the side effects took years to show up (such as long-tailed insurance claims for asbestos liability), and the insurance industry was totally unprepared to manage the risk. Deregulation – often including the activities of formerly public companies – has now brought these concerns to the private sector. In order to reassure the voters, government has brought in a whole array of new regulations ("regulation") to make the newly privatized operations transparent. Another great force for change has been globalization. Companies have been released from the confines of the regional or national markets and have taken a global stake. The largest companies have been doing this for 100 years. Now much of the rest of the economy is following. Some newly privatized businesses-like water supply and treatment-find themselves on the global scene for the first time.

Legal redress is becoming globalization. Ironically, this has been possible for a long time, specifically through the U.S. Alien Tort Claims Act of 1789, which allows foreign nationals to sue American companies in the American courts. This is now being used by diverse groups around the world to sue American companies for damage to their environments. Even if companies and their financial service providers had ignored these developments, they could not ignore the very tangible costs of a poor environmental performance. These have been heavy. The costs of asbestos, inadequate landfill management, and oil spills have had major impacts on their balance sheets. Such cases will be identified in the rest of this book. Management failures have led to huge insurance payments and, in some circumstances, eventually to bankruptcy. Environmental problems have pitted old partners – such as manufacturers, insurers, and bankers – against one another.

A number of companies now understand this change of paradigm very well and have moved to address it. On the positive side there is increasing evidence that the market rewards proactive environmental management.

## 2 Environmental Management and Shareholder Value Creation

Research attempting to link environmental and financial performance reveals a growing sense that sound environmental management can lead to increased shareholder value, which is defined as: Value for shareholders which is created when a business, over time, uses capital at its disposal to earn returns greater than, or equal to, the cost of that capital (Willis and Desjardins 2001).

Traditionally, environmental management has been seen as imposing a cost on a company and a "green penalty" on investors, with no corresponding benefit being conferred. The opposing view holds that environmental performance is compatible with, and perhaps central to, competitiveness and superior financial performance

(Porter and van der Linde 1995). There is strong evidence that improved environmental behavior has a strong impact on shareholder value (Dowell et al. 2000; Sustainability/UNEP 2001; UBS 2000).

A business case can be made that not only dispels notions that environmental initiatives have an adverse effect on profitability, but holds that they contribute to shareholder value creation. Figure 1 demonstrates the linkages between improved corporate environmental performance and the creation of shareholder value. Areas of strategic decision making within a company's product management, operations, capital assets, and finance departments govern the processes that create value for the corporation, through their impacts on revenues, operating costs, and the cost of capital. Improved environmental management decisions at this level influence these value drivers, which in turn generate shareholder value. The following discussion demonstrates how a focus on environmental issues can lead to increased revenues, decreased operating costs, and a lower cost of capital.

*Product management.* A strong environmental focus in the product design can lead to new product development and, in some cases, can redefine markets (U.S. EPA 2000). Considerations of a product's environmental impacts at the design stage can keep a firm in the forefront of market innovation and position it well to reap marketing advantages. From this marketing standpoint, an environmental focus can help improve a company's revenues as its environmentally improved products are differentiated from others, contributing to increased brand recognition and competitive advantage. From a liability perspective, if a company's product has adverse effects on the environment, the company can bear liabilities that strike at the core of its business.

*Operations.* In addition to product management and design, many firms' environmental impacts come from their manufacturing processes. Taking environmental considerations into account in these processes can help firms reduce the energy and raw materials inputs, as well as reduce waste outputs. Process changes that reduce environmental impacts can lead to lower costs and increased operational

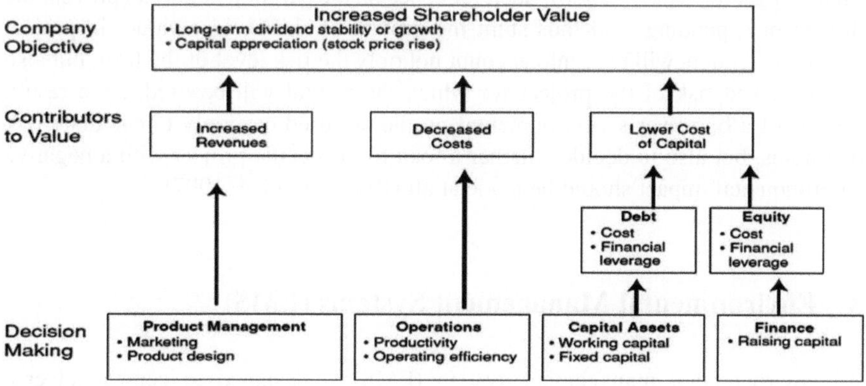


**Fig. 1** Shareholder value creation

efficiency. Insurance specialists have recognized a reduction in risk for firms with strong operational environmental management. Some insurers have created products that translate improved environmental performance into lower premiums.

*Capital assets.* A focus on environmental issues when making capital asset investment decisions also helps to lower a company's costs. Not only do investments in environmentally appropriate fixed capital assets lower production costs and make the operating process more efficient, they also help to improve a firm's environmental profile. As a result, a firm that has invested in environmentally favorable assets will be well positioned to comply with new environmental regulations, and to increase its ability to use those assets that benefit the environment over their full operating lives. In addition, the firm will be less prone to environmental incidents, which lead to costly cleanup charges and legal liability. Lending institutions take into consideration a company's reduction in environmental risk, in considering favorable lending terms.

*Finance.* While increasing revenues and decreasing costs help to improve a firm's income, financing decisions are central to the long-term creation of value in the organization. Financing decisions are crucial to the retention of firm value over time, allowing for expansion or acquisitions as well as having an impact on tax and interest expenses.

A firm's two main choices for raising funds are debt or equity financing, although hybrid instruments do also exist. The cost of capital for a firm is defined in terms of the weighted average of its costs of equity and debt, and reflects the company's marginal costs of raising capital (Damodaran 2001). Firms with poor environmental management, therefore, can be expected to pay higher rates of interest than others, due to the increased risk of environmental liability in the eyes of its investors. This results in a higher cost of debt and larger debt obligations, thus reducing residual earnings that provide a return to equity holders and destroying shareholder value. Lenders may, indeed, view a certain level of risk as too great, and may not be willing to lend to a firm demonstrating poor environmental management. Studies illustrate that the exposure to Superfund liability can decrease the likelihood of loan approval (Schaltegger and Burritt 2000). Such reticence on the part of lenders can prevent the firm from expanding, and thus stunt the growth of shareholder value. In project financing, lenders will take into account not only the risk level of the firm, but also the perceived risk of the project for which the capital will be used. As a result, aspects of a borrower's environmental profile are used, not only to calculate risk premiums, but also to decide whether a loan for a specific project with a negative environmental impact should be made at all (Blumberg et al. 1997).

## 3 Environmental Management Systems (EMS)

An environmental management system (EMS) is designed to control adverse environmental impacts, just as financial management is designed to control a company's economic well-being.

A firm's approach to the development of an environmental management system entails both the formulation of long-term environmental policies and goals as well as the adaptation of current business activities in order to reduce the impacts of the firm's product and process on the environment. Figure 2 outlines the basic components required for the establishment of a comprehensive EMS. The first phase involves the collection and development of evidence of the need for an environmental policy and strategy, followed by top management's commitment and board approval for new environmental management and reporting strategies. Company specifics such as its mission statement and available budget are significant at this stage. Once the commitment has been obtained and the strategy communicated throughout the company, the next phases involve the development of the policy and programs to be implemented and the development of the management system components. Reporting of an environmental policy statement serves to establish the direction the firm is taking as well as to communicate the plan to employees and the broader public. The sixth stage outlined in Fig. 2 involves the actual

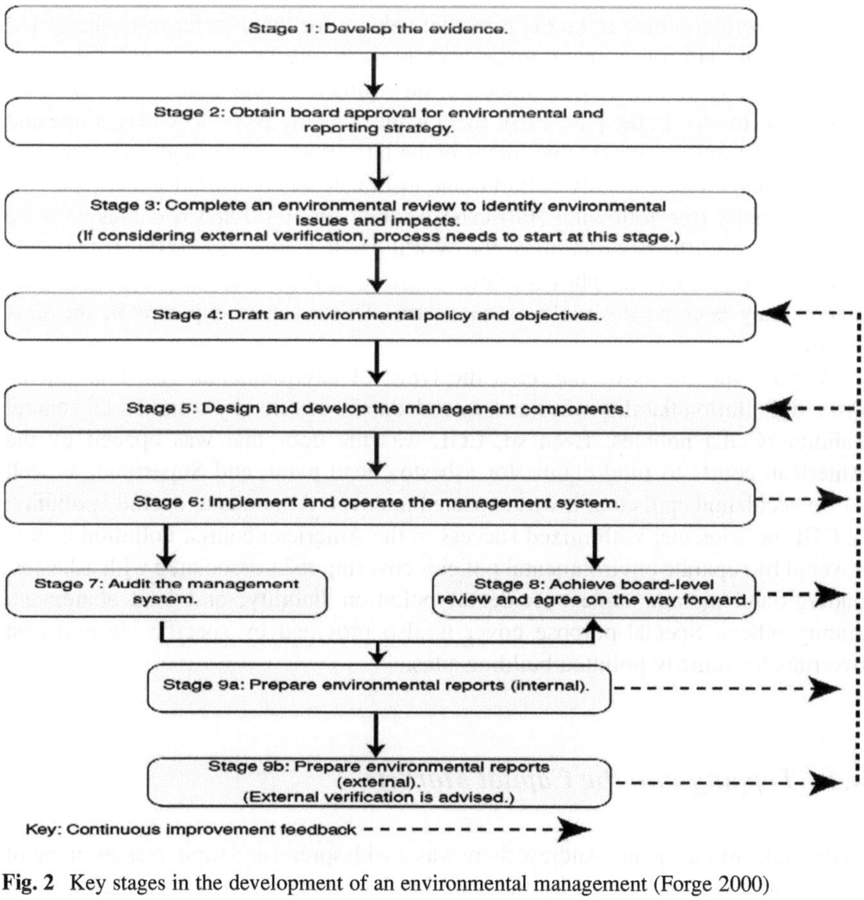


**Fig. 2** Key stages in the development of an environmental management (Forge 2000)

implementation of the plan. As the performance of the EMS is measured, it is also evaluated, with feedback then creating the basis for adjusting the programs and perhaps even adapting the environmental policy. Such feedback implies continual improvement within the EMS framework.

To help the financial services sector meet the requirements of an EMS, a group of British financiers brought out the Forge Report (Forge 2000), which offers practical guidance on the development of an EMS within financial companies. The report pays more attention to the first stage of evidence development and senior management commitment, since financial institutions have historically not seen themselves as a polluting industry.

## 4 Tools for Risk Management

### *4.1 Traditional Insurance Mechanisms*

Insurance will continue to be the principal vehicle for the transfer of business and personal risk. However, there are at least two circumstances in which insurance companies may find themselves unable or unwilling to accept certain risks that they may have covered in the past. First, there is the capacity issue. The magnitude and frequency of major catastrophic losses in the late 1980s and through to the present have challenged the capacity of traditional insurance and reinsurance markets. This was especially true following Hurricane Andrew (in 1992) and becomes an issue again following the destruction of the World Trade Center (in 2001). If the scale and frequency of catastrophes continue to grow, then a wider diffusion of the risk market may become desirable or necessary. This issue is explored in the next subsection.

A more specific issue concerns the types of environmental risk that can be insured. Pollution liability was never intended to be covered by commercial general liability (CGL) policies. Even so, CGL was the door that was opened by the American courts to fund claims for asbestos, lead paint, and Superfund, as well as the accidental spills that it was expected to cover. Attempts to exclude pollution in CGL policies met with mixed success in the American courts. Pollution is now covered by separate environmental policies covering risks associated with asbestos, underground storage tanks, accidental pollution liability, and lead abatement, among others. Special-purpose cover is also provided by specific cleanup cost overruns for remedy polluted building sites.

### *4.2 Tapping into the Capital Markets*

In the wake of Hurricane Andrew there was a widespread and rapid reassessment of just what could be insured by the traditional insurance and reinsurance market.

Whereas the major companies were well prepared, others were vulnerable. There was a real danger that government (especially the elected insurance commissioners in each state) would step in to force the solvent companies to fill the breach by obliging them to join involuntary pools to provide backup cover, as had happened so often in the past. That point forced some consideration of tapping into the capital markets with their much greater volume of transactions and capital base. Thus began an exploration of various off-balance-sheet instruments to make this transition. New products were developed to mimic derivative instruments that had been appearing since the mid-1970s to hedge risks in the financial markets, principally volatility in foreign exchange and interest rates (Smithson 1998). Options and swaps are the instruments most widely used. In the 1990s catastrophe options were designed to provide a flexible infusion of capacity through the Chicago Board of Trade, based on the insurance losses due to catastrophes in the United States. Swaps based on exposure to extreme weather events, such as heavy rainfall and both high and low temperatures, have proven more durable. There is now a growing "weather market," driven, so far, by large energy companies hedging their volume exposures in a deregulated world. Catastrophe bonds, or "cat bonds," have been developed to bring in additional partners to share the financial risk by going directly to the institutional investors. These bonds have the advantage of being quite simple conceptually, compared with the derivative products described in the previous subsection. The downside is that each bond must be configured for each placement, which takes time and therefore carries a higher transactional cost. The market is growing steadily, so it certainly seems to meet a need. Also, the secondary market in cat bonds is developing quickly, which encourages liquidity in the market and hence further growth.

## 5 Conclusion

Environmental finance is developing as a field in response to an acceptance of the idea that sound environmental management is positively correlated with sound economic management. We are no longer tying ourselves to the old assumption that a clean environment is bad for profits. Thus, there is growing confidence that environmental quality is justified by the bottom line. However, because environmental quality cannot be packaged like a physical commodity and sold in a traditional marketplace, innovation has been required to develop new financial instruments that recognize and reward environmental virtue in the private sector. All this is happening at a time when our biggest environmental challenge – climate change – is injecting both uncertainty and urgency into the global situation.

We can admit now that the results have been mixed. This is largely because the development of new financial products can happen only if the regulatory framework is there to make it happen. It requires clear rules that charge the polluters for polluting and reward those who enhance the quality of the environment. Once the regulatory framework has been constructed, then market forces have the potential to provide a dynamic motor for improved environmental performance. To meet that potential we need a trading infrastructure that provides transparency for price discovery and liquidity to allow traders to enter and leave markets.

## References


Blumberg J, Blum G, Korsvold A (1997) Environmental performance and shareholder value. World Business Council for Sustainable Development, Geneva. www.wbcsd.com/ecoeff1

Damodaran A (2001) Corporate finance theory and practice, 2nd edn. Wiley, New York

Dowell G, Hart S, Yeung B (2000) Do corporate global environmental standards create or destroy market value? Manage Sci 46(8):1059–1074

Forge (2000) Guidelines on environmental management and reporting for the financial services sector. Forge Group, London

Porter M, van der Linde C (1995) Green and competitive: ending the stalemate. Harv Bus Rev 73(5):120–134

Schaltegger S, Burritt R (2000) Contemporary environmental accounting: issues, concepts and practice. Greenleaf, Sheffield

Smithson CW (1998) Managing financial risk: a guide to derivatives products, financial engineering and value maximization, 3rd edn. McGraw-Hill, New York

Sustainability/UNEP (2001) Buried treasure: uncovering the business case for corporate sustainability. SustainAbility, London

UBS (2000) Environmental report. Union Bank of Switzerland, Zurich. www.ubs.com/environment

Willis A, Desjardins J (2001) Environmental performance: measuring and managing what matters. Canadian Institute of Chartered Accountants, Toronto

# Research on Urban Water Security Evaluation Based on Technique for Order Preference by Similarity to Ideal Solution Model

**Junfei Chen, Lu Xia, and Huimin Wang**

**Abstract** Urban water safety evaluation is an important content of urban water safety management. In this paper, combining with the characters and influencing factors of urban water safety system, index system of urban water safety evaluation is established. The model of urban water safety evaluation based on Technique for Order Preference by Similarity to Ideal Solution is provided and is used to evaluate water security of Nanjing. The result shows that the model is effective and the state of Nanjing water security is better and better in the future years.

**Keywords** Evaluation · Index system · Technique for order preference by similarity to ideal solution (TOPSIS) · Urban water safety

## 1 Introduction

Water is the most important natural resources and irreplaceable, which is not only basic natural resources but also strategic economic resources and public social resources. However, with fast development of economic and improvement of urbanization, urban water shortage, water pollution, flood, drought and other water safety problems are revealing (Shao 2004). China is a country with serious urban water safety issues. In china, the per capita possession of water resources is few, which is less than one fourth of per capita of the world; spatial and temporal distribution of water resources is extremely uneven; the phenomena of North drought South flood or spring drought summer flood is serious. In addition, water is badly polluted, and water wastage is common. Therefore, how to ensure the sustainable use of water resources and protect the urban water safety is a major subject of great significance currently.

---

J. Chen (✉), L. Xia, and H. Wang
State Key Laboratory of Hydrology – Water Resources and Hydraulic Engineering, Business School, Hohai University, Nanjing 210098, China
e-mail: chenjunfei@hhu.edu.cn; lemonkathy@126.com

Y. Zhou and D.D. Wu (eds.), *Modeling Risk Management for Resources and Environment in China*, Computational Risk Management,
DOI 10.1007/978-3-642-18387-4_38, © Springer-Verlag Berlin Heidelberg 2011

At present, China's urban water safety research is still in its infancy, so it is urgent to discuss deeply in theoretical research, index system of evaluation, evaluation models and methods, emergency plans, etc. Urban water security issue involves multiple factors and indicators, and therefore we must use an effective method for analysis. Now there are some methods, such as AHP method, fuzzy evaluation method (Peng 2000; Jiang 2002). As a multi-objective decision making approach, the TOPSIS method which calculates the relative closeness between ideal point scheme and evaluation scheme as the proof of evaluation, is an effective, convenient and applicative method (Peng 2000). In this paper, the city's water safety evaluation system is established, and we established the urban water safety's assessment model and analyze it to evaluate urban water safety of Nanjing.

## 2 Connotation of Urban Water Security

As for the concepts and contents of water safety, some research results were obtained. Jia believes that water security means the water supply can meet the reasonable needs of water resources. If a regional water supply can meet the reasonable demands for their social and economic development in the long run, then the region's water resources is safe, otherwise it is not safe (Jia et al. 2002). Han thinks that water safety can be understood as: in the present or the future, due to the natural hydrologic cycle fluctuations or unreasonable changes the human make on the water cycle balance, the status of regional water that human live on occur the evolution adverse for mankind, and have a negative impact on all aspects of human society, which present as drought, floods, water shortage and water pollution, water environmental damage and thus could lead to food production, social instability, economic decline and regional conflicts, etc. (Han et al. 2003). Zhang et al. (2005) thinks: water security refers to the presence of water (quantity and quality, physical and chemical characteristics, etc.) and water activities (Government Administration, sanitation, water supply, disaster mitigation, environmental protection, etc.) are not a threat on human society's stability and development, or a threat to some degree, but you can control its consequences within the scope people can bear (Zhang et al. 2005). Chen describe the concept of water safety as one region's (or country) capability of water hazards and sustainable use of water can ensure sustainable development of society, economy, and ecology (Chen 2006). Li defines water security as: the waters remain adequate water quality and safe water conditions to maintain its natural ecosystems and the ecological functions, ensure aquatic life surviving effectively and the surrounding environment in good condition, so that water system can function normally and sustainably, and satisfies the needs of human life and production to a large extent, so that humanity itself and human intergroup relationship is not in the threatened status (Zeng et al. 2004).

Based on the understanding of the safety of water, this paper defines urban water safety as: in a particular area of the city, there is plenty of water to meet the material needs of the human society, economic development and the maintenance of ecological environment; human society attach importance to the limitation of water resources in the process of using water, and exploit appropriately and use scientifically the water resources not beyond the carrying capacity of water resources and the bearing capacity of water environment, so that water resources can be recycled sustainably; the city (region) is not in severe losses caused by floods, drought, water scarcity, water pollution, water environmental destruction.

## 3 Evaluation Model of Urban Water Security Based on TOPSIS

### 3.1 Establishing Evaluation Index System

Establishing index system should follow the four basic principles: systemic principle, comparable principle, scientific principle and practical principle. According to the four principles, we divide urban water safety evaluation system into four layers, as showed in Table 1. The highest layer is the objective layer, and use the comprehensive level to measure the development of city water safety; the second layer is the criterion, composed by relevant indicators reflecting the objective layer, including five aspects such as water supply, water ecological environment, drinking water security, water hazards and water management; the third is the standard layer, which is a further refinement of the criterion layer; the fourth is indexes layer including specific indicators (Han and Ruan 2003).

### 3.2 Computing the Weight of Evaluation Indexes

Suppose the number of water safety system is $M$, and the number of the evaluation indexes is $N$. Each evaluation index expresses by vector, note it as $X_i = (x_{i1}, x_{i2}, \cdots, x_{iN}), i = 1, 2, \cdots, M$, so we can get the primal evaluation matrix $X = (x_{ij})_{M\times N}$.

The index weight is the quantitative performance which reveals the relative importance of each index in the whole index system. Whether the index weight is reasonable or not will affect the comprehensive evaluation results. In this paper, the combining method of expert survey and AHP is used to computing the weight of evaluation index, shown in Table 1.

Evaluation indexes of different evaluation scheme often have different quantity scale and dimension. Some indexes are positive. That is, the value is the smaller the better. Consequently, the primal indexes should be normalized for eliminating the effect of quantity scale and dimension.

**Table 1** Urban water security index system and weight

| Objective | Criterion | Weight | Sub-criterion | Weight | Index | Weight |
|---|---|---|---|---|---|---|
| The comprehensive level of urban water security | Water supply | 0.196 | Water resources conditions | 0.050 | Per capita water quantity | 0.025 |
| | | | | | Per MU water quantity | 0.025 |
| | | | Water supply indicators | 0.095 | Rate of water resources development and utilization | 0.019 |
| | | | | | Groundwater mining rate | 0.018 |
| | | | | | Water investment accounted for ratio of GDP | 0.015 |
| | | | | | Water consumption in GDP per 10,000 | 0.017 |
| | | | | | Rate of industrial water reuse | 0.026 |
| | | | Water demand indicators | 0.051 | Agricultural water quota | 0.015 |
| | | | | | Water consumption of industrial output per 10,000 | 0.019 |
| | | | | | Daily water consumption per capita | 0.017 |
| | Water ecological environment | 0.213 | Water environment security | 0.082 | Emissions of COD | 0.040 |
| | | | | | Standard-reaching rate of water quality in surface water functional zone | 0.042 |
| | | | Water ecological security | 0.131 | Green ratio in built up area | 0.043 |
| | | | | | Standard-reaching rate of waste water emission | 0.043 |
| | | | | | Treatment rate of domestic sewage | 0.045 |
| | Drinking water security | 0.182 | Drinking water shortage | 0.061 | Popularizing of tap water use | 0.061 |
| | | | Drinking water sanitation | 0.121 | Quality standard-reaching rate of centralized drinking water source | 0.060 |
| | | | | | Annual comprehensive qualified rate of urban water supply | 0.061 |
| | Water hazards | 0.179 | Floods and drought disasters | 0.179 | Effective irrigation area | 0.050 |
| | | | | | Ensuring good harvests despite drought or excessive rain area ratio | 0.043 |
| | | | | | Flood damage | 0.043 |
| | | | | | Drought disaster losses | 0.043 |
| | Water management | 0.230 | Engineering technical management measures | 0.040 | Flood embankment length | 0.02 |
| | | | | | Urban water supply network leakage rate | 0.020 |
| | | | Management measures of laws and regulations | 0.190 | Water-saving awareness of regional population | 0.030 |
| | | | | | Integrity of water laws | 0.025 |
| | | | | | Executive force of water laws and regulations | 0.025 |
| | | | | | Level of water-saving technology | 0.040 |
| | | | | | Reasonable expenses of water | 0.030 |
| | | | | | Level of water pollution control technology | 0.040 |

## 3.3 TOPSIS Evaluation Model

The TOPSIS model consists of the following steps:

1. The matrix $Y = (y_{ij})_{M\times N}$ can be got through normalization as follows:
For the positive indexes, the performance-oriented indexes, can be normalized via formula (1),

$$y_{ij} = \frac{x_{ij} - \min_i \{x_{ij}\}}{\max_i \{x_{ij}\} - \min_i \{x_{ij}\}} \tag{1}$$

For the negative ones, the cost-oriented indexes, can be normalized via formula (2),

$$y_{ij} = \frac{\max_i \{x_{ij}\} - x_{ij}}{\max_i \{x_{ij}\} - \min_i \{x_{ij}\}} \tag{2}$$

2. Computing the weighted normalized matrix $Z$,

$$Z = (z_{ij})_{M\times N} = (w_j y_{ij})_{M\times N} \tag{3}$$

3. Ideal solution $V^+$ and inverse ideal solution $V^-$ of the evaluation problem are computed as follows:

$$V^+ = \left\{ z_j^+ \middle| j = 1, 2, \cdots, N \right\} = \left\{ \max_i z_{ij} \middle| j = 1, 2, \cdots, N \right\} \tag{4}$$

$$V^- = \left\{ z_j^- \middle| j = 1, 2, \cdots, N \right\} = \left\{ \min_i z_{ij} \middle| j = 1, 2, \cdots, N \right\} \tag{5}$$

4. Computing the distance from ideal solution $D_i^+$ and inverse ideal solution $D_i^-$ to corresponding evaluation index vector as follows:

$$D_i^+ = \sqrt{\sum_{j=1}^{N} (z_{ij} - z_j^+)^2},\ D_i^- = \sqrt{\sum_{j=1}^{N} (z_{ij} - z_j^-)^2}\ i = 1, 2, \cdots, M. \tag{6}$$

5. Computing relative closeness of ideal solution $D_i^+$ and the evaluation index vector,

$$C_i = \frac{D_i^-}{D_i^- + D_i^+}, i = 1, 2, \cdots, M \tag{7}$$

6. The relative closeness value $C_i$ is as the comprehensive value of the evaluation scheme. According to the relative closeness value, all scheme can be conducted to sort. The relative closeness is the bigger the better.

# 4 Case Study

Follow the above steps and methods, the Nanjing city water security system is evaluated while the time is 2010–2015 and 2020. Through the analysis of those years' situation, it is easy to see the trend of Nanjing city water safety system. During calculation, some data are directly access, some are through the necessary prediction methods. Finally, the results are shown in Table 2 and Fig. 1. From Nanjing city water safety system evaluation score we can seen, Nanjing city water safety system development level is on the rise from 2010 to 2020, while the value from 0.3962 to 0.6761. Water security conditions in Nanjing city water continuously improve overall, which essentially reflect the Nanjing city water safety system trends.

From the results we can see, it can be broadly divided into two phases from 2010 to 2020.

The first phase is 2010–2011, Nanjing water safety system is in a relatively stable stage. In this phase, the economy is relatively mature and stable, but water security issues have yet to be further improved.

**Table 2** Index value of Nanjing city water security from 2010 to 2020

| Index year | 2010 | 2011 | 2012 | 2013 | 2014 | 2015 | 2020 |
|---|---|---|---|---|---|---|---|
| $S_i^*$ | 0.1445 | 0.1369 | 0.1152 | 0.0972 | 0.0897 | 0.0900 | 0.0848 |
| $S_i^-$ | 0.0948 | 0.0870 | 0.0996 | 0.1138 | 0.1263 | 0.1390 | 0.1771 |
| $C_i$ | 0.3962 | 0.3887 | 0.4636 | 0.5391 | 0.5847 | 0.6072 | 0.6761 |

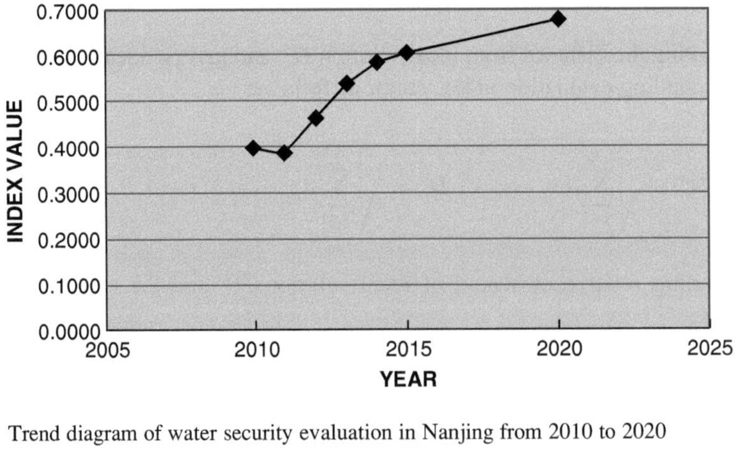


**Fig. 1** Trend diagram of water security evaluation in Nanjing from 2010 to 2020

The second stage is 2012–2020, water security is steadily rising, the situation is better and better, so we can say water safety conditions in Nanjing will continue to improve over the next 10 years. At this stage, the implementation of green Nanjing in-depth strategies, promoting ecological city, the development of recycling economy, and actively promote resource conservation, resource use efficiency, significantly enhanced sustainable development capacity; the strengthening of water environment, planning and construction development zones, focusing on urban waste water treatment plants and sewage treatment facilities in rural areas; to strengthen the focus on industrial pollution, according to the gateway, stopping, and water consumption, transfers, high-polluting enterprises, strictly controlling the emission of pollutants; increasing reuse to promote residential water reuse, pushing the tail water; water management laws and regulations and continuously improve, increasing population awareness of the importance of saving. With the constant implementation of these measures, the water security situation in Nanjing will continually be improved.

## 5 Conclusion

With the rapid development of China's cities, urban water safety issues have become increasingly prominent. Protection of the urban water security is the basic requirement to promote urban sustainable socio-economic development and build a harmonious society. Therefore, the development of region urbanization must adhere to the principle of sustainable development, strengthen the prevention of water pollution, water resources, water environment, and aquatic ecosystems, especially urban drinking water source protection work, so that realize the economic, social and environmental coordinated development. This paper examines the connotation of urban water safety, establishes the urban water security evaluation system, and adopts the TOPSIS model in Nanjing city water safety evaluation. Because of the complexity of the urban water security itself, there are a number of issues for further study and discussion.

**Acknowledgement** This work was supported in partially by the National Society Science Fund of China (Grant No. 09CJY020, 10AJY005), the National Nature Science Foundation of China (No. 90924027), the Fundamental Research Funds for the Central Universities of HoHai University (Grant No. 2009B22114), Public-interest Industry Project of Ministry of Water Resources (200801027) and Yunnan Province Science and Technology Plan Projects (2010).

## References

Chen SJ (2006) Research on evaluation, forecasting and regulation in the water safety system. China Hydraulic Press, Beijing

Han P, Ruan BQ (2003) Research on evaluation index system of water safety. Acta Scientiae Circumstantiae 23(2):267–272

Han YP, Ran BQ, Xie CJ (2003) Multi-objective and multilevel fuzzy optimization model and its application in water security evaluation. Resour Sci 25(4):37–42

Jia SF, Zhang JY, Zhang SF (2002) Regional water resources stress and water resources security appraisement indicators. Prog Geogr 21(6):528–45

Jiang LM (2002) Method of ideal point and its application in the commercial banks operating performance comprehensive evaluating. Syst Eng Theory Methodol Appl 11(3):227–30

Peng YX (2000) Analysis of management decision-making. Science Press, Beijing

Shao YS (2004) To strengthen urban water system planning. China Construction News

Zeng SY, Li GB, Fu H (2004) Study on water environment security and its evaluation index system – a case study of Beijing. S N Water Transferred Water Sci Technol 2(4):31–5

Zhang X, Xia J, Jia SF (2005) Definition of water security and its assessment using water poverty index. Resour Sci 27(3):145–9

# Application of Extreme Value Analysis to Extreme Drought Disaster Area in China

**Lingyan Xu, Huimin Wang, and Junfei Chen**

**Abstract** Recently, the natural balance and harmony of human society has been broken by the frequent occurrence of extreme drought. Extreme value theory is a statistical analysis tool of extreme events on the risk management which provides a good support of theoretical and methodological application. In this paper, we analyze the data of drought disaster losses area from 1949 to 2008 years in China, and establish a G model of extreme value distribution, and then verify the application of extreme value theory can significantly improve the fitting results.

**Keywords** Disaster area · Extreme drought · Extreme value theory · Pareto distribution

## 1 Introduction

During recent decades, drought disaster has frequently occurred in annual meteorological disasters, most accounts for 55% of the disaster affected area (Weng 2010). According to incomplete statistics, 2010 Southwest Drought caused economic losses over 35.186 billion Yuan, 1.01 million Mu of cultivated land area. This study focuses on these extreme droughts which are difficult to predict, largely dangerous, and highly uncertain.

The probability of extreme drought events is very low, but they often break the relative balance and harmony of nature, and cause great loss of human production and life. This paper attempts to describe the extreme events by using a statistical tool – extreme value theory (EVT) to analyze these extreme events. L. von Bortkiewicz (1922) is the first statisticians who clearly put forward extreme value

L. Xu (✉), H. Wang, and J. Chen
State Key Laboratory of Hydrology Water Resources and Hydraulic Engineering of Hohai University, Nanjing 210098, China
and
Management Science Institute of Hohai University, Nanjing 210098, China
e-mail: xulingyan333@yahoo.com.cn

Y. Zhou and D.D. Wu (eds.), *Modeling Risk Management for Resources and Environment in China*, Computational Risk Management,
DOI 10.1007/978-3-642-18387-4_39, 

(EV) (von Bortkiewiez 1992). M. Frechet (1927) published the first articles on the asymptotic distribution of the maximum paper (Frechet 1927). Currently statistics of EV has been widely applied in weather, floods, earthquakes, rainfall, human life, radioactive and other issues, etc. Chen (1973), Mcneil (1997), Reiss (2007), Wei (2008), Xie (2008), Hua (2009), Ma (2010) respectively used EVT for studying flood, earthquake, fire, and financial losses.

However, EVT is still in its infancy, whose theory and applications requires extensive digging. This article mainly applies the data of drought disaster area from 1949 to 2008 in China, and uses EVT in extreme drought disaster area to give threshold, and beyond the threshold of the affected area we want to establish Pareto extreme value distribution model.

# 2 Extreme Value Distribution Model

## 2.1 Some Common Extreme Value Distribution

Consider the following statistic statistical characteristics $M_n = \max\{X_i, X_2, \ldots, X_n\}$, and $X_1, X_2, \ldots, X_n$ is an independent and identically distributed random variables with common distribution function $F$. Usually $X_i$ means value of a unit of time or a process, such as in this article it refers to the year's drought disaster area. $M_n$ represents a maximum in this process during n period of time. To standardize $M_n$, we order $M_n^* = \frac{M_n - b_n}{a_n}$ and $\{a_n > 0\}, \{b_n\}$. So possible distribution functions of $M_n^*$ had been given by Fisher–Tippett, $P\{M_n^* \le x\} \to G(x), n \to \infty$, and $G_x$ must be one of Gumbel distribution, Frechet distribution, and Weibull distribution (Ouyang 2008). Jenkinson (1955) proposed the generalized extreme value (GEV), which defines these three types of distribution functions into a general form, as follows:

$$G_X(x) = \begin{cases} \exp\left\{-\left[1+\xi\left(\dfrac{x-b}{a}\right)^{-1/\xi}\right]\right\} & -\infty < x \le b - a/\xi \quad \xi > 0 \\ & b - a/\xi \le x < \infty \quad \xi < 0 \\ \exp\left\{-\exp\left[-\left(\dfrac{x-b}{a}\right)\right]\right\} & -\infty < x < +\infty \quad \xi = 0 \end{cases}$$

Scale parameter $a > 0$, location parameter $b$, shape parameter $\xi$. When $\xi > 0$, it refers to Frechet distribution; when $\xi < 0$, it refers to Weibull distribution; when $\xi = 0$, it refers to Gumbel distribution.

## 2.2 Threshold Selection and Parameter Estimation

There are two methods to determine the value of $u$ in EVT, one is Block Maxima Method (BMM) (Mcneil 1997), and another is Peaks-Over-Thresholds model

(POT), which is modeled by the data over a large threshold value. Broadly speaking, generalized Pareto distribution (GPD) can be used as the approximate distribution of POT model. In this paper, GPD is used as the threshold selection method.

Calculating of the average excess volume function $e(u) = E(X - u|X > u)$, and drawing scatter plot $\{u, e(u)\}$, $u > 0$. Then selecting sufficiently large value of $u$ to make $e(u)$ linearly when $X \geq u$. If MEF of $u$ is a positive slope, it follows the generalized Pareto distribution whose shape parameter is positive.

After determining the threshold value, parameter estimation is another core issue of extreme value distribution model. Commonly used methods are maximum likelihood method, moment method, moment method probability weights and regression method. Maximum likelihood estimation is the most basic method, which allows parameter estimates to reflect the overall statistics for the sample, and has good statistical properties, thus it has the most common applications.

# 3 Fitting Models of Extreme Value Distribution on China's Extreme Drought Disaster Area Losses

## 3.1 Data Analysis

Due to the lack of unity about drought economic losses measurement and few Statistical Yearbook data, this paper uses arid crop disaster area as samples during 1949 and 2008, and the data is from "China Statistical Yearbook Home", 1949–2000 droughts in Chinese history, unit is 10,000 Mu.

According to Fig. 1, the last 10% of the maximum points of the disaster area is regard as a national reference point loss of maximum, 1961, 1992, 1994, 1997, 2000, 2001. And during these 6 years, the drought disaster area (DDA) all exceeded 250,000 thousand Mu.

General analysis of disaster losses are insured catastrophe losses based on the assumption of normal distribution, however we conclude from the results from

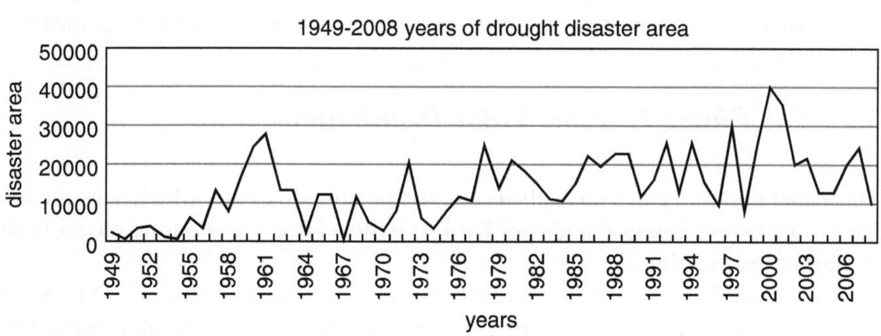


**Fig. 1** Drought disaster area within 60 years

**Table 1** The statistical description of drought disaster area losses in China over the years

| | N | Minimum | Maximum | Standard deviation | Kurtosis | Skewness | JB statistics |
|---|---|---|---|---|---|---|---|
| Disaster area | 60 | 389 | 40,176 | 9,061.712 | 0.542 | 0.029 | 25.0047 |

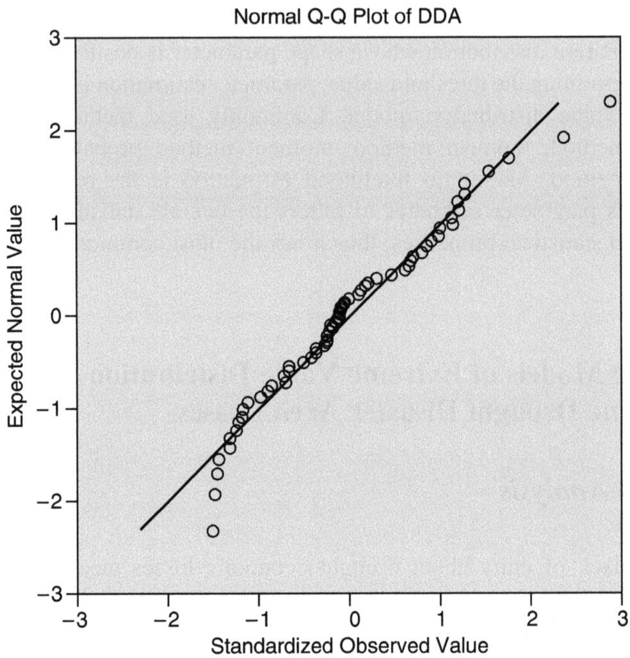


**Fig. 2** Normal Q–Q plot of DDA

Table 1 that the data's kurtosis and skewness of DDA deviated from the normal distribution hypothesis value of 0 and 3, and JB statistical value is greater which also rejected the null hypothesis of normal distribution. Therefore, we cannot use the normal distribution fitting the inundated area of drought losses accurately. After the comparison of Figs. 2 and 3, we find the tail of Fig. 3 is more near the straight line $y = x$. This shows a better fitting effect, and extreme data's effects are significant.

## 3.2 The Fitting Extreme Value Distribution

This paper hopes to place on analysis of extreme drought events which happened in the history of our country, and uses EVT to build a strong asymptotic model to fit the tail of the distribution.

The threshold is the key to modeling extreme value distribution. This paper applies the average excess volume function $e(u)$ to determine the threshold. The formula is $e_n(u) = \frac{1}{N_u} \sum_{i=1}^{n} (X_i - u)^+$, when $X_i \leq u$, $(X_i - u)^+ = 0$. Seen from the

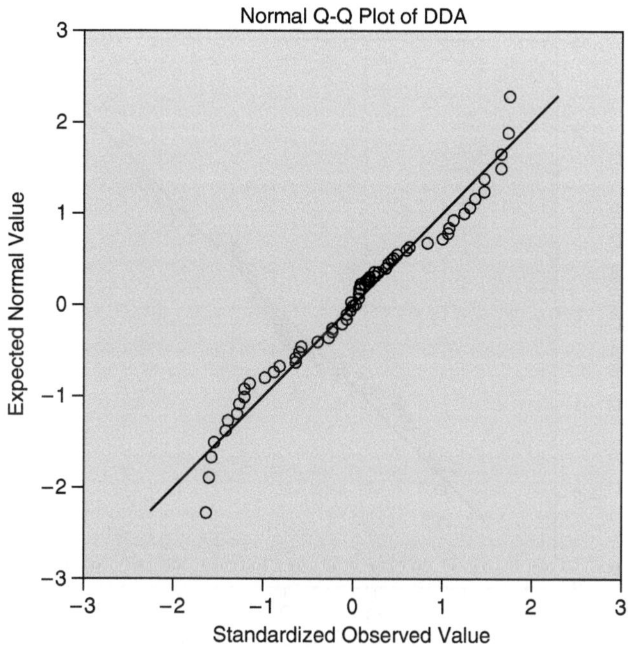


**Fig. 3** Normal Q–Q plot of DDA after removing six EV

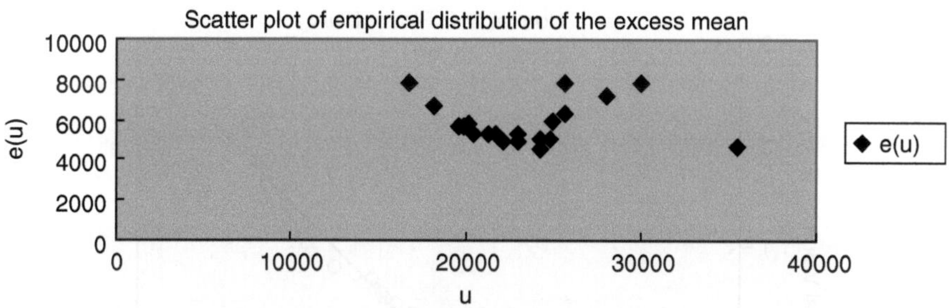


**Fig. 4** Scatter plot of empirical distribution of the excess mean

formula $e(u)$, we can conclude $e(u)$ is a linear function of $u$, so a certain value after which EMEF becomes linear is the threshold.

From Fig. 4, there is a clear upward trend when the sample mean reaches more than 24,255 acres and the slope of the graph is approximately a linear function, so the samples follow the GDP model while the shape parameter $\xi > 0$. From this point we choose the threshold value $u = 242{,}550$ thousand Mu.

When EV is determined, we compare the Q–Q plot of replacing the EV with the mean or moving directly with Fig. 2. It is found that the normal distribution of sample quintiles and the median distance of the experience points of Fig. 5, Fig. 6 are closer. While compared between Figs. 5 and 6, it was found that moving the EV directly fits the tail distribution better.

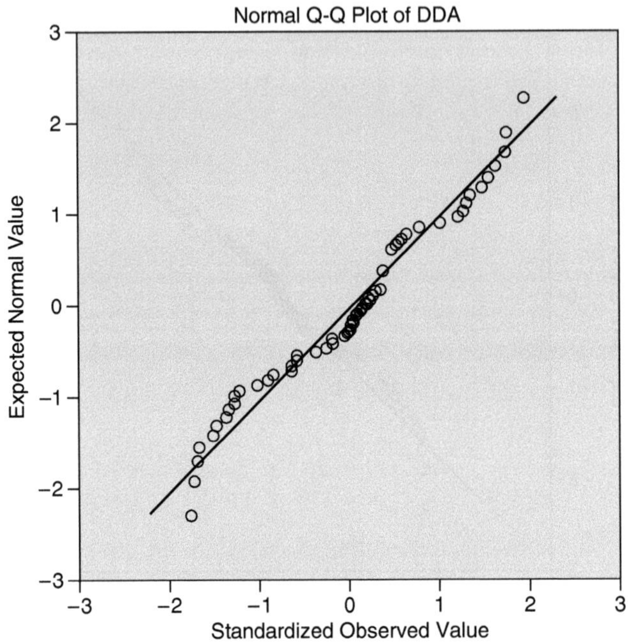


**Fig. 5** Normal Q-Q plot of DDA after EVwith mean substitution

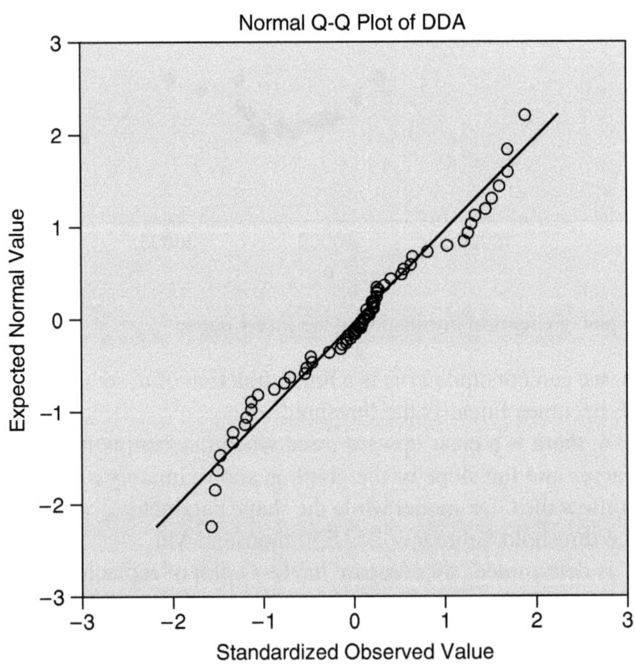


**Fig. 6** Normal Q-Q plot of DDA after removing EV directly

In this paper, maximum likelihood estimation method is used to estimate the parameters of the Pareto distributions' scale and shape parameters. General definition of GDP is $G_{a,\xi}(x) = 1 - (1 + \xi \frac{x}{a})^{-\frac{1}{\xi}}$, and the maximum likelihood function is $L(x; a, \xi) = -n \ln a - (1 + \frac{1}{\xi}) \sum_{i=1}^{n} \ln(1 + \xi \frac{x_i}{a})$. The distribution over $u$ is defined as: $F(x) = (1 - F(x))G_{a,\xi}(x - u) + F(u) = 1 - \frac{N_u}{n}(1 + \xi \frac{x-u}{a})^{-\frac{1}{\xi}}, x > u$. The result calculated by using MATLAB is $a = 51682, \xi = -1$. So $F(x) = 1 - 0.15 (1 + 1.9349 \times 10^{-5}(x - 24255))^{-1}, x > 24255$.

To test the fitting results of GDP about exceeding threshold of DDA loss, the normal Q-Q plot of exceed threshold fitting GDP is given above (Fig. 7). As seen

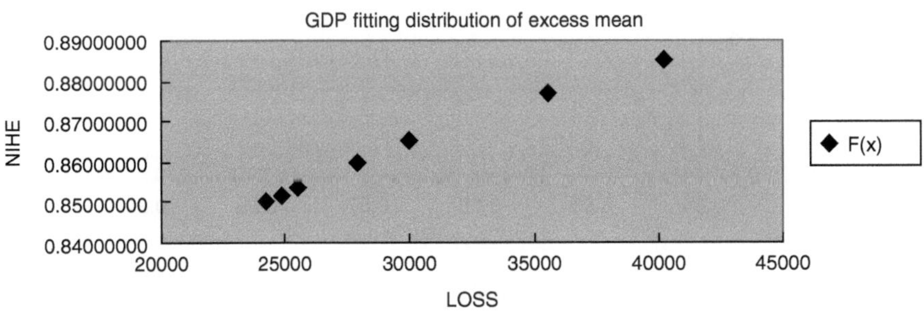


**Fig. 7** GDP fitting distribution of excess mean

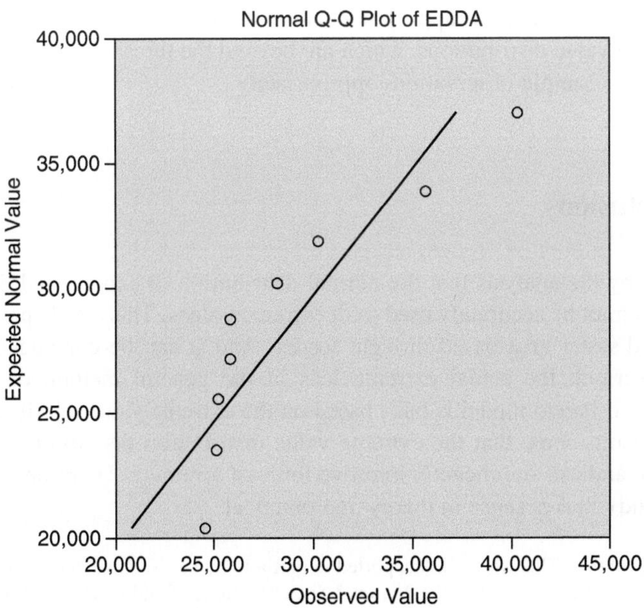


**Fig. 8** Normal Q-Q plot of exceed DDA

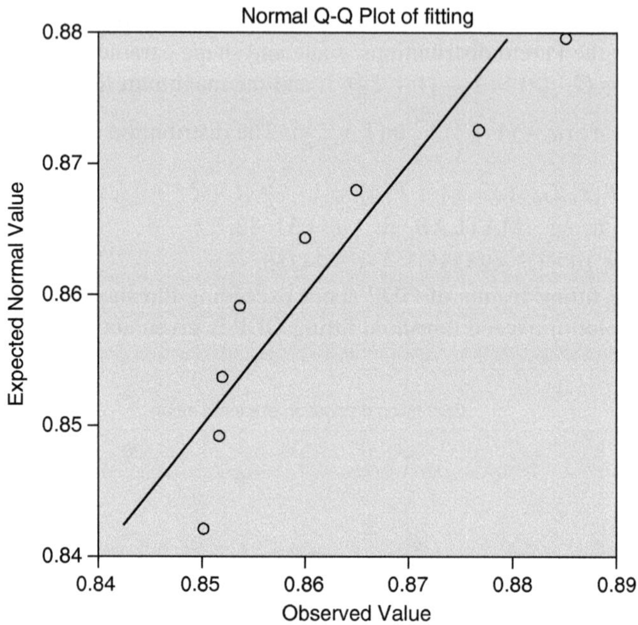


**Fig. 9** Normal Q-Q plot of exceed fitting GDP

from the diagram, the value follows the distribution of a linear function, and graph is approximately a linear function of positive slope, therefore, we believe that threshold $u = 24{,}255$ is appropriate, and it fits well. According to Fig. 8 and 9, nine extreme value distributions, which are beyond the threshold value of the GDP model, fits the sample observations appropriately.

# 4 Conclusions

It is found by the analysis that the normal distribution or exponential distribution approach cannot be accurately used to fit extreme values. This article presents EVT for fitting disaster area-based drought losses. And it can be concluded that it is easy to overlook the actual extreme loss in the general method of modeling. A generalized Pareto model is built based on the extreme value distribution of the data, the results show that the extreme value distribution fits closer to the actual distribution and can significantly improve forecast accuracy. Therefore, we believe that this study makes sense in theory and practical.

**Acknowledgements** This work was supported in partially by the National Nature Science Foundation of China (No. 90924027), the National Society Science Fund of China (No. 09CJY020, 10AJY005), Public-interest Industry Project of Ministry of Water Resources (200801027) and Yunnan Province Science and Technology Plan Projects (2010).

## References


Chen P (1973) Extreme value theory in long-term earthquake prediction. Geophys J 9:7–24

Freehet M (1927) Sur la loi de probabilite de l'ecart maximum. Ann Soe Polon Math Craeovie 6:93–116

Hua Y (2009) Application of extreme value theory in Chinese stock market risk measurement. Ph.D. thesis, Chongqing University

Jenkinson AF (1955) The frequency distribution of the annual maximum (or minimum) values of meteorological elements. Quart J Roy Meteor Soc 81:158–171

Ma S (2010) Research on risk management of extreme flood disaster led by government in China. Ph.D. thesis, Hohai University

Mcneil AJ (1997) Calculating quantile risk measures for financial return series using EVT. Working paper ETHZ

Ouyang Z (2008) Extreme value estimation and its application in finance and insurance. China economic publishing house

Reiss R-D (2007) Statistical analysis of extreme values, 3rd edn. Springer, Berlin

von Bortkiewiez L (1992) Variationsbreite und mittlerer Fehler, Sitzungsber Berli. Math Ges 21:3–11

Wei H (2008) Application of extreme value theory in the catastrophe insurance. Henan University master's thesis

Weng B (2010) China's integrated response of drought under the Changing environment. Resour Sci 2:209–316

Xie Q (2008) Extreme value theory in catastrophe losses fitting (CCISSR). pp 377–383

# Distribution Characteristics of Water Pollution on Hainan Island of China

**Zhong-yuan Yu, Bo Li, Te-sheng Sun, and Hua Bi**

**Abstract** The article uses multiple-factor method, Lorenz curve, comprehensive pollution index and Borda law analyzing the spatial and industrial distribution traits, studying the reasons, discussing strategic ways to further building of eco-province in Hainan. All study is based on the data form 2004 water environmental report of Hainan province and sticks to the national standard of water quality. By calculation and analysis, the article divides the province into three water pollution areas: low pollution area, middle pollution area and high pollution area. And colludes that (1) water pollution distribution on Hainan Island is uneven spatially; (2) waste water is the main source of water pollution; (3) economical development, population distribution, natural environments and the way of land usage have great impact on the format of water pollution distribution on the island. At the end, the article puts forward some strategies of building eco-province in Hainan.

**Keywords** Distributional traits · Hainan Island · Hazard · Water pollution

## 1 Introduction

Most of the researches on surface water pollution focus on the processes and reasons of pollution or on quality analysis. Jizhen studied Kuznets traits of surface water of Xuzhou city in Jiangsu province, analyzing the relationship of surface

Z.-y. Yu (✉)
China Research Center for Assessment of Ecological Assets, College of resources, Beijing Normal University, Beijing 100875, China
and
Institute of Geography and Tourism, Hainan Normal University, Haikou 571158, China
e-mail: yzy123456789@126.com

B. Li and T.-s. Sun
China Research Center for Assessment of Ecological Assets, College of resources, Beijing Normal University, Beijing 100875, China

H. Bi
Institute of Geography and Tourism, Hainan Normal University, Haikou 571158, China

Y. Zhou and D.D. Wu (eds.), *Modeling Risk Management for Resources and Environment in China*, Computational Risk Management,
DOI 10.1007/978-3-642-18387-4_40, 

water quality with economic growth factors (Jizhen 2006). Dong-yajie analyzed spatial and temporal distribution of water system pollution and its change by using a model of pollution barycenter, taking some main pollution characterization factors of a river as examples (Dong-yanjie 2008); Liu yan and Liu-jiaxiang analyzed the dynamic change and its reasons of surface water bodies (Liuyan 2007). More scholars studied the status quo of water pollution in some river basins. In all, few scholars have been studying distribution traits of surface water in a whole geographical region like Hainan Island and put forward its counter measures.

Hainan province is located in the southernmost of China with total area 35,400 $km^2$, for which Hainan Island (34,100 $km^2$) is the main body. It governs 19 affiliated administrative regions of county level with a population of 8,263,100 (China Statistic Bureau 2005). As early as 1999, Hainan put forward the strategy of eco-province construction, being the first experimental eco-province in China. Since then on, Hainan has gone a long way in developing economy, at the same time keeping the environment first-class in the country. However, in some area, there still exist contradiction and discord between development and environmental protection, with some areas more or less polluted.

The research studies the area of Hainan Island of the province (including 18 county-level areas) based on the data of Hainan Provincial sewage outfall census in 2004. According to the nature of water pollution and the reality of water resources of Hainan Province, the research selects flow rate, water temperature, PH, sewage amount entering the rivers, CODcr, $BOD_5$, $NH_3$-N, TP and volatile phenol as factors, uses multiple-factor evaluation method, Board Law, Lorenz curve, comprehensive water pollution index to analyze the spatial and industrial distribution traits as well as their reasons. The research has theoretical value in that it bridges the gap of studying traits of surface water in provincial unit and providing advice and theoretical reference for decision making in building eco-province in Hainan.

## 2 Method of Monitoring Water Pollution

Monitoring water quality: use method of swage analyzing and sampling in comprehensive swage discharge standard (GB8978-1996). Methods of analyzing water quality factors are listed as followed (Table 1). Monitoring flow rate: use current meter method, float method; overflow weir method and volumetric methods.

**Table 1** Method of analyze of water quality

| | Items | No. of used standard |
|---|---|---|
| 1 | Temperature | GB13195-1991 |
| 2 | PH | GB6420-986 |
| 3 | CODcr | GB11914-1989 |
| 4 | $BOD_5$ | GB7488-1987 |
| 5 | $NH_3$-N | GB7479-1987 |
| 6 | TP | GB11893-1989 |
| 7 | Volatile phenol | GB7490-1987 |

# 3 Method of Research

## *3.1 Calculate the Comprehensive Pollution Index of each county and industry*

$$S_j = \sqrt{\frac{1}{n}\sum_{i=1}^{n}\left(X_{ij} - \overline{X}_j\right)^2} \tag{1}$$

$$K_{ij} = \left(X_{ij} - \overline{X}_j\right)/S_j \tag{2}$$

$$K_i = \sum_{j=1}^{6} a_j K_{ij} \tag{3}$$

1. Use formula (1) to get standard deviation of the distribution of each pollutants among different counties or different industries. In the formula, $X_{ij}$ is the value of j pollutant in i county (or industry); $\overline{X}_j$ is average quantity of j pollutant in the province (or all industries); $S_j$ is the standard difference (SD) of j pollutant among counties (or industries). "n" is the number of i (18 counties or 20 industries).
2. Use the formula (2) for standardization of pollutant distribution data to get distribution index. In this formula, $K_{ij}$ is the pollution index of j pollutant in i county (or industry).
3. And follow the formula (3) to get the comprehensive pollution index of each county and industry. In the formula, $K_i$ is the composite pollution index of i county (or industry). $a_j$ is the weighting coefficient of pollutant j. In this research, the weighting coefficients of pollutant factors of yearly swage charge entering the rivers, are set as 0.1, 0.25, 025, 0.15, 0.1 and 0.15 respectively.

## *3.2 Calculate the Borda Index and Rank Water Pollution of Each County*

$$N_i = \sum_{j=1}^{6} N_{ij} \tag{4}$$

(Wang-jiaotuan et al. 2008; ShinjiOhseto 2007; Nitzan and Rubinstein 2002; Xiong-yang and Xu-xiaodong 2005; Yue-chaoyuan 2003)

In formula (4), $N_i$ is the Borda Index; $N_{ij}$ is the number of county which follow i county in the ranking list of j pollutant. For a county, the bigger the Borda index, the more forward it stands in ranking list of water quality and the smaller, the pollution index. Vice versa.

### 3.3 *Divide the Distribution Types of Water Pollution of Each County*

Using the comprehensive water pollution index $K_i$ and Borda index $N_i$ to divide the distribution types. Considering the reality of Hainan Province, the research decides (1) if comprehensive water pollution index $K_i > 1$ and Borda index $\leq 10$, this county is belong to high pollution area; (2) If $0 \leq K_i \leq 1$, and $10 \leq$ Borda $\leq 30$, we call this county middle pollution area; (3) if $K < 0$, Borda $\geq 30$ we call the county low pollution area (Liuyan 2007).

### 3.4 *Calculate Pollutant Spatial Concentration Index and Draw Lorenz Curve*

$$I_j = \frac{C_j - R_j}{M_j - R_j} \tag{5}$$

(Lu-dadao 1991)

In the formula, $\mathbf{I_j}$ is the spatial concentration index of pollutant, $C_j$ is the sum of cumulative percentage of j pollutant of each county. $R_j$ is the concentration index when pollutant distributes evenly (minimum), $M_j$ is the concentration index under extreme situation that all j pollutant of the province concentrates in one county (max). The bigger the $I_{j,}$ the more j pollutant concentrates.

### 3.5 *Check the Feasibility of the Factors Selected and the dependability of Division of Water Pollution Type of Each County by Judgment Index*

Use the difference value between 1 and the ratio value of number of unmoral samples against number of total samples to calculate the Judgment Index.

$$J = 1 - \frac{B + G - |B_m - G_m|}{T} \tag{6}$$

(Yuan-jianping Liu-fuke et al. 2008)

In the formula, J is the Judgment index, B is the total number of factors which present bad in ranking list in low pollution counties, G is the total number of factors which present well in ranking list in high pollution counties. $|B_m - G_m|$ is the absolute value of the difference between the numbers of factors presenting well and factors presenting bad respectively in medium class counties. T is the amount of the samples. If R ≧ 85%, it means that the factors are correctly selected and we can use them to make a judgment. If R < 85%, it means that we have not selected enough factors or what we select are not suitable for making a judgment, need to adjust or add some other factors. In this research, $R = 1 - \{0 + 0 + |6 - 14|\}/90 = 91.11\%$, it indicates that the factors selected have high dependability, can be used to make a judgment.

# 4 The Distribution Traits of Water Pollutants on Hainan Island

## 4.1 *Spatial Distribution Traits*

Water pollutants are relatively concentrated. Water pollutants of Hainan province are unevenly distributed, as demonstrated by the deviation of accumulative curve of pollutants from even-distribution curve in the picture (Fig. 1). Most of the pollutants are concentrated in Haikou, Lin'gao, Tunchang, Baisha, Danzhou and Sanya city, especially Haikou has more than 30–50% of all water pollutants of the province, being the most seriously polluted area. All spatial concentration index of water pollution factors on Hainan island are bigger than 50%, especially the sewage amount entering the rivers, CODcr, $NH_3$-N, and volatile phenol (see Table 2).

Water quality of inland rivers is better than that of coastal rivers; southeastern is better than northern western part of the island. The average index of the western (Lin'gao, Dongfang, Ledong, Cheng'mai, Changjiang, Baisha, Danzhou) of the island is −0.0614; the middle (Haikou, Sanya, Baoting, Qiongzhong, Ding'an, Wuzhishan, Tunchang) 0.29 and the eastern (Wenchang, Linshui, Qionghai, Wanling) −0.41. If we divide the province into north and south parts, the northern has an average comprehensive water pollutant index of 0.332, whereas the

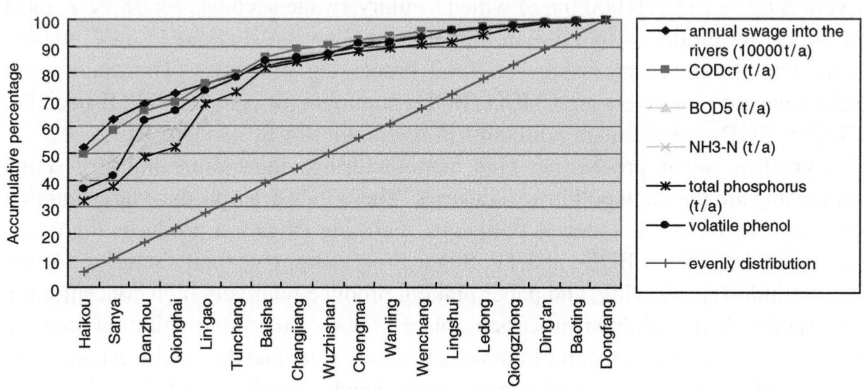


**Fig. 1** Lorenz curve of spatial water pollution in Hainan province

**Table 2** Spatial difference of water pollution of counties in Hainan Province

| Type of water pollution | Counties | Main pollutants | Ki | Ni |
|---|---|---|---|---|
| High pollution area | Haikou, Lin'gao | CODcr, BOD5 | ≥1 | <10 |
| Middle pollution area | Sanya, Danzhou, Baisha, Tunchang | CODcr, TP, $NH_3$-N $BOD_5$, | 0–1 | 10–30 |
| Low pollution area | Dongfang, Qionghai, Ledong, Baoting, Qiongzhong, WanglingWenchang, Lingshui Chengmai Wuzhishan Changjiang Ding'an | | ≤0 | >30 |

southern −0.386. Also, if we divide the province into inland area and coastal area, the inland has an index of −0.033, while the coastal 0.109.

The water quality of big river is better than that of small river, Trunk Stream better than tributary, suburban river better than urban one, lake better than river.82.5% of the stream segments reach or surpass the national water quality standard III, trunk streams of the three main rivers, Nandujiang, Wanquan river, Changhua river reach or surpass the national water quality standard III. Stream segments with standard IV, V are mainly in medium or small rivers and some branch river of Nandujiang. Thirty-three percent of urban rivers are inferior to standard III water quality. Most of lakes surpass national water quality standard III, better than river. Only 6.2% of lakes in the province are inferior to standard III. Big reservoirs have greater anti-staining ability than small ones. In recent years, part of the reservoirs are experiencing eutrophication due to increase of nutritive salts of nitrogen and phosphor in the waters (Hainan Provincial Water Business Bureau 2006).

## *4.2 Analyze the Industrial Distribution Traits*

Sanitary sewage is the main sources of water pollution. Sewage quantity entering rivers is up to 257 270 000 t/a, of which sanitary swage accounts for 78.92%, most of which are produced by civilian life; industrial swage is about 15.84%, mainly from farm product processing and food processing industries. The main water pollutants entering rivers are CODcr, $BOD_5$ ammonia nitrogen and TP. Especially, CODcr, $BOD_5$ are the main pollutants in rivers inferior to standard III.

Cynicism, sugar processing, glue manufacturing, aquaculture industry starch processing are the main polluting industries. These industries produce large number of swage with huge amount of pollutants. Farming irrigation produces high concentration of CODcr, $BOD_5$ and TP. Sugar Processing, glue manufacturing, Aquaculture industry, weaving, and paper making produce relatively high concentration of volatile phenol. Rubber processing plant is one of the main pollution sources, most of them in southern part of the island, Nandu river and the northwest region of the island. Most of the rubber processing plants scattering on the island use inefficient anaerobic method and oxidation pond process to deal with swage, being the main reasons of pollution of some stream segments (Hainan Provincial Water Business Bureau 2006).

Cultivation, rubber processing, aquaculture, butchery and sugar processing are industries with high comprehensive pollutant index and highest concentration of pollutants. Irrigation swage, produces high concentration of CODcr, $BOD_5$, TP and $NH_3$-N. Sugar processing industries give out volatile phenol, and starch processing and sugar processing industries bring relatively high concentration of $BOD_5$, while rubber processing plants volatile phenol, NH3-N and TP and aquaculture, butchery volatile phenol and $NH_3$-N.

Aquaculture industry and livestock and poultry breeding industry are the main areal source of water pollution. In recent year, tropic agriculture and aquaculture develop rapidly, the resulting areal pollution problems are getting more and more serious. This mainly exits in urban area and suburban village, where the pollutants are organic matter, fertilizer, and pesticide. The unscientific fertilizing and drainage and irrigation system, the improper ways of fresh water aquiculture stocking, urban waste, and sanitary swage in the area resulted in serious pollution of the surface water.

## 5 Conclusion and Countermeasures

Water pollution distribution on Hainan Island is uneven spatially and industrially; waste water is the main source of water pollution; economical development, population distribution, natural environments and land usage have great impact on the format of water pollution distribution. Hainan province is the biggest tropic oceanic province and the biggest special economical zone in China, also the first to put forward the strategy of building eco-province. Furthermore, recently, Hainan set up to build international tourist island. Water pollution is an important factor threatening construction of eco-province and international tourist island. Ecological environment, efficient agriculture, ocean industry, vacation tour are the strengths of Hainan province. We should foster the strengths and circumvent the weaknesses, promote developing recycling economic, strengthen industrial adjustment, further develop eco-agriculture, eco-industry and eco-tourism, improve technology of production and swage prevention and treatment; rely on the masses, depend on the support and the concerted consecutive effort of all sides of society to build Hainan a harmonious, ecological civilized, prosperous and beautiful eco-province, paradise for living, production, investing and vocation.

## References

China Statistic Bureau (2005) Hainan provincial statistical year book, 16, 49, 292 China statistic bureau Press

Dong-yanjie (2008) Unbanization's impact on surface water quality of Guangzhou city. J Water Conserv Water Power Chinese Rural Area 2

Hainan Provincial Water Business Bureau (2006) Hainan provincial hydrology and water resource bureau research report on investigation of sewage outfall of hainan province 6

Jizhen K (2006) Traits of surface water in Xuzhou city. J Water Resour Preservation 4:65–67

Liuyan (2007) Evolution characteristic of water quality in Weihe basin in Shanxi province. J Water Resour Preservation 3:30–31

Lu-dadao (1991) Region theory and methods of region study. Science Press, Beijing

Nitzan S, Rubinstein A (2002) A further characterization of Borda ranking method. J Public Choice 36:153–158

ShinjiOhseto (2007) A characterization of the borda rule in peer ratings. J Math Social Sci 54:147–151

Wang-jiaotuan, Mao-zheyong, Zhou-chaowei (2008) Environmental comprehensive assessment, forecast and countermeasure. J Chinese Agri Sci Bull 1

Xiong-yang, Xu-xiaodong (2005) Relation and comparison between Borda point method and ticket authorizing method. J Huazhong Univ Sci Technol Urban Sci (city edition) 5(22): 132–134

Yuan-jianping Liu-fuke, Wang-ping et al (2008) Spatial difference of counties' economical development level and poverty alleviation development. J Hainan Normal Univ 21:94–99

Yue-chaoyuan (2003) Decision theory and method Science Press, Beijing, pp 315-317

# Study on Double Auction Model for Discharge Quota Trading of Water Pollutants

**Huirong Tang, Huimin Wang, and Lei Qiu**

**Abstract** This paper briefly introduces the market structure and operating mechanism of the water pollutant emissions trading in order to increase efficiency, and proposes the double auction model which consists of water pollutant emissions trading transaction costs and trading volume, and the auction price and the exchange price trading mechanism rules. Then a numerical example is given to illustrate the application of this model, and it is significant in water pollutant emissions trading.

**Keywords** Discharge quota trading · Double auction · Water pollution

## 1 Introduction

Water pollutant emissions trading is a market-based water pollution control mode, its experimental work are being carried out in many places of China and achieved some success. Discharge quota of water pollutants is referred to the maximum number of water pollutants in a specific period of time and place for discharge agent who is set by the government regulatory agencies. Relevant government authorities select a different allocation of these emissions, and establishing emissions trading market to legitimate the trading rights. Dischargers of water pollutants decide to buy or sell emission rights on the extent of their pollution control from their own interests (Shi 2003).

The auction is a market system arrangement, in which the bid is based on a series of participants to determine the allocation of resources and the "clearing price"

H. Tang (✉), H. Wang, and L. Qiu
State Key Laboratory of Hydrology Water Resource and Hydraulic Engineering, Nanjing 210098, China
and
Institute of Management Science, Hohai University, Nanjing 210098, China
e-mail: tanghuirong1983@126.com

Y. Zhou and D.D. Wu (eds.), *Modeling Risk Management for Resources and Environment in China*, Computational Risk Management,
DOI 10.1007/978-3-642-18387-4_41, © Springer-Verlag Berlin Heidelberg 2011

(Nicolaisen et al. 2001). Auction is a market-based resource allocation approach, there are two basic functions: first is the revealed information, the second is to reduce agency costs (Zhang 1996). Auction allocation is an effective method for the transfer of rights of scarce resources of the public (McAfee and McMillan 1987). In the double auction, the final offer price is the new price which is based on buyers and sellers, buyers and sellers is an equal supply-demand relationship, which can effectively solve the "conspired" and "malicious quote" problems (Zhan and Wang 2003), so it has been applied in many fields (Fang and Wang 2005; Wang and Wang 2006; Fu et al. 2006; Liu et al. 2007).

This article focuses on water pollutant emissions in the secondary market and considers more buyers and more sellers of emissions trading to take double auction, and builds the double auction model of water pollutant emissions trading.

# 2 Double Auction model for Discharge Quota Trading of Water Pollutants

There are more buyers and more sellers in the water pollutants discharge quota trading market, buyers want to buy emission rights in the market, the seller hopes to sell in the market of emission rights. Buyers and sellers determine the transaction price through bargaining to achieve the optimal allocation of emission rights. The Government is the organizers of market water pollution discharge, and also the designer of the market mechanism. Buyers and sellers abide by market rules, take into account of their own valuation of the emissions at the same time offer, ultimately the market determine the final transaction price in accordance with the relevant rules.

## 2.1 *Basic Assumptions*

Suppose there are $m$ buyers and $n$ sellers in a certain period of trading market, the total demand for emission rights is $Q_d$, the total supply of emission rights is $Q_s$, these constitute a double water pollutant emissions trading auction. Each market participant submit their wish of trade emissions unit price and the number of emission rights, which corresponds to a set of bidding strategies (unit price of emission rights, the number of emission rights). Suppose the buyer's offer is $(d_i, x_i)$ $(i = 1, 2, \cdots, m)$, and the seller's offer is $(s_j, y_j)$ $(j = 1, 2, \cdots, n)$. $d_i$ is the buyers' expected unit price of emissions, $x_i$ is buyers' expected transactions number under the price of emission rights. $s_j$ is sellers' expected unit price of emissions, $y_j$ is sellers' expected transactions number under the price of emission rights. In other words, the buyers wish to purchase $x_i$ units emission rights at the price of $d_i$, and the sellers wish to sell $y_j$ units emission rights at the price of $s_j$.

The value of water pollutant emissions is difficult to quantify, the same emissions value for different participants are different. Therefore, each participant has their own private information. Assume that all buyers' value of emissions per unit of power is $v_i$ $(i = 1, 2, \cdots, m)$. It is the buyers' private information and other market participants only know the value of the probability distribution. Assume that all buyers' $v_i$ are independent and identically distributed, $v_i$ follow the distribution of $F(\cdot)$ during $[0, \alpha]$, probability density is $f(\cdot)$. Assume that any sellers' value of emissions per unit of power is $c_j$ $(j = 1, 2, \cdots, n)$. It is the sellers' private information and other market participants only know the value of the probability distribution. Assume that all sellers' $c_j$ are independent and identically distributed, $c_j$ follow the distribution of $G(\cdot)$ during $[0, \beta]$, probability density is $g(\cdot)$. Therefore, the buyers' bidding strategy is $d_i(v_i)$, and the sellers' bidding strategy is $s_j(c_j)$, then the buyers' unit income is $v_i - d_i(v_i)$ and the sellers' unit income is $s_j(c_j) - c_j$. The buyers and sellers should also consider the overall effectiveness of the auction market, and the number of emission rights trading.

## 2.2 *Model Construction*

Charter Gold and Samuelson established a simple model of the double auction (Chatterjee and Samuelson 1983). In this model there is only one buyer and a seller, both parties decide whether to trade a unit of the commodity. When $d^*(v) \geq s^*(c)$, the transaction occurs and the trade price is $p = [d^*(v) + s^*(c)]/2$. As buyers and sellers expect to maximize their income, we calculate $d^*(v)$ and $s^*(c)$ according to $u_d = v - d(v)$ and $u_s = s(c) - c$. In this case of incomplete information, there are a lot of Bayesian equilibriums of this game. When $d^*(v) < s^*(c)$ the transaction will not occurred.

Buyers are more than one in the water emissions pollutants market, the seller is multiple, it is a "many-to-many" the market structure. Market organizers should not only consider the whole social welfare maximization, but also the concentration of buyers and sellers maximize return to enter transactions. This is equivalent to solving the following optimization problem.

$$\max \sum_{i=1}^{m} E[u_{di}(v_i)] \tag{1}$$

$$\max \sum_{j=1}^{n} E[u_{sj}(c_j)] \tag{2}$$

$$\max \sum_{j=1}^{n} \sum_{i=1}^{m} \{E[u_{di}(v_i)] + E[u_{sj}(c_j)]\} \tag{3}$$

$$s.t.\ \ u_{di}(v_i) = T_i(d)[v_i - d_i(v_i)] \geq 0 \quad (i = 1, 2, \cdots, m) \tag{4}$$

$$u_{sj}(c_j) = T_j(s)[s_j - s_j(c_j)] \geq 0 \quad (j = 1, 2, \cdots, n) \tag{5}$$

$$Q_T = \min\left\{\sum_{i=1}^{m} Q_i(d), \sum_{j=1}^{n} Q_j(s)\right\} = \sum_{i=1}^{m} T_i(d) = \sum_{j=1}^{n} T_j(s) \tag{6}$$

$Q_T$ is the market turnover of water pollutants emission rights, which is the little one between the buyers purchasing number of emission rights $\sum_{i=1}^{m} Q_i(d)$ and all sellers selling number of emission rights $\sum_{j=1}^{n} Q_j(s)$. $T_i(d)$ is buyer $i$' purchasing number, $T_j(s)$ seller $j$'s selling number. $u_{di}(v_i)$ is buyer $i$' income, $u_{sj}(c_j)$ is seller $j$'s income. Formula (1) refers to the maximum expected profit of buyer. Formula (2) refers to the maximum expected profit of seller. Equation (3) refers to the maximum expected profit of two, which means the largest social welfare. Formula (4) refers to the buyer returns is non-negative. Formula (5) refers to the seller returns is non-negative. Formula (6) refers to the buyer purchases and sellers to sell are an equal volume. This is a multi-objective optimization model, the following mechanism design model and trading rules, clearing rules are given below.

## 3 Double Auction Mechanism for Discharge Quota Trading of Water Pollutants

### 3.1 Trading Rules

Nicolaisen (2001)proposed "level matching" trading rules, buyers were arranged in accordance with the quotation from high to low priority, Sellers were arranged in accordance with the offer ranked from low to high. The former high-ranking has priority. Highest priority buyer and seller first trade, then buyers and sellers of the second high-priority trade, and so on. In this article his "level matching" rules are used in auction trade. The buyers' offer set is $D = \{d_{s1}, d_{s2}, d_{s3}, \cdots, d_{sk}, \cdots, d_{sm}\}$, when $d_{s1} \geq d_{s2} \geq d_{s3} \geq \cdots \geq d_{sk} \geq \cdots \geq d_{sm}$. The sellers' offer set is $S = \{s_{s1}, s_{s2}, s_{s3}, \cdots, s_{sl}, \cdots, s_{sn}\}$, when $s_{s1} \leq s_{s2} \leq s_{s3} \leq \cdots \leq s_{sl} \leq \cdots \leq s_{sn}$.

We propose $d_{sk} \geq s_{sl}, d_{sk} < s_{s(l+1)}, d_{s(k+1)} < s_{sl}$, and then this deal set to buy into a collection of quotations is $D' = \{d_{si} | d_{si} \geq d_{sk}\}$, sell into a collection of quotations is $S' = \{s_{sj} | s_{sj} \leq s_{sl}\}$. Under the offer to determine the level of buyers and sellers enter the transaction set does not necessarily have to finally win the emissions trading.

### 3.2 Pricing

After determining the transactions set of buyers and sellers, the buyers select sellers from the transactions set in accordance with the priority order. The buyers will assess the following three factors: each vendor's offer; the number of emission rights sold by the seller; transaction costs. So in this case the buyer's transaction object is not necessarily the lowest offer the seller, but to give the buyer the right to receive units of water to minimize the cost to pay the seller. If a seller and buyer after the transaction to sell emission rights to be left there, you can continue to trade buyers back. Trading continues in accordance with the above principles, until water rights transactions have all completed.

In order to achieve equitable income between buyers and sellers, and to fully mobilize the enthusiasm of both transactions parties, further promote buyers and sellers truly quote, in this paper the two sides of each transaction under the purchase transaction to determine the market price of the transaction method is applied that is no longer a uniform price market transactions .Instead formula (7) is used to calculate the transaction prices.

$$p = \frac{d + s}{2} \tag{7}$$

$p$ is final transaction price, $d$ is buyers' offer, $s$ is sellers' offer. For a buyer and seller transactions over the situation, the seller offer by the seller of the highest offer.

## 4 An Example

Suppose on water pollution emissions trading market, there are ten buyers, six sellers at one point. All participants are ordered to offer the unit emissions, and to report the number of emission rights to be traded. Datum refer to trading are shown in Tables 1 and 2.

The buyers' offer is arranged in accordance with the high-to-low; the sellers' offer is arranged in accordance with the low-to-high. The buyers' transaction set collection is $D = \{D_4, D_9, D_1, D_5\}$ and the sellers' transaction set collection is $S = \{S_2, S_6, S_4, S_5, S_1\}$, the elements of collection $D$ and $S$ is arranged by priority.

**Table 1** Participant transactions information sheet

| Buyers | Price | Number | Sellers | Price | Number |
|---|---|---|---|---|---|
| $D_1$ | 4.1 | 24 | $S_1$ | 3.6 | 30 |
| $D_2$ | 2.5 | 8 | $S_2$ | 2.4 | 15 |
| $D_3$ | 2.3 | 16 | $S_3$ | 4.6 | 45 |
| $D_4$ | 4.5 | 30 | $S_4$ | 2.9 | 37 |
| $D_5$ | 3.8 | 25 | $S_5$ | 3.5 | 35 |
| $D_6$ | 2.7 | 23 | $S_6$ | 2.8 | 23 |
| $D_7$ | 3.5 | 19 | | | |
| $D_8$ | 3.4 | 44 | | | |
| $D_9$ | 4.4 | 20 | | | |
| $D_{10}$ | 3.1 | 30 | | | |

**Table 2** The buyer's and seller's cost of units emissions trading

| Buyer \ Seller | $S_1$ | $S_2$ | $S_3$ | $S_4$ | $S_5$ | $S_6$ |
|---|---|---|---|---|---|---|
| $D_1$ | 0.1 | 0.2 | 0.2 | 0.3 | 0.4 | 0.1 |
| $D_2$ | 0.1 | 0.1 | 0.3 | 0.2 | 0.1 | 0.1 |
| $D_3$ | 0.3 | 0.3 | 0.2 | 0.4 | 0.1 | 0.1 |
| $D_4$ | 0.2 | 0.1 | 0.3 | 0.4 | 0.2 | 0.3 |
| $D_5$ | 0.1 | 0.3 | 0.1 | 0.1 | 0.1 | 0.2 |
| $D_6$ | 0.2 | 0.2 | 0.2 | 0.2 | 0.1 | 0.3 |
| $D_7$ | 0.3 | 0.1 | 0.1 | 0.2 | 0.3 | 0.1 |
| $D_8$ | 0.4 | 0.4 | 0.2 | 0.4 | 0.2 | 0.1 |
| $D_9$ | 0.2 | 0.2 | 0.1 | 0.2 | 0.1 | 0.1 |
| $D_{10}$ | 0.1 | 0.1 | 0.2 | 0.3 | 0.1 | 0.1 |

According to trading rules, the buyer $D_4$ who offers the highest price (highest priority buyer) can first transact, and choose the seller freely. According to five sellers' quote and transaction costs and the number of emission rights to be traded, $D_4$ will select $S_2$, $S_6$ to achieve transaction, the volume are 15 and 15, trade price is $P_1 = (4.5 + 2.8)/2 = 3.65$. At this time there are remaining eight emission trading rights for seller $S_6$. Secondly $D_9$ will trade with $S_6$ and $S_4$, and the volume are 8 and 12, the transaction price is $P_2 = (4.4 + 2.9)/2 = 3.65$. At this time there are remaining 25 emission trading rights for seller $S_4$. Again $D_1$ will trade with $S_4$, the volume is 24 and the transaction price is $P_3 = (4.1 + 2.9)/2 = 3.5$. At this time there are remaining one emission trading rights for seller $S_4$. Lastly $D_5$ will trade with $S_4$ and $S_5$, and the volume are 1 and 24, the transaction price is $P_4 = (3.8 + 3.5)/2 = 3.65$. At this time there are remaining 11 emission trading rights for seller $S_5$. As can be seen from the transaction price, purchasers who offer the highest price will first get into the transaction, but the transaction price is not the lowest. The third buyer who gets into the transaction has the lowest price. From this we can see that the dominant strategy of real offer is effective. It is also proved that the trading mechanism and trading rules to determine the price is reasonable and effective.

## 5 Conclusion

Water pollutant emissions' trading is an important means to achieve optimal allocation of water resources and environment, and study of their trading patterns and mechanism is the basis for promoting emissions trading. This paper proposes the double auction model for Water pollutant emissions' trading which consists of water rights transaction costs and trading volume, and designs trading mechanism rules to ensure the operation of the market. The rules and auction price of double auction market trading have been studied in this paper. And the mechanism is proved reasonable through a simple example. It will have certain practical significance of emissions transactions of water pollutants on the current market.

**Acknowledgments** This work is supported provided by the National Social Science Foundation of China (No. 08CJY02), the Key Project of Chinese Ministry of Education (No. 108064) and Project 333 Foundation of Jiangsu Province.

## References

Chatterjee K, Samuelson W (1983) Bargain under incomplete information. Oper Res 31:835–851

Fang D, Wang X (2005) Electricity market generation companies and large users of both electricity trading bid auction model. Netw Technol 29:32–36

Fu J, Shao P, Yang X (2006) Online double auction simulation of the incomplete information game. J Manage 3:376–380

Liu B, Zeng Y, Li P (2007) Summary of the financial market microstructure based on continuous double auction. J Manage Eng 21:91–100

McAfee P, McMillan J (1987) Auctions and bidding. J Econ Lit 25:699–738

Nicolaisen J, Petrov V, Tesfatsion L (2001) Market power and efficiency in a computational electricity market with discriminatory double-auction pricing. IEEE Trans Evol Computation 5(5):504–523

Shi M (2003) Establishment of water pollutant emissions trading framework. Shanghai Environ Sci 22:639–641

Wang Q, Wang X (2006) Study of the water trading market based on game theory. Water Econ 24:16–18

Zhan W, Wang S (2003) Comment on "Smith mystery" and the double auction Progress. J Manage Sci 6:1–12

Zhang W (1996) Game theory and information economics. Shanghai People's Publishing House, Shanghai

# Analysis on Variation and Factors of Water Resources Consumption Intensity in China

**Jinping Tong, Jianfeng Ma, and Gaofeng Liu**

**Abstract** In China, water resources consumption intensity (WI) was obviously decreased during 1997–2008. Based on complete decomposition model, the paper makes a profound analysis on the variation and factors of WI from the point of industry. The results show that the decline of total WI was caused by both adjustment of industrial structure and increase of water utilization efficiency in various industries. Particularly, the decline of economic proportion of primary industry and the increase of water utilization efficiency of primary and secondary industry played an important role in the decline of WI. And WI has plenty of space to decline, so the water utilization efficiency should be continuously improved.

**Keywords** China · Factor decomposition · Water resources consumption intensity

## 1 Introduction

Water is the important strategic resources of national development and ecological environment in China, and the basis of Chinese national economy as well. But more serious problems are faced to water resources sustainable development in China.

---

J. Tong (✉)
State Key Laboratory of Hydrology-Water Resources and Hydraulic Engineering, HoHai University, Nanjing, China, P.R. 210098
and
School of Economics and Management, Jiangsu Polytechnic University, Changzhou, China 213164
e-mail: tong19781978@126.com

J. Ma
School of Economics and Management, Jiangsu Polytechnic University, Changzhou, China 213164

G. Liu
State Key Laboratory of Hydrology-Water Resources and Hydraulic Engineering, HoHai University, Nanjing, China, P.R. 210098
and
Business school, Hohai University, Changzhou 213022,China

Y. Zhou and D.D. Wu (eds.), *Modeling Risk Management for Resources and Environment in China*, Computational Risk Management,
DOI 10.1007/978-3-642-18387-4_42, © Springer-Verlag Berlin Heidelberg 2011

Water shortage is one of the critical problems. During 2000 to 2008, per capita water resources of China has been decreased from 2,071 $m^3$ per 10,000 yuan to 2,194 $m^3$ per 10,000 yuan, now less than one fourth of the per capita water resources of the whole world. And water resource scarcity becomes increasingly conspicuous. At the same time, with the rapid growth of China economy, socioeconomic development results in an increase of water resources demands, and water shortage has become the main factor to restrict the development of a considerable fraction of regions. Per capita water resources is lower than 2,000 $m^3$ per person considered by the UN Commission on Sustainable Development in 18 provinces, autonomous regions and municipalities. Among them, there are ten areas lower than the minimum standard of 1,000 $m^3$ per person (Sustainable Development Strategy Study Group from the Chinese Academy of Science 2007). Water pollution occurred with the rapid Industrialization and Urbanization has further increased the contradiction between supply and demand of water resources. According to the investigation of the American RAND Corporation, water shortage and water pollution are two of eight factors impacting on China's economic sustainability, and their negative effect on the economic growth rate reached 1.0–2.0% which is more than effect of rising energy prices and declining foreign investment (RAND 2006). It is obvious that water shortage has been one of the serious bottlenecks to restrict the socioeconomic development.

At the same time, it is quiet outstanding of the problems such as extensive water utilization, lower water utilization efficiency, serious water waste and so on. In the current circumstances of supply and demand of water resources, the core of resolving water shortage is to improve the water utilization efficiency. Therefore how to improve the water utilization efficiency has been the key to water saving and water resources sustainable development at present. In order to solve the problem, variation and factors of water resources consumption intensity (WI) should be analyzed, which can correctly control the objective circumstance of water utilization, reasonably appraise our past water-saving policies and find the orientation and countermeasure of improvement of water efficiency. So the paper makes a deep study on the variation and factors of WI based on complete decomposition model of Laspeyres index.

## 2 Approach

### *2.1 Factor Decomposition Model*

The so-called water resources productivity (WP) is the ratio of output (Y) to water input (W) in production process, namely Y/W. Water resources consumption intensity (WI) appeared in many references to evaluate the efficiency of water utilization indirectly. Between WI and WP is reciprocal, so WI is the ratio of W to Y (W/Y). Water usage per unit GDP is usually taken as an typical index to measure (Shadrick et al. 2006; Chen 2008; Li et al. 2008). Considered on the bigger difference of water utilization among primary industry, secondary industry and tertiary industry, variation of WI will be analyzed by decomposition from the point of industry.

Let $W_1$, $W_2$, $W_3$ be the water usage of primary, secondary and tertiary industry respectively, $Y_i$ be industrial economic output, then the $i$-industry WI is $w_i = W_i/Y_i$. $W = \sum_i W_i$ denotes the total water usage of three industries and $Y = \sum_i Y_i$ denotes the total output, then the equation of WI decomposition model from industrial point shows that

$$WI = \frac{W}{Y} = \frac{\sum_i W_i}{Y} = \frac{\sum_i w_i Y_i}{Y} = \sum_i w_i \bullet \frac{Y_i}{Y} = \sum_i w_i \bullet y_i, i = 1,2,3 \qquad (1)$$

where $y_i = Y_i/Y$ shows the ration of each industrial output to the total output.

Let $WI^n$ denote WI of the reference period where n = 0,1,2, ..., N. Let $WI^0$ denote the WI of the base period, then

$$WI^n = \sum_i w_i^n y_i^n, WI^0 = \sum_i w_i^0 y_i^0$$

So variation of WI shows that

$$\Delta WI = WI^n - WI^0 = \sum_i (w_i^n y_i^n - w_i^0 y_i^0) \qquad (2)$$

According to Lasperyres index method, variation of WI is decomposed into two main parts: variation of industrial structure and variation of various industrial departments.

$$\Delta WI = \sum_i w_i^0 (y_i^n - y_i^0) + \sum_i y_i^0 (w_i^n - w_i^0) + \sum_i (w_i^n - w_i^0)(y_i^n - y_i^0) \qquad (3)$$

Where $\sum_i (w_i^n - w_i^0)(y_i^n - y_i^0)$ denotes the residual value of decomposition. Based on the principle of jointly created and equally distributed (Sun 1998), the total WI is decomposed into two parts: one is called structure portion ($\Delta WI_{str}$) caused by adjustment of industrial structure, the other is called efficient portion ($\Delta WI_{eff}$) caused by variation of water utilization efficiency of various industrial departments, namely

$$\Delta WI_{str} = \sum_i [w_i^0 (y_i^n - y_i^0) + \frac{1}{2} (w_i^n - w_i^0)(y_i^n - y_i^0)] \qquad (4)$$

$$\Delta WI_{eff} = \sum_i [y_i^0 (w_i^n - w_i^0) + \frac{1}{2} (w_i^n - w_i^0)(y_i^n - y_i^0)] \qquad (5)$$

Based on (4) and (5), we can obtain two contribution rates to WI variation: one is caused by industrial structure adjustment ($r_{str}$), the other by efficient enhancement ($r_{str}$). Then

$$r_{str} = \frac{\sum_i [w_i^0(y_i^n - y_i^0) + \frac{1}{2}(w_i^n - w_i^0)(y_i^n - y_i^0)]}{\sum_i w_i^n y_i^n - \sum_i w_i^0 y_i^0}. \tag{6}$$

$$r_{eff} = \frac{\sum_i [y_i^0(w_i^n - w_i^0) + \frac{1}{2}(w_i^n - w_i^0)(y_i^n - y_i^0)]}{\sum_i w_i^n y_i^n - \sum_i w_i^0 y_i^0}. \tag{7}$$

## 2.2 Data Sources and Processing

Calculation and decomposition of WI are analyzed with the data during the period of 1997–2008 in the paper. The indexes of water usage and output are isolated from China water resources Bulletin (1997–2008) and China Statistical Yearbook (1997–2008) separately. In order to analyze easily, the industry is divided into primary industry, secondary industry and tertiary industry.

In order to guarantee the data consistency, WI is obtained by the ratio of water usage over the years to GDP over the years at comparable prices (the base year of 1997). WI of various industries will be obtained by the ratio of water usage to added value at comparable prices. By displaying the data, total WI and WI of various industries are shown in Table 1.

**Table 1** Total WI and WI of various industries unit: $m^3$ per 10,000 yuan

| Year | Total WI | Primary industry | Secondary industry | Tertiary industry |
|---|---|---|---|---|
| 1997 | 705 | 2,714 | 341 | 23 |
| 1998 | 638 | 2,520 | 314 | 22 |
| 1999 | 610 | 2,518 | 298 | 22 |
| 2000 | 553 | 2,405 | 267 | 21 |
| 2001 | 517 | 2,365 | 246 | 20 |
| 2002 | 468 | 2,245 | 224 | 19 |
| 2003 | 412 | 2,012 | 205 | 17 |
| 2004 | 390 | 1,977 | 191 | 17 |
| 2005 | 359 | 1,876 | 179 | 16 |
| 2006 | 331 | 1,829 | 166 | 15 |
| 2007 | 294 | 1,732 | 151 | 14 |
| 2008 | 274 | 1,671 | 137 | 14 |

# 3 Empirical Results and Analysis

## *3.1 Variation of WI*

As shown in Table 1, the total WI has taken on an obvious decrease from 705 $m^3$ per 10,000 yuan in 1997 to 274 $m^3$ per 10,000 yuan in 2008, which decreases 61.15%. In the light of concrete circumstances, the characteristics of WI are shown that WI of three industries all decreases year by year. The industrial difference coefficient is broadening gradually. WI of primary industry, secondary industry and tertiary industry was reduced by 38.43%, 59.66% and 41.87% respectively in 2008 as compared to that in 1997, and increased ratio of water efficiency in secondary industry was higher than that in the other two industries. The difference coefficient of WI among three industries is a gradual extension in Fig. 1, increasing from 1.43 in 1997 to 1.52 in 2008.

## *3.2 Industrial Decomposition of WI*

Table 2 shows the results of industrial decomposition of WI in China during 1997–2008. Seen from Table 2, the structure and efficient contribution rates were all positive values to reduction of WI during 1997–2008. The accumulative contribution rate caused is 59.52% by efficiency and 40.48% by structure, which propels a sustainable increase of water utilization efficiency with double impelling forces. Seen from Fig. 2, the contribution rate took on three variable stages:

- The first stage (1999–2003): the structure contribution rate was reduced from 72.74% to 30.7%, but the efficient contribution rate was increased from 27.26% to 69.3%.

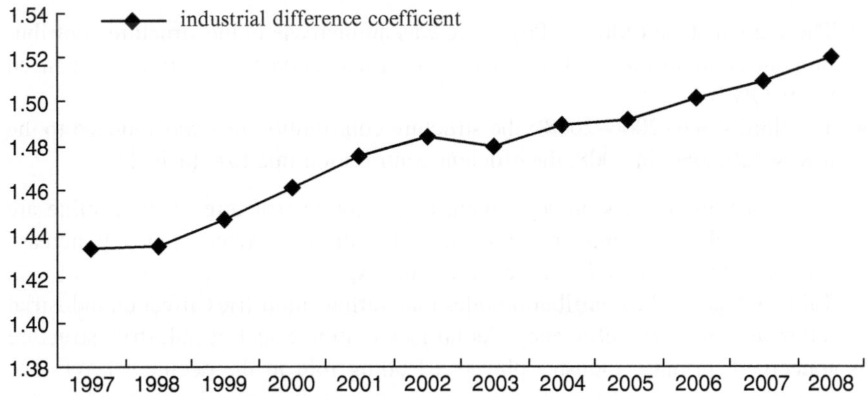


**Fig. 1** Industrial difference coefficient of WI

**Table 2** Results of industrial decomposition of WI

| year | Total variation | Variation effect of industrial structure | | Variation effect of industrial efficiency | |
|---|---|---|---|---|---|
| | | Structure portion (ton per 10,000 yuan) | Contribution rate (%) | Efficient portion (t/10,000 yuan) | Contribution rate (%) |
| 1997–1998 | −67.14 | −17.92 | 26.69 | −49.22 | 73.31 |
| 1998–1999 | −28.17 | −20.49 | 72.74 | −7.68 | 27.26 |
| 1999–2000 | −57.12 | −22.83 | 39.97 | −34.29 | 60.03 |
| 2000–2001 | −35.96 | −19.73 | 54.88 | −16.23 | 45.12 |
| 2001–2002 | −49.05 | −19.64 | 40.03 | −29.41 | 59.97 |
| 2002–2003 | −62.68 | −19.24 | 30.70 | −43.44 | 69.30 |
| 2003–2004 | −20.01 | −8.87 | 44.34 | −11.14 | 55.66 |
| 2004–2005 | −30.46 | −10.97 | 36.01 | −19.49 | 63.99 |
| 2005–2006 | −25.73 | −13.06 | 50.76 | −12.67 | 49.24 |
| 2006–2007 | −36.03 | −16.76 | 46.51 | −19.27 | 53.49 |
| 2007–2008 | −19.64 | −5.87 | 29.89 | −13.77 | 70.11 |
| 1997–2008 | −431.99 | −174.87 | 40.48 | −257.11 | 59.52 |

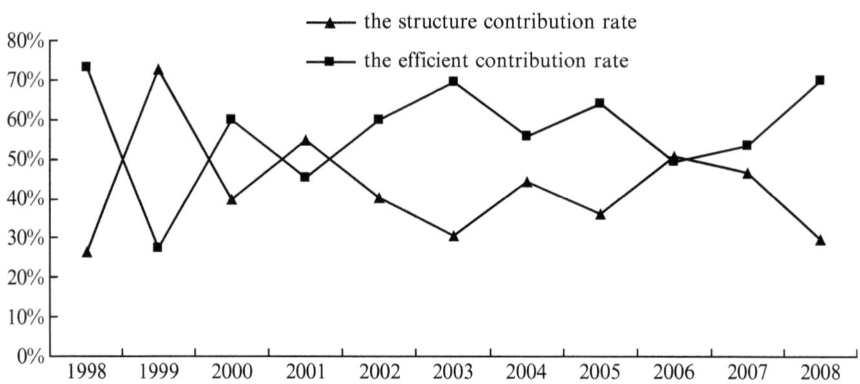


**Fig. 2** Trend of the structure and efficient contribution rate

- The second stage (2003–2006): there was an increase in the structure contribution rate from 30.7% to 50.76%, while the efficient contribution rate was reduced to 29.89%.
- The third stage (2006–2008): the structure contribution rate was reduced to the lowest (29.89%) in 2008, the efficient contribution rate rose to 70.11%.

In a word, the effects on adjustment of economic structure to WI decline are becoming smaller and smaller, while the contributions of water efficiency increase in various industries to WI decline larger and larger.

Table 3 denotes the contribution rates that various industries affect on industrial structure and industrial efficiency. As far as variation effect of industrial structure is concerned, primary industry played a leading role in the decline of WI. The contribution rate was above 100% every year, and the cause lay in the fact that water consumption was increased with the rise of the proportion of secondary

**Table 3** Contribution rate of industrial structure and industrial efficiency in various industries

| Year | Variation effect of industrial structure | | | Variation effect of industrial efficiency | | |
|---|---|---|---|---|---|---|
| | Primary industry (%) | Secondary industry (%) | Tertiary industry (%) | Primary industry (%) | Secondary industry (%) | Tertiary industry (%) |
| 1997–1998 | 109.66 | −9.35 | −0.30 | 75.71 | 23.47 | 0.83 |
| 1998–1999 | 105.49 | −4.91 | −0.58 | 4.04 | 94.95 | 1.01 |
| 1999–2000 | 107.48 | −7.07 | −0.42 | 58.14 | 40.48 | 1.39 |
| 2000–2001 | 102.93 | −2.22 | −0.71 | 40.49 | 57.73 | 1.78 |
| 2001–2002 | 105.37 | −4.87 | −0.49 | 64.33 | 34.01 | 1.66 |
| 2002–2003 | 112.83 | −12.98 | 0.14 | 78.75 | 20.00 | 1.26 |
| 2003–2004 | 112.63 | −12.74 | 0.11 | 43.90 | 55.33 | 0.77 |
| 2004–2005 | 109.54 | −9.43 | −0.10 | 69.42 | 29.12 | 1.46 |
| 2005–2006 | 108.02 | −7.79 | −0.23 | 47.62 | 50.09 | 2.29 |
| 2006–2007 | 107.72 | −7.51 | −0.21 | 59.61 | 37.81 | 2.58 |
| 2007–2008 | 105.54 | −5.17 | −0.38 | 49.20 | 49.25 | 1.56 |
| 1997–2008 | 109.53 | −9.22 | −0.31 | 68.42 | 30.38 | 1.20 |

industry and tertiary industry, which produced negative effects on the WI decline under the circumstance of constant efficiency. So far as variation effect of industrial efficiency is concerned, various industries played active roles in the decline of WI, especially the contribution rate of primary industry and secondary industry reached 68.42% and 30.38% respectively during 1997–2008.

## 4 Conclusion

The paper analyzes the variation and factors of WI by decomposition model of Laspeyres index from industrial point. The empirical results show that the total WI in China takes on an obvious declining trend; that the WI of primary industry, secondary industry and tertiary industry is decreasing respectively, but the differences coefficient among three industries increase gradually. And the decline of total WI in China was caused by both adjustment of industrial structure and increase of water utilization efficiency in various industries; on the variation effect of industrial structure, primary industry played a very important role in the decline of WI; on the variation effect of industrial efficiency, increase of water utilization efficiency in primary industry and secondary industry played active roles in the decline of WI, and their contribution rates reached 68.42% and 30.38% respectively.

Despite that water usage per unit GDP has obviously decreased in China, compared with that in developed countries, WI has plenty of space to decline. For meeting the needs of socio-economic development and easing up the increasingly water shortage problem, we should continuously improve the water utilization efficiency.

**Acknowledgments** Financial supports are acknowledged from Major Project of Chinese National Program s for Fundamental Research and Development (973 Program) (project number: 2010CB951104), National Social Science Foundation of China (project number: 10CGL069), Humanities and Social Science Fund from Ministry of Education (project number: 09YJC790125), China Post-doctoral Science Foundation (project number: 20100471372) and humanities and social science fund from Changzhou University (project number: ZMF08020033).

## References

Chen DJ (2008) Structure share and efficiency of industrial water consumption intensity change in China (in Chinese). China Population Resour Environ 3:211–214

Li SX, Cheng JH, Wu QS (2008) Regional difference of the efficiency of water usage in China (in Chinese). China Population Resour Environ 3:215–220

RAND (2006) China's continued economic progress: possible adversities and obstacles. In: 5th Annual CRF-RAND Conference, Beijing

Shadrick L, Richard M, Jesper S (2006) Index number analysis of Namibian water intensity. Ecol Econ 57:374–381

Sun JW (1998) Changes in energy consumption and energy intensity: a complete decomposition mode (in Chinese). Energy Econ 29:85–100

Sustainable Development Strategy Study Group from the Chinese Academy of Science (2007) China sustainable development strategy report 2007: water: governance and innovation (in Chinese). Science Press, Beijing

# Part V
# Case Study of Risk Management

# The Empirical Study of Liquidity Risk and Closed-End Fund Discounts Based on Panel-Data

**Wenbin Huang**

**Abstract** Within the bounds created by limits to arbitrage and the transaction costs there are many dimensions in which characteristics of CEF shares and its underling portfolios can differ. Both theoretical studies and empirical evidence have shown that liquidity is a factor for capital asset pricing. This paper uses a data-based panel linear regression model to investigate the sensitive analysis of CEF discounts to illiquidity. We find that Chinese CEF discounts and fund spread are significantly and positively affected by illiquidity of fund shares, illiquidity of the market, illiquidity of fund underlying portfolios, and the difference between illiquidity of fund shares and its underlying assets has significantly and positively influence on discounts. Expected and unexpected illiquidity of fund shares are significantly and positively associated with discounts and fund spread. To some extent, these relations between illiquidity and discounts may explain the high volatility of Chinese CEF discounts. Discounts contain liquidity risk premium.

**Keywords** Closed-end fund · Discounts · Illiquidity · Risk

## 1 Introduction

The primarily difference between closed-end funds (CEFs) and open-end funds is that CEFs like stocks are traded only in the secondary market, the stock exchange, whereas open-end funds are directly bought from and sold to the fund companies. Therefore, the trading price of open-end funds are equal to their net asset value (NAV). However, because the trading prices of CEFs are determined by demand and supply, CEFs often sell at a discount or premium to their NAV. CEFs exhibit general discounts to their NAV, and they cannot be eliminated by arbitrage because

W. Huang
Department of Statistics, School of Management, University of Fuzhou, 350108 P.R. China
e-mail: bear_23@163.com

Y. Zhou and D.D. Wu (eds.), *Modeling Risk Management for Resources and Environment in China*, Computational Risk Management,
DOI 10.1007/978-3-642-18387-4_43, 

CEFs cannot be redeemed on demand. Discounts disappear only until CEFs are liquidated or conversed to open-end funds. CEFs sell at a premium to their NAV during the first few months on the listing and then gradually traded at a discount. Moreover, discounts vary widely over time. They are generally smaller in bull market and larger in bear market. The fact has been cited as strong evidence against market efficiency. See, for example, Lee et al. (1991), Malkiel (1977), Pontiff (1997). In fact Lee et al. (1991) claim "few problems in finance are as perplexing as the CEF puzzle," because it seems to go back on the value additivity principle, which is the most important principle of financial market.

If the market is efficient and complete, the market price of the portfolio should be equal to the sum price of all the individual securities. However, the perspective that the price of CEF is just only the sum of its underlying securities may be simplistic. Within the bounds created by limits to arbitrage and the transaction costs there are many dimensions in which the characteristics of the CEF and its underling portfolio can differ. In particular, since the CEF shares and its underlying securities trade independently, they can have different liquidities. In recently, the studies of the liquidity and asset pricing have documented a negative relationship between illiquidity and price. Hence, illiquidity of CEF may depress their price to their NAV.

We all know that liquidity have many dimensions such as the transaction cost, the ability to trade quickly, the ease with which large quantities can be traded, and the impact of trading on prices. Because of the above-mentioned dimensions, the similar financial assets, even with the same payoff, often have different liquidity. Since liquidity is a key feature of the capital market, an important question to ask is how liquidity affects asset prices. Both the theoretical and empirical aspects of the interaction between liquidity and asset prices have been studied extensively. In theoretical studies, Kyle (1985) showed that liquidity conveys information and affects the asset price, and Allen and Gale (1994) also thought that an illiquid asset price is given by the smaller of the asset long-term fundamental value and the amount determined by the supply and demand of cash (liquidity). Further, they show that illiquidity always depresses the asset price to below its fundamental value. In empirical studies, the positive relationship between the stock returns and illiquidity over time is examined and confirmed in Amihud (2002). In addition, Amihud (2002), Brennan and Subrahmanyam (1996), and Brennan et al. (1998) examined a similar relation across different stocks. Pastor and Stambaugh (2003) found that stock returns are related not only to liquidity levels but also to liquidity risk factors. The results in empirical studies show that illiquidity depresses asset prices and raises asset returns.

When liquidity is an important factor in capital asset pricing, the difference of the illiquidity of the CEF shares and their underlying portfolios should influence the discount. When the liquidity risk of the CEF shares is increased relative to the CEF underlying assets, the discount should increase. Therefore, there are two potential liquidity effects. One is the static differences of the liquidity between the CEF and its underlying asset. The other is the liquidity risk differences between the CEF and underlying asset.

This paper tries to examine whether the CEF discount is related to liquidity. That is if the CEF discount is caused by the CEF shares illiquidity, and its underlying asset illiquidity and the liquidity difference between CEF shares and its underlying asset.

## 2 Our Approach

There are several different measures of illiquidity in the empirical studies. In this paper, we use the method in Amihud (2002) to create the measure of illiquidity in each month and for each stock and each CEF. Although this measure of illiquidity is rough, its attraction is that it only requires daily data on trading volume and asset price, which are readily available. It does not need to collect more detailed data such as the bid-ask spread.

The Amihud illiquidity measure is defined as

$$IL_{i,t} = \frac{1}{T_{i,t}} \sum_{k=1}^{T_{i,t}} (|R_{i,t,k}|/VOL_{i,t,k}), \tag{1}$$

where $T_{i,t}$ is the number of trading days for risk asset $i$ in month $t$, $R_{i,t,k}$ and $VOL_{i,t,k}$ are respectively the daily return and volume of the risk asset $i$ in day $k$ of month $t$. The ratio, $|R_{i,t,k}|/VOL_{i,t,k}$, is the absolute proportional price change because of per unit of daily trading volume, also indicates how much the daily price was impacted by the order flow. $IL_{i,t}$ is representative for the measure of illiquidity for risk asset $t$. The market is more illiquid if its value is higher, otherwise the market is more liquid.

The illiquidity measure of fund underlying portfolios is defined as

$$PL_{i,t} = \sum_{k=1}^{10} IL^{i}_{k,t}(PV^{i}_{k,t}/PV_{i,t}), \tag{2}$$

where $PL_{i,t}$ is the illiquidity measure of underlying portfolios for the CEF $i$ at $t$ moment, $PV^{i}_{k,t}$ and $IL^{i}_{k,t}$ are respectively the total tradable value and the illiquidity measure of underlying security $k$ in CEF $i$ at $t$ moment, $PV_{i,t}$ is the sum tradable value of the top ten underlying securities in CEF $i$ at $t$ moment.

The illiquidity measure of the whole market, we replace the whole market with Shanghai stock composite index, is defined as

$$MIL_t = \frac{1}{N} \sum_{i=1}^{N} IL_{i,t}, \tag{3}$$

where $N$ is the number of the stocks that their trading day number is larger than 15 in month $t$.

The CEF discount is defined as

$$D_{i,t} = -(\log(P_{i,t}) - \log(NAV_{i,t})), \tag{4}$$

where $P_{i,t}$ and $NAV_{i,t}$ are respectively the closing price and the net asset value for the CEF $i$ at $t$ moment. When $D_{i,t}$ is positive, it is often referred to as the discount, otherwise it is often referred to as the premium.

The "fund spread" is defined as the change in the discount:

$$\begin{aligned}\Delta D_{i,t} = D_{i,t} - D_{i,t-1} &= -[\ln P_{i,t} - \ln NAV_{i,t}] + [\ln P_{i,t-1} - \ln NAV_{i,t-1}] \\ &= r_{i,NAV,t} - r_{i,P,t},\end{aligned} \tag{5}$$

where $r_{i,NAV,t}$ and $r_{i,P,t}$ are respectively the NAV monthly return and the price monthly return for the CEF $i$. Because the (5) is nearly the difference between the price return and the NAV return, we call it as "fund spread." We can think that it is the return on a zero-investment portfolio where investors are long the CEF and short the underlying assets.

In order to enhance the power of the estimation and hypothesis testing, we use a data-based panels linear regression model approach those coefficients are constrained to be the same. Firstly, we test the relationship between the level of fund discount and illiquidity of fund shares and Shanghai stock composite index. We find that the fund discount is positively associated with the illiquidity of fund shares. Secondly, we examine the association between the level of fund discount and the difference of the illiquidity between the CEF shares and its underlying asset. The result shows that the difference of the liquidity has a significant influence on the discount. Thirdly, because expected and unexpected illiquidity could represent the liquidity risk, we detect that if they affect asset prices and returns. The result indicates that unexpected illiquidity has a significant influence on the fund discount. Finally, we examine the relationship between "fund spread" and the above-mentioned indicators, the result shows that "fund spread" is significant positive with the illiquidity of the CEF underlying assets, unexpected illiquidity and the illiquid difference between of the CEF shares and its underlying assets.

## 3 Data

Chinese daily closing price, daily volume, weekly NAV and weekly closing price were obtained for 19 CEFs from Wind. All Chinese CEFs in this paper are on the listing. The samples are from 01/01/2001 to 04/30/2010. Weiss (1989) found that CEFs usually sell at a premium at first and most of the price decline between 30 and 100 days after they issue. There has always been "speculation Issue" phenomenon in Chinese market. When a new CEF was listed on exchange, there was always the phenomenon of IPO that is lasting a longer time. Hence, all funds included in the

**Table 1** Summary statistics of Chinese CEFs data

| Fund name | Fund spread% | | Discount | | Illiquidity | |
|---|---|---|---|---|---|---|
| | Mean | Std | Mean | Std | Means | Std |
| JinTai | 0.060 | 6.81 | 0.290 | 0.187 | 1.39 | 1.53 |
| TaiHe | 0.103 | 5.83 | 0.284 | 0.196 | 3.07 | 3.48 |
| AnXin | 0.056 | 6.97 | 0.262 | 0.195 | 1.94 | 1.71 |
| HanSHeng | 0.036 | 6.42 | 0.316 | 0.188 | 2.52 | 2.98 |
| YuYang | 0.121 | 7.72 | 0.284 | 0.186 | 2.61 | 2.72 |
| XinHua | 0.034 | 4.79 | 0.242 | 0.165 | 1.49 | 1.58 |
| AnSHun | 0.013 | 6.21 | 0.287 | 0.169 | 1.56 | 1.84 |
| JinXin | 0.012 | 5.56 | 0.312 | 0.197 | 1.83 | 2.06 |
| HanXing | 0.110 | 5.94 | 0.317 | 0.200 | 1.32 | 1.26 |
| XingHua | 0.049 | 5.29 | 0.315 | 0.201 | 1.37 | 1.52 |
| KaiYuan | 0.028 | 7.82 | 0.272 | 0.177 | 3.14 | 3.4 |
| PuHui | 0.104 | 5.44 | 0.326 | 0.199 | 2.88 | 3.00 |
| TongYi | 0.044 | 7.18 | 0.311 | 0.193 | 2.25 | 2.40 |
| JingHong | 0.103 | 6.21 | 0.302 | 0.202 | 2.23 | 2.00 |
| YuLong | 0.035 | 6.88 | 0.321 | 0.203 | 1.38 | 1.35 |
| PuFeng | 0.121 | 5.31 | 0.336 | 0.20 | 1.36 | 1.14 |
| TianYuan | 0.031 | 6.65 | 0.309 | 0.175 | 1.67 | 1.78 |
| TongSHeng | 0.063 | 6.06 | 0.338 | 0.202 | 1.49 | 1.67 |
| JingFu | 0.090 | 5.53 | 0.343 | 0.195 | 1.39 | 1.53 |

paper have a data that is at least 1-year after IPO date. According to this principle, we exclude off CEFs those were issued after 2000. Table 1 contains the summary statistics of the Chinese CEFs monthly data.

Columns 2–3 in Table 1 report the sample mean and sample standard deviation (STD) of the spread (which is defined as the first order difference in the discount). About almost all CEFs have positive mean spread, indicating that if investors are long the CEF shares and short the underlying portfolios during the sample period, they should on average make a profit, otherwise they should make a loss. Columns 4–5 in Table 1 report the sample mean and sample standard deviation of the discount. All 19 CEFs have positive sample mean discounts indicating that positive discounts are common for most of CEFs during this sample period. Most of CEFs have sample average discounts of 28% or higher, even that some CEFs have a discount more than 30%. For example, XingHua Fund and JingFu Fund have mean discounts of 24.2% and 34.3%, respectively. The sample volatility of the CEFs discounts also varies widely during the sample period across different CEFs, ranging from a low of 16.5% per month for XingHua Fund to a high of 20.9% per month for YuLong Fund. The last two columns in Table 1 report the sample mean and sample standard deviation of the illiquidity. The results show that the illiquidity measure varies widely across different CEFs. There are a best illiquidity of 1.32 for HanXing Fund and a worst illiquidity of 3.14 for KaiYuan Fund. During the same sample period, the sample illiquidity of Shanghai stock composite index was 1.96 and its volatility was 1.48. The illiquidity of the CEFs and Shanghai stock composite index varies in the same direction. In bull market, they make a little difference. However, in bear market, the illiquidity of the CEFs is much more than the illiquidity of Shanghai stock composite index, in general. Further, the illiquidity of the CEFs varies much more widely. In this paper, we replace the illiquidity of the whole market with the illiquidity of Shanghai stock composite index.

## 4 Empirical Analysis

In this section, we analyze the relation between illiquidity and the discount of CEFs. Firstly, we examine how the time series variation of discounts and fund spread are associated with the time series variation in the illiquidity of the CEF shares, and the market. Secondly, we examine whether the expected and unexpected illiquidity would have an influence on the discount.

### *4.1 Illiquidity, Discount and Fund Spread*

In order to enhance the power of the estimation and hypothesis testing, we use a data-based panels linear regression model approach those coefficients are constrained to be the same. And in order to distinguish the difference between listed on shanghai stock exchange (SH) and listing on Shenzhen stock exchange (SZ), we put all the CEFs into two parts according to where they were issued. The results show that these two parts have the same results.

Firstly, we use (6) to test how the illiquidity of the CEF shares and the market influence the discount.

$$D_{i,t} = \alpha_1 + \beta_1 IL_{i,t} + \gamma_1 MIL_t + \varepsilon_{i,t}. \tag{6}$$

Secondly, we investigate how the time series variation of the discount is associated with the difference of the illiquidity of the CEF shares and their underlying assets. Because all underlying assets position ratio only has been official public announcement in the semi-annual report and the annual report and this paper use the monthly data, there is a conflict between them. Hence, this paper find the rough substitute variable, $PL_{i,t}$, for the illiquidity measure of CEF underlying portfolios. In Chinese financial market, only up to the 80% of the CEF is invested in stock market, so its underlying portfolios may have a linear relationship associated with the Shanghai stock composite index. Based on the idea, we use Amihud (2002) method to regress the illiquidity of the CEFs on the illiquidity of the Shanghai stock composite index, $IL_{i,t}=a+b*MIL_t+\varepsilon_{i,t}$. We replace the illiquidity of the underlying assets with the fitted value, replace the difference of the illiquidity of the CEF shares and their underlying assets with the errors from regression. Then, we regress the (7).

$$D_{i,t} = \alpha_2 + \beta_2 \hat{IL}_{i,t} + \gamma_2 \hat{\varepsilon}_{i,t} + \mu_{i,t}. \tag{7}$$

**Table 2** Pooled regression of discount on illiquidity

| | $\beta_1$ | $\gamma_1$ | $R^2$ | | $\beta_2$ | $\gamma_2$ | $R^2$ |
|---|---|---|---|---|---|---|---|
| All fund | 0.041 (21.2) | 0.06 (20.6) | 0.276 | All fund | 0.0033 (3.60) | 0.0055 (10.4) | 0.201 |
| Fund in (SH) | 0.043 (17.3) | 0.06 (16.1) | 0.319 | Fund in (SH) | 0.0032 (2.51) | 0.0053 (7.25) | 0.233 |
| Fund in (SZ) | 0.037 (12.7) | 0.06 (13.0) | 0.233 | Fund in (SZ) | 0.0033 (2.54) | 0.0059 (7.51) | 0.182 |

**Table 3** Pooled regression of spread on illiquidity of fund shares and market

| | $\beta_3$ | $\gamma_3$ | $R^2$ |
|---|---|---|---|
| All fund | 0.0038 (8.19) | 0.0029 (4.16) | 0.072 |
| Fund in (SH) | 0.0034 (5.52) | 0.0030 (3.08) | 0.065 |
| Fund in (SZ) | 0.0041 (6.09) | 0.0029 (2.79) | 0.0079 |

Column 2–4 in Table 2 reports the results from the pooled regression of model (6). In three situations, the fitted values of $\beta_1$ are significantly 0.041, 0.043 and 0.037 at 5% level, respectively. The results for $\gamma_1$ are same. It indicates that the discount becomes larger when the illiquidity of the fund shares or the market is larger. Column 6–8 in Table 2 shows that the fitted values of $\beta_2$ and $\gamma_2$ of model (7) are significantly positive. The *t*-statistcs of $\gamma_2$ is larger than the *t*-statistcs of $\beta_2$. It indicates the difference of the illiquidity of the fund shares and its underlying assets is much more significantly effect on the discount. Further, $\gamma_2$ is larger than $\beta_2$. That shows clearly that the difference of the illiquidity of the fund shares and its underlying assets should raise higher discount than the illiquidity of the fund shares does.

We have shown that the discount is positive associated with the illiquidity. Now, we analyze the relationship between the fund spread and the illiquidity of the fund shares and the market.

$$\Delta D_{i,t} = \alpha_3 + \beta_3 IL_{i,t} + \gamma_3 MIL_t + \varepsilon_{i,t}. \tag{8}$$

From Table 3, the results show that the fitted values of $\beta_3$ and $\gamma_3$ are significantly positive. Tables 2 and 3 explain how the illiquidity of the fund shares and the market influence the discount and the spread, respectively. We can explain these two influences as the relationship between the illiquidity and the price and the relationship between the illiquidity and the return. In general, the former is much more distinct than the latter. We can find the illiquidity of the fund shares and the market much more significantly effect on the discount than on the fund spread from the results in Tables 2 and 3.

## 4.2 Expected and Unexpected Illiquidity

Since the illiquidity contains both expected and unexpected illiquidity, and these two components may have different effect on asset returns. Therefore, we decompose the illiquidity into expected and unexpected illiquidity and investigate the relationship between the discount and these two components. We use Amihud (2002) method to create expected and unexpected illiquidity. We take the realized illiquidity as AR (1), $IL_t = a_0 + a_1 * IL_{t-1} + v_t$. Then, we define expected ($ILE_t$) and unexpected ($ILU_t$) illiquidity as the fitted value and the errors from regression, respectively.

We examine how the discount relate to expected and unexpected illiquidity by (9)

$$D_{i,t} = \alpha_4 + \beta_4 ILE_{i,t} + \gamma_4 ILU_{i,t} + \lambda_1 ILE_{z,t} + \lambda_2 ILU_{z,t} + \varepsilon_{i,t}. \tag{9}$$

In Table 4, the fitted values of $\beta_4$, $\gamma_4$, and $\lambda_1$ are significantly positive. The unexpected illiquidity has a stronger influence on the discount than the expected illiquidity. Although the unexpected illiquidity of the market has a positive association with the discount, it is statistically significant for all CEFs sample and is not statistically significant for CEFs listed on SH or listed on SZ separately. The reason may be that the unexpected illiquidity of the market should lower both the fund price and NAV, but there was significant difference of extent of decline so that it is not clearly how the unexpected illiquidity of the market influences on the discount. When investors anticipate the illiquidity of the market, they may demand higher returns, leading to a lower fund price, for holding the CEFs.

We examine how the expected and unexpected illiquidity influence on the fund spread.

$$\Delta D_{i,t} = \alpha_5 + \beta_5 ILE_{i,t} + \gamma_5 ILU_{i,t} + \lambda_3 ILE_{z,t} + \lambda_4 ILU_{z,t} + \varepsilon_{i,t}. \tag{10}$$

The results in Table 5 show that the fund spread is significantly and positively associated with unexpected illiquidity of the CEF shares. The $\lambda_3$ is 0.0054 for the group of all funds. For example by JingHong Fund, its sample spread and volatility was only 0.103% and 6.21%. However, when its shares unexpected illiquidity is increased by one standard deviation (1.183), its spread is increased by 0.666%. On the other hand, the coefficient of expected illiquidity of the fund shares is significant and positive. For example by JingHong Fund, when its shares expected illiquidity

**Table 4** Pooled regression of discount on expected and unexpected illiquidity

| | $\beta_4$ | $\gamma_4$ | $\lambda_1$ | $\lambda_2$ | $R^2$ |
|---|---|---|---|---|---|
| All fund | 0.0033 (5.36) | 0.0046 (6.19) | 0.0037 (3.42) | 0.0020 (2.08) | 0.201 |
| Fund in (SH) | 0.0033 (3.80) | 0.0045 (4.53) | 0.0032 (2.08) | 0.0016 (1.22) | 0.221 |
| Fund in (SZ) | 0.0033 (3.82) | 0.0048 (4.19) | 0.0042 (1.73) | 0.0024 (1.73) | 0.203 |

**Table 5** Pooled regression of fund spread on expected and unexpected illiquidity

| | $\beta_5$ | $\gamma_5$ | $\lambda_3$ | $\lambda_4$ | $R^2$ |
|---|---|---|---|---|---|
| All fund | 0.0023 (4.18) | 0.0053 (6.79) | 0.0054 (5.47) | 0.0009 (1.00) | 0.082 |
| Fund in (SH) | 0.0020 (2.69) | 0.0048 (4.73) | 0.0054 (4.01) | 0.0010 (0.751) | 0.076 |
| Fund in (SZ) | 0.0023 (3.24) | 0.0058 (4.90) | 0.0053 (3.69) | 0.0009 (0.63) | 0.09 |

goes up by one standard deviation (1.70), its spread goes up by 0.6%. The expected illiquidity of the market has a significantly and positively influence on the fund spread, but the unexpected illiquidity of the market has only a positively influence on it.

## 5 Conclusion

This paper mainly tests the impact of the illiquidity on the fund discount and the fund spread. We find that the discount and the spread are significantly and positively associated with the illiquidity of the fund shares, the illiquidity of the market, the illiquidity of the fund underlying assets. And we also find that the difference between the illiquidity of the fund shares and its underlying portfolios has a significantly and positively influence on the discount. The discount goes up when the fund shares has worse illiquidity than its underlying assets. On the other hand, the expected and unexpected illiquidity of the fund shares significantly and positively affect the discount and the fund spread. The expected illiquidity of the market also has the same relationship with them, but the unexpected illiquidity of the market only has a positive influence on them. These results, to some extent, may explain the high volatility of the Chinese CEF discounts. To sum up, the discount is associated with the illiquidity and the illiquidity risk. That is to say the illiquidity and the illiquidity risk is pricing factors for the discount.

## References

Allen F, Gale D (1994) Limited market participation and volatility of asset prices. Am Econ Rev 84:933–955
Amihud Y (2002) Illiquidity and stock returns: cross-section and time-series effects. J Finance 5:31–56
Brennan MJ, Subrahmanyam A (1996) Market microstructure and asset pricing: on the compensation for illiquidity in stock returns. J Finance Econ 41:441–464
Brennan MJ, Chordia T, Subrahmanyam A (1998) Alternative factor specification, security characteristics, and the cross-section of expected stock returns. J Finance Econ 49:345–373
Kyle AS (1985) Continuous auctions and insider trading. Econometrica 53:1315–1335
Lee C, Shleifer A, Thaler R (1991) Investor sentiment and the closed-end fund puzzle. J Finance 46:75–109
Malkiel BG (1977) The valuation of closed-end investment-company shares. J Finance 32:847–858
Pastor L, Stambaugh RF (2003) Liquidity risk and expected stock returns. J Pol Econ 111:642–685
Pontiff J (1997) Excess volatility and closed-end funds. Am Econ Rev 87:155–169
Weiss K (1989) The post-offering price performance of close-end funds. Finance Manage 18(3): 57–67

# Empirical Analysis of Largest Eigenvalue of Leontief Matrix

**Daju Xu and Shitian Yan**

**Abstract** In Input–output Analysis, consumption coefficients play an important role and reflect the production technology of the economic system. The largest eigenvalue of Leontief matrix has some economic meanings. To some degree, it indicates a limit on economic growth, which means the economic system may grow only when the gross input coefficient is bigger than the largest eigenvalue. On the other hand, it has inverse proportion relation with final demands rate. In an empirical analysis for the five countries, China, Japan, Britain, America and Australia, results are calculated for these countries' eigenvalues of Leontief matrices. The outcomes exhibit a large degree of stability across the different years' matrices for a country. Therefore it is confirmed that the largest eigenvalue can reflect a kind of inner regularity. List of these countries according to the average value of largest eigenvalues in a decreasing order are China, Japan, Britain, America and Australia.

**Keywords** Input–output analysis · Largest eigenvalue · Leontief matrix · Sensitive analysis

## 1 Introduction

In 1930s, Input–output Analysis was founded by American economist (Leontief 1936). It was Input–output Analysis that made it possible for some countries in the world to build and develop macroeconomics theory supported by method and data, and to build national account. The theoretical basis and mathematical method for

D. Xu (✉)
Department of Mathematics and Physics, Shandong Jiaotong University, 250023 Jinan, China
e-mail: xudj1968@163.com

S. Yan
Institute of Automation, Qufu Normal University, 273165 Qufu, China
e-mail: shitian_yan@163.com

Y. Zhou and D.D. Wu (eds.), *Modeling Risk Management for Resources and Environment in China*, Computational Risk Management,
DOI 10.1007/978-3-642-18387-4_44, 

Input–output Analysis was due to Leon Walras' general equilibrium models. In Leontief's opinion, the national economic system can be divided into many production sectors according to the categories of their products. In order to make production, one sector needs to consume other sectors' products, including its own products. On the other hand, the sector's partial products should be distributed to other sectors to be utilized. Therefore, Input–output Analysis can exhibit the balance relations of production and distribution, and technical relations among sectors in national economic system (Leontief 1951, 1953). The most important property of Input–output Analysis is considering problem as a whole, and the most important equation is showed as follows:

Material consumption + net production value = gross production value.

The literatures on Input–output Analysis are very much. But, few of them are on eigenvalue and eigenvector of Leontief matrix. Some researchers examine the second eigenvalue of Leontief matrix and provide its economic meanings (Brody 1997; Bialas and Gurgul 1998). In 1985, Luogeng Hua researched on closed and dynamic macroeconomic model and finally provided *Positive Eigenvector Method* (Hua 1984, 1985. Daju Xu generalized Hua's model through adding consumption vector, and the conclusion is that *Positive Eigenvector Method* is correct for open model (Xu et al. 2005). His latest researching results are giving the largest and smallest eigenvalues' economic meanings using linear system theory as a theoretical analysis tool (Xu and Wang 2010).

In this paper, we propose the empirical analysis of the largest eigenvalue of Leontief matrix. In the next section, we prepare some basic definitions and lemma, including Leontief matrix, irreducible matrix, and the famous and very important Perron-Frobenius Theorem. First part in Sect. 3 indicates the economic meaning of the largest eigenvalue, and second part lists some tables in which five countries' eigenvalues are calculated using software Mathematica. Section 4 concludes.

## 2 Preparation: Definitions and Lemma

**Definition 1.** *Eigenvalue, right eigenvector and left eigenvector (Lancaster and Tismenetsky 1985)*

*Let* $A = (a_{ij})_{n\times n}$ *be a real number matrix, there exist a complex number* $\lambda$, *and a nonzero* $n \times 1$ *column vector* $X$ *such that* $AX = \lambda X$, *then we call* $\lambda$ *and* $X$ *be eigenvalue and right eigenvector of matrix* $A$, *respectively. Vector* $V^T$ *is termed left eigenvector of matrix* $A$ *if* $V^T A = \lambda V^T$, *where* $V^T$ *is the transpose of column vector* $V$.

*In fact,* $V$ *is right eigenvector of the matrix* $A^T = (a_{ji})_{n\times n}$ *corresponding to its eigenvalue* $\lambda$.

**Definition 2.** *Irreducible matrix*

*Real number matrix* $A = (a_{ij})_{n\times n}$ *is termed irreducible if there does not exist a permutation matrix* $P$, *such that*

$$PAP^T = \begin{pmatrix} A_{11} & 0 \\ A_{21} & A_{22} \end{pmatrix},$$

*where $A_{11}$ is a $k \times k$ square matrix ($1 \le k \le n-1$), and top right corner is $k \times (n-k)$ zero matrix.*

**Definition 3.** *Leontief matrix*

*In the producing process of a national economy, production manufactured in sector i, one part will be distributed to other sectors, including itself, to manufacture their production; the left part will be consumed as final product. Suppose $x_{ij}$ be the input from sector i in sector j, and $x_j$ be total product of sector j. Let*

$$a_{ij} = x_{ij}/x_j,$$

*then $a_{ij}$ is termed direct consumption coefficient, or technical coefficient. Matrix $A = (a_{ij})_{n\times n}$ be termed Leontief matrix.*

**Lemma 1 (Perron–Frobenius Theorem).** *Suppose $A = (a_{ij})_{n\times n}$ is a nonnegative and irreducible matrix, then (Rothblum 1975)*

1. *Matrix A has a positive eigenvalue $\lambda(A)$ (named Frobenius Root), which is single root, and $|\lambda_j| \leqslant \lambda(A)$, where $\lambda_j$ is other eigenvalues of matrix A. In particularly, $|\lambda_j| < \lambda(A)$ when matrix A is primitive.*
2. *Matrix A has only one positive left and right eigenvectors $V^T$, U corresponding $\lambda(A)$without considering multiples.*
3. *It is correct that $\lambda(B) \le \lambda(A)$ when $B \geqslant A$, where B is an another nonnegative and irreducible matrix.*

For *Perron–Frobenius Theorem*, it should be noted that matrix has unique right and positive eigenvector, and unique left and positive eigenvector only when the matrix is nonnegative and irreducible.

In empirical analysis, all Leontief matrices satisfy the conditions of *Lemma 1*. When the sector number $n$ is smaller, for example $n = 3$, Leontief matrix is positive.

# 3 Results for Empirical Analysis

## *3.1 Economic Meaning of the Largest Eigenvalue*

Beginning with Hua's closed model, Daju Xu has been examining the open, dynamic and macroeconomic models. If Leontief matrix is nonnegative and irreducible, for a kind of open model, the following conclusions are correct (Xu and Wang 2010):

1. To some degree, the optimal input structure and the optimal output structure must be equal to the right and positive eigenvector of Leontief matrix.

Otherwise, some years later, the economic system will lost its equilibrium, which means some sectors' output will be equal to zero or negative number.

2. For a kind of open model, to some degree, the outputs grow only when the gross input coefficient is bigger than the largest eigenvalue of Leontief matrix.
3. Introduced linear system theory, the eigenvalues of Leontief matrix have been given fixed economic meanings. When the gross input coefficient is less than the smallest eigenvalue, the economic system will die out; when the gross input coefficient is bigger than the smallest eigenvalue but less than the largest, the economic will be unstable; when the gross input coefficient is bigger than the largest eigenvalue as well as the input structure is equal to the right positive eigenvector, the economic system will grow year after year.

Therefore, economic growth occurs only when the economic system satisfies two conditions, i.e. aggregate input satisfaction (bigger than the largest eigenvalue), and appropriate input proportion of every sector (related to right and positive eigenvector).

## 3.2 *Different Largest Eigenvalues of Five Countries: Sensitive Analysis*

Using software Mathematica, we calculate the largest eigenvalues of five countries including China, Japan, Britain, America, and Australia, and exhibit in the following tables.

*China's Largest Eigenvalues.* Data come from websites of China Input–output Association, National Bureau of Statistics of China, and China Research Center of Tertiary Industry, Sun Yat-sen University (Table 1).

**Table 1** Largest eigenvalues of China

| Year | Sector number | Largest eigenvalue |
|---|---|---|
| 1987 | 3 | 0.576301 |
| 1990 | 3 | 0.601937 |
| | 33 | 0.598351 |
| 1992 | 3 | 0.626991 |
| | 6 | 0.624415 |
| | 33 | 0.621923 |
| 1995 | 3 | 0.631842 |
| | 33 | 0.630630 |
| 1997 | 6 | 0.630431 |
| | 17 | 0.635346 |
| | 40 | 0.629984 |
| 2000 | 17 | 0.652756 |
| 2002 | 42 | 0.623914 |
| 2005 | 17 | 0.669850 |
| | 42 | 0.689704 |
| 2007 | 42 | 0.684582 |
| | 135 | 0.679020 |

**Table 2** Largest eigenvalues of Japan

| Year | Sector number | Largest eigenvalue | Input–output table's property |
|---|---|---|---|
| 1970 | 36 | 0.564139 | Domestic transactions table, current prices |
| | 36 | 0.615557 | Total (domestic+imported) transactions table, at 1985-based constant prices |
| 1975 | 36 | 0.630592 | Total (domestic+imported) transactions table, at 1985-based constant prices |
| 1980 | 36 | 0.545927 | Domestic transactions table, at 1985-based constant prices |
| | 36 | 0.630396 | Total (domestic+imported) transactions table, at 1985-based constant prices |
| | 36 | 0.625297 | Total (domestic+imported) transactions table, current prices |
| 1985 | 36 | 0.584915 | Total (domestic+imported) transactions table, at 1985-based constant prices |
| 1990 | 36 | 0.574083 | Total (domestic+imported) transactions table, at 1985-based constant prices |
| 2000 | 13 | 0.512979 | Total (domestic+imported) transactions table, producers' prices |
| | 32 | 0.522019 | |

**Table 3** Largest eigenvalues of Britain

| Year | Sector number | Largest eigenvalue | Input–output table's property |
|---|---|---|---|
| 1968 | 36 | 0.421808 | Domestic transactions table, current prices |
| | 36 | 0.553545 | Total (domestic+imported) transactions table, at 1980-based constant prices |
| 1979 | 36 | 0.545036 | Total (domestic+imported) transactions table, at 1980-based constant prices |
| 1984 | 36 | 0.529522 | |
| 1990 | 36 | 0.681719 | |

It should be noted that, for China, there are two or three different largest eigenvalues of the same year because these data are calculated according to different sector numbers' matrices.

*Other Countries' Eigenvalues.* All Input–output Tables come from websites of Organization For Economic Co-operation And Development, and Bureau of Economic Analysis, U.S. Department of Commerce (Tables 2–5).

For Japan, Britain, America, and Australia, there are also two or three different largest eigenvalues of the same year because these data are calculated according to different properties' matrices but same sector numbers with only exception of Japan in 2000.

# 4 Conclusions

According to economic meaning of the largest eigenvalue of Leontief matrix, we analyze the results in tables and provide the conclusions as follows:

1. For China between 1987 and 2007, the largest eigenvalue is becoming bigger and bigger. In detail, it is 0.57630 in 1987, and it is 0.689704 in 2005. Although

**Table 4** Largest eigenvalues of America

| Year | Sector number | Largest eigenvalue | Input–output table's property |
|---|---|---|---|
| 1972 | 36 | 0.465720 | Domestic transactions table, current prices |
| | 36 | 0.497176 | Total (domestic+imported) transactions table, at 1982-based constant prices |
| | 36 | 0.497073 | Total (domestic+imported) transactions table, current prices |
| 1977 | 36 | 0.449758 | Domestic transactions table, current prices |
| | 36 | 0.488350 | Total (domestic+imported) transactions table, at 1982-based constant prices |
| | 36 | 0.490506 | Total (domestic+imported) transactions table, current prices |
| 1982 | 36 | 0.506606 | Total (domestic+imported) transactions table, at 1982-based constant prices |
| | 36 | 0.506606 | Total (domestic+imported) transactions table, current prices |
| 1985 | 36 | 0.441186 | Total (domestic+imported) transactions table, at 1982-based constant prices |
| | 36 | 0.443730 | Total (domestic+imported) transactions table, current prices |
| 1990 | 36 | 0.484748 | Total (domestic+imported) transactions table, at 1982-based constant prices |
| | 36 | 0.484748 | Total (domestic+imported) transactions table, current prices |
| 1987 | 9 | 0.485310 | Commodity-by-industry direct requirements, after redefinitions (1987, 1992, 1997 to 2007) |
| 1992 | 9 | 0.486410 | |
| 1997 | 12 | 0.492656 | |
| 1998 | 16 | 0.486205 | |
| 1999 | 16 | 0.485015 | |
| 2000 | 16 | 0.489227 | |
| 2001 | 16 | 0.484975 | |
| 2002 | 16 | 0.475935 | |
| 2003 | 16 | 0.475014 | |
| 2004 | 16 | 0.477235 | |
| 2005 | 16 | 0.490290 | |
| 2006 | 16 | 0.491158 | |
| 2007 | 16 | 0.486652 | |

**Table 5** Largest eigenvalues of Australia

| Year | Sector number | Largest eigenvalue | Input–output table's property |
|---|---|---|---|
| 1968 | 36 | 0.514020 | Total (domestic+imported) transactions table, at 1989-based constant prices |
| 1974 | 36 | 0.525033 | |
| 1986 | 36 | 0.490586 | |
| 1989 | 36 | 0.394866 | Domestic transactions table, current prices |
| | 36 | 0.481753 | Total (domestic+imported) transactions table, at 1989-based constant prices |

in 2002 the eigenvalue is less than that in 2000, the eigenvalue in 2005 becomes maximum and the increasing proportion is 18.8%. All changes indicate for China's economy, the proportion between intermediate use and aggregate outputs is being bigger and bigger, i.e. the rate of final demands is being smaller and smaller. Therefore, in order to ensure economy growing, China economy must input more and more natural materials consuming in intermediate use.

**Table 6** Average value of largest eigenvalues

| Nation | Time period | Maximum | Minimum | Average value |
|---|---|---|---|---|
| China | 1987–2007 | 0.689704 | 0.576301 | 0.635763 |
| Japan | 1970–2000 | 0.630592 | 0.512979 | 0.580590 |
| Britain | 1968–1990 | 0.681719 | 0.421808 | 0.546326 |
| America | 1972–2007 | 0.506606 | 0.441186 | 0.482492 |
| Australia | 1968–1989 | 0.525033 | 0.394866 | 0.481252 |

2. For Japan between 1970 and 2000, the largest eigenvalue has the tendency of becoming smaller. In detail, it is 0.630592 in 1975, and it is 0.512979 in 2000. The decreasing proportion is about 18.7%. In addition to, the same year such as 1970 or 1980, the largest eigenvalue of Leontief matrix corresponding to total (domestic+imported) transactions table is bigger than that corresponding to domestic transactions table. That characteristic illustrates Japan's imports are the goods with higher intermediate use rate.
3. For Britain between 1968 and 1990, it is same as China that the largest eigenvalue is being bigger and bigger.
4. For United States between 1972 and 2007, the eigenbalue exhibits a large degree of stability. The mean of the largest eigenvalues is 0.482492, and the maximum is 0.506606 in 1982, the minimum is 0.441186 in 1985. The changes range is about −8.5% to 5.0%. it is same as Japan that United States imports the goods with higher intermediate use rate.
5. For Australia between 1968 and 1989, the eigenvalue is being smaller and smaller.

Comparing the five countries' average value in Table 6, the list in a decreasing order is China, Japan, Britain, America and Australia. Note that the time periods are different and the two countries, China and America, have more samples. If we can find the same period and same properties' Leontief matrices of different countries, the conclusions should be more reasonable.

## References


Bialas S, Gurgul H (1998) On hypothesis about the second eigenvalue of the Leontief matrix. Econ Syst Res 10(3):285–289

Brody A (1997) The second eigenvalue of the Leontief matrix. Econ Syst Res 9(3):253–258

Hua LG (1984) Optimum mathematical theory in national planed economy. Chinese Sci Bull 29 (12):705–709, 29(13):769–772, 29(16):961–963, 29(18):1089–1092, 29(21):1281–1282 (in Chinese)

Hua LG (1985) Optimum mathematical theory in national planed economy. Chinese Sci Bull 30 (1):1–2, 30(9):641–645 (in Chinese)

Lancaster P, Tismenetsky M (1985) The theory of matrices with applications, 2nd edn. Academic, New York

Leontief W (1936) Quantitative input-output relations in the economic system of United States. Rev Econ Statistic 18:105–125

Leontief W (1951) The structure of American economy, 1919–1939. Oxford University Press, New York

Leontief W (1953) Studies in the structure of the American economy. Oxford University Press, New York

Rothblum UG (1975) Algebraic Eigen space of nonnegative matrices. Linear Algebra Appl 12:281–292

Xu DJ, Wang WP (2010) Research on application of linear system theory in macroeconomics. In: 8th IEEE International Conference on Control & Automation, China, pp 1843–1847

Xu DJ, Liu Y, Li ZQ, Yu YB, Li AQ (2005) Research on economic growth models in multi-sector macroeconomics. In: Proceedings of the 12th International Conference on Management Science & Engineering, vol 2, pp 1498–1502

# Staff Informal Learning Problems and Influencing Factor Empirical Study

**Rongrong Huang and Ze Tian**

**Abstract** On the basis of studies and theoretical analysis, this paper researches the relationship between personal characteristics, work environment and informal learning, and then draw some hypotheses. The hypotheses are verified and analyzed by applying survey data and SPSS statistical software and sensitive analysis. The results showed that personal characteristics and the work environment had a significant effect on informal learning. Finally, suggestions are made on the basis of the results of empirical studies.

**Keywords** Informal learning · Personal characteristics · Sensitive analysis · Uncertainty analysis · Work environment

## 1 Introduction

People pay more attention to informal learning science the learning organization theory proposed and knowledge was easily achieved by the rapid development of internet. Some researches show that informal learning has significant influence on the performance of organizations and employees. Capital Works investigated hundreds of knowledge workers to find how they become skilled workers. The result shows the effect of learning by doing, collaboration and learning from others are much better than formal training (Xin 2004).

Now, in order to improve staff productivity and to control business risk, more and more enterprises begin to pay more attention to staff informal learning and developing a corporate learning environment. In future, informal learning in the

---

R. Huang (✉)
Business school of Hohai University, Nanjing 210098, China
e-mail: cucumberseven@163.com

Z. Tian
Business School of Hohai University, Changzhou 213022, China
e-mail: tianze21@126.com; tianze21@yahoo.cn

Y. Zhou and D.D. Wu (eds.), *Modeling Risk Management for Resources and Environment in China*, Computational Risk Management,
DOI 10.1007/978-3-642-18387-4_45, 

organization will be an essential part of life. The boundaries between formal learning and informal learning will become more and more blurred and indistinguishable. In this study, we referred to Paul Squires' classification of types of learning to distinguishing formal and informal learning in staff training in work contexts.

This study focused on employee informal learning within the organization or in work contexts to find the factors that affect their informal learning. It is hoped that the research will provide some valuable advice and approaches to improve informal learning so as to control the business risk.

## 2 Theoretical Model and Research Hypothesis

1. Assumption 1: personal characteristics and staff informal learning
   The previous research mainly studied the role individual characteristics played in formal learning, and few people noticed their role in informal learning. Individual differences in cognitive ability have been shown to be strong predictors of job and training performance (Hunter and Schmidt 1996; Jensen 1980). In a meta-analysis of the "big-five" personality dimensions, Barrick and Mount (1991) found that individual differences in openness to experience and extraversion were significantly correlated with learning proficiency. Lievens et al. (2003) did a study of learning proficiency and language acquisition for European managers preparing for job assignments in Japan, a large number of cognitive, leadership, and personality variables were measured. The results indicated that learning proficiency was predicted by assessment center exercise scores for teamwork, communication, adaptability, and organizational and commercial awareness. Although most studies only find the great impact of personal characteristics made on formal learning, we have reason to believe that personal characteristics plays a very important role in informal learning as it is everywhere in our life and work, so we reach the following assumptions:
   H1: Personal characteristics have a significant effect on staff informal learning.
2. Assumption 2: characteristics of the workplace and staff informal learning
   Foreign scholars have done a lot of research about the characteristics of the workplace which can promote informal learning. Eraut (2000) studied the informal learning in the workplace, he found that the informal learning is enhanced when there is sufficient task variation, workers have the opportunity to participate in temporary work teams and opportunity to consult with experts inside and outside of the organization, and the opportunity to change their work duties and roles periodically. Skule (2004) find jobs that supply workers opportunities for problem solving and innovation facilitate informal learning. Programs and incentives for sharing knowledge, job mobility, and autonomy on the job can also facilitate informal learning. In a study of factors that assist or inhibit learning of new job assignments among 604 naval officers, Morrison and Brantner (1992) found that work factors that influenced informal learning

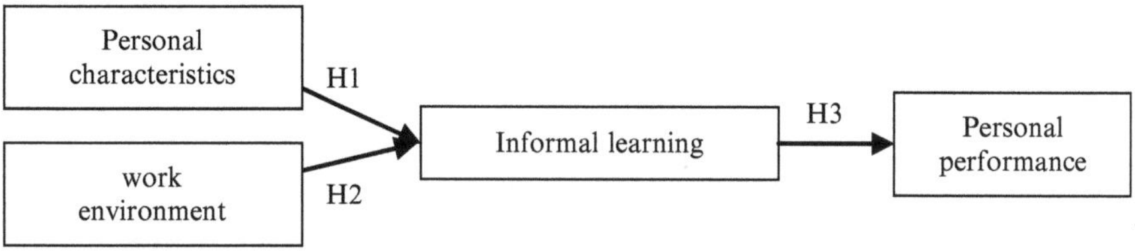


**Fig. 1** Conceptual model

were the organization's climate and pace of work. From the above we can see the influence of characteristics of the workplace on informal learning is obvious, so we draw the following assumptions:
H2:characteristics of the workplace have a significant effect on staff informal learning.

3. Assumption 3: informal learning and employee performance
Informal learning is facilitated when individuals face problems, it's consistent with the laws of learning, learners can immediately put the new knowledge into practice, and this can enrich one's knowledge, and thereby enhancing the ability of solving problems. Informal learning of employees within the enterprise often involves a lot of experience and skills critical, and these experiences and skills are often implicit, unstructured. In order to better adapt to their work and improve performance, employees must master these tacit knowledge. Nowadays, more and more enterprises want to be a learning organization, it indicates that learning is regarded as important as technology and capital to performance. Employee performance is the basis of organizational performance, the impact of learning on organizational performance is realized through the intermediate variable employee performance. Most learning of the staff is informal learning, thus we can draw the following assumptions:
H3: informal learning has a positive correlation with employee performance

Based on the above theoretical analysis, this article proposed the following theoretical research model, shown in Fig. 1.

# 3 Questionnaire Survey and Design of Scale

## 3.1 Data Sources

The main method we use in this survey is called random sampling, interview direct reservation and so on. This paper mainly studies the informal learning of workers in college, enterprise and government. We use method of email and on-site fill to finish this survey. From June 2010 to July, we handed out about 300 surveys, and recovered 172 effective surveys; the recovery percent is 57.3%. The description of the sample date is shown in Table 1.

**Table 1** Description of sample data

| Characteristic variables | Item description | Frequency | Proportion (%) |
|---|---|---|---|
| Type of work | Colleges and universities | 5 | 2.9 |
| | Enterprises or institutions | 78 | 45.3 |
| | Government departments | 37 | 21.5 |
| | Others | 52 | 30.2 |
| Gender | Male | 89 | 51.7 |
| | Female | 83 | 48.3 |
| Age | 20–25 | 132 | 76.7 |
| | 26–30 | 23 | 13.4 |
| | 31–35 | 9 | 5.2 |
| | 36–40 | 6 | 3.5 |
| | >41 | 2 | 1.2 |

### *3.2 Design of Scale and Method of Data Processing*

This study delimits four variables. We try to use existing literature and latest maturity scale as reference during the study. Specifically, about the personal characteristics factors and work environment factors which effects informal learning of workers, we mainly refer to Paul Squires's (2009) scale design of informal learning factors. On informal learning scale, we also refer to Paul Squires's (2009) scale of learning styles classification and the scale of informal learning designed by Yinhuan and Xiaobin (2009). The employee performance scale is designed on the basis of previous research and also considers the operability of the data.

We ask human resource experts for consultation and guidance after scales are drawn up, and we make some changes on the design of some variables and the description of some items according to their suggestions. All items are used Likert Five Points Scale to measure, the value of each variable is the mean of the included options for the variable. The operational definition of scale and Cranach is in Table 2.

The table shows good consistency that the Cronbach's $\alpha$ for each variable is above 0.7, indicating a high correlation of the items of each variable and the variables have good reliability.

This paper use SPSS 17.0 to do correlation and regression analysis, we study the impact of personal characteristics and characteristics of workplace on the informal learning of employee, and the impact of informal learning on employee performance, trying to find out the factors that influence informal learning.

## 4 Results of Correlation Analysis and Regression Analysis

As we can see from Table 3, informal learning has a significant impact on personal performance. In addition, both personal characteristics and working environment are significantly related to informal learning, but in terms of the level of impacts on

**Table 2** The operational definition and Cronbach of variables

| Variables | Operational definition | Cronbach's α |
|---|---|---|
| Personal characteristics (PC) | You are a strong-willed person (PC1) | 0.731 |
| | Are you introverted or extroverted (PC2) | |
| | Do you think your life is controlled by yourself or affected by the external environment and opportunities (PC3) | |
| | You have a strong sense of responsibility (PC4) | |
| | What about your openness to experience (PC5) | |
| | You are confident it will be successful when you taking a task (PC6) | |
| | You will concern the consequences of behavior (PC7) | |
| | You agree your job and believe it can realize your value (PC8) | |
| | You have good teamwork spirit (PC9) | |
| | You have strong communication skills (PC10) | |
| | You are a adaptable person (PC11) | |
| | You have a strong sense of belonging to your organization (PC12) | |
| | You have great commercial awareness (PC13) | |
| Characteristics of the workplace (CW) | The content of your work is complicated or simple (CW1) | 0.749 |
| | You usually get feedback of your work (CW2) | |
| | Your job has high mobility (CW3) | |
| | How about the role autonomy of your work (CW4) | |
| | You often participate in temporary teams (CW5) | |
| | You can often access to experts in and outside (CW6) | |
| | You have many opportunities to use new skills and knowledge (CW7) | |
| | There are incentives in sharing knowledge in your organization (CW8) | |
| | The pace of your work is fast (CW9) | |
| | There is fierce competence of peers (CW10) | |
| | You often get demands from customer and manager (CW11) | |
| | You have to communicate with colleagues frequently (CW12) | |
| Informal learning (IL) | You often observe how a co-worker performs complex job duties (IL1) | 0.835 |
| | You often share experience with a colleague actively (IL2) | |
| | You often ask a colleague for needed information (IL3) | |
| | You often look up information about work on the Internet (IL4) | |
| | You often read the methods and procedure manual on your own initiative (IL5) | |
| | You solve a problem by brainstorming or meeting when participating in team work (IL6) | |
| | You often attend presentations at a experience-sharing sessions (IL7) | |
| | You will learn knowledge and skills about your work in free time by yourself (IL8) | |
| | You often ask for feedback from the supervisor about work on your own initiative (IL9) | |
| | You often discuss your work with friends and family to get advice (IL10) | |
| Performance (P) | You can usually accomplish the set objectives very well (P1) | 0.842 |
| | You can well play the appropriate role and assume the corresponding responsibility in the course of work (P2) | |

**Table 3** Mean, standard deviation and Pearson's correlation coefficient

| Variables | Mean | Standard deviation | 1 | 2 | 3 | 4 |
|---|---|---|---|---|---|---|
| Performance (P) | 6.744 | 1.017 | 1 | | | |
| Informal learning (IL) | 5.189 | 1.415 | 0.220** | 1 | | |
| Personal characteristics (PC) | 6.242 | 0.747 | 0.478** | 0.465** | 1 | |
| Characteristics of the workplace (CW) | 5.112 | 1.216 | 0.241** | 0.611* | 0.384** | 1 |

* means $p < 0.1$, ** means $p < 0.05$, *** means $p < 0.01$

informal learning, working environment is stronger than personal characteristics. Besides, unexpectedly, we discover that both personal characteristics and working environment are significantly related to personal performance and that in terms of personal characteristics, the level of its impacts on personal performance is stronger than the impacts on informal learning, which can be understood that there is a very important relationship between personal characteristics and personal performance.

This paper uses regression analysis to do hypothesis testing, the results are in Table 4. In models 1–4 the dependent variable is informal learning, in model 5 and model 6, the dependent variable is individual performance. To further identify the impact factors of informal learning and individual performance, we analyze the specific factors as independent variables. We use regression analysis to conduct the collinearity diagnosis of variables of each model, all the VIF is less than 2, indicating collinearity among the independent variables will not affect the research obviously.

From model 1, we can see that there is a significant positive correlation between personal characteristics and working environment and informal learning. The regression coefficients of the personal characteristics $\beta = 0.51; p < 0.01$, regression coefficient of the working environment $\beta = 0.590$; $p < 0.01$, which proved that hypothesis 1 and hypothesis 2 were established. It means that the more consistent with personal characteristics and work environment described in the question item, the more informal learning there will be.

From model 2 and model 4, we can see that self-efficacy, adaptability, job involvement and sense of responsibility of personal characteristics have significant positive effects on the informal learning. Among which, the regression coefficient of self-efficacy $\beta = 0.230$; $p < 0.05$, the regression coefficient of adaptability $\beta = 0.121$; $p < 0.1$, the regression coefficient of job involvement $\beta = 0.195$; $p < 0.01$, the regression coefficient of responsibility $\beta = 0.133$; $p < 0.1$. From model 3 and model 4, we can see that among working environment factors, job rotation, access to the experts, opportunities of application of new knowledge, incentives for knowledge sharing and the pace of work have significant positive effects on informal learning. Among which, the regression coefficient of job rotation $\beta = 0.094$; $p < 0.05$, the regression coefficient of access to experts $\beta = 0.180$; $p < 0.01$, the regression coefficient of the opportunities of application of new knowledge $\beta = 0.132$; $p < 0.01$, the regression coefficient of incentives on knowledge-sharing $\beta = 0.109$; $p < 0.1$, the regression coefficient of working pace $\beta = 0.096$; $p < 0.1$. In model 4, the regression coefficient of job rotation $\beta = 0.078; p < 0.1$, the regression coefficient of the opportunities of application of new knowledge $\beta = 0.101; p < 0.1$.

**Table 4** Results of regression analysis

| Variables | Independent Variables: Informal Learning | | | | Independent Variables: Performance | |
|---|---|---|---|---|---|---|
| | Model 1 | Model 2 | Model 3 | Model 4 | Model 5 | Model 6 |
| Constants | −1.025 | −0.187 | $0.926^{*}$ | −0.848 | $5.922^{***}$ | $5.681^{***}$ |
| PC | $0.512^{***}$ | | | | | |
| CW | $0.590^{***}$ | | | | | |
| PC1 | | 0.030 | | −0.007 | | |
| PC2 | | −0.031 | | −0.059 | | |
| PC3 | | 0.080 | | 0.039 | | |
| PC4 | | 0.079 | | $0.133^{*}$ | | |
| PC5 | | 0.122 | | 0.099 | | |
| PC6 | | $0.230^{**}$ | | 0.115 | | |
| PC7 | | 0.055 | | 0.006 | | |
| PC8 | | 0.075 | | −0.012 | | |
| PC9 | | −0.089 | | −0.084 | | |
| PC10 | | 0.016 | | 0.041 | | |
| PC11 | | $0.121^{*}$ | | 0.033 | | |
| PC12 | | $0.195^{***}$ | | 0.067 | | |
| PC13 | | 0.036 | | 0.016 | | |
| CW1 | | | 0.048 | 0.061 | | |
| CW2 | | | −0.013 | −0.01 | | |
| CW3 | | | $0.094^{**}$ | $0.078^{*}$ | | |
| CW4 | | | 0.042 | 0.029 | | |
| CW5 | | | −0.033 | −0.037 | | |
| CW6 | | | $0.180^{***}$ | $0.179^{***}$ | | |
| CW7 | | | $0.132^{***}$ | $0.101^{*}$ | | |
| CW8 | | | $0.109^{*}$ | 0.069 | | |
| CW9 | | | $0.096^{*}$ | 0.068 | | |
| CW10 | | | 0.050 | 0.036 | | |
| CW11 | | | 0.059 | 0.049 | | |
| CW12 | | | 0.080 | 0.084 | | |
| IL | | | | | $0.158^{***}$ | |
| IL1 | | | | | | −0.036 |
| IL2 | | | | | | 0.093 |
| IL3 | | | | | | 0.020 |
| IL4 | | | | | | 0.018 |
| IL5 | | | | | | −0.043 |
| IL6 | | | | | | −0.036 |
| IL7 | | | | | | 0.038 |
| IL8 | | | | | | $0.130^{***}$ |
| IL9 | | | | | | 0.027 |
| IL10 | | | | | | −0.044 |
| F | $65.229^{***}$ | $4.725^{***}$ | $12.902^{***}$ | $6.986^{***}$ | $8.681^{***}$ | $2.618^{***}$ |
| $R^2$ | 0.436 | 0.281 | 0.497 | 0.550 | 0.049 | 0.141 |
| Adjusted $R^2$ | 0.429 | 0.222 | 0.458 | 0.471 | 0.043 | 0.087 |

$^{*}$means $p < 0.1$, $^{**}$means $p < 0.05$, $^{***}$means $p < 0.01$

Hypothesis 3 is verified in model 5, the regression coefficient of informal learning $\beta = 0.158$; $p < 0.01$, showing that informal learning has a significant positive correlation with personal performance. From model 6, we can see that of all ways of informal learning, self-learning of skills and knowledge related to job has a significant impact on personal performance and it's regression coefficient $\beta = 0.130$; $p < 0.01$.

## 5 Conclusions

During the study of the factors affecting the staff informal learning, our three assumptions are all been verified. Based on the above analysis results of data, we can draw the following enlightenments for practice:

1. Organizations should train the self-efficacy, adaptability and sense of responsibility of staff consciously in career training and work. And organization can confer appropriate authorities to the staff to increase the their organizational identities, so we can enhance the staff's performance as well as the organizational performance.
2. In practice, organizations can encourage staff to fully share knowledge, provide opportunities to apply new knowledge, more access to experts, some opportunities for job changing, and speed up the work place appropriately to promote staff informal learning,which can improve organizational performance at a lower cost.
3. Organizations should encourage staff to learn in their working place, and develop policies to support informal learning to promote more staff informal learning. We also found that organizations can most enhance the impact of informal learning through encouraging staff to self-learning job related skills to achieve the goals of improving both personal and organizational performance and controlling business risk.

This article has the following limitations: First, the subjects are not popular. The ages of people who fill in the questionnaire are ranged from 20 to 25 years old and they almost fresh workers, so the results of impact factors may not be comprehensive. Second, the measurement of personal performance is not comprehensive enough and the respondents may not make an objective response for face and other reasons, which may affect the results.

## References

Barrick MR, Mount MK (1991) The big five personality dimensions and job performance: a meta-analysis. Pers Psychol 44:1–26

Eraut M (2000) Non-formal learning and tacit knowledge in professional work. Br J Educ Psychol 70:113–136

Hunter JE, Schmidt FL (1996) Intelligence and job performance: economic and social implications. Psychol Public Policy Law 2:447–472

Jensen AR (1980) Bias in mental testing. The Free Press, New York

Lievens F, Harris MM, Van Keer E, Bisqueret C (2003) Predicting cross – cultural training performance: the validity of personality, cognitive ability, and dimensions measured by an assessment center and a behavioral description interview. J Appl Psychol 88(3):476–489

Morrison RF, Brantner TM (1992) What enhances or inhibits learning a new job? A basic career issue. J Appl Psychol 77(6):926–940

Skule S (2004) Learning conditions at work: a framework to understand and assess informal learning in the workplace. Int J Train Dev 8(1):8–21

Xin J (2004) Management of tacit knowledge – to create a informal learning environment in organizations. Full-text Database of China's Outstanding Master's Degree Thesis (in Chinese)

Yinhuan W, Xiaobing Y (2009) Informal learning's application in corporate training – with Chery automobile Co., Ltd. as an example of enterprise. Research on Modern Distance Education. No. 3, pp 63–65 (in Chinese)

# Developmental Tendency and Empirical Analysis of Staff's Boundaryless Career: Statistic Analysis Based on the Experience in China

**Ze Tian and Jianjun Han**

**Abstract** Boundaryles profession career is a new profession career era, which presents a dynamic, unpredictable and multi-directional development characteristics. In view of the practice of Boundaryless profession career development, we conducted a questionnaire survey and statistical sampling analysis and we applied statistical tools to conduct empirical analysis. The research reveals the current situation, characteristics and developmental trends of staff boundaryless career in China. The analysis is helpful for future expansion of existing theory of boundaryless career orientation, and provide some valuable suggestions and solutions.

**Keywords** Boundaryless career · Empirical analysis · Sensitivity · Uncertainty

## 1 Introduction

With economic globalization, the Internet technology advances, organizational structure flat, Employee's career entered a new era of his career that is boundaryless career. In the boundaryless career background, employment relationship and psychological contract changes made people showing more liquidity, liquidity seems to have become the career development of a specific phenomenon. New career to both the individual employee career management and organizations has produced a significant effect, how the new career under the background of good career management has important practical significance.

Since reform and opening, China's staff turnover rate showed increasing trend. In recent years the annual staff turnover of 15–30%, has exceeded 15% of the

Z. Tian (✉)
Business School of Hohai University, Changzhou 213022, China
e-mail: tianze21@126.com

J. Han
Business School of Hohai University, Nanjing 210098, China
e-mail: freak8866@yahoo.com.cn

Y. Zhou and D.D. Wu (eds.), *Modeling Risk Management for Resources and Environment in China*, Computational Risk Management,
DOI 10.1007/978-3-642-18387-4_46, © Springer-Verlag Berlin Heidelberg 2011

general view that the optimal turnover rate and, especially the turnover rate is as high as 60% in China's IT industry according to U.S. management consultants Smith's research, The high staff turnover has become a major obstacle to Chinese business sustainable development. Of our staff turnover rate so high, it implies that China has advanced into the boundaryless career era? The boundaryless career management issues have a great impact on both staff's career development and organizational career management. Therefore, it is of great practical significance to conduct research on career management in such a new boundaryless career era.

## 2 Literature Review

The concept of the boundaryless career was put forwarded in the early 1990s. boundaryless career was defined as "a series of job opportunities beyond a single employment range"(De Fillipi and Arthur 1994). Baker and Aldrich (1996) enlarged the concept of boundaryless career by putting forward the following idea: number of employers, knowledge accumulation and personal identity are three characteristics for professionals with the tendency of boundaryless career; besides internal movement, transferable skills and personal identity are also important factors to judge whether a person is in the pursuit of the boundaryless career. De fillippi and Arthur put forward the concept of competency-based career. They argue that, under the background of boundaryless career, employees can improve their career by developing professional competency, i.e., they should know who they are, why and how to work (Baruch, 2004).Sullivan and Arthur think that boundaryless career is a multidimensional phenomenon. It is beyond a various of boundaries and includes physical and psychological as well as objective and subjective dimensions (Blau and Lunz 1998).Through the analysis of these concepts, we can summarize that the main characteristics of boundaryless career is that employees are no longer complete their lifelong career in one or two organizations. They achieve their career goals in several organizations, departments, careers and professional positions. Moreover, this mobility contains psychological or subjective mobility.

On research model and empirical analysis, Cardle verified the effect of boundaryless career on personal adaptability, psychological success and performance, etc. from the perspective of personal career. Choosing business class students as research object, Vicenc Fernandez and Mihaela Enache (2008) used the method of qualitative comparative analysis to conducted research on the relationship between emotional commitment and attitude tendency of the boundaryless career and protean career. The results shows that the attitude tendency of boundaryless career and protean career is a good predictor of emotional commitment (consistency coefficient is 0.85). Besides, attitude tendency of boundaryless career and protean career not only has positive effect on personal performance, but also brings positive effect on hiring organizations. Furthermore, highly protean staffs show high emotional commitment without high mobility tendency in organization. The survey also shows that it is more important for employees to be responsible for managing their own career than passively depend on clear career path provided by organizations.

Briscoe et al. (2006) has verified two tendency of the boundaryless career attitude through an empirical research: boundaryless tendency and organizational mobility preference tendency. A survey on 13,655 respondents from European carried out by Jesse Segers et al had indicated more men than women had tendency of psychological movement. Because men had been easier tempted by money, status and promotion, and work security of men had been significantly lower than that of women. On the contrary, women had been rarely influenced by the limits of objective, traditional criteria for career success. They had enjoyed working according to their principles. In the attitude tendency of self-guidance, men and women hadn't showed significant gender differences (Segers et al. 2008). In the framework of protean career model, the study has also showed individuals had not been active in career self-management. The reason was people generally reduce the expectations and they were lack of cognition of future scenarios. While in the aspect of value-driven, age and value-driven dimensions of the protean career showed good positive correlation. Additionally, management experience and high qualifications had a positive correlation with the impact of physical movement, self-career management and incentives of psychological movement, which showed people with higher level of education become manager was the result of high self-directed level.

Another survey on 289 employees carried out by DeVos and Soens (2008) had indicated after receiving career counseling, individual holding the attitude of variability career had higher job satisfaction and perceived ability to be employed. The positive correlation between Variability career attitudes and self-management behavior has also supported Hall (2004) who proposed a view that individual holding variability career orientation would be active in pursuit of professional success and turn this motion into a concrete power to manage their own career operations. Changeability career tendency makes individuals to draw up and guide their own career path. In addition, the results also show that individuals holding the volatility career attitudes are more willing to manage their own career, and therefore produce a series of career achievements. Studies had shown that if the organization wants to motivate more career self-management, only staff training in career self-management behavior is not enough, it is important to enhance the staffs' attitude of career. Besides, the construction of interrelated organizational culture is also important.

## 3 An Empirical Analysis of the Boundaryless Career Development Based on the Data in China

In accordance with the background of Chinese economy, culture and career development, this article combines the research of both boundaryless career and protean career. From the view of demonstration, the SPSS 15.0 statistics software was used to study the population statistics variables which may influence the boundaryless career and protean career. Through the Questionnaires, The empirical analysis was carried out in the paper theory of boundaryless career and protean career, such as the effect that boundaryless career and protean career could make on the tendency of job hopping, or the impact that the organization career management could bring to the probability of job hopping in the age of boundaryless career in China. Providing the basis for development of the theory for the boundaryless career and the volatility of career management of Chinese enterprises.

## 3.1 Research Assumptions

*Assumption 1*: Population statistics variables, such as gender, age, business quality, and position, may be significant difference in the boundaryless career tendency.

The fast development of Chinese economy, the tremendous changes of our society and the recombination of different enterprises have brought some changes to people's everyday life and work. To some degree, these may also affect Chinese employees' career tendency. It is assumed that in accordance with the situation, population statistic variables, such as gender, age, business quality, and position have great impact on the boundaryless career tendency.

*Assumption 2*: The ideology tendency of boundaryless career has a strong relationship with the tendency of the employee's job-hopping.

For the organization, the inner movement (like the job transposition) has little impact on the organization development. What's more, sometimes it is even good for on organization development (like the transposition in order to cultivate the talented). However, to go beyond the organizational boundary, the occurrence of job-hopping, could bring huge negative influences to the enterprise. Based on this, this article focuses more on the relationship between career tendency and employees'job-hopping tendency under the background without boundary Thus the *Assumption 2* is made.

*Assumption 3*: In the age of boundaryless career, organization career management can still effectively avoid employees' job-hopping.

Many factors have proved that organization career management still has effects on such an age of boundaryless career. Organizations can provide better platforms for their employees to express themselves and challenge themselves. Good career management can support self-development therefore the employees would like to stay in the organization. According to the above, Assumption 3 is made.

## 3.2 Demonstration Research Methods

1. Mathematical statistics
   Statistics method and SPSS 15.0 were used to describe and analyze the population statistics variables of the sample, such as gender, age, business quality, and position. The article then compares each variable's division frequency,

attributes, mean value and discrete case, in order to supply some references and proofs for the comparative analysis of the samples. What's more by analyzing the mean value, standard deviation, and coefficient of variation of each scale, we also provide available data for the test of the samples' boundaryless career tendency, organization career management, and the current situation of employees' job-hopping.

2. Regression analysis

This article judges the tightness degree of each variable by analyzing the relevant relationships The linear regression analysis helps to decide the forecasting ability which career tendency of boundaryless career and organization career management have to forecast the possibility of employees' job-hopping.

## 3.3 Assumption Examination and Statistics Analysis

Gap analysis on demographic variables in boundaryless career orientation. To determine whether there are significant differences among them by using T-test and ANOVA test samples on gender, age, education level, marital status, work experience, job level, job type, company type, annual salary level and the relationship between non-border vocational orientation test.

1. Differences on the boundaryless Career Orientation concerning gender
   Because gender is binary variable, so a T-test method was used to analyze differences. Analysis showed that in the absence of career orientation on the boundary, the average score for men than women. But the results of T-test showed that Sig. $= 0.113 > 0.05$, which indicates that gender didn't affect boundaryless Career Orientation to a large degree. Men and women do not have significant differences concerning the aspects of subjective and psychological mobility in organizations. As for the protean career orientation, men's average score is still higher than that of women. While the test results of T-test, Sig. $= 0.034 < 0.05$ (Table 1), is statistically significant, suggesting the protean tendency of his career were significant for two gender and men rates higher in self-guided career than women.
2. The differences concerning different ages in boundaryless Career Orientation
   Age is a multiple sub-variables, so difference should be tested by ANOVA. According to homogeneity test, boundaryless career orientation Sig $= 0.719 > 0.05$, and the protean career orientation Sig $= 0.171 > 0.05$. Proving that it's suitable for the variance test. Upon examination of the results, boundaryless career orientation Sig $= 0.730 > 0.05$, the protean career orientation Sig $= 0.456 > 0.05$ (Table 2), both of them did not reach significance level. And based on the results from the descriptive statistics, there is not great difference for means in the various levels. Therefore, age doesn't lead to obvious differences in the boundaryless professional orientation. Against the major background of

**Table 1** Gender independent sample analysis

| | | Levene's test of equal variance | | T-value equal to the mean test | | | | |
|---|---|---|---|---|---|---|---|---|
| | | F | Sig. | t | df | Sig. | Mean difference | Standard error of difference |
| Boundaryless career orientation | Assuming equal variance | 8.029 | 0.005 | 1.971 | 317 | 0.050 | 1.407 | 0.714 |
| | Assuming unequal variance | | | 1.602 | 76.366 | 0.113 | 1.407 | 0.878 |
| Protean career orientation | Assuming equal variance | 1.123 | 0.290 | 2.134 | 317 | 0.034 | 1.474 | 0.691 |
| | Assuming unequal variance | | | 1.933 | 83.526 | 0.057 | 1.47 | 0.763 |

**Table 2** Test the consistency of variance of different ages on boundaryless career

| | Levene statistics | Freedom 1 | Freedom 2 | Sig. |
|---|---|---|---|---|
| Boundaryless career orientation | 0.523 | 4 | 314 | 0.719 |
| Proteanf career orientation | 1.613 | 4 | 314 | 0.171 |

modern social-economic, people of all ages have similar concept of non-border professional orientation.

3. Differences caused by different levels of education in boundaryless Career Orientation

   Results showed that no borders career orientation Sig. Value = 0.871 > 0.05 and protean career Sig. Value = 0.260 > 0.05, both of which have not reached high level, suggesting that the level of education does not lead to significant difference in career orientation without borders.

4. Differences on the career orientation in the absence of boundaries of different marital status

   Marital status includes married and unmarried and it belongs to binary variables. We use T-test model which shows that boundaryless career orientation Sig. Value = 0.977 > 0.05, protean career Sig. Value = 0.788 > 0.05, both of which did not reach significance level. This result indicates that marital status does not bring about significant difference in boundaryless professional orientation.

5. Analysis for differences on different working life of employees in the boundaryless career orientation

   Protean career Sig.value = 0.528 > 0.05, indicating that they are in homogeneity and are suitable for variance analysis. The results of analysis of variance are shown in the table. Boundaryless career tendency Sig. Value = 0.640 > 0.05, the protean his career tendency Sig. Value = 0.765 > 0.05, indicating the differences in the non-working life of employees does not make much difference on the boundaryless vocational interest (Table 3).

**Table 3** Consistency test for employees of different working life

| | Levene statistics | Freedom 1 | Freedom 2 | Sig. |
|---|---|---|---|---|
| Boundaryless career orientation | 0.437 | 2 | 316 | 0.647 |
| The volatility of career orientation | 0.639 | 2 | 316 | 0.528 |

6. Differences Analysis for different types of staff positions in boundaryless Career Orientation

   Job types include management class, market class, service type, technology class and the party type. According to analysis of variance, boundaryless career orientation Sig. Value $= 0.095 > 0.05$, the protean a career orientation Sig. Value $= 0.411 > 0.05$ the two were not significant, suggesting a different type of staff positions in the absence of career orientation on the boundary causes no significant difference.
7. Different analysis of staff with different salary levels on borderless profession career orientation

By test inspection, we find that borderless profession career orientation Sig. Value $= 0.044 < 0.05$, the protean profession career orientation Sig. Value $= 0.001 < 0.05$, both are significant levels, suggesting staff with different salary levels have significant difference in non-border professional tendencies: the higher the levels, the higher staff knowledge accumulation and skill level they have, the higher potential of being employed, the more terms and self-directed and mobile abilities they will have. Therefore, they will show a higher enthusiasm on the job of bounbaryless mean test of staff with different salary levels in non-border profession tendencies.

## 3.4 *Correlation Analysis of Boundaryless Career Orientation*

This article will make a correlation analysis on the boundaryless career orientation, organizational profession career management and turnover orientation to determine whether there is significant dependency among them.

From the statistical analysis results, there is positive correlation between boundaryless career and the protean profession career and the correlation coefficient is 0.536, showing a strong correlation which is consistent with Sullivan and Arthur's (2006) results. That is, there is the cross-cutting concept between boundaryless career and the protean profession career.

There is a significant positive correlation between boundaryless career, the protean profession career and organizational profession career management. The correlation coefficients are 0.125 and 0.161respectively, which indicates that organizational profession career management, improves staff's knowledge and skills.

Organizational profession career management and turnover show a significant negative correlation and the negative correlation coefficient is 0.425, indicating that organizational profession career management has significant inhibitory effect on staff turnover, which is consistent with the literature's empirical research results on organizational profession career management and turnover (Table 4).

**Table 4** Correlation analysis of boundaryless career orientation, career management

| Factors | | 1 | 2 | 3 | 4 |
|---|---|---|---|---|---|
| Boundaryless career orientation | Pearson correlation | 1 | 0.536* | 0.125** | 0.098 |
| | Sig. | . | 0.000 | 0.026 | 0.080 |
| | N | 319 | 319 | 319 | 319 |
| Protean career orientation | Pearson correlation | 0.536* | 1 | 0.161* | 0.052 |
| | Sig. | 0.000 | . | 0.004 | 0.357 |
| | N | 319 | 319 | 319 | 319 |
| Organizational career management | Pearson correlation | 0.125** | 0.161* | 1 | −0.425* |
| | Sig. | 0.026 | 0.004 | . | 0.000 |
| | N | 319 | 319 | 319 | 319 |
| turnover intention | Pearson correlation | 0.098 | 0.052 | −0.425* | 1 |
| | Sig. | 0.080 | 0.357 | 0.000 | |
| | N | 319 | 319 | 319 | 319 |

*That was significantly related to levels of 0.01 (2-tailed)
**That was significantly correlated to the level of 0.05 (2-tailed)

Besides, from the correlation analyses of boundaryless career, the protean profession career, organizational profession career management and turnover times, we can see that there is no significant correlation between boundaryless career orientation, the protean profession career and turnover times. From the point of view of the correlation coefficient, the correlation coefficient is −0.003 and 0.108, respectively. Both of them are less than 0.2 levels, indicating that there is weak correlation or no correlation among them.

# 4 Conclusions

From the empirical research under China's economic and social background, we draw the following conclusions:

1. Statistical analysis of empirical data validate that our staff's career has been basically in a boundaryless career and a protean profession career era.
2. Our employees of different gender have no significant differences in the boundaryless career. However, in the protean career, they show significant differences and man is significantly better than the females in the self-direction of profession career.
3. Staff of different levels has no significant differences in the boundaryless career orientation, which means that the boundaryless career mindset is a common concept and job degrees have no significant differences. In the protean career, staff of different levels have significant differences and employees of management class have a higher protean career orientation than normal workers, which shows that managers are better in knowing how to self-class career management to enhance their employability, and continuously improve their core competitiveness.
4. Staff of different types of enterprises have significant differences in boundaryless career orientation. Staffs of state-owned enterprises have higher boundaryless career mindset level than Staff of government institutions, the latter have the

lowest boundaryless career mindset level in all Staff of enterprises of different types, which may be related with job position and stability of the government institutions.

5. Staff of different salary levels have significant differences in boundaryless career orientation: Staff of high salary levels have higher orientation of physical and mental movement than Staff of low salary levels, which is related to the former's less material pressure and more resources to broader their space for development. In the protean career, staffs of different salary levels have significant differences. Staff of high salary levels have more motivation than the common staff in self-directed career and cultivating their ability to obtain a higher cause and achievements.
6. Organizational career management can effectively predict staff's turnover. Through correlation and regression analysis, organizational career management has a significant prediction effect on turnover orientation; taking organizational career management as control variables, the results show that boundaryless career orientation and the protean career have a significant correlation with turnover orientation, indicating organizational career management have inhibition effect on turnover orientation caused by boundaryless career orientation and the protean career orientation. And by further regression analysis, it shows that organizational career management and career without boundaries mindset act on turnover orientation together, and have a prediction effect on turnover orientation.

## References

Baker T, Aldrich HE (1996) Prometheus Stretches: building identity and cumulative knowledge in multiemployer careers. In: Arthur MB, Rousseau DM (eds) The boundaryless career. Oxford University Press, New York, pp 132–149

Baruch Y (2004) Managing careers: theory and practice. Prentice Hall, Harlow

Blau G, Lunz M (1998) Testing the incremental, effect of professional commitment on intent to leave one's profession beyond the effects of external, personal and work-related variables. J Vocational Behav 52:260–269

Briscoe JP, Hall DT, DeMuth RLF (2006) Protean and boundaryless careers: an empirical exploration. J Vocational Behav 69:30–47

De Fillipi RJ, Arthur MB (1994) The boundaryless career: a competency-based perspective. J Org Behav 15:307–324

De Vos A, Soens N (2008) Protean attitude and career success: the mediating role of self-management. J Vocational Behav 10:10–16

Fernandez V, Enache M (2008) Exploring the relationship between protean and boundaryless career attitudes and affective commitment through the lens of a fuzzy set QCA methodology. Intangible Capital 4:31–66

Hall DT (2004) The protean career: a quarter-century journey. J Vocational Behav 65(1):1–13

Segers J et al (2008) Protean and boundaryless careers: a study on potential motivators. J Vocational Behav 10:10–16

# Part VI
# Energy Risk Management

# A Preliminary Evaluation of China's Implementation Progress in Energy Intensity Targets

**Yahua Wang and Jiaochen Liang**

**Abstract** China proposed an ambitious goal of reducing energy consumption per unit of GDP by 20% from 2006 to 2010. This paper evaluates the progress of provincial governments implementing energy conservation targets assigned by the central government. The empirical analysis of this paper is divided into two parts, a static analysis and a dynamic analysis. In the static analysis, we established a multiple linear regression model based on provincial cross-sectional data, to explore factors that affect the reduction of energy intensity. In the dynamic analysis, we established a fixed group and time effect model based on provincial panel data, to explain the annual changes in energy intensity. The analysis results show that the framework of the energy conservation policy introduced by the Chinese government is quite robust, and provincial governments respond positively to the instructions from the central government.

**Keywords** China · Climate change policy · Energy intensity · Fixed group and time effect model · Panel data analysis

## 1 Introduction

Global climate change will create enormous challenges to human development in terms of ecological, economic and social disasters. Stern (2007) says, "Climate change will affect the basic elements of life for people around the world – access to water, food production, health and the environment." The Chinese government has taken active actions to address serious domestic energy issues and the challenges from climate change. In March of 2006, the Chinese government propounded the

Y. Wang (✉) and J. Liang
School of Public Policy and Management, Tsinghua University, Beijing 10084, China
e-mail: wangyahua@tsinghua.edu.cn; ljc08@mails.tsinghua.edu.cn

Y. Zhou and D.D. Wu (eds.), *Modeling Risk Management for Resources and Environment in China*, Computational Risk Management,
DOI 10.1007/978-3-642-18387-4_47, 

ambitious targets in its 11th 5-Year Plan from 2006 to 2010: energy intensity per unit of GDP should be reduced by 20%, and total major pollutants emission volume should be reduced by 10%. These have been regarded as obligatory indicators incorporated into the performance appraisal system for local officials. To achieve these goals, the State Council assigned targets for energy conservation and emission reduction to various provinces and released a series of policies to urge the provinces to implement the energy conservation targets.

Due to the energy conservation work done in the first 3 years of the 11th 5-Year period, the energy intensity nationwide has been reduced by an accumulative percentage of 10.1%, equivalent to 50.4% of the goal set in the 11th 5-Year Plan, being still in arrear of the expected schedule. There is a world of difference amongst various provinces in fulfilling the task of reducing energy intensity. Out of 30 provinces excepting Tibet, 16 have not fulfilled the energy conservation targets assigned by the central government. In fact, the overall target completion has not yet reached 60 percent. Beijing has best fulfilled the task of reducing energy intensity, accounting for 87.6% of all tasks assigned in the 11th 5-Year Plan; whereas Qinghai has fulfilled the least of the tasks, making up only 28.7% of all tasks.

If we take a look at the implementation of tasks dynamically, the energy intensity reduction has been accelerating year by year, with intensity down 1.79% countrywide in 2006, 4.04% in 2007, and 4.59% in 2008. And in the past 3 years most of the provinces have reduced their energy intensity at an accelerated pace. Many provinces that had not done well in reducing energy intensity 1 or 2 years before have evidently accelerated the reduction of their energy intensity in the second or third year; these include Ningxia, Qinghai and Shanxi. Nevertheless, several provinces that had done well in fulfilling the tasks in the first 2 years, such as Shanghai and Sichuan, have slowed down their pace in the third year.

The above-mentioned situation suggests that the implementation of the energy conservation target in China is both gratifying and worrying, driving us to make an evaluation of the energy conservation policy framework established during the 11th 5-Year period. In the past 2 years some scholars have commenced probing into-China's energy conservation policy from the public management perspective. For instance, Zhang et al. (2008) have examined the behavioral patterns of Chinese local governments addressing climate change and implementing the energy conservation policy. Wang and Yu (2009) have probed into the interest-driven factors for local government to develop low-carbon economies. But so far there has been a shortage of systematic assessments of China's newly-established energy conservation policy framework. This research is an attempt to move in this direction. What we mainly care about is whether China's policy framework for reducing energy intensity in recent years is effective or workable. In the current policy framework of energy conservation, the assignment of energy intensity targets from the central government to provincial governments is the crucial part.

What merits our further attention is the question: under this policy framework, have provincial governments responded positively to the instructions from the central government? Why is there a disparity of performance for different provinces in implementing the energy conservation instructions? What factors have decided the performances of various provinces in energy intensity reduction? The study of these questions will be conducive to an evaluation of China's energy conservation policy and help us to indentify the characteristics of this initially established policy framework.

## 2 Methodology

This study will make a preliminary analysis of the implementation of the energy conservation policy on the provincial level in China. The analysis will be conducted in both static and dynamic ways. The static analysis is focused on the study of why there is such a world of difference in the energy conservation outcomes of various provinces? What factors affect the outcome of energy intensity reduction in various provinces? The dynamic analysis is centered on how to explain the annual changes in the outcomes of energy intensity reduction of various provinces? Will the provinces adjust their behaviors according to the previous outcomes of energy intensity reduction? These two analyses will jointly reveal the intrinsic mechanisms that provide impetus for the provincial governments to carry out the energy conservation policy.

In the static analysis conducted in Sect. 3, we have established a multiple linear regression model. Taking the rate of energy intensity reduction in each province as the dependent variable and the exogenous variables as the independent variables, we have identified the independent variables that can be used to explain the outcome of energy conservation tasks through econometric analysis. Considering that the exogenous variables are numerous and that multiple collinearity exists among these variables, we have introduced the factor analysis technique to deal with the groups of possible independent variables so as to pick up the main factors and take them as the possible independent variables for the analysis in the regression model.

In the dynamic analysis in Sect. 4, we set up a panel data model. As we consider that there was a general trend of changes in energy intensity in the 30 provinces during the period 2006 to 2008, we have chosen the two-way panel model with fixed group and time effect. The dependent variable of this model is the rate of energy intensity reduction of each province during 2006 to 2008. For independent variables, we have examined the influence from the rate of energy intensity reduction during the previous year as well as trying other possible variables, such as the GDP growth rates of each province in the same year and the growth rate of the added value of the secondary industry.

# 3 A Static Analysis

## 3.1 Model

Here we use the rate in energy intensity reduction $I$, to represent the real outcome of a province's implementation of energy conservation policy, which can be defined as,

$$I = F(y_i) \tag{1}$$

Where $y_i$ are the exogenous variables that may exert influence on $I$. Considering that the effects of the exogenous variables interact with each other, we assume $y_i$s in (1) are in multiplication form, and the expression of $I$ can be written as,

$$I = A\prod_i {y_i}^{\alpha_i} \tag{2}$$

Where $A$ is constant, and $\alpha_i$, the exponential values of $y_i$, are unknown coefficients that need to be estimated. We get logarithm on both sides of (2) to have a linear equation:

$$Ln(I) = Ln(A) + \sum_i \alpha_i Ln(y_i) \tag{3}$$

Next, we need to identify the possible exogenous variables $y_i$ that may influence the dependent $I$, and then to use these $y_i$s to estimate (3) with the Ordinary Least Square Method in order to find out the factors that have significant impact on the rate of a province's energy intensity reduction.

## 3.2 Data

In order to find out the independent variables $y_i$ that have effect on $I$, we have collected the exogenous variables concerned. We have selected the following groups of data as alternative independent variables: GDP, GDP per capita, the percentage of the added value of the secondary industry in the GDP, the percentage of the added value of the heavy industry in the GDP, the initial energy intensity in 2005, and the energy conservation targets assigned to each province. The dependent variable is the accumulative rate of energy intensity reduction of each province during 2006 to 2008.

The energy conservation targets assigned by the central government to each province during the 11th 5-Year period and the initial energy intensity in 2005 come from *the Written Reply of the State Council to the Plan for Energy Intensity Reduction Targets per Unit GDP Allocated to Various Provinces During the*

*Eleventh Five-Year Period.*[1] The data of 2007 of the other four variables, come from the *China Statistics Yearbook 2008*. As dependent variables, the accumulative rate of energy intensity reduction of each province during 2006 to 2008 is calculated based on the energy intensity data of each province released by the National Bureau of Statistics in 2008.[2]

As the data in (3) is in a logarithm form, the logarithms of the above variables have first been taken to set up an SPSS data file. Through analysis, we find that different groups of the alternative independent variables have high correlation. Thus, we are unable to use the above variables to estimate (3) directly. In order to avoid the adverse influence of the multiple collinearity, we use factor analysis in the alternative independent variables before the regression analysis.

### 3.3 *Factor Analysis*

We adopt the method of factor analysis to extract factors from the six groups of alternative independent variables. The testing shows that the KMO indicator of the samples is 0.502, basically suitable for the factor analysis. According to such determining methods as "the Eigen Value greater than 1," "the Scree Plot," and the Accumulative Explaining Ratio of Variance, we have picked up three factors with the aggregate explanatory ratio of 84.87%. In order to better understand the meaning of these factors, we have adopted the Varimax Orthogonal Rotation in data processing to conduct the "Varimax" rotation of the Component Matrix. See Table 1 for the results. The main purpose of the Varimax rotation is to focus each variable's load in one and only one factor.

**Table 1** Rotated component matrix in factor analysis

| Indicators | Factor | | |
|---|---|---|---|
| | 1 | 2 | 3 |
| GDP | 0.828 | 0.424 | 0.006 |
| GDP per-capital | 0.821 | −0.123 | 0.347 |
| Initial energy intensity (2005) | −0.871 | 0.220 | 0.327 |
| Targets of energy intensity reduction | −0.041 | 0.854 | 0.362 |
| Percentage of secondary industry in GDP | 0.005 | 0.329 | 0.777 |
| Percentage of heavy industry in GDP | 0.019 | 0.145 | 0.953 |

Extraction method: principle component analysis; rotation method: varimax with Kaiser normalization.

[1]China's Central Government, *The* Written *Reply of the State Council to the Plan for Energy intensity Reduction Targets per Unit GDP assigned to Various Regions during the 11th 5-year Plan Period.* September 17, 2006.

[2]China's Central Government, *Energy intensity* Targets *per Unit GDP of Various Provinces in 2008*. June 30, 2009.

**Table 2** Descriptions for the factors extracted

| Factors | | Indicators | Descriptions |
|---|---|---|---|
| Capability factor | Factor 1 | GDP<br>GDP per capital<br>Initial energy intensity (2005) | The potential of a province in reducing energy intensity, which mainly presents a province's economic and fiscal capabilities, as well as the abilities to overcome the path dependence of high-carbon economy development. |
| Rules factor | Factor 2 | Targets of energy intensity reduction | The pressure facing by provincial governments in policy implementation of energy conservation policy from the central government. |
| Structure factor | Factor 3 | Percentage of secondary industry in GDP<br>Percentage of heavy industry in GDP | The economic and industrial structure of various provinces. |

Table 2 gives further descriptions to the factors extracted. Factor 1 contains three variables, namely, GDP, GDP per capita, and the Initial Energy Intensity. We call Factor 1 the "Capability Factor," which reflects the potential capacity of a province to reduce energy intensity. On one hand, it has a positive correlation with the GDP and GDP per capita. Where the GDP is larger and the economic development level is higher, more resources can be mobilized to realize the policy goals. On the other hand, Factor 1 has a negative correlation with the Initial Energy Intensity, because the provinces with higher energy intensity are usually the regions that are much more dependent on highly energy-consuming industries and less efficient in utilizing their energy resources, thus making it more difficult to reduce energy intensity. Factor 2 contains one variable, the targets of energy intensity reduction assigned to various provinces by the central government, which reflects the instructions set by the central government, and we interpret it as the "Rules Factor." Factor 3 contains two variables, namely, the percentage of the GDP contributed by the secondary industry and by the heavy industry, respectively, which reflect the economic and industrial structure of various provinces, and we interpret this combination as the "Structure Factor."

## 3.4 *Results of the Multivariate Regression*

By taking the accumulative rate of energy intensity reduction (logarithm value) of the 30 provinces from 2006 to 2008 as the dependent variable and the three factors[3] obtained from factor analysis as the independent variables, we fit (3) and find that the "Structure Factor" is not significant from zero at the 5% level of significance.

[3]With SPSS software, the regression method can be used to work out the scores of the three factors instead of the observations.

**Table 3** Linear regression results *i*th two factors

| | Estimator | Standard Error | T Statistics | Sig. |
|---|---|---|---|---|
| Constant | 2.409 | 0.035 | 69.206 | 0.000 |
| Capability | 0.115 | 0.035 | 3.258 | 0.003 |
| Rules-in-use factor | 0.188 | 0.035 | 5.299 | 0.000 |
| $R^2 = 0.589$, adj-$R^2 = 0.559$, $F = 19.346$ | | | | |

Therefore, we reject it from the equation. By taking the other two significant factors, the "Capability Factor" and the "Rules Factor" as independent variables, we evaluate (3) again with the Ordinary Least Square Method. The results are shown in Table 3.

The $F$ value of this regression model is 19.347, which is significant. And the regression coefficients of the "Capability Factor" and the "Rules Factor" are significantly positive, suggesting that the conditions reflecting potential capability and the pressure from the central government produce a positive influence on the implementation of energy conservation policy. The R-Square of this model is close to 0.589, indicating that these two factors explain quite a large part of the outcomes of policy implementation.

# 4 A Dynamic Analysis

## *4.1 Model*

This section mainly illustrates why annual changes took place in the rate of energy intensity reduction of each province in the first 3 years of the 11th 5-Year period. From the data of energy intensity reduction during 2006 to 2008, it can be seen that there is an overall trend towards a rising rate of energy reduction each year. For this reason, we adopt the fixed group and time effect model as (4) to estimate the dynamic mechanism of implementation of the energy conservation targets.

$$I_{it} = C + \alpha_i + \gamma_t + \mathbf{X_{it}} \bullet \boldsymbol{\beta} + u_{it} \tag{4}$$

Where $I_{it}$ denotes the rate of energy intensity reduction of province $i$ in year $t$; $\alpha_i$ is the intercept of province $i$; $\gamma_t$ is the time fixed effect of year $t$; $X_{it}$ is the vector formed by a number of independent variables of province $i$ in year $t$, to which we have collected three alternative variables: the GDP growth rate in year $t$, the growth rate of the added value of the secondary industry in year $t$, and the completion percentage of the energy intensity reduction by year $t$. (Specifically, that is the ratio of the province's accumulative rate in energy intensity reduction by year $t$ to the expected completion rate according to the target assigned to the province). Finally, $u_{it}$ is the residual.

In panel data analysis, the correctness of the model determines the effectiveness of the estimation. Hence, we should firstly test whether (4) has been established correctly. For the fixed group and time effect model in (4), we use the F statistics to test the following hypothesis (Bai 2008):

$$H_0^3 : \beta = 0 \quad \text{and} \quad \gamma_{2007} = \gamma_{2008} = 0$$

If the hypothesis $H_0^3$ is rejected, we can accept the model established in (4) as correct. The test is carried out through following the F test:

$$F_3 = \frac{(RRSS - URSS)/(N + T - 2)}{URSS/[(N-1)(T-1) - K + 1]} \sim F[N + T - 2, (N-1)(T-1) - K + 1] \quad (5)$$

where the RRSS is the residual sum of squares acquired through the mixed regression model, and the URSS is the residual sum of squares acquired from the regression of (4).

## 4.2 Data and Regression Results

This study uses 90 groups of data from the 30 provinces in Mainland China (not including Tibet) recorded in the first 3 years of the 11th 5-Year period from 2006 to 2008 to conduct econometric analysis. The completion percentage in energy intensity reduction by year $t$ for province $i$ is derived from dividing the accumulative energy intensity reduction by year $t$ by the expected total reduction according to the target assigned to the province. As the data for the first year cannot be calculated, it is assumed to be 100% for all provinces in 2006. In addition, the GDP growth rate and the growth rate of the added value of the secondary industry of various provinces come from *China Statistics Yearbook 2006–2009.*

By using these data to conduct the F test in (5) and the regression results in (4), we can obtain the regression coefficients and the significant level of the independent variables. We will reject the variables with the minimum t value one by one from the independent variables that are not significant at the level of 5%, in order to find out the independent variables that have significant influence on the dependent variable in (4). According to this principle, both the growth rate of the added value of the secondary industry and the GDP growth rate are rejected, leaving the variable of the accumulative completion percentage in energy intensity reduction by year $t$ as significant. It suggests that the two variables (the growth rate of the added value of the secondary industry and the GDP growth rate) do not produce significant influence on the implementation of energy conservation policy in the same year. Based on the analysis above, (4) can be written as:

$$I_{it} = C + \alpha_i + \gamma_t + \beta r_{it} + u_{it} \quad (6)$$

**Table 4** Results of the fixed group and time effect model used panel data 2006–2008

| | Estimator | Standard error | t statistics | Sig. |
|---|---|---|---|---|
| Constant | 5.801645 | 0.464527 | 12.48936 | 0.0000 |
| $\beta$ | −0.021515 | 0.005454 | −3.94506 | 0.0002 |
| Time fixed effect $\gamma_t$ | | | | |
| 2006 – C | −0.91751 | | | |
| 2007 – C | −0.31957 | | | |
| 2008 – C | 1.23708 | | | |

$R^2 = 0.8675$, adj-$R^2 = 0.7931$, D.W. = 2.1078, F = 11.6592

where $I_{it}$ represents the accumulative completion percentage of energy intensity reduction for province $i$ by year $t$. The $F_3$ statistics of model (6) is calculated through (5) to be 12.23, which is greater than the critical value at the 0.5% level of significance. Therefore, the hypothesis $H_0^3$ can be rejected and thus model (6) is deemed to be correct. The results obtained from the regression analysis in model (6) are shown in Table 4.

The estimated result of the parameter $\beta$ in Table 4 is negative, suggesting that $r_{it}$ produced significant negative feedback impact on the implementation of energy conservation policy. That is, the provinces which had fulfilled less of the energy intensity reduction targets suffered from greater pressures and intensified their efforts, which would thus increase the rate of energy intensity reduction in the following year. It can also be found from $\gamma_t$, the estimated results of the time fixed effect in each year, that $I_{it}$ increased in 2007 and 2008 as compared with previous years. As we only have 3 years to look at so far, we are still unable to identify the time fixed effect in a longer period of time. However, in view of the implementation of the energy conservation targets over past few years, we suppose that this trend in the time fixed effect was mainly incurred by the macro political environment in China. During 2007 and 2008, the central government has suffered more and more pressures to promote the energy conservation policy. These pressures include domestic factors such as the slow progress made towards energy conservation goal in 2006 and the threat of an energy supply shortage, as well as international factors like the wild price rise in oil on the international market and the pressure to mitigate carbon dioxide emissions. All these factors may form political pressures on the central government, which drive it to bring out more stringent measures and thereby result in the acceleration in energy intensity reduction of various provinces.

## 5 Conclusion

Based on empirical analysis, this paper conducted an evaluation on the operation of China's initially-established energy conservation policy framework since 2006 at the provincial level. This paper conducted static and dynamic analyses by applying the method of econometric models to analyze the implementation of energy conservation targets by the provincial governments in China. The main conclusions drawn by this study can be summed up as follows.

*Firstly, the framework of the energy conservation policy that China initially established is robust.* Starting from its own national conditions, China has adopted a framework that breaks the responsibilities down to various levels. The empirical analysis of the implementation of this policy at the provincial level shows that the provincial governments have responded positively to the instructions of the central government. Although we are not able to distinguish the difference in the extent of efforts made by different provinces, quantitative analysis shows that the obligatory targets set by the central government significantly influenced the energy intensity reduction in various provinces, and the provincial governments have been intensifying their efforts to implement the energy conservation tasks year by year.

*Secondly, the outcomes of energy intensity reduction on the provincial level are restrained by provincial conditions.* The quantitative analysis of this paper shows that variables such as GDP, GDP per capita, and initial energy intensity of each province had a significant impact on overall energy intensity reduction, which can explain, to a large extent, why there was such variation in the outcomes of energy intensity reduction among various provinces. It shows that the implementation of the energy conservation targets in various provinces not only relies on subjective efforts, but is also limited by the objective factors – the level of economic development, the resources that can be mobilized, and the initial energy intensity. However, some variables, such as economic growth rate and industrial structure, have not had a significant impact on the rate of energy intensity reduction in this study.

*Thirdly, the provincial governments have strong motivations to follow the instructions of the central government for better relative performance.* The quantitative analysis of this paper found that the energy conservation tasks fulfilled by various provinces produce pronounced impacts on subsequent implementation, and the rates of energy intensity reduction of some provinces have been annually increasing. This implies that the provincial governments are facing pressure from the central government. The provincial governments in China have attached importance to and worked hard at their energy conservation tasks, but, in essence, it is an administrative reaction to the call from the central government.

**Acknowledgments** The funding supports come from the National Science Foundation of China (70973064) and the Center for Industrial Development and Environmental Governance, School of Public Policy and Management, Tsinghua University

## References

Bai, Z (2008) Econometric analysis of panel data, the Nankai University Press. Nankai University Press (in Chinese)

Han Z-y, Wei Y-m, Fan Y (2003) Research on change features of Chinese energy intensity and economic structure. Appl Stat Manage 23(1):1–6, In Chinese

He J-k, Zhang X-l (2006) Analysis declining tendency in China's energy consumption intensity during the eleventh five-year-plan period. China Soft Sci Mag 4:33–38 (in Chinese)

Stern NH (2007) The economics of climate change: the Stern review. Cambridge University Press, Cambridge

Wang H, Yu Y-d (2009) An analysis of the interest-driven co-operation in low-carbon economy between the central government and the local governments. Paper presented at the International Symposium of Governmental Governance and Policy in the Low-Carbon Development, Beijing, Tsinghua University, 12 Sept 2009 (in Chinese)

Zhang H-b, Qi Y et al (2008) Analysis of the development and mechanisms for actions in climate change by China's local government. China Public Admin Rev 8:80–97 (in Chinese)

# Analysis on Volatility of Copper and Aluminum Futures Market of China

**Wang Shu-ping, Wang Zhen-wei, and Wu Zhen-xin**

**Abstract** The metal futures market is a typical nonlinear dynamic system. Using R/S method and FIEGARCH model, the paper study nonlinear characteristics and long-term memory of copper and aluminum futures market of China. The empirical results show that: the return series and volatility series of copper and aluminum futures have significant long-term memory, and the volatility leverage effect of copper futures is more obvious than aluminum futures. Furthermore, the copper futures prices respond vehemently to bad news. Testing find that FIEGARCH model is more suitable for the volatility analysis on copper and aluminum futures market of China.

**Keywords** FIEGARCH model · Leverage effect · Long-term memory · R/S method · Risk

## 1 Introduction

With a high development of commodity economy, future markets play a significant role in our capital markets. As a main feature of price behavior, long memory breaks through the efficient market hypothesis, and gives a new study direction for capital pricing and risk management. Therefore, using some nonlinear methods, this paper analyzes empirically the long memory in the returns and volatilities of copper and aluminum futures, which are for well understanding of the volatility behavior in Chinese futures market.

After 1970s, the definition about fractal was given by Mandelbrot (1963). Peters (1999) proposed fractal market hypothesis (FMH), the theory considers that historical information has long-term impact on market volatility. On the long memory of

W. Shu-ping (✉), W. Zhen-wei, and W. Zhen-xin
School of Economics and Management, North China University of Technology, 100144 Beijing, P.R. China
e-mail: dwangshuping@163.com; wang_zhenwei@yahoo.com.cn; wuzx1@ncut.edu.cn

Y. Zhou and D.D. Wu (eds.), *Modeling Risk Management for Resources and Environment in China*, Computational Risk Management,
DOI 10.1007/978-3-642-18387-4_48, 

futures market, Helms et al. (1984) earlier uses the R/S method to analyze soybean futures, the results indicate that there may be the existence of long memory effect. Panas (2001) applies the six LME metals futures price, using R/S, modified R/S method and ARFIMA model, the results show that the presence of long-term memory on aluminum futures.

In addition, Chinese scholars also have done a lot of work, but these work are mainly concentrated on stock markets, interest rates and exchange rates, and so on. Tang et al. (2005), Li et al. (2005) use R/S method to research yields time series of China's futures market, the results show that China's futures market yields have long-term memory. Subsequently, Li and Zou (2007) apply classic R/S and modified R/S method in the copper futures market of Shanghai as well as the soybean futures market of Dalian, and find that all of the return rate and volatility rate of futures exist characteristics of state continuity.

Currently, GARCH family models have been widely used to describe the fluctuation characteristics of stock prices, interest rates, exchange rates and other financial time series, while less deal with the futures market. Ji and Yang (2004) find that it exists significant ARCH effect and leverage effect in Chinese copper futures market, soybean, and wheat futures, but they do not consider the non-linear characteristics of time series.

There are a few literatures use R/S non-parametric statistical method to analyze the volatility of China's futures market. However, the R/S tests only tell us there is a significant long-term memory in time series from the statistics, and memory itself is not made from the long-term measure. Therefore, it is necessary to establish models reflecting the characteristics of long-term memory, such as long memory test and leverage effect. Therefore, from the perspective of model, the paper will discuss the long memory of copper and aluminum futures market of China.

## 2 Models and Methods

### *2.1 R/S Analysis*

R/S analysis was firstly proposed by British water scientist Hurst. Hurst found that the river inflow level is usually assumed to be a random sequence over several years, while the sequence has certain stability-related behaviour. Thus, he proposed a new statistic H to identify the systematic non-random features, namely Hurst index. Mandelbrot and other statisticians proved that the new statistic H is better than the traditional identification methods such as autocorrelation function, variance ratio, etc.

For a time series, when $H = 0.5$, it means that the sequence in all scales are independent for each other. If $H = 1$, it shows that the sequence are related to the characteristics of the system. If $0.5 < H < 1$, it implies that the sequence in the self-similar time scales are relevant. This is a critical fractal characteristics of the market, and it is also a characteristic of nonlinear dynamic system which is sensitive dependence on initial conditions. When 0 < H < 0.5, the sequence in each scale show the anti-persistence.

## 2.2 FIGARCH Model

As the non-linear characteristics of time series, in order to measure the characteristics of long-term memory well, this paper will use FIGARCH model to analyze the volatility of copper and aluminum futures market of China.

Baillie et al. (1996) extended IGARCH to FIGARCH model, this model can measure the sequence's long-term memory according to conditional variance. FIGARCH (p, d, q) model is defined as:

$$y_t = x_t\gamma + \varepsilon_t \tag{1}$$

$$h_t = \beta_0[1-\beta(1)]^{-1} + \left\{1-\left[1-\beta(L)^{-1}\phi(L)(1-L)^d\right]\right\}\varepsilon_t^2 \tag{2}$$

Formula (1) is the mean equation, where $y_t$ is the dependent variable, $x_t$ is the lag item of $y_t$, $\gamma$ is the parameter vector to be estimated,. Formula (2) is the conditional variance equation, where $h_t$ is the conditional variance of $\varepsilon_t$, d $\in$ [0, 1] is the fractional difference operator, it is a measure of the parameters of long-term memory, if d $\in$ (0, 1), it indicates that the sequence have long-term memory. Meanwhile, $\phi(L) = 1-\alpha(L)-\beta(L)$, $\alpha(L) = \alpha_1 L + \alpha_2 L^2 + \ldots + \alpha_q L^q$ $\beta(L) = \beta_1 L + \beta_2 L^2 + \ldots + \beta_p L^p$, where $1-\beta(L)^{-1}$, $\phi(L)$ are polynomial lag operator, respectively, their total roots are located outside the unit circle. We know that when d = 0, FIGARCH (p, d, q) is the GARCH model; when d = 1, FIGARCH (p, d, q) is IGARCH model.

## 2.3 FIEGARCH Model

As the EGARCH model reflects leverage effect of financial time series, Bollerslev and Mikkelsen (1996) further proposed FIEGARCH model, which the mean equation is similar to above FIGARCH model, but conditional variance equation is different, as follows:

$$\phi(L)(1-L)^d \ln h_t = \beta_0 + \sum\nolimits_{i=1}^{q} (\beta_i|x_{t-i}| + \gamma_i x_{t-i}) \tag{3}$$

where $x_t = \varepsilon_t/\sqrt{h_t}$. When $\gamma_i = 0$, it indicates there is no leverage effect; if $\gamma_i < 0$, it suggests that the leverage effect existing, i.e., bad news impact futures prices more intense; if $\gamma_i > 0$, then good news impact futures prices more intense. When 0 < d < 1, FIEGARCH model is stable.

# 3 Empirical Analysis

## *3.1 Sample Selection*

The paper studies the copper and aluminum futures of Shanghai Futures Exchange in China. The prices for each type of futures contracts are trading day's closing prices, the time span is from April 17, 1995 to March 31, 2010, and data are provided by the Wind System.

This method for constructing continuous futures price range is: As the time span of each futures contract is limited, the futures prices are different from the continuity feature of the stock prices. For copper and aluminum futures, each year has 12 futures contracts from January to December, while the trade is more active near delivery month. Therefore, we choose the data of the third month prior to the delivery month as the sample, in which have no data without trading. For example, data in February 2003 are selected from the futures contracts delivering in May 2003, data in May 2004 are selected from the futures contracts delivering in August 2004, and so on. The number of selecting sample for copper and aluminum are 3,836 and 3,839, respectively. The data are not only continuous, but also the futures prices and the spot prices are closer, furthermore, the data are more stable.

## *3.2 Statistical Analysis and Long-Term Memory Test*

The return rate of futures prices is defined as $R_t = \ln(S_t/S_{t-1})$, where $S_t$ is the closing price of futures contracts in period $t$.

When establishing GARCH model, it is necessary to do autocorrelation test, unit root test and ARCH effect test. ADF tests indicate that the return rate series of copper and aluminum futures are stationary. LM tests indicate that the return rate series exist ARCH effects. In addition, the autocorrelation of the return rate series is not obvious according to the ACF diagram, but the square of the return rate series is self-correlative significantly, and Ljung-Box Q statistic tests also obtain the same conclusions. These testing results show that the return rate series of two futures may exist conditional heteroscedasticity which changing with time.

In this paper, MATLAB program is used to compute the Hurst indexes of the return rate and volatility rate for copper and aluminum futures under different frequencies such as day, week and month. To measure volatility rate, we choose the most commonly used indicators: $|r_t - \bar{r}|$ and $|r_t - \bar{r}|^2$, where $r_t$ is the logarithm of the return rate $R_t$, $\bar{r}$ is the average value of the return rate in the observed interval. Here, we only list the daily output of R/S analysis. Table 1 gives the H values about the return rate series of copper and aluminum futures.

Some results can be obtained from Table 1, as follows:

**Table 1** The H values about the return rate series of copper and aluminum futures

| H values | $r_t$ | $\lvert r_t - \bar{r} \rvert$ | $\lvert r_t - \bar{r} \rvert^2$ |
|---|---|---|---|
| Copper | 0.645 | 0.756 | 0.728 |
| Aluminum | 0.606 | 0.875 | 0.806 |

1. The H indexs of the return rate series are not equal to 0.5, which suggests that the metal futures prices have fractal structure and persistence, and this characteristic is significantly different from the behavior of pure random walk, it is a biased random walk.
2. For copper and aluminum futures market, there exists significant long-term memory in the volatility of the return rate. Under two indexes which measuring the volatility of the return rate, H indexes were significantly greater than 0.5, it indicates that there exists significant long-term memory in the volatility of the return rate. This shows when establishing some models to describe the evolution of the futures market, we should consider the long-term memory in the volatility of the return rate.

## 3.3 *FIGARCH and FIEGARCH Modeling*

For the general financial time series, GARCH (1, 1) model has better fitting effect. However, there is a long-term memory in futures prices data according to above analysis. Thus, we select FIGARCH (1, d, 1) to establish model, its form can be simply expressed as:

$$\begin{cases} r_t = c + \varepsilon_t \\ h_t = \beta_0 + \beta_1 h_{t-1} + \left[1 - \beta_1 L - (1 - \phi_1 L)(1 - L)^d\right] \varepsilon_t^2 \end{cases} \tag{4}$$

where $c$ is the mean value of the return series $r_t$, $\phi_1 = \alpha_1 + \beta_1$. The corresponding FIEGARCH (1, d, 1) model may be expressed as:

$$(1 - \phi_1 L)(1 - L)^d \ln h_t = \beta_0 + \beta_1 |x_{t-i}| + \gamma_i x_{t-i} \tag{5}$$

Table 2 gives the parameters estimates and some statistics of FIGARCH (1, d, 1) and FIEGARCH (1, d, 1) model.

From Table 2, we can get the following results:

1. For Shanghai copper futures market, the fractal differential coefficient $d$ of both FIEGARCH and FIGARCH models are located between 0 and 1, furthermore, they are not zero significantly. This indicates that there is significant long-term memory in the volatility of the returns series. In addition, the leverage coefficient $\gamma_1$ of FIEGARCH model is negative, and it is significant at 1% level. This shows that the copper futures prices respond more strongly to the same degree of bad news. In FIGARCH model, the sum of the parameters GARCH(1) and ARCH(1) is equal to 1.1, slightly larger than 1. This implies that FIGARCH

**Table 2** The results of FIGARCH and FIEGARCH model

| | FIGARCH | | | FIEGARCH | | |
|---|---|---|---|---|---|---|
| | coefficient | t statistic | p value | coefficient | t statistic | p value |
| Testing results about Shanghai copper | | | | | | |
| c | $0.0008^{*}$ | 1.328 | 0.0923 | $0.0039^{*}$ | 1.47 | 0.0732 |
| $\beta_0$ | $9.36e{-6}^{**}$ | 1.962 | 0.0251 | $-0.326^{**}$ | −1.69 | 0.046 |
| GARCH(1) | $0.8^{***}$ | 6.15 | 0 | $0.734^{***}$ | 0.449 | 0 |
| ARCH(1) | $0.3^{***}$ | 3.177 | 0.0001 | $0.1390^{***}$ | 2.88 | 0.002 |
| $\gamma_1$ | – | – | – | $-0.0121^{***}$ | −0.717 | 0.2368 |
| *d* | $0.5^{***}$ | 3.296 | 0.0005 | $5.20e{-8}^{**}$ | 2.75E−07 | 0.5 |
| Testing results about Shanghai aluminum | | | | | | |
| c | 3.05E−06 | −0.1161 | 0.4538 | 0.0012 | 1.0682 | 0.1349 |
| $\beta_0$ | 3.50E−07 | 0.7322 | 0.2321 | $-0.3057^{***}$ | −7.5686 | 0 |
| GARCH(1) | $0.8^{***}$ | 11.8991 | 0 | 0.0364 | 0.4137 | 0.396 |
| ARCH(1) | $0.3^{***}$ | 5.0444 | 0 | $0.4035^{***}$ | 7.3686 | 0 |
| $\gamma_1$ | – | – | – | 0.1536 | 4.7058 | 0 |
| *d* | $0.5^{***}$ | 6.6897 | 0 | $0.6458^{***}$ | 16.4532 | 0 |

$^{*}$denotes significance level at 10%, $^{**}$indicates significance level at 5%, $^{***}$indicates significance level at 1%

model has certain instability. In FIEGARCH model, the sum of GARCH(1) and ARCH(1) is slightly less than 1, and the model is stable.

2. For Shanghai aluminum futures market, the fractal differential coefficient ($d = 0.5$) of FIGARCH model is located between 0 and 1, and it is significant at 1% level. This indicates that there is significant long-term memory in the volatility of the returns series for Shanghai aluminum futures market. Furthermore, in FIGARCH model, the sum of the coefficients GARCH(1) and ARCH (1) is slightly larger than 1, which implies the model has some instability. In FIEGARCH model, the sum of GARCH(1) and ARCH(1) is approximately equal to 0.5, less than 1, which implies the model is very stable. At the same time, $d= 0.6458$ is still located between 0 and 1, and it is significant at 1%level. This suggests that the long memory is still significant. The leverage coefficient $\gamma_1 = 0.1536 > 0$, but it is not significant at 10% level. This indicates that there is not obvious leverage effect in Shanghai aluminium futures prices, which further implies that futures prices respond to the positive and negative information symmetrically.

## 4 Conclusions

The empirical analysis shows that there are significant features of nonlinear dynamics in the yields time series of Shanghai metal futures. The futures prices have asymmetry, hysteresis and other nonlinear behaviour when reacting to market information. Specifically, there is significant long-term memory in the volatility of copper and aluminum futures prices. However, the leverage effect is not obvious for

Shanghai aluminum futures, and the copper futures prices respond more strongly to the same degree of bad news, this may be due to the reason that the domestic futures market is vulnerable by the bad news of the foreign markets. Meanwhile, we find that FIEGARCH model is more suitable for the volatility analysis on copper and aluminum futures market of China.

For futures investors, especially institutional investors, the presence of long memory may indicate that the movement trend of futures price may be predictable to some extent in a certain non-cyclical period.

**Acknowledgments** This research is supported by the Humanities and Social Sciences Research Youth Project of Ministry of Education (08JC790004), and the Special Fund of Subject and Graduate Education of Beijing Municipal Education Commission (PXM2010_014212_093659).

## References

Baillie RT, Bollerslev T, Millelsen HO (1996) Fractionally integrated generalized autoregressive conditional heteroscedasticity. J Econometrics 74:3–30

Bollerslev T, Mikkelsen H (1996) Modeling and pricing long memory in stock market volatility. J Econometrics 73:151–184

Helms BP, Kaen FR, Rosenman RE (1984) Memory in commodity futures contracts. J Futures Mark 10:559–567

Ji Guangpo, Yang Junhong (2004) An empirical study on autoregressive conditional heteroscedasticity effect in China's futures market. Econ Rev 5:100–103 (in Chinese)

Li Jiang, Zou Kai (2007) The empirical study on fractal structure of China's futures market. Zhejiang Finance 8:38–39 (in Chinese)

Li Yan, Qi Zhongying, Niu Hongyuan (2005) R/S analysis of time series of copper futures prices of Shanghai futures exchange. J Manage Sci 18:87–92 (in Chinese)

Mandelbrot BB (1963) The variation of certain speculative prices. J Business 36:394–419

Panas E (2001) Long memory and chaotic models of prices on the London Metal Exchange. Resour Policy 27:23–246

Peters EE (1999) Chaos and order in the capital markets. Economic Science Press, Beijing

Tang Yanwei, Chen Gang, Zhang Chenhong (2005) An empirical research on the long-term correlation of the price volatility of the agricultural products futures markets. Syst Eng 23:79–84 (in Chinese)

# The Evaluation of Hydraulic Engineering Scheme Based on Choquet Fuzzy Integral

**Chen Ling and Ren Zheng**

**Abstract** It is often difficult to establish an indicator system to evaluate the hydraulic engineering scheme where the indicators are mutually independent and the project attributes can be revealed comprehensively. In the paper, firstly, probabilistic measures are calculated by diversity between indicators from different and uniform hydraulic engineering scheme based on information entropy and variable weights. Secondly, optimal model of fuzzy measure is built by means of Shapley value definition of multi-people collaborative gambles and Marichal entropy theory, thus the probabilistic measure can be converted to fuzzy measures. Thirdly, on the basis of Choquet integral definition, synthetical evaluation of alternative schemes is calculated according to the known value from bottom to top. The demonstration shows that the method is feasible to array the order of hydraulic engineering scheme, and that computational complexity obviously increases with increasing indictor numbers and application scope of the method will be greatly widened with the improvement of the optimal algorithm.

**Keywords** Choquet fuzzy integral · Hydraulic engineering · Marichal entropy · Variable weights

## 1 Introduction

The purpose of hydraulic engineering is to achieve economic, social, ecological and environmental efficiencies, which is also the ultimate results of the utilization of hydraulic engineering. The evaluation of hydraulic engineering scheme is essential

C. Ling (✉)
College of Hydraulics and Electric Power, Hebei University of Engineering, China
and
College of Economy and Trade, Shihezi University, China

R. Zheng
College of Hydraulics and Electric Power, Hebei University of Engineering, China

Y. Zhou and D.D. Wu (eds.), *Modeling Risk Management for Resources and Environment in China*, Computational Risk Management,
DOI 10.1007/978-3-642-18387-4_49, 

to the choice of the best one relative to others under some indicator system (Ding et al. 2007). It is expected that it will exert considerable effect on the promotion of sustainable utilization of water resources, as well sustainable development of society and economy. Establishing indicator system is the basic premise of evaluation of the hydraulic engineering scheme. However, indicator system that can reflect comprehensive capability of hydraulic engineering and is mutually absolute to different indicator is considerably difficult. For example, high economical return usually attributes to high investment. Therefore, how to deal with these relationships among different indicators directly affects rationality and validity of the choice of the hydraulic engineering scheme. Recent studies are mainly focused on the choice of probabilistic measure weights (Yang et al. 2005; Xue et al. 2005; Nie and Lu 2006), while there are few reports about mutual relationships treatment from indicator system of hydraulic engineering.

With regard to mutual relationships treatment from indicator system, some scholars pay much attention. Sugeno reported the concept of fuzzy measure and considered the destruction of additional rule of indicator weights by mutual relationships. Subsequently, some scholars addressed the concept of $\lambda$ fuzzy measure that would eliminate the handicap of collecting a great deal of resources and reducing the calculational complexity because $2^n$–2 fuzzy measures usually require $n$ indicators. Murofushi and Sugeno introduced the concept of Choquet $\lambda$ fuzzy integral through contacting $\lambda$ fuzzy measure with Choquet integral (Murofushi and Sugeno 1989) .The Choquet $\lambda$ fuzzy integral has been widely applied to risk assessment system (Zhang et al. 2007), handwritten word recognition (Gader et al. 1996), multi-information fusion (Guan and Feng 2004) and route choice of communication technology (Liu et al. 2003), etc.

The purpose of this paper is to apply such method to evaluate hydraulic engineering scheme. The key is to confirm $\lambda$ fuzzy measure. In Sect. 2, basic theories that contribute to the paper are introduced. Fuzzy measure, Choquet integral, determining probabilistic measure weights and fuzzy densities are reviewed. In Sect. 3, an application of hydraulic engineering scheme evaluation is explained. In Sect. 4, the conclusion and further research are given.

## 2 Basic Theories

### *2.1 Fuzzy Measure and Choquet Integral (Sandanee et al. 2002)*

Before we discuss the application of the Choquet fuzzy integral, we would first review the regular fuzzy integral. The fuzzy integral relies on the concept of a fuzzy measure that generalizes the concept of a probabilistic measure. A fuzzy measure over a set $X$ is a function:

$g : 2^X \to [0, 1]$, such that: (1) $g(X) = 1$, $g(\Phi) = 0$; (2) $g(B) \geq g(A)$ if $B \supset A$; (3) if ... $\supset A_m \supset ... \supset A_1$ then $\lim\big(g(A_j)\big) = g\big(\cup A_j\big)$.

A fuzzy measure $g_\lambda$ is called a Sugeno measure if it satisfies the following additional properties for some $\lambda > -1$:

If $A \cap B = \Phi$, then $g_\lambda(A \cup B) = g_\lambda(A) + g_\lambda(B) + \lambda g_\lambda(A) g_\lambda(B)$. If $\lambda = 0$, then $g$ is a probabilistic measure. Let $X$ be a finite set of information sources $X = (x_1, x_2, \ldots, x_m)$ and define $g^j = g_\lambda\left(\left\{x_j\right\}\right)$ (fuzzy densities). The fuzzy density defines the importance of individual information sources. If we know the fuzzy densities, the value of $\lambda$ can be found by:

$$\lambda + 1 = \prod_{j=1}^{m} \left(1 + \lambda g^j\right) \tag{1}$$

Let $h$: $X \to [0, 1]$ be a confidence function, i.e., $h(x_j)$ is the confidence provided by the source $x_j$ that an input sample is from a particular class. The Choquet integral of $h$ over $X$ with respect to a fuzzy measure g is defined by:

$$\int_c h \circ g = \sum_{j=1}^{m} g(X_j) \left[h(x_j) - h(x_{j+1})\right] \tag{2}$$

where $h(x_{m+1}) = 0$, $g(X_0) = 0$, $h(x_1) \geq h(x_2) \geq \ldots \geq h(x_m)$ and $X_j = \left\{x_1, x_2, \ldots, x_j\right\}$. For particular values of the measure, the Choquet integral can implement all linear combinations of order statistics, as well many general forms of combination. This numeric form of the Choquet integral has been widely used in combining feature and algorithm confidence values.

## 2.2 *Determining Probabilistic Measure Weights*

The criteria for the evaluation can be classified into a positive efficacy (the sample is better with an increase in the indicator, such as guarantee rate of water supply) and a negative efficacy (the sample is better with a decrease in the indicator, such as inundation loss) depending upon its attributes. In order to make the result more accessible, an improved efficacy function was adopted, which is widely used in related studies (Li et al. 2006).

To avoid subjective opinion, the model is built from the transformed value where the entropy weight coefficients are determined by Shannon entropy theory. The specific procedure is as follows (Zhang et al. 2005).The average weighted model is applied to integrate different transformed values, which will result in a wrong decision due to ignoring mutual information of indicator of the same scheme. Therefore, the concept of variable weights can be used further to retrieve discrete information from different indicators of uniform schemes. The specific procedure is as follows (Li and Li 2004). The key to calculate variable weights is to choose the vector of state variable weights.

To take fully the advantage of single indicator into account, we will give more weights to these indicators in possession of higher transformed value. Therefore, the element of vector can be defined as follows.

$$S_j(X) = (a_j(X))/\left(\sum_{k=1}^{m} a_k(X)\right) \tag{3}$$

## 2.3 Determining Fuzzy Densities

Probabilistic measure weights have described importance of different among both indicators and schemes. However, mutual relationships of different indicators remain poorly understood. The following natural question arises: how to bridge between the probabilistic measure and the fuzzy measure. Referring to multi-persons cooperative game theory, the relation can be described in virtue of Shapley value definition. Therefore, the above relation can be defined as follows (Kelly 2007; Lu and Wu 2006).

$$w_j(X) = \sum_{t=0}^{m-1} \frac{(m-t-1)!\,t!}{m!} \sum_{T \subseteq X \backslash x_i, |T|=t} \left(g(T \cup x_j) - g(T)\right) \tag{4}$$

where $t$ is potential of indicator set. $w_j(X) = g^j$ if all indicators are mutually absolute.

While probabilistic measure weights $w_j$ are clear, $m$ equations similar to (4) can be built. According to Sect. 2.1, equation group contains $m+1$ unknown variables. Therefore, there is uncertainty in calculating fuzzy densities when probabilistic measure weights are known. The outcome is from infinite solution using (4). For general Choquet capacities, it seems that no definition of generalized entropy was available until recently when three proposals were introduced successively by Marichal (2002), Yager (1994) and Dukhovny (2002). All three proposals can be regarded as direct extensions of the Shannon entropy since they coincide with the latter when the capacity is additive. The generalized entropy proposed by Marichal is defined by:

$$\max_{\lambda,g} H_M(g) = \sum_{j=1}^{m} \sum_{S \subseteq X \backslash x_j, |S|=s} \gamma_s(m) \cdot h\left[g(S \cup x_j) - g(S)\right] \tag{5}$$

$s \cdot t \cdot$

$$\begin{cases} w_j(X) = \sum_{t=0}^{m-1} \frac{(m-t-1)!\,t!}{m!} \sum_{T \subseteq X \backslash x_j, |T|=t} \left(g(T \cup x_j) - g(T)\right) \\ \lambda + 1 = \prod_{j=1}^{m} (1 + \lambda g^j) \\ g(A \cup B) = g(A) + g(B) + \lambda g(A) g(B) \\ g(X) = 1 \end{cases}$$

# 3 Empirical Analyses

## 3.1 Evaluation Indicator

An indicator is a measure of evaluation for hydraulic engineering efficiency. The following principles should be observed when choosing indicators: (1) scientific: indicators and their choices must be scientifically based; (2) comprehensive: the efficiency should be comprehensively and objectively represented; (3) comparable: the meaning of indicators should be as uniform as possible; (4) obtainable: the indicator data should be accessible and credible; (5) accurate: the efficiency should be clear and exactly reflected; (6) dynamic: indicators should be able to reflect future prospects and accommodate changes.

We established 14 evaluation indicators for reservoir use efficiency from the aspect of total engineering investment, water supply benefit, inundation control efficiency, ecological benefit, power generation benefit, social influence and technical difficulty according to the principles listed. The evaluation indicator system and hierarchical structure are listed in Fig. 1.

## 3.2 Application Process of the Method

The indicator data of different schemes of planning the reservoir are listed in Table 1.

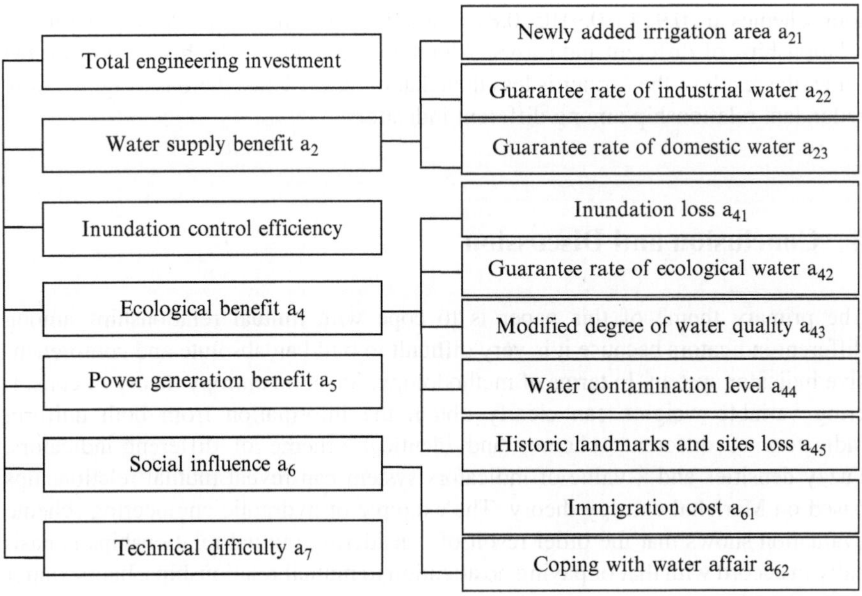


**Fig. 1** Evaluation indicator system and hierarchical structure of planning reservoir

**Table 1** Bottom layer indicator data of different scheme of planning reservoir

| Sch | $a_1$ | $a_{21}$ | $a_{22}$ | $a_{23}$ | $a_3$ | $a_{41}$ | $a_{42}$ |
|---|---|---|---|---|---|---|---|
| $u_1$ | 26.907 | 1.67 | 95 | 95 | 0.92 | 1.637 | 100 |
| $u_2$ | 33.568 | 2.02 | 95 | 100 | 1.03 | 2.056 | 95 |
| $u_3$ | 26.214 | 1.36 | 90 | 90 | 0.92 | 1.637 | 80 |
| $u_4$ | 20.221 | 1.07 | 70 | 75 | 0.55 | 0.858 | 70 |
| Sch | $a_{43}$ | $a_{44}$ | $a_{45}$ | $a_5$ | $a_{61}$ | $a_{62}$ | $a_7$ |
| $u_1$ | 0.5 | 0.2 | 4,200 | 0.9575 | 13.269 | 1 | 0.3 |
| $u_2$ | 0.5 | 0.5 | 4,800 | 0.9622 | 18.224 | 1 | 0.5 |
| $u_3$ | 0.3 | 0.2 | 4,200 | 0.9000 | 13.260 | 0.5 | 0.3 |
| $u_4$ | 0.2 | 0.2 | 4,000 | 0.8600 | 8.569 | 0.2 | 0.1 |

**Table 2** Fuzzy densities and $\lambda$ value of different schemes of planning reservoir

| Sch | $g\{a_1\}$ | $g\{a_2\}$ | $g\{a_3\}$ | $g\{a_4\}$ | $g\{a_5\}$ | $g\{a_6\}$ | $g\{a_7\}$ | $\lambda$ |
|---|---|---|---|---|---|---|---|---|
| $u_1$ | 0.054 | 0.068 | 0.067 | 0.110 | 0.090 | 0.070 | 0.016 | 3.713 |
| $u_2$ | 0.058 | 0.099 | 0.096 | 0.135 | 0.12 | 0.102 | 0.017 | 1.777 |
| $u_3$ | 0.072 | 0.084 | 0.100 | 0.140 | 0.118 | 0.062 | 0.038 | 1.881 |
| $u_4$ | 0.083 | 0.048 | 0.033 | 0.100 | 0.080 | 0.075 | 0.078 | 3.291 |

According to According to Sect. 2.2,we can calculate probabilistic measure weights of these indicators from bottom to top .Then, fuzzy densities and $\lambda$ value of these indicators can be obtained by Eq. (8),and these results are shown in Table 2.

In the same way, according to these values of Table 2 and definition of Choquet integral, compositive evaluation values of four schemes are 0.8822, 0.8701, 0.7729 and 0.8345. The optimal scheme is the first scheme ($u_1$). The results of evaluation of four schemes are 0.9143, 0.9105, 0.8151 and 0.8889 without considering the mutual relationships of different indicators. Although the orders in both methods are the same, the result of the former is less than that of latter. Thus, there is a significantly redundant relationship among different indicators.

## 4 Conclusion and Discussion

The primary theme of this paper is to cope with mutual relationships among different indicators because it is very difficult to build an absolute and comprehensive indicator system. In terms of methodology, Shannon entropy theory of considering variable weights can clearly obtain the information from both uniform indicators of different schemes and identical scheme of different indicators. Fuzzy densities and $\lambda$ value of indicators system can reveal mutual relationships based on Marichal entropy theory. The example of hydraulic engineering scheme evaluation shows that the order result of considering mutual relationships is basically in accord with that of paying no attention to mutual relationships, but the latter is usually higher. So, the method of the paper can reflect the diversity of different schemes, which will drive a broader application for the method. However, the process of the example synchronously shows that the complexity of calculation is swiftly rising with the dimension of evaluation indicators, especially in Eq. (8). Therefore, the application of advanced optimal arithmetic will further promote the scope of the method.

## References


Ding Y, Liang CY, Fang BH (2007) Application of multi-reservoir flood control systems based on evidence theory [J]. Adv Water Sci 18(4):591–597

Dukhovny AD (2002) General entropy of general measures. Int J Uncertainty, Fuzziness Knowledge-Based Syst 10(3):213–225

Gader PD, Monhanmed MA, Keller JM (1996) Dynamic-programming-based handwritten word recognition using the Choquet fuzzy integral as the match function. J Electronic Imag 5(1): 15–24

Guan T, Feng BQ (2004) Roughness of Choquet fuzzy integral and information fusion. J Xi'an Jiaotong Univ 38(12):1251–125

Kelly A (2007) Decision making using game theory: an introduction for managers. Peking University Press, Beijing

Li DQ, Li HX (2004) Analysis of variable weights effect and selection of appropriate state variable weights vector in decision making. Cont Decision 19(11):1241–1245

Li HL, Wang XG, Cui YL et al (2006) Comprehensive evaluation methods for irrigation district. Adv Water Sci 17(4):543–548

Liu YX, Li X, Zhuang ZW (2003) Decision-level information fusion for target recognition based on Choquet fuzzy integral. J Electronics Inform Technol 25(5):695–699

Lu YY, Wu XQ (2006) Evaluation for data fusion system based on generalized relative entropy. J Syst Simulation 18(5):1283–1285

Marichal JL (2002) Entropy of discrete Choquet capacities. Eur J Oper Res 137(3):612–624

Murofushi T, Sugeno M (1989) An interpretation of fuzzy measure and the Choquet integral as an integral with respect to a fuzzy measure. Fuzzy Sets Syst 29(2):201–227

Nie XT, Lu YW (2006) Water conservancy project construction scheme selection based on fuzzy decision making [J]. Water Resour Power 24(3):46–48

Sandanee AW, Keller JM, Paul D (2002) Generalized Choquet fuzzy integral fusion. Inform Fusion 3(7):69–85

Xue CS, Jin JL, Wei YM (2005) Projection pursuit method for optimizing water resources project schemes. J Yangtze River Scient Res Inst 22(4):80–83

Yager RR (1994) A class of fuzzy measures generated from a Dempster-Shafer belief structure. Int J Intelligent Syst 14(12):1239–1247

Yang HJ, Li N, Du ZX (2005) The ideal scheme method applied in the choice of water project. J N China Inst Water Conservancy Hydroelectric Power 26(4):66–68

Zhang XQ, Liang C, Liu HQ (2005) Application of attribute recognition model based on coefficient of entropy to comprehensive evaluation of groundwater quality. J Sichuan Univ 37(3): 28–31

Zhang CY, Wang ZF, Xing HG (2007) Risk assessment system for bidding of construction projects based on Choquet fuzzy integral. China Civil Eng J 40(10):98–104

# Early-Warning Framework of China's Energy Security

**Zhang Minghui, Song Xuefeng, and Li Yongfeng**

**Abstract** Rapid development of society and economy is inseparable from the support of energy. The contradictions between China's energy supply and energy demand of socio-economic development and between energy consumption and environment capacity become increasingly prominent, determined by China's energy occurrence conditions and energy consumption characteristics. Energy is a time-space category. It relates to energy supply and demand, energy occurrence, environment capacity and so on. In this paper based on the comprehensive analysis of energy security implication, the factors that affect China's energy security are analyzed; energy security evaluation index system and energy security early-warning framework suitable to our country is established. Measures to guarantee the effective implementation of the early-warning system are put forward. This research offers important foundation for enforcing energy forecasting and early-warning. In light of timely prediction of the contradiction between energy supply and demand, energy crisis can be abated so that promoting the scientific development of social-economy.

**Keywords** Early-warning · Energy security · Risk · Time-space

## 1 Introduction

Energy is the original driving force of social economic development of human being, and the material base for the survival of human race. However, energy brings great benefit to human, such as economic development, scientific and technological

---

Sponsored by national natural science foundation of China (number: 70971129)

Z. Minghui (✉) and S. Xuefeng
School of Management, China University of Mining and Technology, Xuzhou Jiangsu 221116, China
e-mail: zhangmh02@163.com

L. Yongfeng
Jiangsu Key Laboratory of Resources and Environmental Information Engineering, Xuzhou Jiangsu 221116, China

Y. Zhou and D.D. Wu (eds.), *Modeling Risk Management for Resources and Environment in China*, Computational Risk Management,
DOI 10.1007/978-3-642-18387-4_50, 

progress and so on. At the same time, energy brings series inevitable energy security problems that threaten the survival and development of human race, such as energy shortage, scramble for resources, environmental pollutions caused by overuse of energy and so on. Especially since Industrial Revolution, the amount of energy consumption increases rapidly, energy security issues become increasingly prominent due to the limited one-time energy resources and the regional imbalance of energy occurrence. Currently, energy security has risen to the height of the national strategy. Energy security problems of one country are not only economic problems, but also political and military issues (Jiang 2008; Research Group of China National Energy Development Strategy and Policy Analysis 2004). With the acceleration of economic globalization, nations have set up energy policies taken energy supply security as the core. In order to ensure energy security and stable social economic development, it is necessary to establish energy security early warning system (He 2009; Zhang 2009).

## *1.1 Energy Security and Energy Early Warning*

Energy security refers to the state or ability of a country or region to obtain energy sustainable, steadily, timely, sufficiently and economically (Li and Liu 2009; Li and Lu 2009), including energy economic security and energy ecological environment security. Energy security should cover energy occurrence, energy development, energy supply, energy consumption and energy reserves (see Fig. 1). There are mutual effects between energy security and other factors that affect nation security.

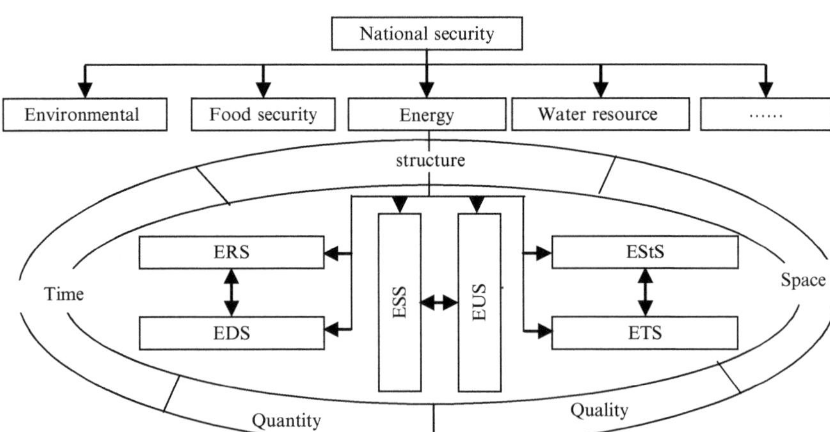


**Fig. 1** Energy security framework. ERS: Energy reserves security. EDS: Energy development security. ESS: Energy supply security. EUS: Energy utilization security. EStS: Energy storage security. ETS: Energy transportation security

Therefore firstly, energy security refers to in any case the energy need of social economic development and people's life should be ensured. Secondly, it means that in all kinds of emergencies effective energy supply should be ensured. At the same time, from long-term perspective, the current energy supply should be guaranteed, and the future energy requirement for social economic development should be guaranteed as well. In addition, energy security also includes ecological security of energy development and utilization, that is to ensure that energy developing and utilizing does not beyond the capacity of regional ecological environment, does not threaten the environment in which human race survive and develop. So that to keep sustainable development of the society.

Obviously, energy security has become an important part of national security. It will not only affect the stability and development of social economy, and will lead to other crisis. It also put accelerating effect on the social economic development. With the rapid development of China's social economy and the accelerated globalization process, the influence of energy reinforces continuously and the meaning of energy security is constantly enriched (Diao 2009; Sun and Pan 2009).

Energy supply security is only a fundamental part of energy security. Comprehensive energy security is the objective requirement of the sustainable development of social economy. Energy plays a fundamental role in the complex social economic system. All aspects of energy reserves, energy supply, energy consumption and environmental protection not only promote each other but also affect one another. The relationship between them must be clarified and managed to ensure the energy security. Therefore combined with social economic development trend, to forecast energy supply and demand according to the relationship between social economic development with energy, and to carry out energy early-warning based on energy occurrence, energy reserves and environmental capacity is conducive to foresee the energy crisis and resolve it, thereby overcoming the constraints on the scientific development from energy insecurity.

## 1.2 Present Energy Early Warning Research in China

Early warning is to send out emergency signal to the relevant department before the crisis comes, based on the law summarized in the past or the warning signs observed. So that to avoid the crisis occurring under the condition that people be unwitting or inadequately prepared, and in this way reduce the loss caused by crisis to the greatest degree (Kang et al. 2004; Nikander and Eloranta 2001; Huang et al. 2003).

After the 1973 energy crisis, people realized the importance of energy issues. The energy situation and its future trends was more and more concerned by all of the world, relevant energy policies were developed, energy crisis and energy early warning studies were carried out. Especially the United States, Japan and other

Western countries, attaches great importance to energy early warning research, and established corresponding energy information monitoring and analyzing agencies. After its establishment, International Energy Agency (IEA) has done a lot of energy forecasting and energy early warning work. The world energy early warning system established by IEA plays a major role in helping member countries to meet the energy crisis. For example in January 17, 1991, IEA launched a preventive contingency plan to cope with the possible shortage of energy supply caused by the Gulf crisis. The move guaranteed that there was no shortage of oil supply before and after the Gulf War and the oil price volatility period was also very short.

In China energy early warning studies are mainly focused on oil, coal and electricity. And the studies give more regard to early warning methods (Chi 2006; Li 2007). The Energy Research Institute of National Development and Reform Commission developed a model for China's energy and environment comprehensive evaluation. This model can be used to predict energy demand. However, there is no one comprehensive energy forecasting and early warning system. This is not conducive to understanding energy security situation and is also difficult to provide basic support for energy strategies and policies development. Therefore it is necessary to establish energy early warning system suitable to our country, according to China's social economic development trends, combined with the characteristics of China's energy structure.

If you have more than one surname, please make sure that the Volume Editor knows how you are to be listed in the author index.

# 2 Energy Supply and Demand Status in China

## *2.1 Energy Status in China*

Since reform and opening, China's society and economy has developed considerably. However, the initial social economic development was at the expense of high energy consumption and serious environmental impact. So energy supply gap increased and environmental burden nearly overwhelmed. Until the past few years economic development model changes gradually from the extensive to intensive. But energy security is increasingly prominent (Zhang 2009).

1. Energy resource reserves drop sharply; the difficulty of resources development is greater and greater. With the increase of energy demand and advancement of technology, energy resources development intensity increase sharply and recoverable reserves decrease rapidly. The contradiction between development and reserves become conspicuous. Take coal resources for example, coal geological exploration degree is low. In proved reserves, the precise reserves are less than 15%. In precise reserves, 68% has been occupied by productive mine and construction mine. The remaining 38%, restricted by

mining conditions, environmental capacity, traffic and other factors, there are only about 300 million tons of reserves can be developed under present technology condition.

2. The contradiction between energy supply and demand is long-existing and becomes increasingly intense. Our country is in the period of accelerating the industrialization and urbanization. So energy consumption intensity is great. Energy demand is increasing and the gap between energy supply and demand is widening. With the expansion of the difference between energy supply and demand, dependency of domestic energy consumption on import expands rapidly. The energy security situation could not be neglected.
3. Energy consumption structure contradiction is serious.

   Although with the social and economic development, China's energy consumption structure is improving, clean and renewable energy consumption is also increasing. However, coal resource reserve is rich in China, which determines the leading position of coal in primary energy consumption. The consumption of coal accounts for about 60% of primary energy structure, which is far higher than the international average level 24.3%. According to China's energy development plan, even by 2020, the share of coal in energy consumption will reach 55%. Therefore, the share of coal in energy consumption in China is equal to the proportion of oil and natural gas in the world. While the share of oil and natural gas in energy consumption in China is equal to the proportion of coal in the world.
4. Energy use efficiency is low and the contradiction between the development and use of energy and environmental protection becomes increasingly sharp.

   Due to the impact of technology, energy consumption of per unit GDP in China is about 2.2 times higher than the average level in the world. Energy output efficiency in China is far lower than the international advanced level. Compare the level of energy use in China with the international advanced level, there is still a big gap. Energy technologies especially energy exploration and utilization technology, ultra high voltage transmission technology, renewable energy, clean energy and alternative energy technologies are lagging behind. Energy efficiency coefficient in China is only about 10%, less than half of the developed countries. In the process of energy processing, transportation, storage and end use, the energy loss and waste is serious and about 90% of the energy is no longer effective.

   The development and utilization of energy has both positive and negative effects on the economic development. On one hand it ensures the economic development. On the other hand it puts negative impact on the environment. In China the development and utilization of energy has caused serious impact on the eco-environment, due to the objective energy resource occurrence structure and low energy efficiency. China is facing the great pressure in fulfilling international convention. Energy development and utilization security is an inevitable problem.
5. The dependency of domestic oil consumption on import is rising.

In the recent 20 years, the annual growth rate of China's oil consumption is 6.3%, which is obviously higher than the annual growth rate of China's oil production. The data from National Bureau of Statistics show that China imported crude oil 118,750 thousand tons and finished oil 17,420 thousand tons in 2005. With the rise of the per-capita income and the popularity the of cars, oil consumption will continue to increase significantly. But the occurrence of domestic oil is limited, more than half of the oil will be dependent on international resources, which will lead to serious oil security problems.

## 2.2 The Main Factors That Affect the China's Energy Supply and Demand Security

Energy security itself is a systematic and complex project, and the factors affecting energy security are more complex (Li and Liu 2009). There are mainly seven interactive factors affecting energy supply and demand security (see Fig. 2)

1. Energy resources factor. The objective occurrence of energy resources is the primary factor affecting energy security. The abundance degree of energy resources is directly proportional to the support degree of energy to social economic development. Therefore energy reserves security is the foundation of energy security.
2. Economic factors. The influences of economy on energy security are shown in two aspects. On one hand, rapid economic development requires the guarantee of adequate energy. On the other hand, the enhancement of economic strength can provide reliable funds for energy resource development, new energy development and technology improvement, thus to promote the rational development and utilization of energy. The cyclical fluctuation of energy supply and demand is conformable to the business cycle (Hu and Wang 2006).

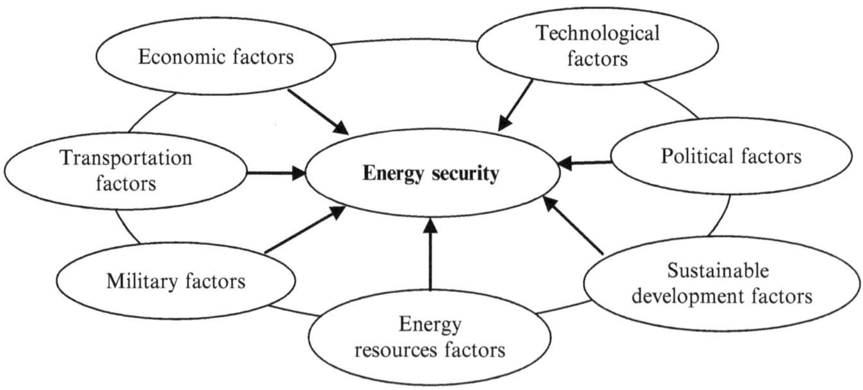


**Fig. 2** Factors that affect energy security

3. Technological factors. Energy is the power source of social economic development, but the energy resources are limited, especially the one-time fossil energy reserves. Therefore social and economic development must be based on the progress of science and technology. All the energy activities, such as energy resource exploration, energy development and utilization, energy reserve, energy structure readjustment, alternate energy development and utilization and so on, depend on the development and use of science and technology. Therefore science and technology is the most effective means in solving the energy problem fundamentally.
4. Political factors. Political factors mainly influence the recreations of energy. In the process of developing international energy market and making full use of the international energy, domestic politics not only affect the relations between trading nations, but also affect the domestic energy supply and demand. For instance the oil development in oil-rich Middle East countries is closely related to their internal political factors.
5. Transportation factors. The occurrence of energy resources is independent of human consciousness. So the space variation of resource occurrence inevitably leads to the space variation of energy development and utilization. The spatial transfer of energy must rely on energy transportation. Of course, the transport distance and transportation mean directly affect the energy transportation security, thereby affecting energy supply security.
6. Military factors. In China, with the exploring of international market, the use of international energy gradually increased, especially oil. China's oil resources are far from enough to meet its oil demand, so a large quantity of imported oil need long distance transportation. Strong, rapid reaction military forces can protect the energy transportation and carry on military intervention to the main energy production base, so that ensure energy supply effectively.
7. Sustainable development and other factors. To develop and utilize energy must take the environment carrying capacity into consideration. Under no circumstances should we seek temporary energy security at the expense of the environment. It is objective requirement of the coordinated development of the population, resources and environment to execute energy green development and low carbon use.

## 3 Early Warning Framework of China's Energy Security

China's energy security possesses its own characteristics determined by factors of rapid and sustained development of China's economy, energy occurrence feature, structure of energy development and utilization and its geopolitical relations. It need take all the factors that affect energy security into consideration to establish comprehensive energy early warning system conforming to the actual conditions of our country from time, space and structure perspectives.

### 3.1 Early Warning Framework of China's Energy Security

Energy security is a space-time category determined by its connotation. Therefore energy security early warning framework suitable to the reality of China should be a multi-dimensional system. From the time dimension, it includes short-term and long term energy early warning. From the space dimension, it includes energy early warning of different regions. From the structure dimension, it includes energy supply and consumption structure early warning. This system combines energy early warning with the development of national economy, and can carry out all directional energy security early warning according to the actual situation of social economy and energy. Thereby ensuring the energy demand of social economic development and promoting coordinated and scientific development of social economic (see Fig. 3).

Energy early warning system is an open, adaptive and dynamic system. In this system information is the basis, and it includes the relative energy information and social economic information, environmental carrying capacity information, technology information and other information of relative factors that affect energy security. Energy security assessment is the direct basis of energy early warning, and with the social economic development energy security situation should be evaluated dynamically to ensure the reality of the warning situation information. The theory system and method system of early warning is the key of energy early warning. The accuracy and timeliness of early warning results depend on the scientific of early warning theory and method. Therefore, early warning method system should be established according to the specific conditions from time, space and structure dimensions. From the time dimension, take short-term and long term

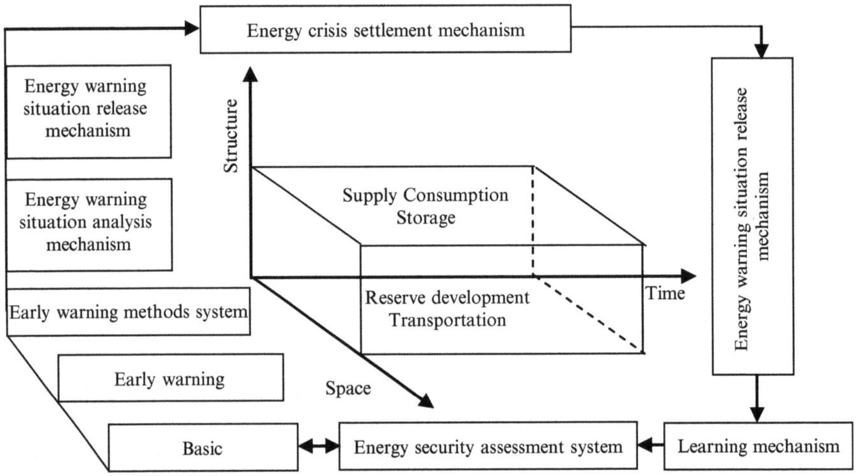


**Fig. 3** Energy early warning framework

emery early warning into consideration. From the space dimension, take the space distribution of energy occurrence, energy supply and energy consumption into consideration. From the structure dimension, take energy supply and consumption structure into consideration. Energy warning situation analysis and release is the process and window of energy early warning system. To ensure the completeness of energy security analysis, warning situation analysis begins with unidimension early warning analysis first and then carry out multidimensional comprehensive analysis. At the same time it is necessary to establish perfect energy warning situation release mechanism, to ensure the standardization and publicity of energy information so as to meet the needs of different policy-makers. Energy crisis settlement mechanism is the realization mechanism of energy early warning goal and the terminal requirement of energy early warning mechanism as well. Energy crisis settlement is to take effective measures to ensure energy security and promote the scientific development of social economy in advance according to energy early warning information. Energy early warning feedback mechanism is the guarantee of energy early warning system's self-improvement and self-adaptation to the development of social economy and technology. It is ensured that energy early warning system responds with the time through learning mechanism.

Therefore, energy early warning system is a complex adaptive system. It should be consistent with the factors of energy security, and can be used to carry out energy security early warning under different time, space and structure conditions.

## 3.2 Energy Security Early Warning Process

Energy security early warning should be timely and accurately. This is the guarantee for resolving energy crisis. Therefore energy security early warning system must be effective and rigid. To this end, first a sound energy statistics mechanism is needed, to ensure the daily energy information be collected and processed completely and accurately. Based on this, it is necessary to carry out comprehensive energy security analysis. Otherwise, the collection and process of relevant information would be meaningless. It is the key of energy early warning to carry out energy early warning according to the results of energy security analysis. In this process, using early warning methodology, based on warning source and warning sign analysis the deep reasons that cause the change of energy security situation are explored, so that to provide basis for putting forward the corresponding settlements. And based on the analysis and reasonable classification of warning situation, energy security should be released through warning situation information release mechanism, so as to remind relevant departments to take active measures to settle the energy crisis in advance. Meanwhile, energy crisis is cyclical. Therefore the experience should be summed up after every effective settlement of energy security issues. In the light of past experiences and combining with new analysis method, to improve the energy early warning system (see Fig. 4).

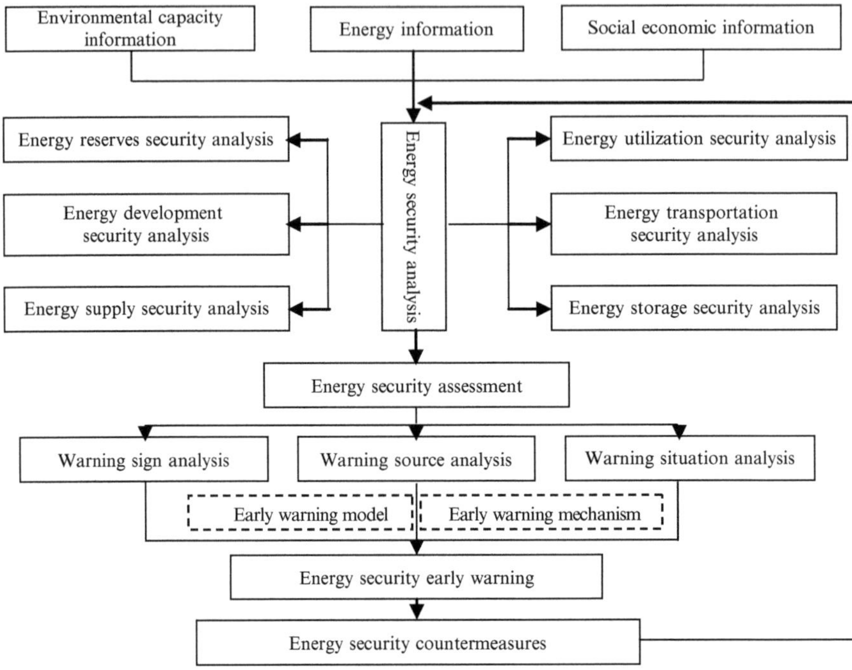


**Fig. 4** Energy early warning process

# 4 Proposals for Establishing and Improving China's Energy Security Early Warning System

1. Strengthen the basic work of energy statistics so that provide reliable information for energy forecasting and early warning.

   Energy forecasting and early warning system is one of the most important measures that are used to realize energy security. Strengthen energy forecasting and early warning is conducive for policy-making departments to grasp energy supply and demand situation timely and correctly, thereby arranging the scale and pace of energy development reasonably. However, energy forecasting and early warning must rest on accurate and reliable basic information. At present China's energy statistics is weak. Related energy space-time information is missing and the caliber and time range of energy statistics are different. It is an important guarantee for timely and accurate energy early warning to strengthen the cooperation among government departments, industry associations, enterprises and research institutions and to regulate energy statistical system, thus to provide basic information for energy forecasting and early warning (Wang)·
2. Using international experiences for reference to establish energy forecasting and early warning system suitable to our country.

Western countries pay more attention on energy and have stronger risk awareness. China should enhance international exchanges and cooperation of energy information with the International Energy Agency (EIA), Eurostat, OPEC and other international organizations. And energy forecasting and early warning system suitable to our country should be established based on the systematic analysis of China's actual situation, and using international experiences and advanced research results for references. At the same time energy early warning and economic early warning should be combined organically, and energy risks should be evaluated rationally, so that to avoid the waste of social resources caused by economic huge fluctuations and to promote the scientific development of social economy.

3. Establishing special research institutes, and improving the energy early warning mechanism constantly

   At the beginning of 2008, China had issued the view of strengthening energy forecasting and early warning. The view puts forward clearly that it is necessary to establish and improve statistical system promptly, to push forward the construction of energy forecasting and early warning information system steadily, to improve the ability and level of energy forecasting and early warning and to establish energy forecasting and early warning information release system. Energy forecasting and early warning information must be serious, accurate, reliable and authoritative to ensure its instruction significance. To this end, perfect energy early warning mechanism need to be established to regulate the actions of energy information's collecting, processing, analyzing and using, so that to ensure the reliability. At the same time special energy security coping mechanism should be set up to ensure reasonable resolving of energy crisis.

4. From one point to the whole, establishing perfect energy early warning system gradually

   Energy security covers a wide range of contents and energy security warning should include all of these contents. At present the focus of energy early warning is energy supply and demand security. Therefore take energy supply and demand security as breakthrough, energy supply and demand security early warning system should be established first. On this basis, according to the connotation of energy security, integrated energy early warning system can be formed by bring energy reserves security early warning, energy mining security early warning, energy transporting security early warning and energy storage security early warning into the energy security early warning system.

## References

Chi C-j (2006) Research on energy security early warning [J]. Stat Decis 11:29–31

Diao X (2009) Present situation, characteristic and countermeasures of China's energy security. J Dongbei Univ Finance Econ 3:50–5

He Q (2009) Discussion and strategy about the energy security of China [J]. China Saf Sci J 19(6):52–7

Hu J, Wang S-C (2006) Total factor energy efficiency of region in China [J]. Energy Policy 34(17):3206–17

Huang J-h, Lei Z-b, Ling C (2003) A survey for early warning system of economics [J]. Syst Eng 21(2):64–70

Jiang Z-m (2008) Reflections on energy issues in China [J]. J Shanghai Jiaotong Univ 13(3): 257–74

Kang X-f, Wang H-T, Huang J-h (2004) Study on farly-warning system of enterprise with quantitative [J]. Sci Sci Manage S& T 7:134–7

Li J-z (2007) Establishment of Chinese energy early warning model and indicators. J China Univ Petrol (Editon of Natural Science) 31(6):161–6

Li G, Liu Y (2009) Report on China's energy safety: early warning and risk settlement [M]. Hongqi Press, p. 3

Li Y-f, Lu G (2009) Sustainable development strategy for energy in China [J]. China Min Mag 18(9):1–5

Nikander IO, Eloranta E (2001) Project management by early warnings [J]. Int J Project Manage 19:385–99

Research Group of China National Energy Development Strategy and Policy Analysis (2004) China National Energy Development Strategy and Policy Analysis [M]. Economic Science Press

Sun X, Pan G (2009) Energy geopolitics in the Middle East and China's energy security strategy [J]. Arab World Stud 4:38–45

Wang S-q Methodology of energy forecasting and early warning [M]. Tsinghua University Press, 2040, 2

Zhang G-b (2009) Report on China's energy development for 2009 [M]. Economic Science Press, p. 3

# The Asymmetrical Analysis of the International Crude Oil Price Fluctuation on Chinese Economy

Xiang Wu, Yanhong Wang, and Yan Pan

**Abstract** In this paper, we applied the method of asymmetric cointegration to analysis the non-symmetry impact of the international crude oil price volatility on the economic of China. The empirical results show that asymmetric cointegration relationship exists, even though there is no long-term cointegration between the international crude oil prices fluctuations and GDP in China. This shows that the rise in international crude oil price on China's economic role played by the obstacles is greater than the price drop on the economy played a stimulating role, but the non-symmetry relationship is not obvious. This paper puts forward policy recommendation.

**Keywords** Sensitive analysis · The asymmetric cointegration · The international crude oil prices fluctuation · Uncertainty

## 1 Introduction

In recent years, due to the financial crisis and global warming, fluctuations in international crude oil prices have been more violent. At present, China's foreign oil dependence has been more than 50%, so the fluctuations in international crude oil prices will undoubtedly affect China's economic growth. Therefore, to explore the relationship between the volatility of oil prices and China's economic growth has important practical significance.

In the study of the relationship between Crude oil price volatility and economic growth, the majority of literature studies by use of cointegration methods, such as Hamilton (1983), Brown and Yucel and so on. These literatures suggest that oil price volatility on a country's economy has a symmetry, namely, crude oil prices

X. Wu (✉), Y. Wang, and Y. Pan
School of Economics and Management, Northeast Dianli University, Jilin 132012, China
e-mail: dianyuanwuxiang@163.com

Y. Zhou and D.D. Wu (eds.), *Modeling Risk Management for Resources and Environment in China*, Computational Risk Management,
DOI 10.1007/978-3-642-18387-4_51, 

pulling effect on the economy and rising oil prices hindered the economy's size is symmetrical. However, 80 of the last century empirical test shows that the international crude oil prices affect the economy with non-symmetry. That is, the decline in oil prices since the economy is less than positive driving role played by prices hinder. These documents are mainly: Mory (1993), Ferderer (1996), Brown and Yücel (2002) and so on.

The aforementioned literature is mainly carried out studies for the OECD countries. Whether there is the asymmetric effect on China's economy from the fluctuations in international crude oil prices? The degree of asymmetry? Currently there are fewer cares. The main contribution of this paper is to use asymmetric cointegration methods to empirically study of crude oil price volatility on the economic impact of China's non-symmetry, and accordingly the corresponding policy recommendations. The main contribution of this paper is to use asymmetric cointegration methods to empirically study of crude oil price volatility on the economic impact of China's non-symmetry, and accordingly the corresponding policy recommendations are put forward.

## 2 Our Approach

In this paper, a non-symmetric cointegration method is used to study the fluctuations in international crude oil price impact on China's economic role of non-symmetry. This method firstly distinguishes the positive and negative values of the increase in volume of the time series, and decomposes this time sequence into its initial value and its value on the negative and positive of the accumulated value. Next, using asymmetric cointegration approach to study the relationship between the combination of multiple variables which has been decomposed.

Schorderet starts the analysis by decompose a time series $Y_t$ into two parts: the positive part and the negative part, then, consider two integrated time series $Y_{1t}$ and $Y_{2t}$, and define $Y_{jt}^+$ and $Y_{jt}^-$, $j = 1, 2$. We assume that there exist a linear combination between $Y_{jt}^+$ and $Y_{jt}^-$, i.e.

$$L_t = \beta_0 Y_{1t}^+ + \beta_1 Y_{1t}^- + \beta_2 Y_{2t}^+ + \beta_3 Y_{2t}^- \tag{1}$$

Following Schorderer, if there exists a vector $\beta' = (\beta_0, \beta_1, \beta_2, \beta_3)$ and $\beta_0 \neq \beta_1$, or $\beta_2 \neq \beta_3$ (and $\beta_0$ or $\beta_1 \neq 0$, $\beta_2$ or $\beta_3 \neq 0$), $Y_{1t}$ and $Y_{2t}$ are asymmetric or we say cointegrated. Thus, the $L_t$ in (1) is a stationary stochastic process. We can explain this thought in a simple way: the relation between two variables will be different when the value of them increase or decrease. With simplification and generalization, we make the assumption that only one component of every series exists in the cointegration relation described in (1), that is,

$$L_{1t} = Y_{1t}^+ - \beta^+ Y_{2t}^+ \quad \text{or} \quad L_{2t} = Y_{1t}^- - \beta^- Y_{2t}^- \tag{2}$$

Due to the nonlinearity properties of $L_{jt}$, $j = 1, 2$, the estimation result of (2) obtained by OLS may be biased. Thus, Schorderet (2004) suggests applying OLS to estimate the auxiliary models. With defining $L_{jt}(j = 1,2)$ as the outcome of the disturbance $\varepsilon_{1t}(j = 1,2)$, we define

$$\Delta Y_{1t}^{-} = \begin{cases} \min\left[0; -\beta^{+}\Delta Y_{2t}^{+} + \varepsilon_{1t}\right] & t = 1 \\ \min\left[0; \beta^{+}\Delta Y_{2t}^{+} - Y_{1t-1}^{+} + \varepsilon_{1t}\right] & t = 2, ..., T \end{cases} \tag{3}$$

Then, under some specific conditions, the equation can be written as $\Delta Y_{1t}^{-} = \varepsilon_{1t} - L_{1t}$. Then we combine (3) with (2), we can obtain the auxiliary model:

$$\varepsilon_{1t} = Y_{1t}^{-} + \Delta Y_{1t}^{+} - \beta^{-} Y_{2t}^{-} \quad \text{or} \quad \varepsilon_{2t} = Y_{1t}^{+} + \Delta Y_{1t}^{-} - \beta^{+} Y_{2t}^{+} \tag{4}$$

As proved by West, there exists a linear time trend in the regressor, so the estimates of (4) obtained by OLS is asymptotic normal distribution and usual statistical inference can be done. In order to test the null hypothesis of no cointegration against the alternative one of asymmetric cointegration, we apply the traditional Engle and Granger procedure to (4).

## 3 Empirical Evidence

In order to test the asymmetric relation between the fluctuation of the oil price and China's economic growth, we apply the asymmetric cointegration method to obtain the empirical results. Before doing this, we have to consider the problem about data selection.

### 3.1 Data Selection

In order to study the long-term relationship between GDP and international crude price, the monthly oil price is the Brunt Spot Price FOB (Dollars per Barrel), which has been deflated by CPI and then we calculate the mean value to obtain quarterly data. The time span of the data is from 1992:1 to 2010:1. We also deflated the GDP quarterly data by CPI in the same time span. We denote *LOIL* and *LGDP* as the natural logarithm of oil price and *GDP* quarterly time series, respectively.

### 3.2 Unit Root Test and Standard Cointegration Test

At first, we apply the ADF test on the LGDP and LOIL series and the results are shown in Table 1.

**Table 1** The ADF result of series LGDP and LOIL

| Level | ADF statistics | First order differences | ADF statistics |
|---|---|---|---|
| GDP | −0.987 (2) | ΔLGDP | $-22.367^{**}$ (1) |
| OIL | −1.542 (3) | ΔLOIL | $-8.687^{**}$ (2) |

Δdenote the difference operator. +, $^{*}$, $^{**}$ denote the rejection of null hypothesis at 10%, 5% 1% significance level respectively. (1) Model without intercept, or trend. (2) Model without trend. (3) Model with intercept and trend. As follow so on.

**Table 2** The cointegration test between LGDP and LOIL

| Null | Eigenvalue | Trace statistics | Critical value | Max-Eigenvalue statistics | Critical value |
|---|---|---|---|---|---|
| $r = 1$ | 0.1902 | 13.347 | 15.496 | 14.323 | 14.265 |
| $r \leq 1$ | 0.002 | 0.352 | 3.837 | 0.245 | 3.841 |

**Table 3** Asymmetric cointegration test for $X_{2t}$ and *LOIL*

| Null | Eigenvalue | Trace statistics | Critical value | Max-Eigenvalue statistics | Critical value |
|---|---|---|---|---|---|
| $r = 1$ | 0.249 | 14.237 | 15.495 | 12.788 | 14.367 |
| $r \leq 1$ | 0.087 | 1.087 | 3.867 | 1.342 | 3.856 |

Table 1 shows that, as to LGDP series, when we use the ADF test with the model which is with intercept and without trend. We cannot reject the null hypothesis at 5% significance level. This means that the level value of LGDP is non-stationary series while its difference is stationary series at 1% significance level. In the same way, we can also know that the LOIL series is non-stationary at level value according to the ADF test, while its difference is stationary at 1% significance level. From above, we know that the LGDP series and LIOL series both are I(1) series, and then we apply the cointegration test to these two series, the results are displayed in Tables 2 and 3.

The test results show that there's no cointegration relationship between *LGDP* and *LOIL* series. However, as the majority of the extant literature mentioned, a lot of evidence show that there exist asymmetric relationships between crude oil price and *GDP* in many countries. In order to study whether there's such relationship in China or not, it's necessary for us to give up the strongly restricted cointegration frame and consider the possibility of asymmetric relationships in China.

### 3.3 Asymmetric Cointegration Test Between International Oil Price and GDP

In order to test the asymmetric cointegration relationship between oil prices and GDP, we estimate the two following auxiliary models, we construct the auxiliary model according to (44) as follows:

$$LGDP_t^- + \Delta LGDP_t^+ = \alpha^- + \beta^- LOIL_t^- + \varepsilon_{1t} \quad (5)$$

$$LGDP_t^+ + \Delta LGDP_t^- = \alpha^+ + \beta^+ LOIL_t^+ + \varepsilon_{2t} \quad (6)$$

We set $LGDP_t^- + \Delta LGDP_t^+ = X_{1t}$ in (5); Meanwhile, we set $LGDP_t^+ + \Delta LGDP_t^- = X_{2t}$, and then, we test the stationarity of $\Delta X_{1t}$,$X_{2t}$,$LOIL_t^-$, and $LOIL_t^+$ these four series. The results showed that the four time series are I(1) series.

After that, we test the asymmetric cointegration relationship between $X_{1t}$ and $LOIL_t^-$, the latter is the sum of $LGDP_t^-$ and $\Delta LGDP_t^+$ in (5). Table 4 reports the testing result. As it shows, we cannot reject the null hypothesis at 5% significance level according to the trace statistics. This result informs us that there's no cointegration relationship between these two series. By combining the two results above, we know that no significant cointegration relationship between international oil price and China's GDP, that's to say, the descending of oil price does not have a stable long term relationships with the descending of China's GDP.

We also test the asymmetric cointegration relationship between $X_{2t}$ and $LOIL_t^+$, the latter is the sum of $LGDP_t^-$ and $\Delta LGDP_t^+$ in (6). The testing results are given in Table 4. As it shows, we cannot reject the null hypothesis at 5% significance level according to the trace statistics and Max-Eigen Statistics. This result informs us that there exists cointegration relationship between these two series. This shows that there exists a significant asymmetric cointegration relationship between these two variables. That's to say, the ascending of oil price has a stable long term asymmetric relationship with *GDP* growth.

Table 5 reports the long term relationship between the estimates of (5) and (6). From the results, we can see that the estimate for $\beta^+$ is 3.086, which is bigger than $\beta^-$(2.648).

Different estimates imply that the asymmetric phenomenon exists. However, this phenomenon is not so significant as western industrialized countries (Lardic and Mignon 2008). It should be pointed out that these relationships are described in terms of partial sums of time series rather than the sums of the variables themselves. Thus, the coefficients $\beta^+$ and $\beta^-$ cannot be explained in the usual way. According to

**Table 4** Asymmetric cointegration test for $X_{2t}$ and *LOIL*

| Null | Eigenvalue | Trace statistics | Critical value | Max-Eigen statistics | Critical value |
|---|---|---|---|---|---|
| r = 1 | 0.193 | 16.427* | 15.495 | 15.032* | 14.265 |
| r ≤ 1 | 0.007 | 0.865 | 3.841 | 0.487 | 3.841 |

**Table 5** Coefficients in (5) and (6)

| | | | | |
|---|---|---|---|---|
| Equation 5 | $\alpha^-$ | $\beta^-$ | $\overline{R^2}$ | Std. deviations |
| | −0.378 (−2.89) | 2.648 (35.49) | 0.921 | 0.579 |
| Equation 6 | $\alpha^-$ | $\beta^-$ | $\overline{R^2}$ | Std. deviations |
| | 2.326 (10.35) | 3.086 (39.88) | 0.948 | 0.721 |

some relative literatures, $\beta^+$ is generally higher than $\beta^-$, which means the ascending of oil price has greater impact on GDP than the descending one.

Why could the impact of international crude oil price on economic growth be asymmetric? Exactly as Brown and Yücel (2002), classic supply theory is unable to explain this kind of asymmetry. It has some plausible explanation, such as the monetary policy, the adjustment cost derived from different sectors, the adverse effect of uncertainty on the investment environment (Ferderer 1996) and the asymmetry in the oil products, especially gasoline.

The first factor we consider here is the monetary policy, suppose the price is nominally sticky downward. Then, a increment in the oil price will lead to *GDP* partial loss, if the monetary authority maintain the nominal *GDP* only by unanticipated inflation. On the contrary, after the decrement of the oil price, the standard of wage has to be levelled up in order to clear the market. Just like this, the monetary policy has the asymmetric effect (Bernanke et al. 1997). If according to the explanation concerning the adjustment cost of different sectors (Hamilton 1988), this cost induced by the fluctuation of oil price retard the economic activity. Such cost arise from the imbalances among different sectors (Lilien 1982; Hamilton 1988), coordination failures between firms, or because of the different energy-to-output ratios embedded in the capital stock. Finally, plenty of studies argues that the oil product prices respond asymmetrically to the crude oil prices, for example, the gasoline prices increase more quickly when the crude oil prices ascending than they decrease when the crude oil prices descending. Thus, there are many kinds of reasons to explain the asymmetric relationship between international oil price and GDP and the reasonability of the existence of this kind of relationship.

## 4 Conclusion and Policy Suggestions

The method, which was applied to study whether there's a long term relationship between international crude oil price and GDP or not, is the usual cointegration frame in the majority of extant literatures. However, the researches on many countries' empirical evidence by large amount of scholars shows that there exists an asymmetric cointegration relation between these two variables. The effect of hindering economic growth, which is from price increasing actually, is much more significant than that of prompting economic growth, which is from price decreasing. The same empirical evidence exist in China, but this is not as significance as industrialized countries. The main reason is that the government pays fiscal subsidy on the domestic oil and natural gas prices, which make the price running at a low level.

The corresponding policy implications are listed as follows. Firstly, the government should straighten out the oil and gas prices, promote and perfect the oil price reform. Secondly, the reform of financial system should be deepened; besides, the oil future market should be made up as soon as possible. These policies are made in order to avoid the result that the fluctuation of international oil price impact on domestic economy. Thirdly, the government should build up the strategic oil reserve system, with encouraging private capital to construct the reserves system and the transition, storage and transportation system, gradually unloosen the oil market, and introduce competition mechanism into the market. Fourthly, we should make the energy conservation technology more widespread, foster the conscious of energy saving, diversify the utilization of energy, and realize the objective of reducing the dependence on crude oil.

**Acknowledgments** This research was supported by Scientific Research Foundation for the Dr. project of Northeast Dianli University of 2009, under Grant BSJM-200910.

## References

Bernanke BS, Gertler M, Watson M (1997) Systematic monetary policy and the effects of oil price shocks. Brookings Pap Econ Activity 1:91–157
Ferderer JP (1996) Oil price volatility and the macroeconomy: a solution to the asymmetry puzzle. J Macroecon 18:1–16
Hamilton JD (1983) Oil and the macro economy since World War II. J Polit Econ 91:228–248
Hamilton JD (1988) A neoclassical model of unemployment and the business cycle. J Polit Econ 96:593–617
Lardic S, Mignon V (2008) Oil price and economic activity: An asymmetric cointegration approach. Energy Econ 30:847–855
Lilien D (1982) Sectoral shifts and cyclical unemployment. J Polit Econ 90:777–793
Mory JF (1993) Oil prices and economic activity: is the relationship symmetric? Energy J 14:151–161

# Building Optimal Operation Model of Cascade Hydropower Stations Based on Chaos Optimal Algorithm

**Liang Wei, Xu Kan Xu, Zheng-hai Xia, and ShanShan Song**

**Abstract** After the status is analyzed on Wujiang River hydropower stations, a model of the mid-long term cascade hydropower station reservoir optimal operation is put forward and using Chaos Optimization Algorithm (COA) to solve the mid-long term cascade hydropower station reservoir optimal regulation problem. The main principle of article is using the randomness of chaos movement. First random chaos series are produced by Logistic mapping, career random series to feasible field S which include hydropower station's objective function, by using of properties of randomness, periodicity and regularity to search optimal solution in global space, it can obtain point belong to the feasible field S. Then by comparison, iteration and secondary career wave, the paper get optimal scheduling graph of cascade hydropower station reservoir optimal regulation. Through the example of the scheduling model is validated, and the result shows that COA can solve nonlinear cascade hydropower station reservoir optimal regulation problem, which has complex constrain conditions. This algorithm not only makes the solution more accurate, converges faster, but also is an effective way to solve the problem of cascade hydropower station reservoir optimal regulation.

**Keywords** Cascade hydropower station · Chaos optimal algorithm · Optimal regulation · Risk · Sensitive analysis · Stochastic · Uncertainty

## 1 Introduction

With the formation of large-scale cascade hydropower stations and the deepening and regulation of the reform in power system. It is a key issue for each generation company that how to develop the optimal dispatch operation of reservoirs. This paper uses the scientific theories and method of COA to research the optimal scheduling problem of the cascade hydropower stations along the upstream of

L. Wei, X.K. Xu (✉), Z.-h. Xia, and S. Song
Department of Information Management, HoHai University, ChangZhou 213022, China
e-mail: xxkwh@hhu.edu.cn

Y. Zhou and D.D. Wu (eds.), *Modeling Risk Management for Resources and Environment in China*, Computational Risk Management,
DOI 10.1007/978-3-642-18387-4_52, © Springer-Verlag Berlin Heidelberg 2011

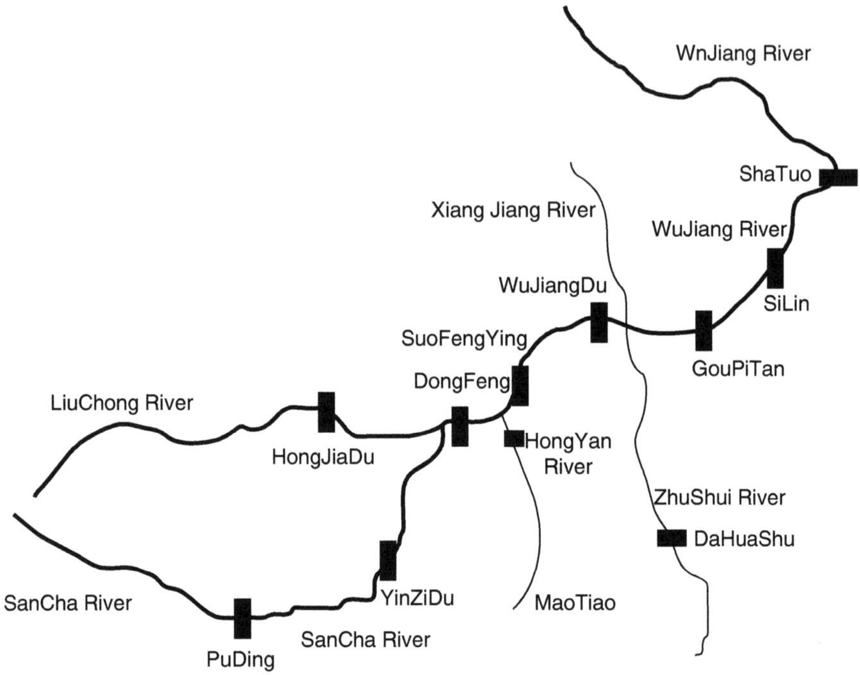


**Fig. 1** Cascade hydropower stations in WuJiang River

WuJiang River, which has the great theoretical and practical significance in catching the operation scheduling rules and improving the composite utilization efficiency of those stations (Huang et al. 2002).

WuJiang River, also known as QianJiang, is the biggest upstream of the Yangtze River. It originates in the foothills of the WuMeng, YunNan-GuiZhou Plateau. It crosses the central part of GuiZhou Province, and flows into Yangtze River in Fuling City of the Chongqing. WuJiang River has the natural large gap, and abundant water resources. The water resources mainly concentrated in the mainstream river, Here are all 11 large hydropower stations: HongJiaDu, DongFeng, SuoFengYing, WuJiangDu, GouPiTan, SiLin, Land, PengShui, PuDing, YinZiDu, XiKou. But in this paper, we will pay main attention to the four Cascade Hydropower Stations: HongJiaDu, DongFeng, SuoFengYing and WuJiangDu (commonly called as Cascade Hydropower Stations in WuJiang upper reaches), as the Fig. 1 shows.

## 2 Building Model of Optimal Operation Model of Cascade Hydropower Stations in Wujiang River

The work of optimal operation model of cascade hydropower stations should organize the scheduling operation reasonably to get the best whole economic benefits, based on meeting the comprehensive demands of Water-power System and downstream water using etc. (Sun and Shi 1995). That is to maximum the total cascade power through the reasonable assignment of water, under the condition that calculate the initial and termination water level by giving the forecasting inflow process line and water-using process line (Ruzic et al. 1994). Combine the feature of the problem, and build the Chaos Optimization Algorithm (COA) model as following:

## 2.1 Objective Function

$$E = \max \sum_{j=1}^{N} \sum_{i=1}^{T} N_i^j \Delta t_i \tag{1}$$

Where: T: the total using time of a year (when calculating, T = 12); N: the all stations (N = 4); $N_i^j$: when i, the power from j (KW); E: the whole year's power (kWh); $\Delta t_i$: the using time (Yang and Chen 1989).

## 2.2 Constraints

(a) Reservoir capacity (level) constraints:

$$V_{i,\min}^j \le V_i^j \le V_{i,\max}^j \quad (j = 1, 2, \cdots, N; i = 1, 2, \cdots, T) \tag{2}$$

(b) Output constraints:

$$N_{i,\min}^j \le N_i^j \le N_{i,\max}^j \quad (j = 1, 2, \cdots, N; i = 1, 2, \cdots, T) \tag{3}$$

(c) Water capacity balance constraints:

$$V_i^j = V_{i-1}^j + (q_i^j + Q_i^{j-1} + S_i^{j-1} - Q_i^j - S_i^j)\Delta t_i \tag{4}$$

$$Q_i^o = S_i^0 = 0 \quad (j = 1, 2, \cdots, N; i = 1, 2, \cdots, T)$$

(d) Outflow capacity constraints:

$$Q_{i,\min}^j \le Q_i^j \le Q_{i,\max}^j \quad (j = 1, 2, \cdots, N; i = 1, 2, \cdots, T) \tag{5}$$

(e) Nonnegative constraints

$$Q_i^j > 0, \quad S_i^j \ge 0 \quad (j = 1, 2, \cdots, N; i = 1, 2, \cdots, T) \tag{6}$$

(f) Cascade output assured constraints:

$$\sum_{j=1}^{N}\sum_{i=1}^{T}N_i^j \geq N \tag{7}$$

Where: $V_{i-1}^j$,$V_i^j$ separately stands for the j reservoir capacity (billion $m^3$) at the beginning and end of the time i; $V_{i,\min}^j$, $V_{i,\max}^j$ shows the min capacity, max capacity of reservoir j separately when i; $N_{i,\min}^j$, $N_{i,\max}^j$ means the min output and max output separately of reservoir j when time i; $Q_i^j$ shows the power flow $(m^3/s)$, of reservoir j at time i; $Q_{i,\min}^j$, $Q_{i,\max}^j$ separately shows the min outflow capacity and max outflow capacity allowed of reservoir j when i; $q_i^j$: the average inflow capacity of reservoir j when i; $S_i^j$ shows the abandoned water flow capacity $(m^3/s)$ from j when i, N means the cascade assured output.

## 3 Chaos Optimization Algorithm

Logistic model is one of the most typical models in the study of chaos. This paper chooses the chaotic variables generated by Logistic model to optimize the search. The equation is

$$x_{k+1} = \lambda \cdot x_k \cdot (1 - x_k), \tag{8}$$

where $\lambda = 4$. For optimization of n parameters, then n different initial values should be set Arbitrarily in the range of (0,1) (except for the fixed point 0.25, 0.5, 0.75 in the forum (8)), after that, we can get n chaotic variables of different track, then convert them into ergodic chaos variable in the solution space of optimization, finally, through searching optimization to find the optimal solution.

The nonlinear programming problem is dealt with equality or inequality constraints of a target function to determine the optimal solution. Generally expressed as:

$$\begin{cases} \min f(X) \\ s.t \quad \begin{matrix} g_i(X) \geq 0 \; i = 1,2,\cdots,m \\ h_i(X) = 0 \; j = 1,2,\cdots,n \end{matrix} \end{cases} \tag{9}$$

where: $X \in E^n$,$f(X)$ is a target function, $g_i(X)$, $h_j(X)$ are two constraint functions, at least one nonlinear function exists in these functions. Constraints sometimes expressed as sets, for example (Ma et al. 1996).

$$S = \{X | g_i(X) \geq 0, i = 1,2,\cdots,m; h_j(X) = 0, j = 1,2,\cdots,n\} \tag{10}$$

Then $S$ is called the feasible set or feasible region, the point is called feasible points (Xu and Ma 2005).

According to the previous described mathematical model of cascade hydropower station reservoir optimal operation, the objective function is to seek the biggest annual energy production in the $N*T$ decision variables, which are $Q_1^1, Q_2^1, \cdots, Q_T^1, Q_1^2, Q_2^2, \cdots, Q_T^2, \cdots\cdots, Q_1^N, Q_2^N, \cdots, Q_T^N$.

To solve the problem of optimal regulation of cascade hydropower stations in the algorithm. First, select a $N*T$-dimensional initial vector randomly. Second, make use of the randomness of chaotic motion, then generate chaotic sequence $\{X_k\}_{k=1,2,\cdots}$ by the Logistic equation randomly $X_{k+1.i} = 4 \cdot X_{k.i} \cdot (1 - X_{k.i})$, $j = 1, 2, \cdots NT$, Then carry its math wave into one region, which includes the feasible region S of cascade hydropower station objective function. The S is the collection meeting the constraint condition of (a)–(b) of the function; Use randomness, ergodicity and regularity to seek the most optimization in the whole region, from which we search the point based in the set S; and then through comparison, iteration and the second carrier, finally we can calculate the most optimal result and the corresponding point, thereby, we will conclude the most optimal operation curve. During the calculation, we should consider the connection from cascade water power. When we find the varying point at the searching, it is the time that the reservoir capacity changes at some period of one cascade reservoir, we should calculate the objective function again at the same time for next level reservoir (ignoring the time of interval water flow).

The basic process of the chaos optimization algorithm is as follows, which is used to solve the problem of optimal regulation of cascade hydropower stations:

Step 1: Initialization: Set the dimension of objective function E in the (1) for $M = N \bullet T$, and $E = \max f(V), V^* = (V_k^1, V_k^2, \cdots, V_k^M), f^* = f(V^*)$, then generate the following M initial values randomly.

$(X_0 = (X_{01}, X_{02}, \cdots, X_{0NT})\ X_{0j} \in [0\ 1], j = 1, 2, \cdots, NT)$

Step 2: Chaotic map: Generate $k$ chaotic variables ($\{X_k\}_{k=1,2,\cdots}$) by the above Logistic map[8], then carry wave as the formula (11), using the above chaotic variables () separately, and amplify them to the range of constraint (Storage capacity variable).

$$V_k^j = V_{k,\min}^j + X_{k,j}(V_{k,\max}^j - V_{k,\min}^j) \tag{11}$$

Step 3: Iteration: Set $k = 0$, then use chaos variables for Iterative search, calculate the objective function $f = f(V_k)$, which meet the constraints (a)–(b) (here $V_k = (V_k^1, V_k^2, \cdots, V_k^M)$), detailed calculation is as follows:

1. Calculate $N*T$ decision variables $Q_1^1, Q_2^1, \cdots, Q_T^1, Q_1^2, Q_2^2, \cdots, Q_T^2, \cdots\cdots,$ $Q_1^N, Q_2^N, \cdots, Q_T^N$, which meet the constraints (a), (d), (e), by the water balance (4). Otherwise, recalculate.
2. Calculate $N*T$ output, which meet the constraints (b),(f), by the output equation, or recalculate into (1).
3. Calculate the objective function $f = f(V_k)$, if $f > f^*$, then

$f = f^*, V^* = V_{k+1}, k = k + 1$, when $k$ meet the maximum number of iterations, turn the next step, otherwise continue iteration.
Step 4: Secondary carrier: Set $k = 0$, given an arbitrarily small positive number $\varepsilon$ and $Z = \lambda Z, \lambda \in [0.9\ ,\ \ 0.999]$, which $\alpha \in (0, 0.5)$, the initial value of $Z$ comes from $Z_0 = \min(V_k^J - V_{k,\min}^J, V_{k,\max}^j - V_k^J) \cdot \alpha$, and according to

$$Z_{kj} = V_{kj} + \alpha(t_j - 0.5) \tag{12}$$

To press the secondary-type carrier. $t = (t_1, t_2, \ldots, t_M)$ is an initial sequence generated by Logistic chaotic map.$V_{kj}$ is the optimal solution for the current. Calculate the objective function $f = f(Z_k)$, which meet the constraints (a)–(f), if $f > f^*$, then $f = f^*, V^* = Z_k$, $k = k + 1$, repeat Step 3, until

$$|f(Z_{k+1}) - f(Z_k)| < \varepsilon \tag{13}$$

Or, meet the maximum number of iterations, then output the optimal solution $E = f^*$.

## 4 Empirical Analyses of Reservoir Scheduling Model Based on COA Arithmetic

In order to prove the feasibility and availability of COA arithmetic in cascade hydroelectric station optimized scheduling mentioned in this paper, we calculate the data of cascade hydropower station in WuJiang Basin in 2009. There are four hydropower stations in WuJiang basin. They are HongJiaDu, DongFeng, SuoFengYing and WuJiangDu. Among them, the HongJiaDu hydropower station is main to generate electricity, flood control, irrigation, tourism, shipping, improving ecological environment of the reservoir ambient area, and other comprehensive benefits. The hydropower will be equipped with three machines, each of them has the power of 200 MW, the whole power can arrive 600 MW, the whole capacity of the reservoir is 4.947 billion cubic meters, adjusted capacity 3.361 billion cubic meters, the common water level is 1,140 m, dead level 1,076 m, and the dead capacity of reservoir is 1.137 billion cubic meters; The DongFeng is the second level hydropower in the trunkstream of WuJiang River, the available machine capacity can arrive 510 MW, and the whole storage capacity is 1.016 billion cubic meters, adjusted capacity 0.491 billion cubic meters, the common water level is 970 m, capacity 0.864 billion cubic meters; The SuoFengYing hydropower is the third level cascade hydropower in WuJiang trunkstream, the available machine capacity is 600 MW, normal reservoir water level is 837 m, and the reservoir water capacity 0.168 billion cubic meters; The hydropower WuJiangDu is the fourth one, belonging to the incompletely year-adjusted reservoir of WuJiang trunkstream, the whole equipped machine capacity is 1,250 MW, common water level reaches 760 m, reservoir water capacity is 2.14 billion cubic meters (Zhao and Jiang 1997).

## *4.1 Methods for Solving This Model*

The randomness of chaotic motion is the main idea when we use COA arithmetic to solve this model while Logistic equation generates chaotic sequences randomly. Then we load these sequences into an area of the feasible region S which contains the target function of this cascade hydroelectric station model. Using randomness ergodicity and regularity for global optimization and search for points belong to the feasible region S. Then through comparison, iterations, second carrier wave we ultimately make out the optimal scheduling line of cascade hydroelectric station's scheduling. The practice shows that COA arithmetic is an effective method in solving high dimension issues (Ma et al. 1996; Zhao 2006).

## *4.2 Calculation Results of COA and Analysis*

According to the calculation methods and optimized scheduling model which is aimed at getting maximum capacity of the Cascade hydropower stations in WuJiang Basin mentioned above, using MATLAB language to realize the chaos optimization algorithm. Meeting the demands of die capacity, normal capacity, minimum output and the greater ability to flow as far as possible While getting the maximum capacity is the ultimate goal. To achieve this goal, making optimized scheduling. Computing this arithmetic ten times and choose the optimum value. The results are shown in table from 1 to 4, the total of output power is 164.839 (100 million kilowatts hour) (Zhou and Liu 2006) (Figs. 2 and 3; Tables 1–4).

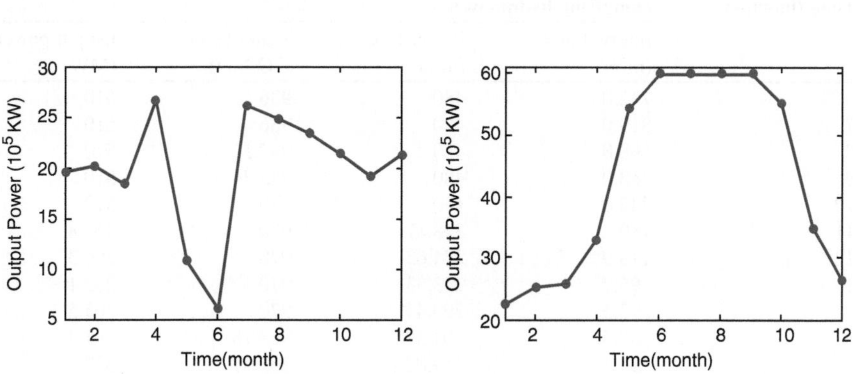


**Fig. 2** Optimal dispatch curve on HongJiaDu and DongFeng Hydropower

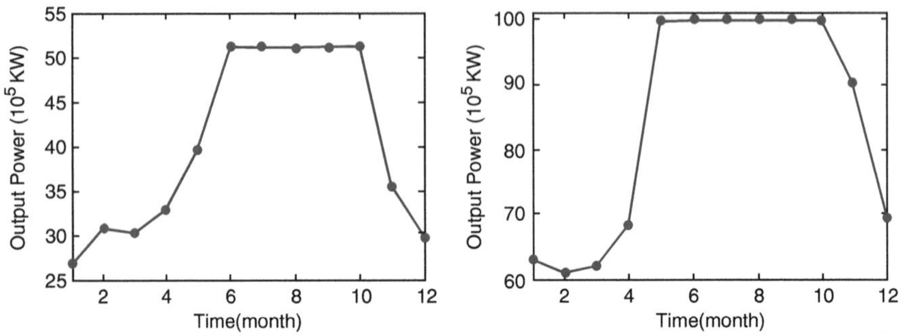


**Fig. 3** Optimal dispatch curve on SuoFengYing and WuJiangDu hydropower

**Table 1** Calculation results of COA results on HongJiaDu hydropower

| Time (months) | HongJiaDu hydropower | | | |
|---|---|---|---|---|
| | Interval flow ($m^3/s$) | Power flow ($m^3/s$) | Water level E.O.M. (m) | Output power (MW) |
| 6 | 296.0 | 46.64 | 1,119.9 | 60.1 |
| 7 | 342.0 | 197.21 | 1,129.7 | 262.5 |
| 8 | 279.0 | 182.47 | 1,135 | 249.8 |
| 9 | 219.0 | 169.13 | 1,138.3 | 235.3 |
| 10 | 154.0 | 154 | 1,140 | 215.6 |
| 11 | 89.1 | 138.97 | 1,140 | 193.8 |
| 12 | 59.2 | 155.73 | 1,138.3 | 213.6 |
| 1 | 50.1 | 146.63 | 1,135 | 197 |
| 2 | 50.3 | 153.48 | 1,131.5 | 201.5 |
| 3 | 47.4 | 143.93 | 1,127.8 | 184.4 |
| 4 | 67.8 | 217.42 | 1,124 | 267.9 |
| 5 | 137.0 | 88.74 | 1,117.7 | 109 |

**Table 2** Calculation results of COA on DongFeng hydropower

| Time (months) | DongFeng hydropower | | | |
|---|---|---|---|---|
| | Interval flow ($m^3/s$) | Power flow ($m^3/s$) | Water level E.O.M. (m) | Output power (MW) |
| 6 | 742.0 | 480 | 936 | 510 |
| 7 | 819.0 | 480 | 936 | 510 |
| 8 | 608.0 | 480 | 947.51 | 510 |
| 9 | 468.0 | 480 | 953.34 | 510 |
| 10 | 342.0 | 480 | 970 | 510 |
| 11 | 189.2 | 328.17 | 970 | 352.8 |
| 12 | 118.9 | 274.63 | 970 | 296.3 |
| 1 | 96.7 | 243.33 | 970 | 263.1 |
| 2 | 93.5 | 290.44 | 970 | 305.5 |
| 3 | 89.5 | 304.57 | 963.78 | 297.7 |
| 4 | 131.1 | 359.02 | 951.47 | 327 |
| 5 | 319.0 | 468.72 | 949.61 | 393.7 |

**Table 3** Calculation results of COA on SuoFengYing hydropower

| Time (months) | SuoFengYing hydropower | | | |
|---|---|---|---|---|
| | Interval flow ($m^3/s$) | Power flow ($m^3/s$) | Water level E.O.M. (m) | Output power (MW) |
| 6 | 916.0 | 990 | 834.97 | 600 |
| 7 | 910.0 | 990 | 837 | 600 |
| 8 | 691.0 | 990 | 837 | 600 |
| 9 | 547.0 | 990 | 837 | 600 |
| 10 | 400.0 | 896 | 837 | 553.7 |
| 11 | 226.0 | 554.17 | 837 | 351.2 |
| 12 | 140.0 | 414.63 | 837 | 265.8 |
| 1 | 112.0 | 355.33 | 837 | 229 |
| 2 | 109.0 | 399.44 | 837 | 256.4 |
| 3 | 99.9 | 404.47 | 837 | 259.5 |
| 4 | 160.0 | 519.02 | 837 | 329.9 |
| 5 | 418.0 | 890.87 | 837 | 542.7 |

**Table 4** Calculation results of COA on WuJiangDu hydropower

| Time (months) | WuJiangDu Hydropower | | | |
|---|---|---|---|---|
| | Interval flow ($m^3/s$) | Power flow ($m^3/s$) | Water level E.O.M. (m) | Output power (MW) |
| 6 | 1,126.0 | 1,080 | 720 | 1,000 |
| 7 | 1,150.0 | 1,080 | 741.22 | 1,000 |
| 8 | 850.0 | 1,080 | 749.42 | 1,000 |
| 9 | 661.0 | 1,080 | 747.28 | 1,000 |
| 10 | 490.0 | 1,080 | 743.25 | 1,000 |
| 11 | 283.0 | 837.2 | 760 | 903.3 |
| 12 | 178.0 | 649 | 760 | 697.5 |
| 1 | 145.0 | 613.2 | 756.78 | 633.4 |
| 2 | 137.0 | 626.9 | 749.42 | 611.6 |
| 3 | 146.0 | 691.5 | 743.25 | 624 |
| 4 | 223.0 | 858.6 | 731.56 | 684.4 |
| 5 | 530.0 | 1,080 | 720 | 1,000 |

# 5 Conclusions

The research about the optimized scheduling of Cascade hydropower station in WuJiang Basin is not only the base to realize the optimized scheduling of hydro power systems (Zhang 1988), but also an important measure to provide reliable operating data and basic method of operation for the scheduling of GuiZhou power system. In this paper we build a long time optimized scheduling model which is aimed at getting maximum capacity of the Cascade hydropower station in WuJiang Basin and we use COA to solve the problem about the dimension of dynamic programming. The results indicate that the model we build is not only reduce the surplus water released from the cascaded reservoirs but also increase the economic benefits of the Cascade hydropower station in WuJiang Basin. It also proves that COA is a new feasible and effective method to solve such an issue.

**Acknowledgements** This research was supported by the Jiangsu University Social Science Foundation of China under Grant 09SJD870001 and the Fundamental Research Funds for the Central Universities.

## References

Huang W, Yuan L, Lee C (2002) Linging genetic algorithms with stochastic dynamic programming to the long-term operation of a multi-reservoir system. Water Resour Res 38(12):41–49

Ma G-w, Wu L, Waiters GA (1996) FP GA in hydropower stations' optimal dispatch. Theory Pract Syst Proj 11:77–81 (in Chinese)

Ruzic S et al (1994) A flexible approach to short-term hydro-thermal coordination. IEEE paper 94 SM 563-7 &564-5 PWRS. Presented at the IEEE/PES 1994 Summer Meeting

Sun X-d, Shi X-c (1995) Joint optimal scheduling of two libraries of parallel hydropower station. Water Resour Sci 13:167–172 (in Chinese)

Xu G, Ma G-w (2005) Cascade hydropower stations optimal operation based in ant colony algorithm. Hydropower Acad J 5:7–10 (in Chinese)

Yang JS, Chen N (1989) Short-term hydrothermai cordination using multi-pass dynamic programming. IEEE Trans Power Syst 4(3):1050–1056

Zhang Y-z (1988) Hydropower principles of economic operations. China Water Conservancy and Hydroelectric Power Press, Beijing (in Chinese)

Zhao Q (2006) Improved chaos optimal algorithm and its application. Autom Instrum 125(3): 90–92 (in Chinese)

Zhao B, Jiang W-s (1997) Chaos optimal algorithm and its application. Control Theory Appl 14(4):613–615 (in Chinese)

Zhou J, Liu K-z (2006) Operation and optimal scheduling of water resource system. Metallurgy Industry Press, Beijing (in Chinese)

# The Sensitive Analysis of Spatial Inverted U Curve Between Energy Efficiency and Economic Development of the Provinces in China

**Aijun Sun**

**Abstract** This thesis aims to study the correlation between energy consumption efficiency and economy development of inter-provinces in China. The Spatial econometrics model is set up, considering geographical influence. The empirical study is conducted. The Moran's I index shows that there is a spatial correlation between the inter-provincial GDP per capita and economy consumption per GDP. The spatial error model (SEM) confirms the spatial inverse U curve between energy usage efficiency and economic development of the provinces in China. If there is not sufficient economic development, the energy usage efficiency is decreasing, though there are some degrees of inter-provincial economic increase. After the turning point of the diagram, we find that energy usage efficiency increases with the further development of economy. Finally, some proposals are put forward.

**Keywords** Energy efficiency · GDP per capita · SEM · Sensitive analysis

## 1 Introduction

As the life line of industrial economic development, energy is vital to economy in the aspects of its overall investment and its consumption efficiency. More heed is being paid to the consumption efficiency due to the increasing scarcity of the global energy. Energy possession per capita in China is 40% of the average world level, but its gross energy consumption stands second in the world. The Chinese energy efficiency is only about 33%, but its energy consumption intensity is about three times that of America and about 7.2 times that of Japan.

A. Sun
School of Economics and Management, Huaiyin Normal University, Huaian, Jiangsu 223300, China
and
School of Economics, Nanjing University, Nanjing 210093, China

Y. Zhou and D.D. Wu (eds.), *Modeling Risk Management for Resources and Environment in China*, Computational Risk Management,
DOI 10.1007/978-3-642-18387-4_53, 

Many scholars, such as Dasgupta, S. gets similar conclusion since Grossman and Krueger bring forward the famous curve. The time series data or cross data is usually adopted by domestic researchers to analyze the connection between energy efficiency and economical development. Liu et al. (2007) investigates the dynamic interactions between economic growth and energy consumption by utilizing VAR model. It is time for China to realize sustainable economic growth under the condition of energy restriction. Wang et al. (2008) explores the dynamic relationship between economic growth and China energy,furthermore, examined the long-term equilibrium relationship between the two variables by means of Co-integration analysis. Growing at a reasonable speed and conducting energy saving continuously is essential to reach the goal of reducing energy intensity for 20%. There are many empirical studies after the concept of environmental Kuznets curve starts from empirical study, which initially refers to the inverted U-type relationship between economic growth and the level of environmental pollution. Based on resource curse hypothesis, Shao Shuai carries out an econometric analysis on the relationship and its transmission mechanism between energy development and economic growth with cross-province panel data. Shu and Wang (2008) measures the relationship between regional resource consuming and economy growth by index of product-consuming. Wei and Shen (2007) takes a measurement based on the panel data to compare with traditional energy productivity. Qi and Luo (2007) models the relationship between the energy intensity and per capita GDP together with other regression variables. He got the findings and policy implication after the coefficients by panel data of 30 provinces from 1995 to 2002 were estimated. Li and Cheng (2008) measures the energy efficiency of China by using panel data. The results show that the energy efficiency is still very low and there are significant differences across provinces and regions as well. Wang Huo geng established and compared the simple linear panel regression model and the spatial linear panel regression. The result indicated that the spatial linear panel regression model was better than the simple linear panel regression. Zhou Yan-fen analyzed the regional characters of china 31 provincial economic development and energy effective through the spatial statistical model and Moran I index to Panel Data.

In fact, there are outstanding differences of inter-provinces between different regions; the efficiency of resource consuming and economic development in developed areas is higher. We seldom find documents studying the spatial panel autocorrelation of the energy efficiency and the Chinese provincial economic growth. In this paper, we set spatial econometric model after the Moran's I index of economic development and energy efficiency is computed. We study the spatial character of energy efficiency based on the analysis of the inverted U curve between the two variables. We suppose that:

Hypothesis 1: There is the spatial inverted U curve between the energy efficiency and the Chinese provincial economic growth;

Hypothesis 2: More and more provinces' GDP Per Capita exceed the turning point value with the economic development.

## 2 Theory and Method

If there is not sufficient economic development, the energy usage efficiency is decreasing, though there are some degrees of inter-provincial economic increase. After the turning point of the diagram, we find that energy usage efficiency increases with the further development of economy. Formally, we write a relationship depicting this as:

$$Y_i = C + \alpha_1 X_i + \alpha_2 X_i{}^2 + u_i \tag{1}$$

Where i indexes observations collected at i = 1,..., n points in space, Xi represents a (1 × k) vector of explanatory variables with an associated set of parameters $\alpha_i$, In this formula, $Y_i$ is the dependent variable at observation (or location) i, and $u$ denotes a stochastic disturbance in the relationship. The method to judge whether there is inverted U curve between two variables is as follows:

Firstly, there exits the inverted U curve if the regression conclusion is significant besides $\alpha_1 > 0$ and $\alpha_2 < 0$; secondly, there exits not any inverted U curve though the regression conclusion is significant for $\alpha_1 > 0$ and $\alpha_2 > 0$. Lastly, there exits not any inverted U curve if the regression conclusion is not significant.

Spatial inverted U curve model is necessary for geographical relations. The Spatial auto-correlation is firstly taken into accounted.

The macrocosm index of spatial auto-correlation is exercised to check up variables' spatial correlation of different variables. Moran's I is the popular index, and it is defined as follows:

$$\text{Moran's I} = \frac{\sum_{i=1}^{n}\sum_{j=1}^{n} W_{ij}(Y_i - \bar{Y})(Y_j - \bar{Y})}{S^2 \sum_{i=1}^{n}\sum_{j=1}^{n} W_{ij}} \tag{2}$$

In this formula, $S^2 = \frac{1}{n}\sum_{i=1}^{n}(Y_i - \bar{Y})^2$, $\bar{Y} = \frac{1}{n}\sum_{i=1}^{n} Y_i$, $Y_i$ denotes the corresponding value of i region, Letting $W_{ij}$ represents an n*n diagonal matrix containing distance-based weights for observation i that reflects the distance between observation i and all other observations, Define $W_{ij} = 1$ for entities that share a common edge to the immediate right or left of the region of interest. Otherwise we would have all $W_{ij}= 0$ in this paper.

There are two kinds of spatial linear panel regression:

The first is spatial lag model (SLM):

$$y = \rho W y + \alpha_1 X + \alpha_2 X^2 + \varepsilon \tag{3}$$

$W_{ij}$ represents an n*n diagonal matrix, the lag variable is introduced to explain geography influence from neighbor region. $\rho$ represents the orientation and character from spatial effect. If $\rho > 0$, it denotes competition relationship among variables, and there is simulation action between some region and its neighbor. There is substitution when $\rho < 0$.

The second is spatial error model (SEM):

$$y = \alpha_1 X + \alpha_2 X^2 + \varepsilon$$

$$\varepsilon = \lambda W + \mu \tag{4}$$

Where y contains an n × 1 vector of cross-sectional dependent variables and X represents an n × k matrix of explanatory variables. Λ represents coefficient of spatial error, and W is known n × n spatial weight matrices, and ε is stochastic error.

# 3 Empirical Analysis

## *3.1 Variables and Data*

Energy efficiency depends on these industrial course including resource exploitation, procession, conversion, utilization, and so on. It can be measured by energy consumption per unit output value, energy consumption of unit product, energy consumption per unit architecture area. There is difference in the energy efficiency and economy development among the 31 provinces in China. It is usually described by energy consumption per capita GDP. According to the cross-sectional data from 2005 to 2008 of the 30 provinces in China except Xi Zang, E represents energy consumption per capita GDP in this paper. Its unit is the standard coal consumption per 10,000 yuan output value. Gdpp denotes economy development which is described by per capita GDP, and its unit is production value per 10,000 yuan. The data come from China Statistical Yearbook from 2006 to 2009.Xi Jiang is neglected.

## *3.2 Statistic Case and Moran's I Index*

The following formula is usually applied:

$$E = \alpha_0 + \alpha_1 gdp + \alpha_2 gdpsq \tag{5}$$

*E* represents energy efficiency, *gdp* and *gdpsq* are GDP per capita and its squared value. We compute these data from the year 2005 to 2008. Take the year 2008 for example (the other years are omitted, since the computing method is similar.):

The Table 1 shows that the coefficients of variables do not pass the statistic test. Therefore there is not environmental Kuznets curve between energy efficiency and economy. We should not ignore spatial influence among inter-provinces if there

**Table 1** OLS result of 2008

| Statistical variable | Coefficient | Std. dev | $P$-value |
|---|---|---|---|
| Constant | 1.4487 | 0.4747 | 0.0049 |
| GDP | 0.1188 | 0.3154 | 0.7093 |
| GDPSQ | −0.0367 | 0.0417 | 0.3861 |
| $R^2$ | 0.1172 | | |
| F | 1.8588 | | 0.1746 |

**Table 2** The Moran's I index of variables

| Variable | Moran's I | | | |
|---|---|---|---|---|
| | 2005 | 2006 | 2007 | 2008 |
| Per capita GDP | 0.3279 ($p = 0.002$) | 0.1277 ($p = 0.073$) | 0.1301 ($p = 0.0520$) | 0.1757 ($p = 0.0300$) |
| Energy efficiency | 0.3885 ($p = 0.003$) | 0.3074 ($p = 0.005$) | 0.5211 ($p = 0.008$) | 0.4385 ($p = 0.0020$) |

exits spatial correlation. In the view of spatial effect we validate if or not exiting the spatial inverted U curve between the two variables. According to (2), the Moran's I index of every variable is computed as Table 2:

Table 2 shows that it pass the statistic test at 5% level. It accounts for there exits spatial auto-correlation concerning with energy efficiency and economy development.

The overall energy efficiency in China is still very low and there are significant differences across provinces and regions as well. To some degree, the energy efficiency may be low for developing economy lag behind some level in despite of abundant resources. The index of resource consuming and economic development in some areas is higher, so are its neighbors'. The main reasons may include similar industry structure and technology feature, as well as the simulated economical policy and energy supervise measurement. Of course, the policy system, technological progress and price mechanism, etc. in some province benefit its neighbors in the improvement of energy efficiency. There is something wrong with the model (5) in the lack of spatial influence. It is necessary to consider the geography effect when we study the spatial inverted U cure.

## 3.3 Parameter Estimation

We only give the 2008 conclusion to save length although the data from 2005 to 2008 is computed separately with the tool of GEODA soft. Table 3 shows that spatial lag model (SLM) is not appropriate, but spatial error model (SEM)passes the statistical test. Furthermore, Likelihood Ratio Test, Akaike info criterion and Schwarz criterion attest to SEM which reflect the real correlation.

The conclusion of parameter estimation from 2005 to 2008 is as follows:

2005:

$$Y = 1.5166X - 0.2929X^2 + \varepsilon \quad \varepsilon = 0.0743W + \mu \tag{6}$$

**Table 3** Spatial regression result of 2008

| Variable | SLM | | | | SEM | | | |
|---|---|---|---|---|---|---|---|---|
| | Coefficient | Std.E | *z*-value | *P*-value | Coefficient | Std.E | *z*-value | *P*-Value |
| Constant | 1.0647 | 0.5891 | 1.8075 | 0.0707 | | | | |
| Gdpp2008 | 0.1941 | 0.2989 | 0.6492 | 0.5162 | 0.89046 | 0.152 | 5.8757 | 0 |
| Gdp2008sq | −0.044 | 0.0394 | −1.119 | 0.2632 | −0.118141 | 0.025 | −4.7814 | 0 |
| $\rho/\lambda$ | 0.1818 | 0.2268 | 0.8016 | 0.4228 | 0.4183 | 0.192 | 2.1821 | 0.029 |
| Stat. test | Sta. value | p-value | | | Stat. value | | | |
| LogL | −33.3155 | | | | −36.1073 | | | |
| LR ratio | 0.6339 | 0.4259 | | | 3.9534 | 0.047 | | |
| AIC | 74.6309 | | | | 74.2148 | | | |
| SC | 80.3669 | | | | 79.0827 | | | |

2006:

$$Y = 1.3120X - 0.2313X^2 + \varepsilon \quad \varepsilon = 0.0694W + \mu \tag{7}$$

2007:

$$Y = 1.1776X - 0.1799X^2 + \varepsilon \quad \varepsilon = 0.0743W + \mu \tag{8}$$

2008:

$$Y = 0.89046X - 0.11814X^2 + \varepsilon \quad \varepsilon = 0.4183W + \mu \tag{9}$$

Y represents energy efficiency, its unit is ton (standard coal consumption) every ten thousand Yuan output value. The regression conclusion is believable in model (6) to (9) .The coefficient of linear item is positive number, but the squared item' is negative. All these validate Hypothesis one that says there is the spatial inverted U curve between the energy efficiency and provincial economic development.

We pay attention to a positive $\lambda$ in model. Take 2008 for example, t means there is competition to develop economy and to improve energy efficiency among provinces. The energy consuming index published by every government is looking up. Affect by GDP, which is tightly associated with government performance evaluation, developing economy and wasting resources is the most important and popular. It is accounted for those governments at all levels would rather enlarge the power of attracting investment and to promote the local economic development.

## *3.4 Analysis of Result*

There is inverted U curve between energy efficiency and per capita GDP of provinces in China. How to compute the turning point value? It is as follows:

$$y = -c_1/(2 * c_2) \tag{10}$$

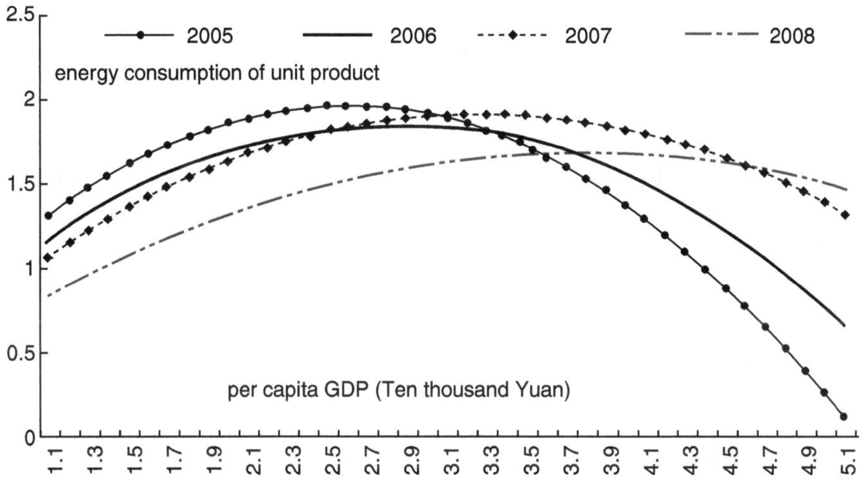


**Fig. 1** 2005–2008 energy efficiency and economy development

The turning point value is 2.5889, 2.8361, 3.2729, 3.7687 ten thousand Yuan from 2005 to 2008, separately based on the SEM model. The Fig. 1 shows that there is competition and smooth shift effect among every province because of spatial influence. Especially, more attention should be paid to innovation, and human capital input, energy saving, and energy consumption standard elevation, etc. The inverted U curve is wholly declining. The energy efficiency is improving year by year.

The result of SEM model discover that the per capita GDP of Shang Hai, Beijing, Tianjin in 2005 is separately higher than the turning point value 2.5889 ten thousand Yuan. Zhejiang, Jiangsu and Guangdong is near the turning point. The rest 24 province are all is under the value. It is similar in 2006 and 2007 except some small changes. When it comes to 2008, there is 20% provinces, such as Shang hai, Beijing, Tianjin, Zhejiang, Jiangsu, whose economy development have exceed the turning point. Guangdong is neat 3.7687 ten thousand. So, hypothesis two is correct.

## 4 Conclusion

The result indicates that the SEM is better than the simple regression model. It is concluded that the provincial energy efficiency and economy development have obvious spatial correlation and cluster in the geographical space. The SEM model is the better one to describe the inverted U curve between the two variables. The turning point value is near 3 ten thousand Yuan. It is from three to five provinces which exceed the turning point value from 2005 to 2008.

Certainly, to improve energy efficiency is not only the most urgent task, but also long and arduous . Each provincial social obligation compels them to obey the agreement signed by the Development and Reform Department in 2007. it is as follows: to improve industry structure, to restrict factories with energy-consuming resources, to popularize energy-saving technologies, to set appropriate price mechanisms, to regard energy efficiency as assessment index, and so on. What is more, openness, innovation and human capital input are all important. It is essential to grow at a reasonable speed and conduct energy saving continuously in order to reach the goal of reducing energy intensity by 20%, which is determined in the 11th 5-Year Program.

## References

Liu F-c, Liu Y-y, Pan X-f (2007) Dynamics of economic growth and energy consumption in China. Resour Sci 29(5):63–68

Li S-x, Cheng J-h (2008) Study on the energy efficiency of China and its determinants. Stat Res, 10:18–27

Qi S, Luo W (2007) Regional Economic growth and differences of energy intensity in China. Econ Res J 7:74–81

Shu Y-l, Wang H-m (2008) A study on relationship between regional resource consuming and economy growth. Ecol Econ 1:111–113

Wei C, Shen M-h (2007) Energy efficiency and energy productivity: a comparison based on the panel data by province. J Quant Tech Econ 9:110–121

Wang Y, Guo J-e, Xi Y-m (2008) Dynamic relationship between economic growth and China energy based on cointegration analysis and impulse response function. China Population Resour Environ 18(4):56–61

# Part VII
# Risk Management Modeling

# Sample Size Determination via Non-unity Relative Risk for Stratified Matched-Pair Studies

**Hui-Qiong Li and Liu-Cang Wu**

**Abstract** A stratified study is often designed for adjusting several independent trials in modern medical research. In this paper, we consider approximate sample size formulas for a non-unity relative risk in stratified matched-pair studies. To evaluate the accuracy and usefulness of these sample size formulae developed in this paper, we further calculate their simulated powers. Our empirical results confirm sample sizes formulae based on the constrained maximum likelihood estimation method can provide a sample size estimate that guarantees pre-specified power of a test at a given significance level. A real example from clinical studies is used to illustrate the proposed methodologies.

**Keywords** Sample size determinations · Score test · Sensitivity · Stratified matched-pair studies

## 1 Introduction

Assessment of non-inferiority is a popular issue in comparative studies. One often uses matched-pair non-inferiority trials to evaluate whether the effectiveness of a less toxic, easier to administer and/or inexpensive new diagnostic method is not inferior in terms of efficacy to that of the standard one. For example, Nam (1997) proposed a one-sided Wald-type statistic for testing non-inferiority via non-zero risk difference based on restricted maximum likelihood estimates of parameters under a null hypothesis of inferiority in a matched-pair design; Tango (1998) derived a score statistic for testing non-inferiority via relative risk with a re-parameterized model in a matched-pair design; Tang et al. (2002) derived approximate sample size

H.-Q. Li (✉)
Department of Statistics, Yunnan University, Kunming 650091, China
e-mail: lhuiqiong2002@yahoo.com.cn

L.-C. Wu
Faculty of Science, Kunming University of Science and Technology, Kunming 650093, China

Y. Zhou and D.D. Wu (eds.), *Modeling Risk Management for Resources and Environment in China*, Computational Risk Management,
DOI 10.1007/978-3-642-18387-4_54, 

formula for establishing equivalence/non-inferiority of two treatments via relative risk on the basis of Tang et al.'s (2003) score statistic; However, all the above mentioned works were confined to a single $2 \times 2$ table, and did not consider confounding effects.

In some clinical studies, ignoring confounding effects may lead to incorrect statistical conclusions. In this case, several statistical methods for testing non-inferiority/equivalence of two treatments were proposed in multiple independent $2 \times 2$ tables. For example, Nam (1995) considered a series of independent binomial variates relative risk or risk difference, and presented asymptotic power and sample size formula of the score test; Nam (2003) gave a homogeneity score test procedure for the interclass version of the Kappa statistics, and derived sample size formula for stratified studies; Recently, Nam (2006)considered the statistical testing for non-inferiority of two treatments via non-zero risk difference under matched-pair setting in a stratified study, and presented a efficient scores and a Mantel–Haenszel procedure with restricted maximum likelihood estimators of nuisance parameters. However, little work has been done for non-inferiority assessment of relative risk in stratified matched-pair designs. Motivated by the above mentioned work, the main purpose of this paper is to propose reliable methods for calculating sample sizes for equivalence/non-inferiority studies in stratified matched-pair designs.

This paper is organized as follows. Section 2 presents several sample sizes formulas of relative risk based on stratified data. Simulation studies are conducted to investigate the performance of various sample sizes formulas in Sect. 3. In Sect. 4, a real example is used to illustrate the proposed methodologies. Finally, some concluding remarks are given in Sect. 5.

## 2 Sample Size Calculation Based on Significance Test Approach

Consider a stratified matched-pair design in which two diagnostic methods (a new diagnostic method and a standard one) are respectively conducted for the same $n_j$ subjects in the *jth* stratum ($j = 1, 2 \cdots, J$). Let $x_{11j}, x_{10j}, x_{01j}$ and $x_{00j}$ be the observed numbers of pairs (1, 1), (1, 0), (0,1) and (0, 0) in the *jth* stratum, respectively; and let $p_{11j}, p_{10j}, p_{01j}, p_{00j}$ be their corresponding probabilities of the pairs (1, 1), (1, 0), (0, 1) and (0, 0) in the *jth* stratum, respectively.

$$0 < p_{ikj} < 1, p_{11j} + p_{01j} = p_{0j}, p_{11j} + p_{01j} = p_{0j}, p_{10j} + p_{00j} = q_{0j}, p_{1j} + q_{1j} \\ = 1.0, p_{0j} + q_{oj} = 1.0,$$

Let $\delta_j = p_{1j}/p_{0j}$ which is the relative risk between the probability of positive of the new diagnostic procedure and that of the standard diagnostic procedure in the *jth* stratum. In this paper, we consider a common relative risk between two marginal probabilities across J strata, i.e., $\delta_j = \delta$ for $j = 1, 2, \cdots, J$. Under this assumption, non-inferiority of the new diagnostic procedure compared with standard one can be expressed by the following one-sided hypothesis:

$$H_0 : \delta = \delta_0 \leftrightarrow H_1 = \delta > \delta_0$$

where $\delta_0$ is a clinical pre-specified acceptable value of inferiority, and assume $\delta_0 < 1$ without loss of generality.

## 2.1 Sample Size Calculation for the Score Statistic on the Basis of Stratified Data

In this article, we consider the score statistics for testing $H_0$

$$T_s = \frac{\sum_{j=1}^{J} \frac{x_{10j} + x_{11j} - (x_{11j} + x_{01j})\delta_0}{\delta_0(2\tilde{p}_{01j} + \tilde{p}_{0j}(\delta_0 - 1))}}{\left\{ \sum_{j=1}^{J} \frac{n_j}{\delta_0(2\tilde{p}_{01j} + \tilde{p}_{0j}(\delta_0 - 1))} \right\}^{\frac{1}{2}}} \tag{1}$$

which is asymptotically distributed as standard normal distribution under $H_0$ as $n_j \to \infty$ for $j = 1, 2 \cdots, J$ . The test rejects the null hypothesis $H_0$ in favor of $H_1$ at the nominal level $\alpha$ if $T_s \geq z_\alpha$ (see, Tang et al.'s 2003).

For score test on the basis of stratified data, the score function of log-likelihood under $H_0$ is

$$S = \sum_{j=1}^{J} \frac{x_{10j} + x_{11j} - (x_{11j} + x_{01j})\delta_0}{\delta_0(2\tilde{p}_{01j} + \tilde{p}_{0j}(\delta_0 - 1))} \tag{2}$$

Variance of $S$ under $H_0$ can be approximated by

$$V_{ar}(S|H_0) \approx \sum_{j=1}^{J} \frac{n_j}{(1 + \delta_0)\bar{p}_{01j} + (\delta_1 p_{0j} + p_{01j})(\delta_0 - 1)} = v_0$$

where $\bar{p}_{01j}$ is the asymptotic limit of $\tilde{p}_{01j}$. Asymptotic expectation and variance of $S$ under $H_1 : \delta = \delta_1$ are given by

$$E(S|H_1) \approx \sum_{j=1}^{J} \frac{n_j(\delta_1 - \delta_0)p_{0j}}{(1 + \delta_0)\bar{p}_{01j} + (\delta_1 p_{0j} + p_{01j})(\delta_0 - 1)} = E_1$$

$$V_{ar}(S|H_1) \approx \sum_{j=1}^{J} \left\{ \frac{n_j[2\delta_0 p_{01j} - \delta_0(1-\delta_0)p_{0j}}{[(1+\delta_0)\bar{p}_{01j} + (\delta_1 p_{0j} + p_{01j})(\delta_0 - 1)]^2} + \frac{(\delta_1 - \delta_0)(1-(\delta_1-\delta_0)p_{0j})p_{0j}}{[(1+\delta_0)\bar{p}_{01j} + (\delta_1 p_{0j} + p_{01j})(\delta_0 - 1)]^2} \right\} = v_1$$

respectively. Therefore, the approximate sample size required for power of $1-\beta$ based on the score statistic $T_s$ at the nominal level $\alpha$ can be obtained by solving the following equation: $E_1 = z_\alpha v_0^{\frac{1}{2}} + z_\beta v_1^{\frac{1}{2}}$. Define the design parameter $t_j$ for allocation among strata as $t_j = \frac{n_j}{N}$. Thus, follows from the above equation that the approximate sample size required for a desired power $1-\beta$ based on the score statistic $T_s$ for testing hypothesis $H_0$ at the nominal level $\alpha$ is given by

$$N_s = \frac{[G_0^{\frac{1}{2}} z_\alpha + G_1^{\frac{1}{2}} z_\beta]^2}{D^2}, G_0 = \sum_{j=1}^{J} \frac{t_j}{(1+\delta_0)\bar{p}_{01j} + (\delta_1 p_{0j} + p_{01j})(\delta_0 - 1),}$$

$$D = \sum_{j=1}^{J} \frac{t_j(\delta_1 - \delta_0)p_{0j}}{(1+\delta_0)\bar{p}_{01j} + (\delta_1 p_{0j} + p_{01j})(\delta_0 - 1),}$$

$$G_1 = \sum_{j=1}^{J} \left\{ \frac{t_j[2\delta_0 p_{01j} - \delta_0(1-\delta_0)p_{0j}}{[(1+\delta_0)\bar{p}_{01j} + (\delta_1 p_{0j} + p_{01j})(\delta_0 - 1)]^2} + \sum_{j=1}^{J} \frac{(\delta_1 - \delta_0)p_{0j}(1-(\delta_1-\delta_0)p_{0j})]}{[(1+\delta_0)\bar{p}_{01j} + (\delta_1 p_{0j} + p_{01j})(\delta_0 - 1)]^2} \right\}$$

## 2.2 *Sample Size Calculation for the Score Statistic Based on Pooling Data*

Similar to Nam (1995), we consider one case that $J$ $2 \times 2$ tables are pooled into a single $2 \times 2$ table, i.e., ignoring the confounding effect. Denote $x_{01} = \sum_{j=1}^{J} x_{01j}$, $x_{11} = \sum_{j=1}^{J} x_{11j}, x_{00} = \sum_{j=1}^{J} x_{00j}$, $x_{10} = \sum_{j=1}^{J} x_{10j}, N = \sum_{j=1}^{J} n_j$

Let $p_{10}, p_{01}, p_{11}, p_{00}$ be their corresponding probability, respectively. According to the arguments of Tang et al. (2003), we obtain that the score statistic based on pooling data is given by

$$T_{sp} = \frac{x_{11} + x_{10} - \delta_0(x_{11} + x_{01})}{\sqrt{N(1+\delta_0)\tilde{p}_{01} + (x_{11} + x_{10} + x_{01})(\delta_0 - 1)}} \tag{3}$$

where $\tilde{p}_{01}$ is the ML estimate of $p_{01}$ under $H_0$. Statistic $T_{sp}$ is asymptotically distributed as standard normal distribution under $H_0$ (Tang et al. 2003). Test rejects the null hypothesis of inferiority at the nominal level $\alpha$ if $T_{sp} \geq z_\alpha$.

Hence, the approximate sample size required for score statistic $T_{sp}$ on the basis of the pooling data to achieve the desired power of $1-\beta$ at $\delta = \delta_1$ with the nominal level $\alpha$ is given by

$$N_{sp} = \frac{\left[G_{sp0}^{\frac{1}{2}} z_\alpha + G_{sp1}^{\frac{1}{2}} z_\beta\right]^2}{D_{sp}{}^2} \tag{4}$$

where

$$G_{sp0} = (1+\delta_0)\bar{p}_{01} + (\delta_1 p_0 + p_{01})(\delta_0 - 1), D_{sp} = (\delta_1 - \delta_0)p_0,$$
$$G_{sp1} = 2\delta_0 p_{01} - \delta_0(1-\delta_0)p_0 + (\delta_1 - \delta_0)p_0(1 - (\delta_1 - \delta_0)p_0).$$

## 2.3 Sample Size Calculation for the Wald-Type Statistic

A Wald-type weighted statistic for testing $H_0$ can be expressed by

$$T_w = \sum_{j=1}^{J} \frac{w_j(\hat{\delta}_j - \delta_0)}{\hat{\sigma}_j}, \text{ Where } \hat{\sigma}_j{}^2 = \sum_{j=1}^{J} w_j{}^2 \frac{(x_{11j} + x_{10j})(x_{10j} + x_{01j})}{(x_{11j} + x_{01j})^3} \tag{5}$$

According to theories given in Bishop et al. (1975), it is easily shown that $T_w$ is asymptotically distributed as standard normal distribution under $H_0$. Test rejects the null hypothesis of inferiority at the nominal level $\alpha$ if $T_w \geq z_\alpha$.Thus, the approximate sample size required for the Wald-type weighted statistic $T_w$ to achieve a given power $1-\beta$ at a significance level $\alpha$ is given by

$$N_w = \frac{(z_\alpha + z_\beta)^2}{(\delta_1 - \delta_0)^2} \sum_{j=1}^{J} w_j{}^2 \delta_1 \frac{2p_{01j} + (\delta_1 - 1)p_{0j}}{t_j p_{0j}{}^2}. \tag{6}$$

Here, we consider three weights, i.e.,
(1) $w_j = n_j{}^{-1} / \sum_{j=1}^{J} n_j{}^{-1}$, (2) $w_j = \sigma_j{}^{-2} / \sum_{j=1}^{J} \sigma_j{}^{-2}$, (3) $w_j = \frac{1}{J}$.

# 3 Monte Carlo Simulation Studies

To investigate the validity of the asymptotic sample size formulae for controlling pre-specified power, we first calculate the approximate sample sizes for a power of 80% at the nominal level 5% with the above configuration and the alternative

**Table 1** Sample size and empirical power based on 10,000 trials for two strata

| $p_{02}$ | $p_{012}$ | $t_1$ | $t_2$ | $N_S$ | $N_{Sp}$ | $N_{W1}$ | $N_{W2}$ | $N_{W3}$ |
|---|---|---|---|---|---|---|---|---|
| 0.5 | 0.2 | 0.75 | 0.25 | 168 | 170 | 572 | 198 | 309 |
| | | | | 79.3 | 80.6 | 87.1 | 82.3 | 85.4 |
| | | 0.5 | 0.5 | 174 | 182 | 216 | 213 | 216 |
| | | | | 80.0 | 81.4 | 84.0 | 81.2 | 84.9 |
| | 0.3 | 0.75 | 0.25 | 180 | 194 | 850 | 174 | 433 |
| | | | | 81.2 | 79.7 | 85.9 | 81.0 | 87.2 |
| 0.8 | 0.1 | 0.75 | 0.25 | 110 | 114 | 124 | 136 | 110 |
| | | | | 80.5 | 81.3 | 82.5 | 89.5 | 85.7 |
| | | 0.5 | 0.5 | 82 | 85 | 117 | 85 | 117 |
| | | | | 81.7 | 83.6 | 85.5 | 83.9 | 85.6 |

hypothesis for J $= 2$; and then compute the corresponding empirical powers of the above obtained sample sizes based on 10,000 random samples. In this simulation study, we take $p_{01} = 0.5, p_{011} = 0.15$, $\delta_0 = 0.8$. The results are reported in Table 1.

In general, sample size formulae based on CMLE method provide fairly accurate sample size estimates in the sense that the empirical power of the estimated sample size is usually rather close to the pre-specified power level (e.g., $80\% \pm 2.2\%$) under the considered cases. We also observe that the sample sizes $N_s$ is generally smaller than those of others. All the above results demonstrate that the stratified score statistic $T_s$ outperform other statistics in stratified matched-pair designs.

## 4 A Worked Example

In this section, an example is used to illustrate the above proposed methods. First we consider an example from Tsai et al.'s (1989) paratuberculosis data analysis that has even been analyzed by Nam (2006). Here we also consider assessment of the two serological methods. Sensitivity is the proportion of positive serological results in cattle with positive fecal results, and specificity is the proportion of negative serological results in cattle with negative culture results. The sensitivities for DIA and ELISA are given by 0.68 and 0.65, respectively. Similarly, the specificities for DIA and ELISA are given by 0.84 and 0.86, respectively. Thus, the relative risk of DIA and ELISA for sensitivities is $\hat{\delta}_1 = 1.05$, whilst the relative risk of DIA and ELISA for specificities is $\hat{\delta}_2 = 0.97$. Now we want to know whether DIA is non-inferiority to ELISA for screening of suspected paratuberculosis infected cattle in positive and negative culture groups. In this case, we regard the positive and negative culture groups as two different strata. Sensitivity and specificity are those proportions of consistent serological results with the corresponding positive and negative cultures. Since estimates of $\delta_1$and $\delta_2$ have a little difference, we assume relative risk between DIA and ELISA is the same across strata. Thus, that DIA is non-inferiority to ELISA in terms of specificity and sensitivity can be expressed as the following one-sided hypothesis

$$H_0 : \delta = \delta_0 \leftrightarrow H_1 = \delta > \delta_0.$$

If we choose the significant level of the one-sided test to be 0.05, $\delta_0 = 0.9$, the results of the one-sided test are $T_s = 4.201$, $T_{sp} = 4.196$, $T_{w1} = 4.968$, $T_{w2} = 4.782$, $T_{w3} = 4.967$. This shows that all tests reject the null hypothesis and claims that the DIA is non-inferiority to ELSIA in terms of sensitivity and specificity jointly at 0.05.

Suppose a researcher wishes to undertake a study similar to that carried out by Tsai et al. (1989) in another clinical center. He may want to know how many subjects are required to achieve 80% power using the stratified score test, the unstratified score test, and Wald-type test at $\alpha = 5\%$ for testing $H_0 : \delta_0 = 0.9$ against $H_1 : \delta_1 = 1.0$ when the design parameters $t_1 = 0.51$, $t_2 = 0.49$, $p_{01} = 0.7$, $p_{011} = 0.09$, $p_{02} = 0.92$, $p_{012} = 0.03$. In this case, we have $N_s = 109$, $N_{sp} = 128$, $N_{w1} = 130$, $N_{w2} = 152$, $N_{w3} = 134$ which indicates that sample size based on the score statistic is significantly smaller than those based on other statistics.

## 5 Conclusion

In this article, we consider non-inferiority test of a new diagnostic procedure compared with the standard one in stratified matched-paired designs. Meanwhile, we also consider approximate sample size formulas for a non-unity relative risk in stratified matched-pair studies. Our empirical results show sample sizes $N_s$ is generally smaller than those of others. In most cases, the stratified score statistic $T_s$ outperform other statistics in stratified matched-pair designs. In view of the above reasons, we recommend the usage of the stratified score test.

**Acknowledgements** This work is fully supported by grants from Natural Science Foundation of Yunnan University (2008YB025), Program of Educational Commission of Yunnan Province (09Y0046), Natural Science Foundation of Yunnan (2009ZC039M) and Doctoral Foundation of Kunming University of Science and Technology (2009–024).

## References

Bishop YM, Fienberg SE, Holland PW (1975) Discrete multivariate analysis: theory and practice. MIT Press, Cambridge

Nam J (1995) Sample size determination in stratified trials to establish the equivalence of two treatments. Stat Med 14:2037–2049

Nam J (1997) Establishing equivalence of two treatment and sample size requirements in matched-pairs design. Biometrics 53:1422–1430

Nam J (2003) Homogeneity score test for the intraclass version of the kappa statistics and sample size determination in multiple or stratified studies. Biometrics 59:1027–1035

Nam J (2006) Non-inferiority of new procedure to standard procedure in stratified matched-pair design. Biom J 48:966–977

Tang ML, Tang NS, Chan ISF, Chan BPS (2002) Sample size determination for establishing equivalence/noninferiority via ratio of two proportions in matched-pair design. Biometrics 58:957–963

Tang NS, Tang ML, Chan ISF (2003) On tests of equivalence via non-unity relative risk for matched-pair design. Stat Med 22:1217–1233

Tango T (1998) Equivalence test and confidence interval for the difference in proportions for the paired-sample design. Stat Med 17:891–908

Tsai SJ, Hutchinson LJ, Zarkower A (1989) Comparison of dot immunobinding assay, enzyme-linked immunsorbent assay and immunodiffusion for serodiagnosis of paratuberculosis. Can J Vet Res 53:405–410

# The Portfolio Risk Analysis Based on Dynamic Particle Swarm Optimization Algorithm

**Qin Suntao**

**Abstract** Risk prediction about investor portfolio holdings can provide powerful test of asset pricing theories. In this paper, we present dynamic Particle Swarm Optimization (PSO) algorithm to Markowitz portfolio selection problem, and improved the algorithm in pseudo code as well as implement in computer program. Furthermore in order to prevent blindness in operation and selection of investment, we tried to make risk least and seek revenue most in investment and so do in the program. As used in practice, it showed great application value.

**Keywords** Dynamic particle swarm optimization · Financial investment selection · Investment combinations · Uncertainty

## 1 Introduction

It is really full of risks and complexities in financial investment, most investors know theory that they should not place all the eggs in one basket, they always use investment combination or multi-invest to disperse risks. Dr. Harry M.Markowitz, a Nobel Laureate and the father of Modern Portfolio Theory, studying the effects of investment and portfolio variability, return, and correlation. Famous for his work in economics, Dr. Markowitz has made equally significant strides in the field of technology. He was awarded the prestigious Von Neumann Prize in Operations Research Theory for his work in portfolio theory, sparse matrix techniques and the SIMSCRIPT programming language. Dr. Markowitz was also recently given the "Man of the Century" award by Pensions and Investments magazine for his life's work in the field of investments. He provided a comprehensive theoretical

Q. Suntao
Department of Information Management, ZheJiang University of Finance and Economics, TX 310018, China
e-mail: qst@zufe.edu.cn

Y. Zhou and D.D. Wu (eds.), *Modeling Risk Management for Resources and Environment in China*, Computational Risk Management,
DOI 10.1007/978-3-642-18387-4_55, 

framework for analysis of the investment portfolio. After his research there are lots of stock investors and mathematicians have done many works by tools of mathematics and statistics, they investigate kinds of investment strategies problem, portfolio of securities is an integrated whole, each security complementing the other, those researchers built a series math models and kinds of algorithms. Stoneb used a linear programming formulation to solve the general portfolio selection problem (Stoneb 1973), Best and Graner did some works sensitivity analysis for mean variance portfolio problem (Best and Graner 1991), Merton analgised optimal consumption and portfolio rules in a continuous time model (Merton 1971). In 1676 Sir Isaac Newton wrote his friend Robert Hooke, 'If I have seen further it is by standing on the shoulders of giants' and that is pretty true of those researchers as well. Large scale portfolio selection computation is always a challenge, besides those traditional optimize algorithm, the recent development of modern intelligent algorithm has far speed and can parallel compute, in this paper we will use a optimized algorithm based on iterative, dynamic particle swarm optimization algorithm, I tried this method to solve portfolio selection problem.

As we know all the investors like return and dislike risk, goal of portfolio is provides the minimum risk for every possible level of return, most portfolio model around the total expected return (average) $E$ and variance $V$. we assume that $X = (x_i)$, $i \in \mathrm{N}$, $1 \geq \mathrm{x_i} \geq 0$, $i \in \mathrm{N}$ is a possible portfolio selection, $x_i$ is the racial to invest stock I, then the basic portfolio model is as below:

$$\begin{cases} \max R(x) = (r_1, r_2, \ldots, r_n)(x_{1,}, x_{2,}, \ldots, x_n) \\ \min V(x) = (x_{1,}, x_{2,}, \ldots, x_n)^T Q(x_{1,}, x_{2,}, \ldots, x_n) \\ s.t.(x_{1,}, x_{2,}, \ldots, x_n)^T I = 1, (x_{1,}, x_{2,}, \ldots, x_n) \geq 0 \end{cases} \quad (1)$$

$\mathrm{R} = (\mathrm{r_1}, \mathrm{r_{2,}} \ldots, \mathrm{r_n})$ is the return of portfolio, it random as joint distribution, $X = (X_1, \ldots, X_n)$ is the proportion vector of every kinds of stocks invested, here we assumed that $\mathrm{X_i} \geq 0$, $(\mathrm{i} = 1,2,\ldots,\mathrm{n})$, $Q$ is covariance matrix, $Q = [\delta_{ik}]$, $\delta_{ik} = E[(r_i - \mu_i)(r_k - \mu_k)]$, E is expected return of a portfolio, it is the weighted sum of the expected return from each of those securities.

After analysis, such expect return and variance discussion way, we can use efficient function as $\mu = f(E(R), V)$, for defined risk plane, investor prefer higher return, that is $\partial\mu/\partial(E(R)) > 0$; for certain expected return, investor like lower risk, that is $\partial\mu/\partial(V) < 0$.

But recently there are more and more people are suspicious of Markowitz's taken variance as risk measure factors. They think that it should have to strict hypothesis when use variance to compute risk, it need expect return spread as normal school; and at same time the assumption of binomial efficient function is not realistic; and in Markowitz's algorithm is should be symmetry that negative and positive of warp, it is also not match investor real psychology feeling, in fact investor always endow with big weight for the lost risk.

For prevent those shortcomings in practical compute, we use real data to this return and risk model by Markowitz's theory frame. As continually training this model by PSO, take Chinese stock market as example, we analysis portfolio selection problem with transaction cost.

We assume that investor have 1 unit bankroll, he want buy n kinds of stocks $S_i$, $i \in N = \{1, 2, \ldots, n\}$, their prices are: $d_i$, $i \in N$. We assume the investment redound of each stock is $r_i$, $i \in N$, $r_i$ can stand by real data in past time, for future time $r_i$ can be used by expectation. The transaction cost to $x_i$ unit stock $S_i$ can be defined as follow:

$$c_i(x_i) = \begin{cases} 0 & x_i = 0 \\ p_i u_i & 0 < x_i < u_i \\ p_i x_i & x_i \geq u_i \end{cases}$$

Here $u_i > 0, p_i \geq 0, p_i u_i$ is initial transaction cost, whenever there is transaction, there should be cost at least. $u_i$ is the threshold of initial transaction cost, when the exchange quantity is more than that threshold, then transaction fee will be counted by the exchange quantity. $p_i$ is the rate transaction fee, it is the cost of one unit transaction.

$$\sum_{i=1}^{n} [d_i x_i + c_i(x_i)] \leq 1$$

So the net return equals investment expected return minus transaction cost:

$$R = \sum_{i=1}^{n} [r_i x_i - c_i(x_i)]$$

$$E = \sum_{i=1}^{n} x_i \mu_i ,$$

here

$$\mu_i = E(r_i)$$

From Markowitz's theory, the variance of capital portfolio return is:

$$V(x) = \delta^2 = \sum_{i=1}^{n} x_i^2 \delta_i^2 + \sum_{i=1}^{n} \sum_{\substack{k=1 \\ k \neq i}}^{n} x_i x_k \delta_{ik}$$

Here: $\delta_{ik} = E[(r_i - \mu_i)(r_k - \mu_k)]$

$$\delta_i^2 = \delta_{ii} = E(r_i - \mu_i)^2$$

In fact, in this model, for each stock $i$, $\delta_i^2$ is the risk that can be separated, and $\delta_{ik}$ is market risk that can not be.

We can find from those functions that this is a searching process for optimize combination. As we know there are so many ways for colony intelligent optimization searching problem, in this paper we will use Particle Swarm Optimization (PSO) in this case.

## 2 Particle Swarm Optimization Algorithm

In 1995, by elicitation of birds' looking for food, Kennedy and Eberhart brought forward Particle Swarm Optimization (PSO) (Eberhart and Kennedy 1995; Eberhart and Shi 2001), this is the presentation of colony intelligent optimization. In this algorithm, it search optimization by society information of agent in colony, all the particles will be decided by the fitness of an optimized function, the searching process will adjust movement contrail by each particle local optimization and whole optimization of all colony. Initial set of PSO is a cluster of random particles, each particle has its speed $v$ and position $X$. We can get optimization by iterative training. Here $X$ is an array of $n$ dimension, we can take it as proportion array of portfolio of different stocks that investor chosen. In each iterative, particles adjust themselves by these two extremums as below: one $P_i$ is the partial optimization searched by particle itself from beginning till now, and the other g is optimization of whole colony. Particles will update its speed and position by the expressions (2) and (3):

$$V_i = w \times V_i + c_1 \times Rand() \times (p_i - X_i) + c_2 \times rand() \times (g - X_i) \tag{2}$$

$$X_i = X_i + v_i \tag{3}$$

Rand( ) is function can release a random data from 0 to 1, $C_1$ and $C_2$ is positive fixed data, as matter of fact they are learning factors, we take $C_1$ as perceive coefficient and $C_2$ as society coefficient, usually it needs adjust $C_1$ and $C_2$ when real computation to get efficient constringency, w is weight of inertia.

The basic steps of our optimized particle swarm algorithm are as follow:

*Step 1*. Abstract optimized goal, define fitness function as inhibit and object formula (1). At same time we can choose some particles as initial particle swarm by experience. Include initialized speed and position for each particle of swarm;

*Step 2*. Compute the fitness of all particles by fitness function;

*Step 3*. Update extremum for each particle according its fitness, compare the best of the swarm with object of *pbest* (partial optimization), if current fitness is better than quondam opt *pbest*, then take current position and speed as *pbest*;

*Step 4*. Update exremum for the swarm, compare current best computation one in the swarm with the best one of swarm *gbest* (global optimization) in the record, if the current better than the record, than get them exchange, we also get the best *gbest* so far;

*Step 5*. Iterative compute the speed and position according to formula (2) and (3);
*Step 6*. Repeat step 2 to step 5, till iterative computation satisfied condition of cease the process, then output optimized solution *gbest* and its fitness.

By MATLAB programming, we can get our application of the algorithm model, we training the model with the real data, and found that the model can work efficiently.

## 3 Practical Application

As usual operation, we tried to choose some open funds in now Chinese stock market as our sample, initially *Zhongyou Hexin* (590002), *Jiashi Celue* (070011), *Nanfang Jiyou* (202003), *Baokang Linghuo* (240002), *Guangfa Jufu* (270001), *Jiashi Fuwu* (070006), *Boshi Zhuti* (160505), *Jiashi Zhuti* (070010), *Changcheng Xiaofei* (200006), *Guotou Ruifu* (121007), *Boshi Sanchan* (050008), *Boshi Jiazhi* (050001), *Jiashi Zengzhang* (070002), *Zhongyou Zhuti* (590005), *Zhongyin Zengzhang* (163803), *Huifeng Jinxin* (540003), *Shenwan Jingji* (310358), *Hua'an Baoli* (040004), *Shenwan Shengli* (310318), *Gongyin Dapan* (481008), these 20 open funds with not much relation are as our first candidates , we want to decide a fine portfolio strategy to gain a better return. In real application, we always aim a few of funds in the range of vision, and then decide ideally some of them to invest, we did as usual application.

In practical computation by program, $C_1$ and $C_2$ could be defined in range 0–4, of cause, we also set the biggest cycle times and least error limit, so the program can be end in proper time. In this case, the biggest cycle time is 2000, and the least error is 1, whatever we also can adjust these pausing condition at anytime. To check up the way we discuss above, we choose past time data period from 16 October to 28 November 2007 with net value each day. Our target is to find some potential revenue funds from those 20s.

After computer program running no more than 20 cycles, we restricted the limit of funds kinds in 3, then we got our chosen: *Guotou Ruifu* (121007), *Shenwan Yingli* (310358) and *Hua'an Baoli* (040004), this is really approach to the practice result, it is a least to bad scheme in that dull environment of all stocks in china, owe to world economical crisis.

## 4 Evaluation and Experimental Results

Compare with Genetic Algorithm (GA), the predominance of PSO is simple easy to realize, and there are not too much parameters to adjust. In practical computation, we still found that it is difficult to define inertia weight w, cognize coefficient $C_1$ and society coefficient $C_2$ and also difficult to combine them together, sometime even slight difference would cause giant aftermath. Especially for those high dimensions function with infinitum partial extremum points, it is very hard to get optimization with this algorithm. After our practical debug and test, we found out the main reasons that there are wrong combinations of those kinds of parameters and target function might have infinitude partial extremum, always in that situation algorithm would get into partial optimize. To prevent the algorithm run into partial convergence too early, we introduce particle position mutation to enlarge the searching scale of solution, to increase the constringency possibility of entire.

After all kinds of ameliorate, it's still not efficient when more than 20 dimensions, always difficult to approach to entire optimization, so we should obviate situation more than 20 dimensions in this approach.

## References

Best M, Graner RR (1991) The analytic of sensitivity analysis for mean variance portfolio problem. Int Rev Financial Anal 1:17–37

Eberhart RC, Kennedy J (1995) A new optimizer using particle swarm theory. In: Proceedings of the sixth international symposium on micro machine and human science, IEEE Service Center, Piscataway, NJ, Nagoya, Japan, pp 39–43

Eberhart RC, Shi Y (2001) Particle swarm optimization: developments, applications and resources. In: Proceedings of the congress on evolutionary computation 2001 IEEE service center, Piscataway, NJ, Seoul, Korea

Kennedy J, Eberhart RC (1995) Particle swarm optimization. In: Proceedings of the IEEE International Conference on Neural Networks, vol IV, IEEE Service Center, Piscataway, pp 1942–1948

Merton JL (1971) Optimal consumption and portfolio rules in a continuous time model. Econ Theory 3:771–802

Stoneb BK (1973) A linear programming formulation of the general portfolio selection problem. J Financial Quant Anal 4:621–638

# Modelling Uncertainty in Graphs Using Regime-Switching Model

**Fengjing Cai, Yuan Li, and Huiming Wang**

**Abstract** We introduce the Markov regime switching model to describe the uncertainty in graphs and design the algorithm by the Markov chain Monte Carlo method. The regime-switching graphical model is applied to the stock market of Shanghai in China to study the conditional dynamic correlation of five segments of the stock market. Empirical results show that the two regimes reflect high and low correlation and the persistent probability of regime is comparatively large. Our results have potential implication for portfolio selection.

**Keywords** Bayesian · Graphical Model · Regime Switching · Uncertainty

## 1 Introduction

Estimating the variance–covariance structure matrix of multivariate data is of fundamental importance in statistical analysis. In empirical finance, the statistical estimation of the covariance matrix is central to asset pricing, portfolio optimization, and various investment strategies based on market data. However, thousands of stocks, bonds, and other assets are traded on the market, which makes the estimation of the variance–covariance matrix particularly challenging.

---

The research was supported by the National Natural Science Foundation of China (NSFC 10971042) and the Project of Wenzhou Science & Technology Bureau (R2010030).

F. Cai
School of Mathematics & Information Science, Wenzhou University, 325035, Zhejiang Province, China
e-mail: cyclie@163.com

Y. Li (✉)
School of Mathematics & Information Science, Guangzhou University, 510006, Guangdong Province, China
e-mail: yli57@hotmail.com

H. Wang
Business School, Hohai University, 210098, Jiangsu Province, China
e-mail: huimingwang@hotmail.com

Y. Zhou and D.D. Wu (eds.), *Modeling Risk Management for Resources and Environment in China*, Computational Risk Management,
DOI 10.1007/978-3-642-18387-4_56, © Springer-Verlag Berlin Heidelberg 2011

Many scholars seek to reduce the dimension of the parameter space by borrowing from economic theory, or mining for structure directly from the data. The newly-developed graphical model is an alternative to the existing structural models. The graphical model is to merge the probabilistic concept of conditional independence with graph theory. For an introduction to graphical model, we refer to Whittaker (1990), Cox and Wermuth (1993), Edwards (2001). A mathematical rigorous treatment can be found in Lauritzen (1996).

As noted recently by Plerou et al. (2000), correlations among financial variables may evolve through time. Recent research shows that correlation differs between volatile and tranquil periods (Longin and Solnik 2001; Goetzmann et al. 2005).The time-varying variance-covariance matrix has been widely accepted. Previous work has focused on modelling those changes using multivariate stochastic volatility models or auto-regressive heteroskedasticity models (Engle 2002).

Talih (2003)) presents an alternative to these models that focuses instead on the latent graphical structure related to the precision matrix. They developed a graphical model for sequences of Gaussian random vectors when changes in the underlying graph occur at random times, and a new block of data is created with the addition or deletion of an edge. The method is applied to study multivariate financial data of US industry portfolios. However, they assume that the graph structure changes slowly over time, which violates the assumption that the parameters of each model in each segment are independent.

An alternative approach, the Markov regime switching model (Hamilton 1989) is proposed to provide the time-varying graphs to describe model uncertainty in this thesis. We assume that the graph structure follows a regime switching model. The transitions between the regimes are governed by a Markov chain. Given parameterization, the algorithm is designed by MCMC methods. At last, the regime-switching graphical model is applied to the stock market in China.

## 2 Our Approach

Let $\mu$ denote the mean vector, and $K = \sum^{-1}$the precision matrix, of the d-dimensional multivariate assets return vector. In the work of Markowitz (1952), investors seek linear combinations of assets that achieve the prescribed return level and minimize the incurred risk. Such portfolios are called mean-variance efficient. The optimal allocation $w_\lambda$ of assets can be written as $w_\lambda = \lambda K \mu$, where the Lagrange multiplier $\lambda$ depends on the investor's preference, measuring the trade-off between the investor's expected return and the risk that investor is willing to incur.

Let returns $Y = (Y_i)_{i \in V}$ are multivariate Normal random vector, indexed by the vertices of an undirected graph $G = (V, E)$. The undirected graphs we consider are so-called conditional independence graphs, in that $(i, j) \notin E$ if and only if $Y_i \perp\!\!\!\perp Y_j | Y_{V \setminus \{i,j\}}$.For graphical Gaussian models, $(i, j) \notin E$ if and only if $k_{ij} = 0$.

We follow Talih (2003) and Zhang (2008), and parameterize the precision matrix $K = (k_{ij})_{d\times d}$ as follows:

$$k_{ii} = \frac{v_i}{\sigma_i^2}, \quad k_{ij} = -\frac{\theta}{\sigma_i \sigma_j} \times I_{\{(i,j)\in E, i\neq j\}}$$

where $I_{\{\}}$ denotes a indicator function, $(i,j)$ represents the undirected edge between vertices $i$ and $j$, $v_i = \max(1, \#\{j : (i,j) \in E\})$.In order to keep $K$ positive definite, we have to limit $|\theta|<1$.

We assume that the time-varying graphs are denoted by $G(S_t)$, where $S_t$ is an unobserved state. The probability law governing $S_t$ is defined by its transition probability matrix, denoted by $P$. The probability of going from regime $i$in period $t$ to regime $j$ in period $t+1$ is denoted by $p_{ij}$.

For simplicity, we assume $\theta$ and $\sigma$ are constant. The likelihood function for $Y$ is given by

$$f(Y|K,S) = \prod_{t=1}^{T} (2\pi)^{-\frac{d}{2}} \cdot |K(S_t)|^{\frac{1}{2}} \cdot exp\left\{-\frac{1}{2} trace(K(S_t)(Y_t-\mu)'(Y_t-\mu))\right\}$$

The joint posterior distribution of parameter is given

$$\begin{aligned} f(G,\theta,\sigma,S,P|Y) &\propto f(Y|K,S) \cdot f(S,P|K) \cdot f(K) \\ &\propto f(Y|G,\theta,\sigma,S,P) \cdot f(S,P|G,\theta,\sigma) \cdot f(G) \cdot f(\theta) \cdot f(\sigma) \end{aligned} \tag{1}$$

# 3 Algorithm

It is difficult to get the solution from (1) directly. We consider to draw the sample from the joint posterior using Monte Carlo simulation. For this purpose, The Gibbs sampling technique and Metropolis-Hastings iterative method are employed for estimation of the model. Given the conditional posterior distributions, we implement the sampling to generate sample draws. The following steps can be replicated until convergence is achieved.

Step 1: Specify starting values for the parameters: $\tilde{S}_T = \{S_1, S_2, \cdots, S_T\}$, $G$, $\sigma$ and $\theta$.

For the prior distribution of $\theta$ and $\sigma^2$, we follow Talih (2003) and Zhang (2008) to assume that

$$p(\theta) \propto \frac{\pi/2}{\cos^2(\theta \cdot \pi/2)}, \quad p(\sigma_i^2) \propto \frac{1}{\sigma_i^2} (i = 1, \cdots, d)$$

For undirected graph, there are at most $d(d-1)/2$ edges. We assume that each graph has the same prior probability, i.e., the prior distribution is described as follows:

$$p(G_j) = 2^{-d(d-1)/2} \quad (j = 1, 2, \cdots, m)$$

where $m$ denotes the number of regime.

Step 2: Generate $G^{new}$, $\theta^{new}$ and $\sigma^{new}$ from $p(G|\tilde{S}_T,Y), p(\theta|\tilde{S}_T,Y)$ and $p(\sigma|\tilde{S}_T,Y)$ using Metropolis-Hastings algorithm.

1. Metropolis-Hastings designs for graph $G$
We consider M-H designs of $G_j$ by changing graph from $G_j^{old}$ to $G_j^{new}$ and keep $\theta$, $\sigma$, $G_{-j}$, $P$ and $\tilde{S}_T$ unchanged, where $G_{-j}$ represents the remain graph except $G_j$. We assume that at most one edge is changed from $G_j^{old}$ to $G_j^{new}$.With the assumption of equal probability, the designed conditional probability density $q(G_j^{new}|G_j^{old})$ is as follows:

$$q\left(G_j^{new}|G_j^{old}\right) = 2^{-d(d-1)/2} \quad (j = 1, 2, \cdots, m)$$

Similarly,

$$q\left(G_j^{old}|G_j^{new}\right) = 2^{-d(d-1)/2} \quad (j = 1, 2, \cdots, m)$$

The acceptance ratio of iterative for graph $G_j$ is given by

$$\begin{aligned}\alpha(G_j^{old}, G_j^{new}) &= \min\left\{1, \frac{f(Y, G_j^{new}, G_{-j}, \theta, \sigma, \tilde{S}_T, P)}{f(Y, G_j^{old}, G_{-j}, \theta, \sigma, \tilde{S}_T, P)} \times \frac{q(G_j^{new}|G_j^{old})}{q(G_j^{old}|G_j^{new})}\right\} \\ &= \min\left\{1, \frac{f(Y|G_j^{new}, G_{-j}, \theta, \sigma, \tilde{S}_T, P)}{f(Y|G_j^{old}, G_{-j}, \theta, \sigma, \tilde{S}_T, P)}\right\}\end{aligned}$$

2. Metropolis-Hastings designs for $\theta$
Keeping $G$, $\sigma$ unchanged, we assume that $\tan(\theta\pi/2)$ follows the random walk:

$$\tan(\theta^{new}\pi/2) = \tan(\theta^{old}\pi/2) + \varepsilon, \quad \varepsilon \sim N(0, 0.9^2)$$

Simple computation shows that the acceptance ratio for parameter $\theta$ is given by

$$\alpha(\theta^{old}, \theta^{new}) = \min\left\{1, \frac{f(Y|G, \theta^{new}, \sigma, \tilde{S}_T, P)}{f(Y|G, \theta^{old}, \sigma, \tilde{S}_T, P)}\right\}$$

3. Metropolis-Hastings designs for $\sigma$
Keeping $G$, $\theta$ and $\sigma_{-i}$ unchanged, where $\sigma_{-i}$ represents the vector left by removing $\sigma_i$ from $\sigma$. We assume that $\log(\sigma_i^2)$ follows the random walk:

$$\log\left((\sigma_i^{new})^2\right) = \log\left((\sigma_i^{old})^2\right) + \varepsilon, \quad \varepsilon \sim N(0, 0.25^2)$$

Then the acceptance ratio for parameter $\sigma_i$ is given by

$$\alpha((\sigma_i^{old})^2, (\sigma_i^{new})^2) = \min\left\{1, \frac{f(Y|G, \theta, \sigma_{-i}^2, (\sigma_i^{new})^2, \tilde{S}_T, P)}{f(Y|G, \theta, \sigma_{-i}^2, (\sigma_i^{old})^2, \tilde{S}_T, P)}\right\}$$

Step 3: Generate $P^{new}$ from $p(P|\tilde{S}_T)$ using Gibbs sampling algorithm.
The transition matrix prior is assumed to be row-wise Dirichlet

$$p(P_{i\cdot}') \sim D(\alpha_{i1}, \alpha_{i2}, \cdots, \alpha_{im}) \quad (i = 1, 2, \cdots, m)$$

where $D$ denotes Dirichlet distribution with parameters $\alpha_{i1}, \alpha_{i2}, \cdots, \alpha_{im}$.
The posterior distribution is also Dirichlet

$$p(P_{i\cdot}'|\tilde{S}_T) \sim D(\alpha_{i1} + n_{i1}, \alpha_{i2} + n_{i2}, \cdots, \alpha_{im} + n_{im})$$

where $n_{ij}$ denotes the number of the transition from the regime $i$ to $j$, that can be counted from given $\tilde{S}_T$.

Step 4: Generate $\tilde{S}_T^{new}$ from $p(\tilde{S}_T|\Theta, Y)$ using multi-move Gibbs sampling algorithm, where $\Theta = \{G, \theta, \sigma, P\}$.

To sample the regime variable $\tilde{S}_T$, we employ the multi-move Gibbs sampling method, which was originally proposed by Carter and Kohn (1994) and applied to a Markov switching model by Kim and Nelson (1998).

The multi-move Gibbs sampling refers to simulating $\tilde{S}_T$, as a block from the following conditional distribution:

$$p(\tilde{S}_T|\Theta, Y) = p(S_T|\Theta, Y) \cdot \prod_{t=1}^{T-1} p(S_t|S_{t+1}, \Theta, Y)$$

To draw $S_t$ conditional on $S_{t+1}$, $\Theta$ and $Y$, we use the following results: $p(S_t|S_{t+1}, \Theta, Y) = \dfrac{p(S_{t+1}|S_t, \Theta, Y) \cdot p(S_t|\Theta, Y)}{p(S_{t+1}|\Theta, Y)} \propto p(S_{t+1}|S_t) \cdot p(S_t|\Theta, Y)$ where $p(S_{t+1}|S_t)$ is the transition probability, and $p(S_t|\Theta, Y)$ can be obtained from the Hamilton filter.

Step 5: Go to Step 2. Step 2 through Step 5 can be iterated N times to obtain the posterior densities. Note that the first L times iterations are discarded in order to attenuate the effect of the initial values.

## 4 Simulation

We generate the data using the graph shown in Fig. 1. Then we have two regimes in our data here, and the transition probability is given by $p_{11} = p_{22} = 0.7$.

Our sample consists of 400 independent observations drawn from a multivariate Normal distribution with mean vector zero, and a precision matrix constrained by the graph in Fig. 1, and parameterized by $\theta = 0.95$, $\sigma = \{0.6, 0.7, 0.6, 0.55, 0.6\}$.

We would like our proposed method to recover that fact. We use 10,000 replications in MCMC algorithm and discard first 1,000 draws of the chain. The posterior mean of $\theta$ and $\sigma$ is 0.9691 and {0.5393,0.6505,0.5357,0.5075,0.5565},

**Fig. 1** The original two-regime graphical structure

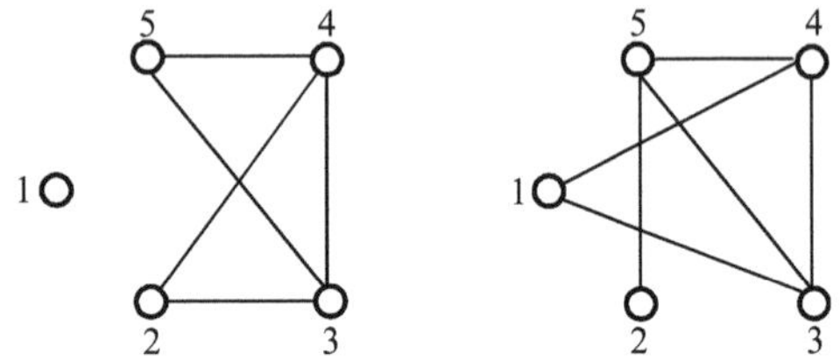


The posterior mean of $p_{11}$ and $p_{22}$ is 0.6674 and 0.7081. The mean is close to the original parameter. It is more important that both the posterior graphs are exactly the true graph.

## 5 Application

For the purpose of this thesis, we choose to work at the portfolio level using five industry index: Real States, Industries, Commerce, Public Utilities and Integration in the stock market of Shanghai in China. We select sample from May 15th 1999 to December 28th 2005, providing 315 weeks of data. The dataset is obtained from by data library of CSMARS in China. We follow Zhang (2008) and assume weekly log returns are multivariate Normality, and center the observations around their sample mean in order to focus only on the precision matrix.

In application, we assume there are two different regimes about graphs. We use 10000 replications in MCMC algorithm and discard first 1000 draws of the chain to find posterior distributions of the parameters. The posterior mean of $\theta$ and $\sigma$ is 0.9733 and $\{0.0248, 0.0171, 0.0198, 0.0160, 0.0195\}$. The posterior mean of $p_{11}$ and $p_{22}$ is 0.9726 and 0.7215.

Figure 2 shows that there are the different graphs between different regimes. We name the first regime as "stronger correlation", as there is seven edges in the first graph. The second regime is named as "weaker correlation", as there is four edges in the second graph. The probability that the stronger correlation will be followed by the stronger correlation is 0.9726, so the regime will persist on average for $1/(1\text{-}p_{11}) = 36.5$ weeks. The probability that the weaker correlation will be followed by the weaker correlation is 0.7215, so the regime will persist on average for $1/(1\text{-}p_{22}) = 3.6$ weeks.

A probabilistic inference can be calculated for each date $t$ in the sample given the posterior mean of the parameters. The smooth probability of stronger correlation regime is shown in Fig. 3. From May 1999 to September 2000, the weaker correlation occupy most of the time. From October 2000 to December 2005, it is associated with the first regime, stronger correlation. In fact, from 1999 to 2001, there is the bull market in China, while from then on China enters into a span of four years of bear market. The result is consistent with the view that the bear market reflects high correlation, while there is low correlation in the bull market.

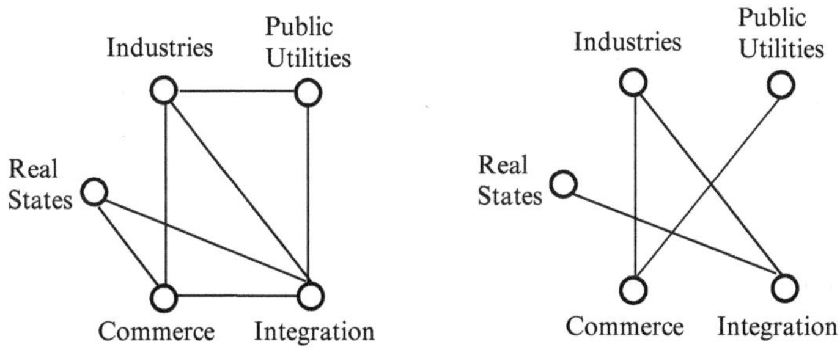


**Fig. 2** The two-regime graphical structure of industry index in China

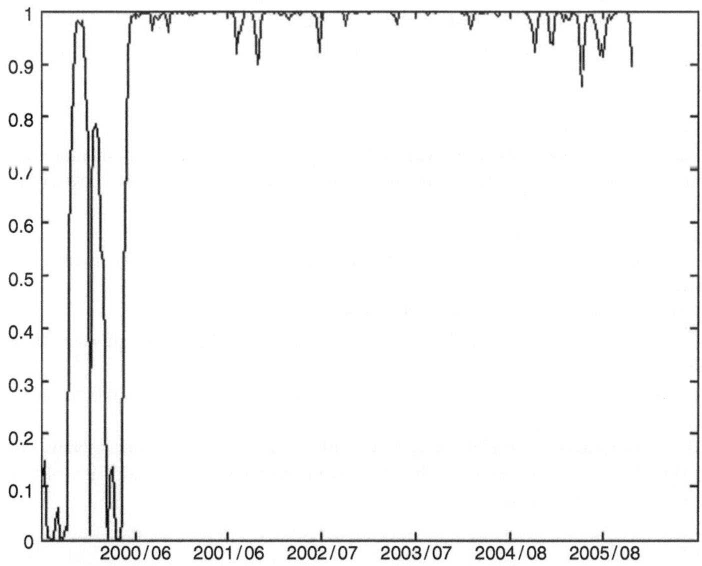


**Fig. 3** The smooth probability of the stronger correlation regime

# 6 Conclusion

The Markov regime switching for the graph structure is proposed in the paper. The Markov chain Monte Carlo method is applied to design the algorithm. The numerical simulation demonstrates the algorithm is effective. The model and algorithm are applied to study the conditional dynamic correlation of five industry index of the stock market in China. Empirical results show that the two regimes reflect high and low correlation and the persistent probability of regimes is comparatively large, denoting the regimes are stable. The smooth probability is calculated and shows that bear markets associate with the stronger correlation, while there is the weaker correlation in bull markets.

In fact, supposing the returns are Gaussian, the method can be applied to the optimal allocation of assets and value at risk measures. Our empirical distinction between bear and bull markets has potential implications for asset allocation and portfolio construction.

## References


Carter CK, Kohn P (1994) On Gibbs sampling for state space models. Biometrika 81:541–553

Cox DR, Wermuth N (1993) Linear dependencies represented by chain graphs. Stat Sci 8:204–218

Edwards D (2001) Introduction to graphical modeling. Springer, New York

Engle R (2002) Dynamic conditional correlation: a simple class of multivariate generalized autoregressive conditional heteroskedasticity models. J Bus Econ Stat 20:339–350

Goetzmann WN, Li L, Rouwenhorst KG (2005) Long-term global market correlations. J Business 78:1–38

Hamilton JD (1989) A new approach to the economic analysis of nonstationary time series and the business cycle. Econometrica 57:357–384

Kim CJ, Nelson CR (1998) Business cycle turning points, a new coincident index and tests of duration dependence based on a dynamic factor model with regime switching. Econ J 80:188–201

Lauritzen SL (1996) Graphical models. Clarendon, Oxford

Longin F, Solnik B (2001) Extreme correlation and international equity markets. J Finance 56:649–676

Markowitz H (1952) Portfolio selection. J Finance 7:77–91

Plerou V, Gopikrishnan P, Rosenow B, Amaral L, Stanley H (2000) Econophysics: financial time series from a statistical physics point of view. Phys A 279:443–456

Talih M (2003) Markov random fields on time-varying graphs with an application to portfolio selection. Yale University, New Haven

Whittaker J (1990) Graphical models in applied multivariate statistics. Wiley, New York

Zhang, F (2008) Detection of structural changes in multivariate models and study of time-varying graphs. Guangzhou University

# A Study on Dynamic Spatial Fixed Effect Model Based on Endogenous Initial Value

**Penghui Guo**

**Abstract** This paper proposes a dynamic spatial fixed effect model. First, simultaneously introduce both temporally and spatially lagged factors. Second, analyze both observable and unobservable spatial effects by taking the initial value as endogenous. Third, derive and prove the asymptotic properties and distributions of estimators, and undertake a Monte Carlo simulation. The simulation results show that the estimators improve as the sample size increases. Moreover, the degree to which the estimation results improve seems more sensitive to temporal dimension than to spatial dimension.

**Keywords** Asymptotic characteristics · Dynamic modeling · Endogenous initial value · Sensitive analysis

## 1 Introduction

The spatial and temporal dimensions of the panel data model depict the characteristics of variables in time and space. However, panel data model emphasizes heterogeneity in depicting spatial characteristics, it neglects the spatial correlation, a more important aspect of spatial analysis. In comparison, spatial econometric model takes account of both the spatial correlation and heterogeneity, and maximizes the "dimensions" of the econometric model by introducing a spatial weight matrix and spatial lag operators. Research on the spatial panel data model can be traced back to that of Anselin. In his spatial stochastic effect model, spatial correlation exists only in the error term. Baltagi generalized the model of Anselin and derived the maximum likelihood estimators (MLEs) and their corresponding LM and LR test statistics. Elhorst (2003) divided the models into eight categories

Supported by the National Social Science Foundation of China under Grant "10CTJ002".
P. Guo
Department of Statistics, School of Economics, Xiamen University, Xiamen 361005, China
e-mail: gpengfly@163.com

Y. Zhou and D.D. Wu (eds.), *Modeling Risk Management for Resources and Environment in China*, Computational Risk Management,
DOI 10.1007/978-3-642-18387-4_57, © Springer-Verlag Berlin Heidelberg 2011

including the spatial fixed model, the spatial stochastic effect model, and the spatial fixed coefficient model.

One drawback common to all of the above studies is that the models they discuss are static. A real economic system requires that the model control for two types of effects: the spatial effect and the temporal effect. The former reflects the impacts of location-varying (but not time-varying) background changes (e.g., changes in the economic structure and in natural endowment) on the level of stability. The latter represents the impacts of time-varying (but not location-varying) background changes (e.g., changes in the business cycle and temporary shocks) on the level of stability. Elhorst (2001) was the first to discuss the dynamics of spatio-temporal data, though his analysis is purely cross-sectional in nature and does not extend to the panel data model. Elhorst (2005) subsequently proposed a dynamic spatial error autocorrelation fixed effect model and suggested that unconditional maximum likelihood estimators (UMLEs) be obtained using either the Bhargava and Sargan approximation or the Nerlove and Balestra approximation. Yu et al. (2006) proposed a dynamic spatial autocorrelation fixed effect model where quasi maximum likelihood estimators (QMLEs) are derived when the number of spatial units $N$ and the time periods $T$ are large. Su and Yang (2007) considered the spatial effect in the error term and constructed a dynamic spatial model that combines both the fixed effect and the stochastic effect, in which QMLEs are derived under taking into account both endogenous and exogenous initial values for $Y_0$ when $N$ is large and $T$ is fixed.

However, it is difficult to determine whether a spatial correlation exists in the explained term or in the error structure, it is vital to construct a model that simultaneously considers both. This paper proposes a dynamic spatial fixed effect model, an integrated model that includes both dynamic temporal and spatial characteristics. It takes into account both spatial autocorrelation and spatial error autocorrelation simultaneously. The rest of this paper proceeds as follows. The next section proposes our model. The third section discusses the asymptotic characteristics of the QMLEs, and undertakes an assessment based on a Monte Carlo simulation. In the final section, we conclude our main findings.

## 2 Model Construction and Estimation

In this paper, we pay ample attention to spatial heterogeneity by considering only the individual effect and also assume the existence of the spatial effect that includes both observable and unobservable components of an economic system. Our model is constructed as follows:

$$\begin{aligned} Y_t = \mu_N + \tau Y_{t-1} + \lambda W_N Y_t + \gamma W_N Y_{t-1} + X_t \beta + U_t, U_t \\ = \rho M_N U_t + \varepsilon_t, \varepsilon_t \sim iid(0, \sigma_0^2 I_N) \end{aligned} \quad (1)$$

where $Y_t = (Y_{1t}, Y_{2t}, \ldots, Y_{Nt})$; $N$ denotes the spatial unit; $t = 1, 2, \ldots, T$ denotes time unit; $\mu_N$ denotes the non-stochastic individual effect; $X_t$ is an $N \times k$ exogenous independent variable matrix, $\beta$ is the corresponding coefficients; $W_N, M_N$ are $N \times N$ spatial weight matrices corresponding to the dependent variable and the error term, respectively, where the observable component is measured by $W_N Y_t$ and the unobservable component by $M_N U_t$, $\lambda, \gamma, \rho$ are their corresponding spatial correlation coefficients; and $\tau$ is the temporal auto-correlation coefficient.

Model (1) offers a convenient way of providing for model degeneration given some restrictive conditions. Specifically, when $\lambda = \gamma = 0$, (1) is degenerated into the dynamic spatial error autoregressive fixed effect model of Elhorst (2005); when $\rho = 0$, (1) is degenerated into the dynamic spatial autoregressive fixed model of Yu et al. (2006); when $\tau = \gamma = 0$, (1) is degenerated into the general static spatial fixed model of Lee and Yu (2010). Also, when impose zero constraints on corresponding parameters in (1), we obtain the spatial models of such as Elhorst (2003) and Baltagi et al. (2007).

For simplification, we define:

$Z_t = (Y_{t-1}, W_N Y_{t-1}, X_t); \theta = (\delta', \lambda, \rho, \sigma^2)'; \delta = (\tau, \gamma, \beta')', \varphi = (\delta', \lambda, \rho)'; S_N(\lambda) = I_N - \lambda W_N, B_N(\rho) = I_N - \rho M_N$ 1 can then be transformed into:

$$Y_t = S_N^{-1}(\lambda)\mu_N + S_N^{-1}(\lambda)Z_t\delta + S_N^{-1}(\lambda)B_N^{-1}(\rho)\varepsilon_t(\varphi) \tag{2}$$

Estimation results in dynamic regression usually depend on the initial value $Y_0$. This paper discusses only the condition that the initial value is endogenous and the exogenous condition is discussed in our forthcoming paper. When the initial value is endogenous and therefore requires to consider the joint probability distribution between the initial value and the other observations before constructing the likelihood function. Elhorst (2005) suggested that either the Bhargava and Sargan approximation or the Nerlove and Balestra approximation be used. When the initial value $Y_0$ is endogenous, it has the same data generating process (DGP) as $Y_t (t = 1, \ldots, T)$ and also has the same bounds, where $Y_t (t = 1, \ldots, T)$ is a stable time series. Under this assumption, neither (1) nor (2) is suitable for constructing a likelihood function directly. Instead, it is necessary to derive the joint probability distribution function first. Following the usual method for estimating the fixed effect model, we first make a first-order differentiation on (1) to eliminate $\mu_N$ and obtain:

$$\Delta Y_t = S_N^{-1}(\lambda)A_N(\tau, \gamma)\Delta Y_{t-1} + S_N^{-1}(\lambda)\Delta X_t\beta + S_N^{-1}(\lambda)B_N^{-1}(\rho)\Delta\varepsilon_t(\varphi), (t = 2, 3, \ldots, T) \tag{3}$$

where $A_N(\tau, \gamma) = \tau I_N + \gamma W_N$. The other items in (3) defined as (2). Iterating (3), we obtain:

$$\Delta Y_t = [S_N^{-1}(\lambda)A_N(\tau,\gamma)]^m \Delta Y_{t-m} + \sum_{i=0}^{m-1} [S_N^{-1}(\lambda)A_N(\tau,\gamma)]^i S_N^{-1}(\lambda)\left\{\Delta X_{t-i}\beta + B_N^{-1}(\rho)\Delta\varepsilon_{t-j}(\varphi)\right\} \quad (4)$$

$$\Delta Y_1 = [S_N^{-1}(\lambda)A_N(\tau,\gamma)]^m \Delta Y_{1-m} + \sum_{i=0}^{m-1} [S_N^{-1}(\lambda)A_N(\tau,\gamma)]^i S_N^{-1}(\lambda)\left\{\Delta X_{1-i}\beta + B_N^{-1}(\rho)\Delta\varepsilon_{1-j}(\varphi)\right\} \quad (5)$$

Note that because $\Delta X_{-i}(i = 1,\ldots)$ is unobservable, the probability function of $\Delta Y_1$ is also undetermined. Hsio et al. (2002) proposed two assumptions on the determination of $\Delta Y_1$:

1. $E(\Delta Y_1) = \pi_0 1_N$, where $1_N$ is an $N$ dimension column series and $\pi_0$ is an estimated coefficient. The assumption holds that the DGP begins from a past period that is not far from a zero period and the expected change in the given initial value is identical in each space unit.
2. $E(\Delta Y_1) = 0$. The assumption holds that the DGP begins from an unlimited past period from a zero period, which requires $m \to \infty$ and $|\tau|<1$.

Obviously, Assumption 2 is subject to a greater constraint and is a special case of Assumption 1. When $|\tau|<1$ and $\pi_0 = 0$, assumption 1 is the same as Assumption 2. Our discussion therefore continues on the basis of the less stringent first assumption.

In determining $\Delta X_{-i}(i = 1,\cdots)$, Bhargava and Sargan (1983) suggested that the prediction be based on all of the exogenous independent variables $\Delta X_i(i = 1,\ldots,T)$. Hsio et al. (2002) suggested using an approximation method based on a first-order differential fixed effect model. Following Bhargava and Sargan (1983), we have:

$$\sum_{i=0}^{m-1} [S_N^{-1}(\lambda)A_N(\tau,\gamma)]^i S_N^{-1}(\lambda)\Delta X_{1-i}\beta = \pi_0 1_N + \sum_{i=1}^{T} \Delta X_i \pi_i + \xi \quad (6)$$

then $E(\sum_{i=0}^{m-1} [S_N^{-1}(\lambda)A_N(\tau,\gamma)]^i S_N^{-1}(\lambda)\Delta X_{1-i}\beta|\Delta X_i, i = 1,\ldots,T) = \pi_0 1_N + \sum_{i=1}^{T} \Delta X_i \pi_i$, where $\xi \sim iid(0,\sigma_\xi^2 I_N)$, $E(\xi, \Delta X_i|_{i=1,\ldots,T}) = 0$. Note that (6) contains $(1+Tk)$ estimated coefficients which increase as $T$ increases, where the incidental parameter problem arises. Specifically, when $T\to\infty$, $(1+Tk)\to\infty$, then no estimation can be realized. Therefore, it is necessary to consider reducing the estimated coefficients in (6). One way to do this is to use the mean of $\Delta X_t$ as suggested by Su and Yang (2007). This gives us:

$$\sum_{i=0}^{m-1} [S_N^{-1}(\lambda)A_N(\tau,\gamma)]^i S_N^{-1}(\lambda)\Delta X_{1-i}\beta = \pi_0 1_N + \Delta\bar{X}\pi + \xi \quad (7)$$

where $\overline{\Delta X} = \frac{1}{T-1}\sum_{i=2}^{T} \Delta X_i$. Note that there are only $(1+k)$ coefficients to be estimated in (7), where $\pi = (\pi_1, \ldots, \pi_k)'$ is a $k \times 1$- series. Substituting (7) into (5), we have

$$\Delta Y_1 = \pi_0 1_N + \overline{\Delta X}\,\pi + e_1(\cdot) \tag{8}$$

where $e_1(\cdot) = \xi + \sum_{j=0}^{m-1} [S_N^{-1}(\lambda)A_N(\tau,\gamma)]^j [B_N(\rho)S_N(\lambda)]^{-1}\Delta\varepsilon_{1-j}(\varphi)$. According to (3), we have

$$\begin{aligned}
e_t(\varphi) &= [B_N(\rho)S_N(\lambda)]^{-1}\Delta\varepsilon_t(\varphi), (t = 2,3,\ldots,T)\\
E(e_t) &= 0, t = 1,\ldots,T; E(e_1(\cdot)e_t(\varphi)') = 0, (t = 3,\ldots,T)\\
E(e_1(\cdot)e_1(\cdot)') &= \sigma_\xi^2 + \sigma^2(B_N(\rho)S_N(\lambda))^{-1}V_N((B_N(\rho)S_N(\lambda))')^{-1}\\
&= \sigma^2(B_N(\rho)S_N(\lambda))^{-1}R_N((B_N(\rho)S_N(\lambda))')^{-1}\\
E(e_{t-1}(\cdot)e_t(\varphi)') &= -\sigma^2(B_N(\rho)S_N(\lambda))^{-1}((B_N(\rho)S_N(\lambda))')^{-1}, (t = 2,\ldots,T)\\
E(e_t(\cdot)e_t(\varphi)') &= 2\sigma^2(B_N(\rho)S_N(\lambda))^{-1}((B_N(\rho)S_N(\lambda))')^{-1}, (t = 2,\ldots,T)
\end{aligned}$$

where

$$\begin{aligned}
V_N(\tau,\gamma,\lambda) &= 2[(I_N + (S_N^{-1}(\lambda)A_N(\tau,\gamma))^{2m-1}) \times (I_N + S_N^{-1}(\lambda)A_N(\tau,\gamma))^{-1}]\\
R_N(\psi,\tau,\gamma,\lambda,\rho) &= \psi^2(B_N(\rho)S_N(\lambda))(B_N(\rho)S_N(\lambda))' + V_N, \psi^2 = \sigma_\xi^2/\sigma^2
\end{aligned}$$

Defining $e = (e_1', e_2', \ldots, e_T')'$,$\omega = (\psi,\tau,\gamma,\lambda,\rho)'$, we have

$$\operatorname{cov}(e,e) = \sigma^2((I_T \otimes B_N(\rho)S_N(\lambda))^{-1})R_{NT} \times (I_T \otimes ((B_N(\rho)S_N(\lambda))')^{-1}) = \sigma^2\Omega(\omega) \tag{9}$$

We can obtain the joint probability distribution function of $\Delta Y_t (t = 1,\ldots,T)$ using (9) and then construct the log-likelihood function of (3) as follows:

$$l(\theta) = \ln L(\theta) = -\frac{NT}{2}\ln(2\pi) - \frac{NT}{2}\ln(\sigma^2) - \frac{1}{2}\ln|\Omega(\omega)| - \frac{1}{2\sigma^2}e(\upsilon)'\Omega^{-1}(\omega)e(\upsilon) \tag{10}$$

where $\theta = (\pi_0,\pi',\beta',\tau,\gamma,\lambda,\rho,\psi,\sigma^2)'$, $\upsilon = (\tau,\gamma,\lambda,\beta')'$, and $Z_t = (Y_{t-1}, W_N Y_{t-1}, X_t)$, $(t = 2,3,\ldots,T)$. Defining $\varsigma = (\pi_0,\pi',\beta')'$, $\phi = (\tau,\gamma,\lambda,\rho)'$, then $\theta = (\varsigma',\phi',\psi,\sigma^2)'$. Given $\phi = (\tau,\gamma,\lambda,\rho)'$ and $\psi$, we maximize (10) to obtain the QMLEs of $\sigma^2$ and $\varsigma = (\pi_0,\pi',\beta')'$.

$$\sigma^2(\hat{\phi}',\psi)=\frac{1}{NT}e(\hat{\upsilon})'\Omega^{-1}(\omega)e(\hat{\upsilon}),\varsigma(\hat{\phi}',\psi)=(\tilde{X}'\Omega^{-1}(\omega)\tilde{X})^{-1}\tilde{X}'\Omega^{-1}(\omega)\tilde{Y}$$

$$\tilde{X}=\begin{pmatrix} 1_N & \Delta\bar{X} & 0_{N\times k} \\ 0_{N\times 1} & 0_{N\times k} & S_N^{-1}(\lambda)\Delta X_2 \\ \vdots & \vdots & \vdots \\ 0_{N\times 1} & 0_{N\times k} & S_N^{-1}(\lambda)\Delta X_T \end{pmatrix}_{NT\times(2k+1)}$$

$$\tilde{Y}=\begin{pmatrix} \Delta Y_1 \\ (\Delta Y_2-\tau S_N^{-1}(\lambda)\Delta Y_1-\gamma S_N^{-1}(\lambda)W_N\Delta Y_1) \\ \vdots \\ (\Delta Y_T-\tau S_N^{-1}(\lambda)\Delta Y_{T-1}-\gamma S_N^{-1}(\lambda)W_N\Delta Y_{T-1}) \end{pmatrix}_{NT\times 1} \tag{11}$$

Substituting the estimators of (11) into (10), we obtain the concentrated log-likelihood function as follows:

$$\begin{aligned} l(\theta|\hat{\sigma}^2,\hat{\varsigma})=\ln L(\theta)=&-\frac{NT}{2}(\ln(2\pi)-\frac{1}{2}\ln|\Omega(\omega)|+\ln(\sigma^2(\hat{\phi}',\psi))) \\ &-e(\hat{\upsilon})'\Omega^{-1}(\omega)e(\hat{\upsilon})/(2\sigma^2(\hat{\phi}',\psi) \end{aligned} \tag{12}$$

Maximizing function (12), we obtain $\hat{\phi}$ and $\hat{\psi}$, the QMLEs of the estimated parameters, and finally $\hat{\theta}$, the QMLEs of $\theta$.

## 3 Asymptotic Characteristics of QMLEs

In order to prove the asymptotic properties and distributions of the QMLEs $\hat{\theta}$, we propose the following hypotheses:

H1: Error term $\{\varepsilon_{it}\}(i=1,\ldots,N;t=1,\ldots,T)$ follows an independent identical distribution $(i.i.d)(\forall i,t)$ with mean zero and variance $\sigma_0^2$, and $E|\varepsilon_{it}|^{4+\zeta}<\infty\,(\zeta>0)$.
H2: Explanatory variable $\{x_{it}\}(i=1,\ldots,N;t=1,\ldots,T)$ is strictly exogenous and non-stochastic. It is also consistent and bounded. $(\forall i,t)$; $\{x_t\}(t=1,\ldots,T)$ is a stable series.
H3: Each element in the spatial weight matrices $W_N$ and $M_N$ is non-stochastic, and $w_{ii}=m_{ii}=0(\forall i)$; $W_N$ and $M_N$ have calculable and real characteristic roots and the row and column sums of $W_N$ and $M_N$ are consistent and bounded.
H4: $S_N(\lambda)$, $B_N(\rho)$ and $A_N(\tau,\gamma)$ are arbitrarily invertible for $\lambda\in\Lambda_\lambda$, $\rho\in\Lambda_\rho$, $\tau\in\Lambda_\tau$ and $\gamma\in\Lambda_\gamma$. $\Lambda_\kappa(\kappa=\tau,\gamma,\lambda,\rho)$ are compact coefficient spaces and their true values $\tau_0$, $\gamma_0$, $\lambda_0$ and $\rho_0$ lie inside the spaces $\Lambda_\kappa(\kappa=\tau,\gamma,\lambda,\rho)$. The row and column sums of $A_N^{-1}(\tau,\gamma)$, $S_N^{-1}(\lambda)$, $B_N^{-1}(\rho)$ are consistent and bounded in space $\Lambda_\kappa(\kappa=\tau,\gamma,\lambda,\rho)$.

H5: the initial value $Y_0$ is endogenous and its DGP is the same as that of $Y_t(t = 1, \ldots, T)$ and is bounded. $\{Y_t\}(t = 1, \ldots, T)$ is a stable series.
H6: The true value of $\theta$, $\theta_0$ lies inside the compact parameter space $\Theta$ and is consistent and bounded in $\Theta$.
H7: $T \to \infty$, $N$ is limited or $N \to \infty$.

H1 is a common hypothesis in econometrics. The method used in this study is QMLE and is based on the i.i.d error. It's unable to obtain consistent estimators when the error is heterogeneous or auto-correlated. It is often assumed exogenous variables are strictly exogenous, consistent, and bounded, as in H2. H3 is a standard hypothesis in spatial econometrics and $w_{ii} = m_{ii} = 0(\forall i)$ can avoid the spatial self-impact. The calculable real characteristic roots ensure that the log-likelihood function is solvable. H4 ensures (9) be tenable. The consistent bounds with the row and column sums of $A_N^{-1}(\tau, \gamma), S_N^{-1}(\lambda), B_N^{-1}(\rho), W_N$ and $M_N$ ensure that the spatial correlation can be confined to a manageable range, as suggested by Kelejian and Prucha (1998, 2001) and adopted by Lee (2004). H5 avoids the occurrence of an unstable dependent variable, we consider only the dynamic spatial regression given a stable series here. H6 is a regular hypothesis in model estimation. H7 ensures that the large-sample asymptotic properties of QMLEs are proved.

In order to prove completely, we need to introduce some additional hypothesis.
H8: Define $EH = \frac{1}{NT}\sum\limits_{t=1}^{T}(B_N\widetilde{Z}_t, B_N G_N \widetilde{Z}_t\delta)'(B_N\widetilde{Z}_t, B_N G_N \widetilde{Z}_t\delta)$,[1] $EH$ is non-singular. H8 is a sufficient but not necessary condition to ensure the information matrix is non-singular. If H8 is unsatisfied, then $|EH| = 0$, we introduce an alternative H9.
H9: $\Phi_N = \frac{1}{2N^2}[tr(C_N^S C_N^S)tr(D_N^S D_N^S) - tr^2(C_N^S D_N^S)] > 0$.

In line with the approach of White (1994), we propose the following White hypothesis (WH) to prove the asymptotic consistency of QMLEs.

WH: $l(\theta)$ in (4) has a sole and identifiable maximum solution $\theta_0$ in $\Theta$.
Claim 1: When the hypothetical conditions on H1 to H7 are tenable, we have $\frac{1}{NT}l(\theta) - \frac{1}{NT}E[l(\theta)] \xrightarrow{p} 0$ and $Q(\theta) = \frac{1}{NT}E[l(\theta)]$ has a consistent continuity in $\Theta$.

**Theorem 1.** *Define $\hat{\theta} \xrightarrow{p} \theta_0$. When the hypothetical conditions on H1 to H7, and H8 or H9 (and also WH) are tenable, we have $\hat{\theta} \xrightarrow{p} \theta_0$.*

*To prove $\hat{\theta} \xrightarrow{p} \theta_0$, according to White (1994), two conditions must be satisfied:*

1. $\frac{l(\theta)}{NT} - \frac{E[l(\theta)]}{NT} \xrightarrow{p} 0$, *and $Q(\theta) = \frac{1}{NT}E[l(\theta)]$ has a consistent continuity in $\Theta$;*
2. $\theta_0$ *where $N_\zeta^C(\theta_0)$ is an open neighboring region of $\theta_0$ in space $\Theta$.*

*Condition 1 is satisfied according to Claim 1. Condition 2 is equal to an identical definition of $\theta_0$ in space $\Theta$ and is convenient to be proven.*

Claim 2: When the hypothetical conditions on H1 to H7 are tenable, we have $\frac{1}{\sqrt{NT}}\frac{\partial l(\theta_0)}{\partial \theta} \xrightarrow{d} N(0, \Sigma_{\theta_0}^e + \Omega_{\theta_0}^e)$.

[1] Where $G_N(\lambda) = W_N S_N(\lambda), \widetilde{Z}_t = (\tilde{Y}_{t-1}, W_N \tilde{Y}_{t-1}, \tilde{X}_t)$.

For the log-likelihood function (10) and its first and second order derivatives, we have $E\left[\frac{1}{\sqrt{NT}}\frac{\partial l(\theta_0)}{\partial \theta}\frac{1}{\sqrt{NT}}\frac{\partial l(\theta_0)}{\partial \theta'}\right] = \Sigma_{\theta_0,NT} + \Omega_{\theta_0,NT} + O(\frac{1}{T})$, define $\Sigma_{\theta_0} = \lim_{T\to\infty} \Sigma_{\theta_0,NT}$, $\Omega_{\theta_0} = \lim_{T\to\infty} \Omega_{\theta_0,NT}$, where the former term is the so-called information matrix.

**Theorem 2.** *When the hypothetical conditions on H1 to H7, and H8 or H9 are tenable, have* $\sqrt{NT}(\hat{\theta} - \theta_0) \xrightarrow{d} N(0, \Sigma_{\theta_0}^{e-1}(\Sigma_{\theta_0}^{e} + \Omega_{\theta_0}^{e})\Sigma_{\theta_0}^{e-1})$. *When the error term* $\varepsilon_t$ *follows a normal distribution, have* $\sqrt{NT}(\hat{\theta} - \theta_0) \xrightarrow{d} N(0, \Sigma_{\theta_0}^{e-1})$.

*To prove the asymptotic distribution of* $\hat{\theta}$, *we conduct a Taylor series expansion on* $\frac{\partial l(\hat{\theta})}{\partial \theta}$ *of (12) at* $\theta_0$:

$$0 = \frac{\partial l(\hat{\theta})}{\partial \theta} = \frac{\partial l(\theta)}{\partial \theta}\bigg|_{\theta_0} \cdot + (\hat{\theta} - \theta_0)\frac{\partial^2 l(\theta)}{\partial \theta^2}\bigg|_{\bar{\theta}}$$

*where* $\bar{\theta}$ *lies between* $\hat{\theta}$ *and* $\theta_0$. *Thus, when* $\hat{\theta} \xrightarrow{P} \theta_0, \bar{\theta} \xrightarrow{P} \theta_0$. *Because* $\bar{\theta} \xrightarrow{P} \theta_0$, *we have* $\frac{1}{NT}\frac{\partial^2 l(\bar{\theta})}{\partial \theta^2} - \frac{1}{NT}\frac{\partial^2 l(\theta_0)}{\partial \theta^2} = o_p(1)$. *Based on the information matrix* $\Sigma_{\theta_0}^{e}$, *we have:*

$$\frac{1}{NT}E\frac{\partial^2 l(\theta_0)}{\partial \theta^2} \xrightarrow{p} \Sigma_{\theta_0}^{e}, \frac{1}{NT}\frac{\partial^2 l(\theta_0)}{\partial \theta^2} - \frac{1}{NT}E\frac{\partial^2 l(\theta_0)}{\partial \theta^2} \xrightarrow{p} 0, \frac{1}{\sqrt{NT}}\frac{\partial l(\theta_0)}{\partial \theta} \xrightarrow{d} N(0, \Sigma_{\theta_0}^{e} + \Omega_{\theta_0}^{e})$$

and because $\Sigma_{\theta_0}^{e}$ is non-singular, therefore we have $\sqrt{NT}(\hat{\theta} - \theta_0) \xrightarrow{d} N(0, \Sigma_{\theta_0}^{e-1}(\Sigma_{\theta_0}^{e} + \Omega_{\theta_0}^{e})\Sigma_{\theta_0}^{e-1})$. when error term $\varepsilon_t$ follows a normal distribution, $\sqrt{NT}(\hat{\theta} - \theta_0) \xrightarrow{d} N(0, \Sigma_{\theta_0}^{e-1})$. Therefore, Theorem 2 is proven.

In the final of this section, we design a Monte Carlo simulation to assess the asymptotic properties of QMLEs in our dynamic spatial fixed effect model (1). Assume that independent variables $X_t(t = 1, \ldots, T)$, error term $\varepsilon_t(t = 1, \ldots, T)$ and $\xi$, initial value $(Y_0)$ in the endogenous model all follow independent standard normal distribution. The sample generating based on (3) and (8) follows:

$$\begin{aligned}\Delta Y_1 &= \pi_0 1_N + \Delta\bar{X}\pi + e_1(\cdot), \Delta Y_t \\ &= S_N^{-1}(\lambda)A_N(\tau,\gamma)\Delta Y_{t-1} + S_N^{-1}(\lambda)\Delta X_t\beta + S_N^{-1}(\lambda)B_N^{-1}(\rho)\Delta\varepsilon_t(\varphi)(t = 2, 3, \ldots, T)\end{aligned}$$

We choose the true values of $\theta_0 = (\sigma_0^2, \beta_0, \tau_0, \gamma_0, \lambda_0, \rho_0)'$ and $(\pi_0, \pi_1, \psi_0)'$ as (1, 2, 0.2, 0.3, 0.4, 0.5)′ and (1, 2, 1)′, respectively. The spatial weight matrices $W$ and $M$ are generated on the basis of the so-called rook continuity matrix type. Spatial unit $N$ and temporal unit $T$ take (9, 25, 49) and (5, 20, 40), respectively. The number of simulations is 1,0000. We record results and calculate the corresponding means and RMSEs. The simulation results are shown in Table 1.

From Table 1, we find that the largest bias reflected by the largest RMSE (e.g., RMSE = 0.247 for $\sigma^2$) when the sample size is the smallest ($N = 9$, $T = 5$). However, the RMSEs improve steadily with a further increase in the sample size. When the sample size is at its largest ($N = 49$, $T = 40$), the RMSE reaches its

**Table 1** Monte-Carlo simulation result

| (N,T) | $\sigma_0^2 = 1$ | | $\beta_0 = 2$ | | $\tau_0 = 0.2$ | | $\gamma_0 = 0.3$ | | $\lambda_0 = 0.4$ |
|---|---|---|---|---|---|---|---|---|---|
| | Mean | RMSE | Mean | RMSE | Mean | RMSE | Mean | RMSE | Mean |
| (9,5) | 1.082562 | 0.246511 | 1.99346 | 0.275596 | 0.170521 | 0.063493 | 0.309119 | 0.103684 | 0.39981 |
| (9,20) | 1.027077 | 0.113369 | 2.00629 | 0.138198 | 0.194709 | 0.026123 | 0.302814 | 0.043901 | 0.39259 |
| (9,40) | 1.010253 | 0.075196 | 2.00091 | 0.095528 | 0.196927 | 0.018477 | 0.301812 | 0.029277 | 0.40121 |
| (25,5) | 1.069739 | 0.158696 | 2.0095 | 0.153387 | 0.172037 | 0.039299 | 0.322268 | 0.064304 | 0.39592 |
| (25,20) | 1.02005 | 0.070562 | 1.99781 | 0.074254 | 0.193286 | 0.015968 | 0.304849 | 0.026449 | 0.39831 |
| (25,40) | 1.011766 | 0.046913 | 1.99606 | 0.051821 | 0.197064 | 0.010784 | 0.302384 | 0.018496 | 0.39909 |
| (49,5) | 1.064728 | 0.12025 | 2.00047 | 0.103754 | 0.179161 | 0.026608 | 0.313326 | 0.044681 | 0.40585 |
| (49,20) | 1.020968 | 0.050673 | 1.99909 | 0.04841 | 0.194705 | 0.011524 | 0.303793 | 0.019707 | 0.39897 |
| (49,40) | 1.010796 | 0.034783 | 2.00008 | 0.035653 | 0.197502 | 0.008097 | 0.302592 | 0.014348 | 0.39995 |
| (N,T) | $\pi_0 = 1$ | | $\pi_1 = 2$ | | $\psi_0 = 1$ | | $\rho_0 = 0.5$ | | $\lambda_0 = 0.4$ |
| | Mean | RMSE | Mean | RMSE | Mean | RMSE | Mean | RMSE | RMSE |
| (9,5) | 0.990829 | 0.149375 | 2.00025 | 0.150804 | 1.001963 | 0.219882 | 0.498798 | 0.287704 | 0.299917 |
| (9,20) | 0.997388 | 0.073166 | 2.00182 | 0.076165 | 1.0024 | 0.10497 | 0.506592 | 0.145853 | 0.144486 |
| (9,40) | 0.995782 | 0.053198 | 1.99929 | 0.051663 | 1.003033 | 0.075258 | 0.500728 | 0.101218 | 0.102359 |
| (25,5) | 0.995565 | 0.087228 | 1.99882 | 0.092055 | 0.992905 | 0.124311 | 0.505011 | 0.191558 | 0.17912 |
| (25,20) | 0.999425 | 0.04272 | 2.00042 | 0.046042 | 1.001623 | 0.064055 | 0.500702 | 0.084099 | 0.065807 |
| (25,40) | 1.00011 | 0.031112 | 1.99957 | 0.031494 | 0.999033 | 0.044257 | 0.501673 | 0.061949 | 0.083667 |
| (49,5) | 1.000206 | 0.062038 | 1.99711 | 0.063292 | 1.000806 | 0.091728 | 0.496384 | 0.132007 | 0.13452 |
| (49,20) | 0.998093 | 0.031243 | 1.99922 | 0.031966 | 1.000052 | 0.044722 | 0.497137 | 0.059918 | 0.059898 |
| (49,40) | 1.000655 | 0.021968 | 1.9999 | 0.022458 | 0.999552 | 0.032892 | 0.502129 | 0.041804 | 0.040735 |

smallest value (0.035), i.e., $\sigma^2$ is very close to its true value $\sigma_0^2$. This phenomenon reflects the large-sample asymptotic properties of QMLEs. Looking at the sensitivity of the change in asymptotic properties to both $N$ and $T$, we find that when $N$ is fixed, the asymptotic properties of estimators improve rapidly when $T$ becomes large. Take $\beta$ as an example. When $N = 9$ and $T = 5$, RMSE is 0.276 (bias rate is 13.8%). When $T$ is increased to 20 and 40, respectively, RMSE decreases significantly to 0.138 (bias rate is 6.9%) and 0.096 (bias rate is 4.8%), respectively. On the other hand, when $T$ is fixed, the asymptotic properties of estimators improve to a different extent as $N$ increases. However, the sensitivity of change under this scenario (in which $T$ is fixed but $N$ is not) is not so strong as in the previous scenario in which $N$ is fixed and $T$ is not. Again, take $\beta$ as an example. When $T = 5$ and $N = 9$, the RMSE is 0.275 (bias rate is 13.8%), and the RMSE decreases to 0.153 (bias rate is 7.7%) and 0.103 (bias rate is 5.2%), respectively, when $N$ is increased to 25 and 49, respectively. As can be seen, $\beta$ improves to a lesser extent than in the scenario in which $T$ is increased and $N$ is fixed. Other parameter estimators also indicate a similar pattern.

We conclude that the asymptotic properties of QMLEs depend on a large sample size, i.e., QMLEs do not perform well when the sample size is small. When the sample size ($N \times T$) lies in the range of 125 to 180, the RMSE can basically be controlled at under 0.1, but bias rate is still comparatively large. Take $\tau_0$ as an example. When $N = 25$ and $T = 5$, bias rate rises to nearly 40%. Clearly, the estimation results for such sample size are basically unacceptable. When the sample size lies in the range of 180–500, the RMSE is controlled at under 0.08 and the corresponding bias rate falls by 10% on average. Therefore, the estimation results are basically acceptable. When the sample size exceeds 500, the asymptotic properties of the estimation results improve significantly, with the RMSE basically being controlled at under 0.05 and its corresponding bias rate being below 5% on average. The implications of our simulation results are that using different combinations of $N$ and $T$ effectively allows us to improve the quality of the estimation results if it is possible to increase $T$ based on the existing $N$.

## 4 Conclusion

This paper contributes to the extant literature in following ways. First, the introduction of a temporal lag operator means that our model represents a significant improvement on the static spatial fixed effect model of Lee and Yu (2010). Second, our model represents a major advance on the previous dynamic spatial fixed models of Elhorst (2005) and Yu et al. (2006) as it provides a path for moving from a "general" to "specific" model selection approach. Because Elhorst (2005) assumes in his model that spatial correlation exists but is unobservable, and simply places the measure of spatial correlation in the error term. Yu et al. (2006) assume that spatial correlation not only exists, but is also directly observable. Therefore, they simply place the measure of spatial correlation in the explanatory variables. It is true that both directly observable and directly unobservable spatial effects may exist in a real economy and that the assumptions made in the two predecessor models are therefore reasonable to a certain extent. However, the problem they create is that in the model construction process, only one scenario is considered and the possible existence of the other scenario is ignored. In other words, the models of Elhorst (2005) and Yu et al. (2006) are mutually complementary, but are not compatible. However, such assumptions become untenable when both directly observable and directly unobservable scenarios coexist in a real economy and will thus lead to erroneous conclusions. In contrast, our model has the advantage of being built by relaxing the respective assumptions proposed by the previous researchers. In other words, the assumptions made in building our model are that a spatial correlation exists and includes both directly observable and directly unobservable scenarios. Specifically, we build a more generalized model by relaxing the spatial correlation made in building previous models and then determine the degeneracy of specific models using parameter estimators and significance level results.

## References


Bhargava A, Sargan JD (1983) Estimation dynamic stochastic effects models from panel data covering short time periods. Econometrica 51:1635–1659

Elhorst JP (2001) Dynamic models in space and time. Geogr Anal 33:119–140

Elhorst JP (2005) Unconditional maximum likelihood estimation of linear and log-linear dynamic model for spatial panels. Geogr Anal 37:85–106

Hsio C, Pesaran MH, Tahmiscioglu AK (2002) Maximum likelihood estimation of fixed effects dynamic panel data models covering short time periods. J Econometrics 109:107–150

Kelejian Harry H, Prucha I (1998) A generalized spatial two-stage least squares procedure for estimating a spatial autoregressive model with auto-regressive disturbance. J Real Estate Finance Econ 17(1):99–121

Kelejian Harry H, Prucha I (2001) On the asymptotic distribution of the moran i test statistic with applications. J Econometrics 104:219–257

Lee LF (2004) Asymptotic distributions of quasi-maximum likelihood estimators for spatial econometric models. Econometrica 72:1899–1925

Lee LF, Yu J (2010) Estimation of spatial autoregressive panel data models with fixed effects. J Econometrics 154(2):165–185

Su LJ, Yang ZL (2007) QML estimation of dynamic panel data models with spatial errors. Working Paper, Singapore Management University

White H (1994) Estimation, inference and specification analysis. Econometric Society Monographs, No. 22, Cambridge University Press, pp 27–28

Yu J, de long R, Lee LF (2006) Quasi-maximum likelihood estimators for spatial dynamic panel data with fixed effects when both n an T are large. Working Paper, The Ohio State University

# Correlation Analysis of Yield and Volatility Based on GARCH Family Models

**Yue Xu and Sulin Pang**

**Abstract** This paper conducts an empirical study on the daily return rate of Shanghai composite index from July 3, 2000 to July 1, 2009. The result shows that China financial market daily returns possess remarkable ARCH effects. The result shows that Shanghai stock market daily returns has the features like leptokurtosis, heavy-tailed and possesses remarkable ARCH effects. Then based on three types of distribution, we establish the GARCH (1, 1) models of Shanghai Composite Index daily yield series. By comparing the results, GARCH (1, 1) Model based on GED distribution performs the best. By conducting ARCH-LM test on residual error, the results show that this model eliminates the ARCH effects effectively. At last the leverage effect of Shanghai stock market is checked by using TARCH and EGARCH Model.

**Keywords** ARCH-LM test · GARCH family model · Leverage effect

## 1 Introduction

Stock market returns and volatility has long been a concern of researchers. Engle (1982) proposed autoregressive conditional heteroskedasticity (ARCH) model which was considered to be a new method to research these issues. Then Bollerslev (1986) extended it to generalized ARCH model (GARCH). The model depicts the conditional second order moment property of error by linear form, and describes time varying and clustering of volatility through the change of conditional heteroskedasticity. GARCH family model is now widely used in Econometrics. Akgiray (1989)

---

Y. Xu and S. Pang (✉)
Department of Mathematics & Institute of Finance Engineering, School of Management, Jinan University, Guangzhou, Guangdong 510632
e-mail: pangsulin@163.com

Y. Zhou and D.D. Wu (eds.), *Modeling Risk Management for Resources and Environment in China*, Computational Risk Management,
DOI 10.1007/978-3-642-18387-4_58, 

forecasted and analyzed monthly volatility of American stock market by using GARCH (1,1) model. Bollerslev et al. (1992) and Panorska et al. (1995) had researched ARCH and GARCH models based on stationary time series. Buhlmann (2002) had researched the algorithm of nonparametric GARCH (1,1) model. Garcia and Contreras (2005) provided an approach to predict next-day electricity prices based on the Generalized Autoregressive Conditional Heteroskedastic (GARCH) methodology that was already being used to analyze time series data in general. Wilhelmsson (2006) investigated the forecasting performance of the Garch (1, 1) model when estimated with different error distributions on Standard and Poor's 500 Index Future returns. The result found that allowing for skewness and time variation in the higher moments of the distribution does not further improve forecasts. Floros (2008) examined the use of GARCH-type models for modeling volatility and explaining financial market risk. They used daily data from Egypt (CMA General index) and Israel (TASE-100 index). Various time series methods were employed, including the simple GARCH model, as well as exponential GARCH, threshold GARCH, asymmetric component GARCH, the component GARCH and the power GARCH model.

From many domestic empirical research, it is believed that China's financial time series share common characteristics as follows: First of all, asset prices is generally non-stationary and has a unit root, while the return series are usually stationary; Secondly, return series is hardly autocorrelative but the square of return sequence has a strong autocorrelation, which revealing that the relationship between the observations at different periods is nonlinear. The fluctuation of returns has the feature which is clustering. that is to say, sometimes the series has similar large fluctuations, and some time has the accordant little fluctuations; Moreover, the distribution of return series is significantly different from normal distribution, and generally present the features like leptokurtosis, heavy-tailed which are related to conditional variance. Last, the impact on capital market often has asymmetric effect, that is the impact on the capital market caused by bad news is not equal to it caused by equivalent good news. This phenomenon is known as leverage (Leverage effect).

This article focuses on an empirical study on the daily return rate of Shanghai stock market in China by using GARCH family model to find the best GARCH model based on the appropriate distribute. And we analyze leverage effect of Shanghai stock Market at a specific time quantum by using TARCH model and EGARCH model.

## 2 Theoretical Basis and Model

Conditional heteroskedasticity is one significant feature of financial time series. ARCH model and GARCH model describe the conditional second moment character of error term in the linear form. They take the change of conditional heteroskedasticity to describe the time varying and the clustering of the fluctuations. Now GARCH family model family is widely applied in quantitative finance field.

## *(1). Mean equation and variance equation of GARCH (1,1) model:*

The mean equation is:

$$Y_t = \beta' X_t + \varepsilon_t \tag{1}$$

The variance equation is:

$$h_t = Var(\varepsilon_t|\psi_{t-1}) = \beta_0 + \beta_1\varepsilon_{t-1}^2 + \beta_2 h_{t-1} \tag{2}$$

In the mean equation, $Y_t$ denotes the dependent variable, $X_t$ denotes the row vectors of explanatory variables, $\beta'$ denotes the column vectors of explanatory variables, $\varepsilon_t$ denotes stochastic disturbance term. In the variance equation, $\psi_{t-1}$ denotes the available information set at time t-1, $h_t$ denotes the conditional variance. For GARCH model is an extended form of ARCH model, it has some same characteristics as ARCH model. However, the conditional variance of GARCH model is a linear function of lagged squared residuals and lagged conditional variance.

## *(2). Variance function of TARCH (1,1) model is set to be as the following equation:*

$$h_t = \alpha_0 + \alpha_1\varepsilon_{t-1}^2 + \gamma\varepsilon_{t-1}^2 d_{t-1} + \alpha_2 h_{t-1} \tag{3}$$

Where

$$d_{t-1} = \begin{cases} 1, \varepsilon_{t-1} < 0 \\ 0, \varepsilon_{t-1} \geq 0 \end{cases} \tag{4}$$

In the stock market, another well-known characteristic is leverage effect. It means that different kinds of news will have an asymmetric influence to the stock price volatility. If we take good news as the positive impact to the stock price and bad news as the negative one, the stock price usually has different degrees of responses while encountering positive impact and equivalent negative one. This asymmetry can be described by TARCH model. TARCH model was put forward by Zakoian for the first time. In its variance equation, $d_{t-1}$ is taken as a dummy variable. If $\gamma \neq 0$, the influence caused by positive impact and negative impact would be different. For example, the influence of the positive impact is $\alpha_1$ and the counterpart of the negative one is $(\alpha_1 + \gamma)$ while $\gamma > 0$.

***(3). In order to overcome some weaknesses of GARCH model in dealing with the financial time series, Nelson (1991) put forward EGARCH model. The form of its variance equation is:***

$$\ln(h_t) = \alpha_0 + \alpha_1 \left| \frac{\varepsilon_{t-1}}{\sqrt{h_{t-1}}} \right| + \gamma \frac{\varepsilon_{t-1}}{\sqrt{h_{t-1}}} + \alpha_2 \ln(h_{t-1}) \tag{5}$$

Similar to TARCH model, if $\gamma$ is significantly negative, then bad news will have a greater impact on the stock market. Because conditional variance is denoted in the natural logarithm form, whether the coefficient is negative or not, the value of equation is non-negative. In the same way, the value of residual cannot affect the non-negativity of equation.

# 3 Garch Family Model Building and Analysis

## 3.1 Data Selection

We select Shanghai composite index as a sample of Shanghai stock market for our study. We take daily closing price as sample and the range is from 3rd July 2000 to 1st July 2009. Data are selected in WIND financial database which include 2,175 samples in total. We use Eviews 5.0 software to analyze the data. Logarithmic return rate of Shanghai composite index are given by following formula:

$$r_t = \ln(P_t) - \ln(P_{t-1}) \tag{6}$$

where $r_t$ is daily yield; $P_t$ is daily closing price. RSH denotes daily yield of Shanghai composite index.

## 3.2 Descriptive Statistical Analysis

According to formula (6), we calculate logarithmic yield in order to observe the main trend of Shanghai composite index daily yield. The trend is shown in the Fig. 1

It can be observed directly from the Fig. 1 that the fluctuation of daily yield in Shanghai stock market has the phenomenon of clustering. The Large fluctuations are often followed by the large fluctuations, while the little fluctuations often come after little fluctuations. Then, we conduct a statistical analysis on the Shanghai composite Index and the features are shown in the follow figure:

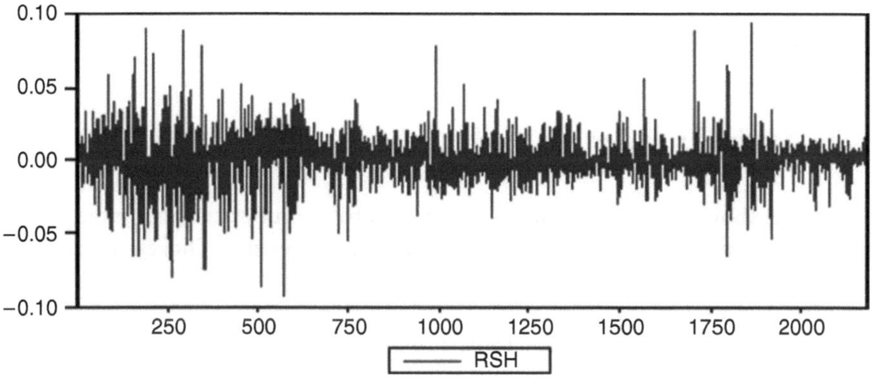


**Fig. 1** The fluctuation of Shanghai composite index daily yield

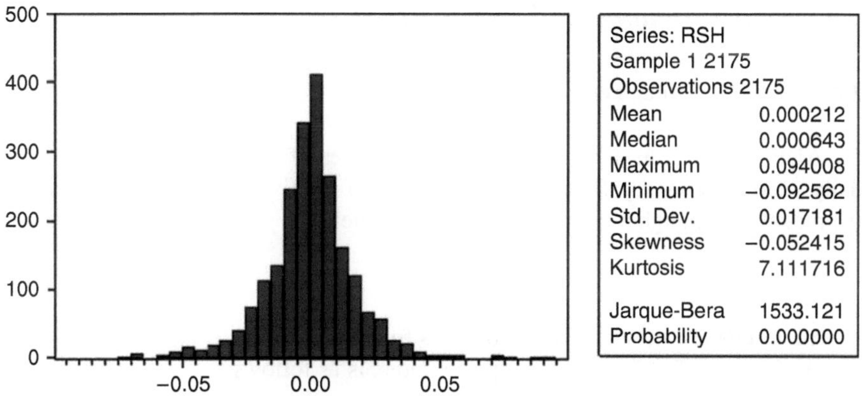


**Fig. 2** Histogram and statistic of Shanghai stock market

The leptokurtosis and heavy-tailed characteristics are shown directly in the Fig. 2. We can also observe yield distribution is slightly left skewed. The mean value of daily yield is 0.000212. The maximum is 0.09401 while the minimum is −0.0926. Skewness is −0.0524 which also proves the result of observation which is left-skewed distribution of yield. Kurtosis is 7.1117, which is significantly higher than the kurtosis of the normal distribution which is 3. Therefore, it can be drawn a conclusion that Shanghai stock market has the phenomenon like leptokurtosis, heavy-tailed. Then, we apply Jarque-Bera test to overall distribution:

$$JB = n\left[S^2 + (K-3)^2/4\right]/6 \tag{7}$$

where $S$ is skewness; $K$ is kurtosis and $n$ is sample size. It is illustrated in the Fig. 2 that the Jarque-Bera value is 1,533.121. The probability of accepting that the distribution is normal distribution is close to 0. So we can draw a conclusion that the distribution of return series is significantly different from normal distribution.

## 3.3 Stationarity Test

From the above statistical analysis we can probably guess that the return series fluctuates around the mean. Therefore, we put an Augmented Dickey–Fuller Test (only with intercept and the lag length is 4) on the return series of Shanghai stock market. And we obtain:

According to the result of the Table 1, the conclusion is obvious. We reject the null hypothesis: RSH has a unit root (1% level). That means the return series is stationary and we can use it to build a model directly without differencing it.

## 3.4 Model Building and ARCH Effect Test

### 3.4.1 ARCH Effect Test

By ADF test, we can conclude that the time series of Shanghai composite index daily yield is stationary. So after analyzing its self-relative chart, we establish a AR (4) model for Shanghai composite index daily yield:

$$r_t = \alpha_0 + \alpha_1 r_{t-4} + \varepsilon_t \tag{8}$$

The results are shown in the Table 2 as follows:

AR(4) model for Shanghai composite index is as follows:

$$r_t = 0.0002 + 0.0611 \times r_{t-4} \tag{9}$$

We calculate auto-correlation coefficients of squared residuals and we get the following table.

The figure above (Fig. 3) shows the series of squared residuals has a significant auto-correlation. The result illustrates that there is a nonlinear relationship among

**Table 1** The result of Augmented Dickey–Fuller Test for Shanghai stock market

| ADF test statistic | | t-Statistic | Probability |
|---|---|---|---|
| | | −19.4807 | 0.000 |
| Test critical values | 1% level | −3.4332 | |
| | 5% level | −2.8627 | |
| | 10% level | −2.5674 | |

**Table 2** Results of Shanghai composite index model

| Parameter | Coefficient | t-Statistic | Probability |
|---|---|---|---|
| $\alpha_0$ | 0.0002 | 0.5087 | 0.6110 |
| $\alpha_1$ | 0.0611 | 2.8517 | 0.0044 |

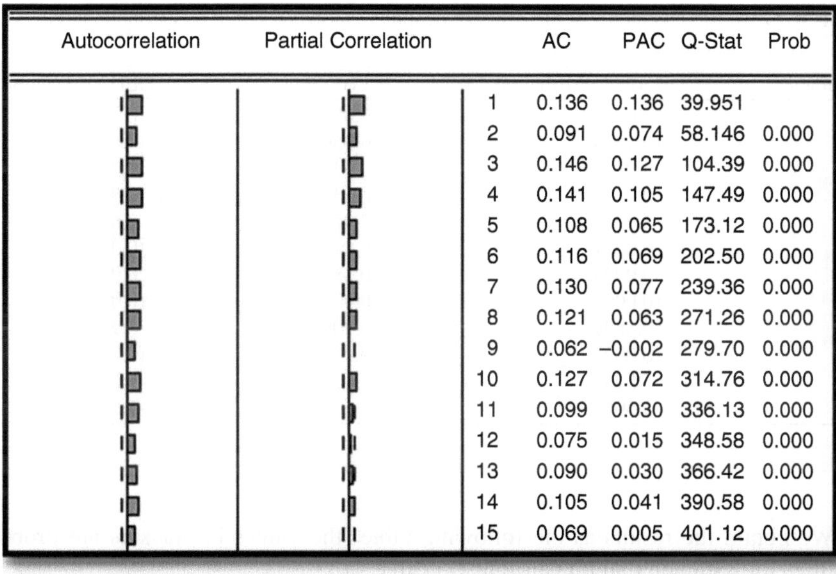


**Fig. 3** Correlogram squared residuals

**Table 3** ARCH-LM test result of model residual

| Lags | 1 | 2 | 3 | 4 | 5 |
|---|---|---|---|---|---|
| F- statistic of residual | 40.63 | 26.45 | 29.80 | 28.65 | 24.83 |
| Prob | 0.00 | 0.0000 | 0.00 | 0.00 | 0.00 |

the observations of different periods, which supports the clustering of volatility with a preliminary evidence.

Then we conduct ARCH-LM test on Shanghai composite index daily yield model, outcome is shown in the Table 3.

From Table 3 we can find that ARCH-LM lags test shows that the probability is close to zero. Then we can conclude that residual series of the model have a high order ARCH effect. Therefore, we should establish GARCH model to eliminate it.

### 3.4.2 GARCH Model Building

We establish GARCH model for Shanghai composite index daily yield as follows:

$$r_t = \alpha_0 + \alpha_1 r_{t-4} + \varepsilon_t \tag{10}$$

$$\varepsilon_t = \sigma_t \sqrt{h_t} \tag{11}$$

$$h_t = \beta_0 + \beta_1 \varepsilon_{t-1}^2 + \beta_2 h_{t-1} \tag{12}$$

**Table 4** GARCH (1,1) model of Shanghai stock market based on different distribution

| GARCH (1,1) model of Shanghai stock market | | | |
|---|---|---|---|
| Parameter | Normal distribution | T distribution | GED distribution |
| $\alpha_0$ | 0.000386 | 0.000503 | 0.000631 |
| | (0.1954) | (0.0592) | (0.0110) |
| $\alpha_1$ | 0.04542 | 0.04229 | 0.04069 |
| | (0.0431) | (0.0381) | (0.0346) |
| $\beta_0$ | $5.97 \times 10^{-6}$ | $2.92 \times 10^{-6}$ | 3.65E-06 |
| | (0.000) | (0.0065) | (0.0022) |
| $\beta_1$ | 0.0923 | 0.086775 | 0.08684 |
| | (0.000) | (0.000) | (0.000) |
| $\beta_2$ | 0.8899 | 0.9097 | 0.9039 |
| | (0.000) | (0.000) | (0.000) |
| AIC | −5.519 | −5.618 | −5.615 |
| SC | −5.506 | −5.602 | −5.599 |

We obtain the results as the following table (the values in brackets are probabilities for parameter significance test) (Table 4).

The results based on three different distribution show that the sum of $\beta_1$ (ARCH coefficient) and $\beta_2$ (GARCH coefficient) of either GARCH (1,1) model based on different distribution is less than 1. That meets the hypothesis of GARCH modeling. And the sum is close to 1, which indicated that impact to Shanghai stock market caused by outer factors will last for a long time. Comparing to AR model, the log-likelihood of GARCH (1, 1) model increases, meanwhile the value of AIC and SC become smaller. The statistics indicate that GARCH (1,1) model can fit the data in a better way. From the comparison of three distribution model in the Table 3 which based on AIC criterion, SC criterion and the significance of estimated parameters, we can conclude that GARCH (1,1) model based on the GED distribution is better than that based on the student t distribution or the normal distribution. So we have chosen GARCH (1, 1) model based on GED distribution for the Shanghai Stock Market.

The GED-GARCH (1,1) model of Shanghai composite index. The mean Equation:

$$rsh_t = 0.000631 + 0.04169 \times rsh_{t-4} \tag{13}$$

The variance Equation is:

$$h_t = 3.65 \times 10^{-6} + 0.08684 \times \varepsilon_{t-1}^2 + 0.9039 \times h_{t-1} \tag{14}$$

Then we conduct ARCH-LM test on residuals of newly established GARCH (1, 1) model. The results obtained are as follows:

The results from the Table 5 show that through the multi-order ARCH-LM test, the probabilities are much greater than 0.05. That indicates the GARCH (1, 1) model can effectively eliminate ARCH effect.

**Table 5** ARCH-LM test results of GARCH (1, 1) model residual

| Lags | 1 | 2 | 3 | 4 | 5 |
|---|---|---|---|---|---|
| F- Statistic of residual | 0.4956 | 0.2553 | 0.2560 | 0.2012 | 0.1625 |
| Probability | 0.4815 | 0.7747 | 0.8571 | 0.9378 | 0.9762 |

### 3.4.3 Leverage Effect Test

The form of variance equation for TARCH (1,1) model:

$$h_t = \alpha_0 + \alpha_1 \varepsilon_{t-1}^2 + \gamma \varepsilon_{t-1}^2 d_{t-1} + \alpha_2 h_{t-1} \tag{15}$$

where

$$d_{t-1} = \begin{cases} 1, \varepsilon_{t-1} < 0 \\ 0, \varepsilon_{t-1} \geq 0 \end{cases} \tag{16}$$

Then we build TARCH (1, 1) model for Shanghai composite index daily yield. We obtain the results as follows:

TARCH (1,1) model of Shanghai composite index daily yield:

$$\begin{aligned} h_t(RSH) = 7.03 \times 10^{-6} + 0.1564 \times \varepsilon_{t-1}^2 \\ -0.0982 \times \varepsilon_{t-1}^2 d_{t-1}(RSH) + 0.8753 \times h_{t-1}(RSH) \end{aligned} \tag{17}$$

The form of variance equation for EGARCH (1,1) model:

$$\ln(h_t) = \alpha_0 + \alpha_1 \left| \frac{\varepsilon_{t-1}}{\sqrt{h_{t-1}}} \right| + \gamma \frac{\varepsilon_{t-1}}{\sqrt{h_{t-1}}} + \alpha_2 \ln(h_{t-1}) \tag{18}$$

Now we build EGARCH (1,1) model for Shanghai composite index daily yield: EGARCH (1,1) model for Shanghai composite index daily yield:

$$\ln(h_t) = \alpha_0 + \alpha_1 \left| \frac{\varepsilon_{t-1}}{\sqrt{h_{t-1}}} \right| + \gamma \frac{\varepsilon_{t-1}}{\sqrt{h_{t-1}}} + \alpha_2 \ln(h_{t-1}) \tag{19}$$

We can see from the Tables 6 and 7, for both $\gamma$ (Leverage Effect) of TARCH model and EGARCH model of Shanghai stock market, the p values are close to 0. Statistically speaking, bad news and good news at the same level will impact Shanghai stock market differently. It means that Shanghai stock market has significant asymmetry of good news and bad news, and the leverage effect is significant.

**Table 6** The results of variance equation for TARCH (1,1) model

| TARCH (1,1) model variance equation | | | |
|---|---|---|---|
| Parameter | Coefficient | Z-statistic | Probability |
| $\alpha_0$ | −0.3882 | −8.8705 | 0.000 |
| $\alpha_1$ | 0.1987 | 10.4015 | 0.000 |
| $\gamma$ | 0.01177 | 4.8157 | 0.000 |
| $\alpha_2$ | 0.9710 | 255.0571 | 0.000 |

**Table 7** The results of variance equation for EGARCH (1,1) model

| EGARCH (1,1) model variance equation | | | |
|---|---|---|---|
| Parameter | Coefficient | Z-statistic | Probability |
| $\alpha_0$ | $7.03 \times 10^{-6}$ | 9.0564 | 0.000 |
| $\alpha_1$ | 0.1564 | 9.7707 | 0.000 |
| $\gamma$ | −0.0982 | −5.0341 | 0.000 |
| $\alpha_2$ | 0.8753 | 75.237 | 0.000 |

# 4 Conclusion

In this paper, we applied financial econometric analysis and statistical test to fit Shanghai composite index in order to analyze the characteristics of yield volatility in Shanghai stock market. In general, the conclusions are as follows:

1. Shanghai stock market presents the phenomenon of clustering and it has the characteristics like leptokurtosis, heavy-tailed. After stationarity test, we can draw a conclusion that the return series is stationary. The result is similar to the characteristic of mature stock market. That is to say, the price of financial assets generally has a unit root and it is non-stationary, but yield series are usually stationary. Homogeneity of the stock market rules is validated to a certain extent.
2. In the GARCH (1, 1) model of Shanghai composite index yield, the sum of ARCH coefficient and GARCH coefficient is less than 1 indicating that the conditional variance of yield is stationary and convergent and the model has the trait of predictability. Meanwhile, the sum of ARCH coefficient and GARCH coefficient is close to 1.This means that the impact on the volatility of Shanghai stock market which caused by external shocks will sustain a long time and the persistence feature is obvious.
3. Shanghai stock market has significant leverage effect, that is the impact on the Shanghai stock market caused by bad news is not equal to the impact caused by equivalent good news. The empirical evidences reveal that Shanghai stock market has a negative leverage effect which is the impact on the Shanghai stock market caused by bad news is stronger than it caused by equivalent good news. This result is the same with foreign research.

**Acknowledgements** The paper is supported by the National Natural Science Foundation (70871055); the New Century Talents plan of Ministry of Education of China (NCET-08-0615); the Key Programs of Science and Technology Department of Guangdong Province (2010).

## References

Akgiray V (1989) Conditional heteroskedasticity in time series of stock returns: evidence and forecasts. J Business 62:55–80

Bollerslev T (1986) Generalized autoregressive conditional heteroskedasticity. J Economitrics 31:307–327

Bollerslev T, chou RT, Kroner KF (1992) ARCH modeling in finance. J Economitrics 52:1–59

Buhlmann M (2002) An algorithm for nonparametric GARCH modeling. Comput Statistic data Anal 40:665–683

Engle RF (1982) Autoregressive conditional heteroskedasticity with estimates of the variance of UK. inflation. Economitrica 50:987–1008

Floros C (2008) Modelling volatility using GARCH models: evidence from Egypt and Israel. Middle East Finance Econ 2:1450–2889

Garcia R, Contreras J (2005) A GARCH forecasting model to predict day-ahead electricity prices. IEEE Trans Power Syst 20(2):867–874

Nelson D (1991) Conditional heteroskedasticity in asset returns: a new approach. Econometrica

Panorska AK, Mittnik S, Rachev ST (1995) Stable GARCH models for financial time series. Appl Math Lett 8(5):33–37

Wilhelmsson A (2006) Garch forecasting performance under different distribution assumptions. J Forecasting 25:561–578

# A Unifying Approach to the Ruin Problems Under the Compound Binomial Model

**Li-juan Sun and Yi-Hau Chen**

**Abstract** In this paper, the aggregate claims are modeled as a compound binomial process and the individual claim sizes are integer-valued. Taking advantage of the expected discounted penalty function, we derive, when a discount factor$v$is taken into account, the recursive formulas, generating functions, defective renewal equations, asymptotic expression and explicit expressions for some quantities related to the ruin. We indicate that the maximal aggregate loss of the surplus process can be expressed as a compound geometric random variable, whose tail is exactly the generating function of the ruin time.

**Keywords** Compound binomial model · Expected discounted penalty function · Maximal aggregate loss · Ruin probability

## 1 Introduction

In actuarial risk models, the compound binomial model was first proposed by Gerber (1988), in which the aggregate claims are modeled as a compound binomial process. Since this work, ruin problems in the compound binomial model has become a topic of interest and further studied by Shiu (1989), Willmot (1993), Dickson (1994), DeVylder and Marceau (1996), and Cheng et al. (2000), Cheng and Zhu (2001), Yuen and Guo (Yuen et al. 2005), Xiao and Guo (2007).

In recent year, the compound binomial model has found wide spread applications in the field of financial risk, especially, for the measurement of risk in credit risk and operational risk. CreditRisk+ is a credit risk model that was originally developed by Credit Suisse Financial Products in 1997. The key idea of CreditRisk + is based on the compound binomial model, and has quickly become one of the

L.-j. Sun (✉)
School of Insurance, University of International Business and Economics, Beijing 100029, China

Y.-H. Chen
Institute of Statistical Science, Academia Sinica, Taipei 11529, Taiwan

Y. Zhou and D.D. Wu (eds.), *Modeling Risk Management for Resources and Environment in China*, Computational Risk Management,
DOI 10.1007/978-3-642-18387-4_59, 

financial industry's benchmarks in the field of credit risk modeling. Loss Distribution Approach (LDA) is a popular technique in financial institution for quantitative modeling of operational risk, which also refers to the compound binomial model. See Giese (2003), Haaf et al. (2003), Frachot et al. (2001), for more information on these subjects.

The set-up of a compound binomial model in ruin problems is as follows. Let $\xi_n$ indicate whether or not a claim occurs in the nth time interval (n − 1, n], $p = P(\xi_n = 1)$ and $1 - p = P(\xi_n = 0)$, for $n = 1,2, \cdots, 0 < p < 1$. The number of claims happening up to time $t$ hence $N(t) \sim Binomial(t, p)$. Assume that the occurrences of claims in different time periods are independent events, and the individual claim amounts $\{X_i, i \geq 1\}$ are mutually independent, identically distributed with a common probability function $p(x) = P(X_i = x)$, where $x = 1, 2, 3, \cdots$, and $p(0) = 0$. Let $\{X_i, i \geq 1\}$ be independent of the binomial process $N(t)$, and $S(t) = X_1 + X_2 + \cdots + X_{N(t)}$ be the aggregate claim amounts up to time $t$. The insurer's surplus process $\{U_t\}_{t \geq 0}$ is then given by

$$U_t = u + t - S(t) = u + t - \sum_{i=1}^{N(t)} X_i, \tag{1}$$

where $u \geq 0$ is the integer initial surplus. The premium collected in a unit time is assumed to be one, which contains a positive security loading $\theta$, that is, $1 = (1 + \theta)\, p\mu$, where $\mu = EX_i$ is the mean claim size.

Let $T = \inf\{t \geq 1 : U_t < 0\}$ be the time of ruin, and $\psi(u) = Pr(T < \infty | U_0 = u)$ the probability of ultimate ruin from initial surplus $u$. Note that here the definition of ruin follows with Shiu (1989) and Willmot (1993), rather than from Gerber (1988), Dickson (1994), and Cheng et al. (2000). Two nonnegative random variables related to the time of ruin are the surplus immediately before ruin, $U_{T-1}$, and the deficit at ruin, $|U_T|$ (or $-U_T$). Define $f(x, y, t \mid u) = Pr(U_{T-1} = x, |U_T| = y, T = t | U_0 = u)$ to be the joint probability function of $(U_{T-1}, |U_T|, T)$. Let $f_\upsilon(x, y|u) = \sum_{t=1}^{\infty} \upsilon^t f(x, y, t|u)$ $(x = 0, 1, 2, \ldots, y = 1, 2, \ldots)$ be the discounted joint probability function of $U_{T-1}$ and $|U_T|$ with the discount factor $0 < \upsilon \leq 1$.

By analogy to the continuous time Poisson model (see Gerber and Shiu 1998, Sun 2005), let $w(x_1, x_2)$ be a nonnegative and bounded function, where $0 \leq x_1, x_2 < \infty$ are integers. For a discount factor $0 < \upsilon \leq 1$, define the expected discounted penalty function

$$\phi_\upsilon(u) = E[\upsilon^T w(U_{T-1}, |U_T|) I(T<\infty) | U_0 = u], \tag{2}$$

where $I(\cdot)$ is the indicator function, the function $w(x_1, x_2)$ can be viewed as the penalty at the time of ruin, and $\upsilon$ is a discount factor.

In this paper, we will discuss $\phi_\upsilon(u)$ in detail. By first giving the recursive formula and the renewal equation satisfied by $\phi_\upsilon(u)$, we prove that $\phi_\upsilon(u)$ can be expressed in terms of the probability function of a related compound geometric random variable. The asymptotic expression for $\phi_\upsilon(u)$ and maximal aggregate loss of the surplus process are further studied.

## 2 Recursive formula and renewal equation

In this section, we will derive the recursive formula and renewal equation satisfied by $\phi_\upsilon(u)$. Throughout this paper, the value of a series $\sum_{k=0}^{u} a(k)$ is set to 0 when $u < 0$, and $a(0)$ when $u = 0$.

**Theorem 1.** *The recursive formula of* $\phi_\upsilon(u)$ *is*

$$\phi_\upsilon(u+1) = \frac{1}{q\upsilon}\phi_\upsilon(u) - \frac{p}{q}\sum_{x=1}^{u+1}\phi_\upsilon(u+1-x)p(x) - \frac{p}{q}\sum_{k=1}^{\infty}w(u,k)p(k+u+1), \tag{3}$$

*with initial value* $\phi_\upsilon(0)$

$$\phi_\upsilon(0) = \frac{p}{q}\sum_{u=0}^{\infty}\sum_{y=1}^{\infty}w(u,y)\rho^{u+1}p(y+u+1), \tag{4}$$

*where* $0 < \rho \leq 1$ *is the solution of Lundberg fundamental equation (7).*

*Proof.* By consider whether or not a claim occurs in the first period, we have

$$\phi_\upsilon(u) = q\upsilon\phi_\upsilon(u+1) + p\upsilon\sum_{x=1}^{u+1}\phi_\upsilon(u+1-x)p(x) + p\upsilon\sum_{x=u+2}^{\infty}w(u,x-u-1)p(x). \tag{5}$$

Making the substitution $k = x - u - 1$, we thus obtain the recursive formulas:

$$\phi_\upsilon(u+1) = \frac{1}{q\upsilon}\phi_\upsilon(u) - \frac{p}{q}\sum_{x=1}^{u+1}\phi_\upsilon(u+1-x)p(x) - \frac{p}{q}\sum_{k=1}^{\infty}w(u,k)p(k+u+1).$$

The initial value $\phi_\upsilon(0)$ is derived in the following. Let $\phi_\upsilon^*(s) = \sum_{u=0}^{\infty}s^u\phi_\upsilon(u)$ be the generating function of $\phi_\upsilon(u)$. Multiplying the two side of (5) by $s^u$ ($|s| \leq 1$, *and* $s \neq 0$), and summing it over 0 to $\infty$, we get

$$\phi_\upsilon^*(s) = q\upsilon s^{-1}[\phi_\upsilon^*(s) - \phi_\upsilon(0)] + p\upsilon s^{-1}\phi_\upsilon^*(s)G_X(s)$$
$$+ p\upsilon\sum_{u=0}^{\infty}\sum_{k=1}^{\infty}w(u,k)s^u p(y+u+1)$$

where $G_X(s) = E[s^X] = \sum_{n=1}^{\infty}p(n)s^n$ ($|s| \leq 1$, *and* $s \neq 0$) denotes the probability generating function of the random variable $X$, we then have

$$\phi_{\upsilon}^{*}(s)=\frac{-q\upsilon\phi_{\upsilon}(0)+p\upsilon\sum_{u=0}^{\infty}\sum_{k=1}^{\infty}w(u,k)s^{u+1}p(k+u+1)}{s-q\upsilon-p\upsilon G_X(s)}. \tag{6}$$

Observing the denominator of (6), and letting

$$s-q\upsilon-p\upsilon G_X(s)=0, \tag{7}$$

the (7) is just the Lundberg fundamental equation, who has at most two roots, and one of them, say $\rho$, satisfies $0<\rho\leq 1$. If some regularity conditions for the tail of the probability function $p(x)$ is satisfied, the other root, say $R$, exists, and is greater than 1 (see Cheng et al. 2000).

Since $\rho$ is a zero of the denominator of (6), it must also be a zero of the numerator, so

$$\phi_{\upsilon}(0)=\frac{p}{q}\sum_{u=0}^{\infty}\sum_{y=1}^{\infty}w(u,y)\rho^{u+1}p(y+u+1).$$

We thus conclude the theorem.

Return to (6). Notice that for any two real numbers $r$ and $s$, with $|\,r\,|\leq 1, |\,s\,|\leq 1$ and $r\neq 0$, $s\neq 0$,

$$\frac{G_X(r)-G_X(s)}{r-s}=\sum_{k=0}^{\infty}\left[\sum_{n=k+1}^{\infty}p(n)r^{n-k-1}\right]s^k\equiv\sum_{k=0}^{\infty}A_r(k)s^k, \tag{8}$$

where $A_r(k)=\sum_{n=0}^{\infty}p(n+k+1)r^n$. The denominator of (6) can thus be rewritten as

$$s-\upsilon[q+pG_X(s)]=(s-\rho)\left[1-p\upsilon\sum_{k=0}^{\infty}A_{\rho}(k)s^k\right],$$

and the numerator of (6) can also be rewritten as

$$p\upsilon\sum_{u=0}^{\infty}\sum_{k=1}^{\infty}(s^{u+1}-\rho^{u+1})w(u,k)p(k+u+1)=p\upsilon(s-\rho)\left[\sum_{k=0}^{\infty}B_{\rho}(k)s^k\right],$$

where

$$B_{\rho}(k)=\sum_{n=k}^{\infty}\sum_{j=1}^{\infty}w(n,j)p(n+j+1)\rho^{n-k}. \tag{9}$$

Plugging the expressions of numerator and denominator into (6), we obtain the generating function of $\phi_{\upsilon}(u)$.

**Corollary 1.** *The generating function of* $\phi_\upsilon(u)$ *is*

$$\phi_\upsilon^*(s) = \frac{p\upsilon\sum_{k=0}^{\infty}B_\rho(k)s^k}{1 - p\upsilon\sum_{k=0}^{\infty}A_\rho(k)s^k}, \textit{for } |s| \le 1, \textit{ and } s \neq 0. \tag{10}$$

*The definitions of* $A_\rho(k)$ *and* $B_\rho(k)$ *are given in (8) and (9), respectively.*

In the following, we will derive the renewal equation for $\phi_\upsilon(u)$, from which the explicit expression and approximate results can be obtained further.

**Corollary 2.** *For* $u = 0, 1, 2, \ldots$, $\phi_\upsilon(u)$ *satisfies the defective renewal equation*

$$\phi_\upsilon(u) = p\upsilon\sum_{t=0}^{u}\phi_\upsilon(u-t)A_\rho(t) + p\upsilon B_\rho(u). \tag{11}$$

*Proof.* We start from (10). Notice that

$$\phi_\upsilon^*(s) \quad p\upsilon\phi_\upsilon^*(s)\sum_{k=0}^{\infty}A_\rho(k)s^k = p\upsilon\sum_{k=0}^{\infty}B_\rho(k)s^k.$$

By using the uniqueness of the generating function, we have

$$\phi_\upsilon(u) = p\upsilon\sum_{t=0}^{u}\phi_\upsilon(u-t)A_\rho(t) + p\upsilon B_\rho(u).$$

Utilizing the fact that $\rho$ is the solution of (7), we have

$$p\upsilon\sum_{u=0}^{\infty}A_\rho(u) = \frac{p\upsilon - p\upsilon G_X(\rho)}{1-\rho} = \frac{\upsilon-\rho}{1-\rho} < 1,$$

so the renewal equation is defective, concluding the theorem.

Rewrite the renewal equation (11) as

$$\phi_\upsilon(u)[1 - p\upsilon A_\rho(0)] = p\upsilon\sum_{y=1}^{u}\phi_\upsilon(u-y)A_\rho(y) + p\upsilon B_\rho(u).$$

Let $g(y) = \dfrac{p\upsilon A_\rho(y)}{1 - p\upsilon A_\rho(0)}$ for $y = 1, 2, \ldots$, then

$$g(y) = \frac{p}{q}\sum_{x=0}^{\infty}\rho^{x+1}p(x+y+1), \tag{12}$$

and the renewal equation (11) can be written as

$$\phi_\upsilon(u) = \sum_{y=1}^{u}\phi_\upsilon(u-y)g(y) + \frac{p\upsilon B_\rho(u)}{1 - p\upsilon A_\rho(0)}. \tag{13}$$

From (11), we see that $\phi_\upsilon(0) = \frac{\upsilon p B_\rho(0)}{1-p\upsilon A_\rho(0)}$. Taking $w(x_1, x_2) = I_{\{x_1=x, x_2=y\}}$, it follows that $f_\upsilon(x, y|0) = \frac{p}{q}\rho^{x+1} p(x+y+1)$, which implies

$$g(y) = \sum_{x=0}^{\infty} f_\upsilon(x, y|0), \quad y = 1, 2, \ldots.$$

$g(y)$ defined in (12) can be interpreted as the discounted probability that the surplus will ever fall below its initial value $u$ and will be $u - y$ when it happens for the first time. Note that here $g(y)$ is different from that in Cheng et al. (2000), which has the form $g(y) = p\upsilon \sum_{x=0}^{\infty} p(x+y+1)\rho^x$. This discrepancy exists because they defined ruin as the event that the surplus $U(t)$ becomes "non-positive" rather than "negative" as we defined here.

To normalize the equation (13), let

$$b = \frac{1}{\sum_{y=1}^{\infty} g(y)} - 1 = \frac{\rho(2-\rho-p\upsilon) - q\upsilon}{\rho(p\upsilon - 1) + q\upsilon} \ (0<\upsilon<1), \ \ so \ \frac{1}{1+b} = \sum_{y=1}^{\infty} g(y).$$

Define

$$\eta(y) = (1+b)g(y) \ with \ \eta(0) = 0, M(u) = (1+b)\frac{p\upsilon B_\rho(u)}{1-p\upsilon A_\rho(0)},$$

then the equation (13) may be expressed in terms of $b$, $\eta(y)$, and $M(y)$ as

$$\phi_\upsilon(u) = \frac{1}{1+b}\sum_{y=1}^{u} \phi_\upsilon(u-y)\eta(y) + \frac{1}{1+b}M(u). \tag{14}$$

An explicit solution of $\phi_\upsilon(u)$ can be derived from (14).

**Theorem 2.** *The solution of the renewal equation (14) can be expressed as*

$$\phi_\upsilon(u) = \frac{1}{b}\sum_{x=0}^{u} M(u-x)\kappa(x), \tag{15}$$

*where*

$$\kappa(x) = \sum_{n=0}^{\infty} \frac{b}{(1+b)^{n+1}}\eta^{*n}(x) \ (x = 1, 2, \ldots), \ \ with \ \ \kappa(0) = \frac{b}{1+b} \tag{16}$$

*is a compound geometric probability function, $\eta^{*n}(x)$ denotes the n-fold convolutions of $\eta(x)$.*

*Proof.* Taking the generating function of (14) gives

$$\phi_\upsilon^*(s) = \frac{M^*(s)}{1+b-\eta^*(s)}.$$

Note that $\kappa^*(s) = \frac{b}{1+b-\eta^*(s)}$, and by the uniqueness of the generating function, we have

$$\phi_\upsilon(u) = \frac{1}{b}\sum_{x=0}^{u} M(u-x)\kappa(x).$$

# 3 The asymptotic formula

In this section, we will derive the asymptotic expression of $\phi_\upsilon(u)$.

**Theorem 3.** *If $\bar{K}(u) = 1 - \kappa(u)$ is the tail of the associated compound geometric distribution function defined in (16), then $\bar{K}(u)$ also satisfies a defective renewal equation*

$$\bar{K}(u) = \frac{1}{1+b}\sum_{x=1}^{u}\bar{K}(u-x)\eta(x) + \frac{1}{1+b}\sum_{x=u+1}^{\infty}\eta(x)\,(u\geq 0),\ \textit{with}\ \bar{K}(0) = \frac{1}{1+b}. \quad (17)$$

*Proof.* Note that the $\kappa(x)$ given in (16) can be interpreted as a compound geometric probability function. Specifically, let $\{V_1, V_2, \ldots\}$ be an i.i.d. sequence of positive random variables with a common probability function $\eta(x)$ $(x = 1, 2, \ldots)$, and are independent of $N$, which is a geometric random variable with probability mass function $P(N = n) = \left(1 - \frac{1}{1+b}\right)\left(\frac{1}{1+b}\right)^n$, $n = 0, 1, 2, \ldots$. So,

$$\kappa(u) = P\left(\sum_{i=1}^{N} V_i = u\right) = \sum_{n=0}^{\infty}\frac{b}{(1+b)^{n+1}}\eta^{*n}(u),\ \textit{for}\ u = 0, 1, 2, \ldots.$$

Also, $\bar{K}(u) = P(\sum_{i=1}^{N} V_i > u) = \sum_{n=1}^{\infty}\frac{b}{(1+b)^{n+1}}\bar{L}^{*n}(u)$ is the tail of the associated compound geometric distribution function, where $\bar{L}(u) = \sum_{x=u+1}^{\infty}\eta(x)$, and $\bar{L}^{*n}(u)$ is the tail of the n-fold convolutions of $L(u)$. Since when $u \geq 0$,

$$\begin{aligned}\bar{K}(u) = P(\sum_{i=1}^{N} V_i > u) &= \frac{1}{1+b}P(\sum_{i=1}^{N} V_i > u | N \geq 1)\\ &= \frac{1}{1+b}\sum_{x=1}^{u}\bar{K}(u-x)\eta(x) + \frac{1}{1+b}\sum_{x=u+1}^{\infty}\eta(x),\end{aligned}$$

proving the equation (17).

**Corollary 3.** *The associated compound geometric tail function $\bar{K}(u)$ can be expressed as*

$$\bar{K}(u) = E[\upsilon^T I\{T < \infty\} | U(0) = u],\ u = 0, 1, 2, \ldots.$$

*Further, if* $\upsilon = 1$, $\bar{K}(u)$ *is exactly the probability of ruin,*

$$\bar{K}(u) = \psi(u) = E[I(T<\infty)|U(0) = u]. \tag{18}$$

*Proof.* When $w(x, y) = 1$, $\phi_\upsilon(u)$ in (2) reduces to $E[\upsilon^T I\,(T < \infty)|U(0) = u]$, and $B_\rho(u)$ in (13) is $B_\rho(u) = \sum_{x=u+1}^{\infty} A_\rho(x)$. Comparing (14) with (17), we then conclude the result.

The fact that $\bar{K}(u)$ is the generating function of the ruin time has been observed by Lin and Willmot (1999) for the Poisson process. We can further obtain the asymptotic expressions for $\bar{K}(u)$ and $\phi_\upsilon(u)$ through the behaviour of $g(y)$ and $A_\rho(y)$. Note that, under sufficient regularity conditions for the tail of the individual claim amount probability function $p(x)$, the Lundberg's fundamental equation (7) has a solution $R > 1$, and the probability generating function $G_X(R)$ exists. From $R - q\upsilon - p\upsilon\, G_X(R) = 0$, and $\rho - q\upsilon - p\upsilon\, G_X(\rho) = 0$, follows that

$$p\upsilon \frac{G_X(\rho) - G_X(R)}{\rho - R} = \sum_{u=0}^{\infty} R^u \left[ p\upsilon \sum_{x=0}^{\infty} p(x+u+1)\rho^x \right] = p\upsilon \sum_{u=0}^{\infty} R^u A_\rho(u) = 1. \tag{19}$$

We now apply the key renewal theorem (see, Feller (1968), p. 331) to (11) and use (19), an asymptotic result for $\phi_\upsilon(u)$ then follows.

**Theorem 4.** *Suppose the Lundberg's fundamental equation (7) has a root* $R > 1$, *then* $\phi_\upsilon(u)$ *has an asymptotic expression*

$$\phi_\upsilon(u) \sim \frac{\sum_{y=0}^{\infty} R^y B_\rho(y)}{\sum_{y=1}^{\infty} y R^y A_\rho(y)} R^{-u}, \quad as\ u \to \infty.$$

*The definitions of* $A_\rho(y)$ *and* $B_\rho(y)$ *are given in (8) and (9), respectively.*

**Theorem 5.** *If Lundberg's fundamental equation (7) has a root* $R > 1$, $\bar{K}(u)$ *satisfies the approximate formula*

$$\bar{K}(u) \sim \frac{\sum_{y=0}^{\infty} \sum_{x=y+1}^{\infty} g(x)}{\sum_{y=1}^{\infty} y g(y)} R^{-u}, \quad as\ u \to \infty.$$

## 4 The maximal aggregate loss and associated compound geometric distribution

In the risk model (1), $S(t)$ is the aggregate claim amounts up to time $t$. Define the maximal aggregate loss of the surplus process (see Bowers et al. 1997) as

$$U = \max_{t \geq 0}\{S(t) - t\}.$$

It is well known that $\psi(u) = P(U > u)$, and $U = \sum_{i=1}^{N_1} U_i (U = 0 \; if \; N_1 = 0)$, where $N_1$ is the number of ladder epoch and has a geometric distribution with $P(N_1 = n) = [1 - \psi(0)]\psi(0)^n$, $n = 0, 1, 2, \ldots$. The $\{U_i, i \geq 1\}$ are i.i.d. random variables with $U_1$ denoting the amount of the claim by which the surplus fall below the initial value for the first time, given that this ever happens. The random variables $N_1$ and $\{U_i, i \geq 1\}$ are mutually independent, and the common distribution function of $\{U_i, i \geq 1\}$ is $L_1(u)$, where $L_1(u)$ is $L(u)$ evaluated at $\upsilon = 1$, and $L(u) = \sum_{x=1}^{u} \eta(x)$, with $\eta(x) = (1 + b)g(x)$. So $\psi(u)$ may be expressed as

$$\psi(u) = \sum_{n=1}^{\infty} [1 - \psi(0)]\psi(0)^n \bar{L}_1^{*n}(u), \; u \geq 0. \tag{20}$$

Note that the tail of the associated compound geometric distribution function, $\bar{K}(u)$, given in (17) can be also expressed as

$$\bar{K}(u) = \sum_{n=1}^{\infty} [1 - \bar{K}(0)]\bar{K}(0)^n \bar{L}^{*n}(u), \; with \; \bar{K}(0) = \frac{1}{1+b} = \sum_{y=1}^{\infty} g(y), \tag{21}$$

where $\bar{L}(u) = 1 - L(u)$. When $\upsilon = 1$, $\bar{K}(u)$ reduces to $\psi(u)$, and (21) reduces to (20) again. Also we have $\psi(0) = \sum_{y=1}^{\infty} g_1(y)$, where $g_1(y)$ is $g(y)$ evaluated at $\upsilon = 1$.

From the proof of Theorem 3, let $V = \sum_{i=1}^{N} V_i$, where $N$ has a geometric distribution with $P(N = n) = [1 - \bar{K}(0)]\bar{K}(0)^n$, $n = 0, 1, 2, \ldots$, the common distribution function for $\{V_i, i \geq 1\}$ is given by $L(y) = \frac{1}{\bar{K}(0)} \sum_{z=1}^{y} \sum_{x=0}^{\infty} \sum_{t=1}^{\infty} \upsilon^t f(x, z, t|0)$. Therefore $V_i$ can be viewed as the discounted version of $U_i$ at the time when the surplus process falls bellow its initial level for the first time. Comparing (20) with (21), it is then interesting to note that $\bar{K}(u) = P(V > u)$. Based on the above relationships, the generating functions of $U$ and $V$ can be further expressed in terms of $g(y)$.

**Theorem 6.** *The generating functions of U and V are given by*

$$G_U(r) = \frac{1 - \psi(0)}{1 - \psi(0)G_{U_i}(r)} = \frac{1 - \sum_{y=1}^{\infty} g_1(y)}{1 - \sum_{y=1}^{\infty} r^y g_1(y)},$$

$$G_V(r) = \frac{1 - \bar{K}(0)}{1 - \bar{K}(0)G_{V_i}(r)} = \frac{1 - \sum_{y=1}^{\infty} g(y)}{1 - \sum_{y=1}^{\infty} r^y g(y)}, \; for \; r \leq 1 \; and \; r > 0.$$

*Where* $g_1(y) = \frac{p}{q}[1 - P(y)]$ *is* $g(y)$ *evaluated at* $\upsilon = 1$, *which can be interpreted as the probability that the surplus will ever fall below its initial value u and will be* $u - y$ *when it happens for the first time, and* $g(y)$ *is given in (12).*

The proof of Theorem 6 see Sun (2005) for the detail.

**Acknowledgments** Li-juan Sun gratefully acknowledges the grant of 211 supported by University of International Business and Economics.

## References

Bowers NL, Gerber HU, Hickman JC, Jones DA, Nesbitt CJ (1997) Actuarial mathematics, 2nd edn. The Society of Actuaries, Schaumburg

Cheng S, Zhu R (2001) The asymptotic formulas and Lundberg upper bound in fully discrete risk model. Applied Matgematics. J Chinese Univ Series A 16:348–358

Cheng S, Gerber HU, Shiu EW (2000) Discounted probabilities and ruin theory in the compound binomial model. Insurance: Math Econ 26:239–250

De Vylder FE, Marceau E (1996) Classical numberical ruin probabilities. Scand Actuarial J 2:191–207

Dickson DCM (1994) Some comments on the compound binomial model. ASTIN Bull 24:33–45

Feller W (1968) An introduction to probability theory and its applications, vol 1, 3rd edn. Wiley, New York

Frachot A, Georges P, Ro T (2001) Working paper, Crédit Lyonnais, Groupe de Recherche Opérationelle. Loss distribution approach for operational risk. http://ssrn.com/abstract=1032523

Gerber HU (1988) Mathematical fun with the compound binomial process. ASTIN Bull 18:161–168

Gerber HU, Shiu ESW (1998) On the time value of ruin. N Am Actuarial J 2(1):48–78

Giese G (2003) Enhancing CreditRisk+. Risk 16(4):73–77

Haaf H, Reiss O, Schoenmakers J (2003) Numerically stable computation of CreditRisk+. J Risk 6(4):1–10

Lin X, Willmot GE (1999) Analysis of a defective renewal equation arising in ruin theory. Insurance: Math Econ 25:63–84

Shiu ESW (1989) The probability of eventual ruin in the compound binomial model. ASTIN Bull 19:179–190

Sun LJ (2005) The expected discounted penalty at ruin in the Erlang (2) risk process. Stat Prob Lett 72:205–217

Willmot GE (1993) Ruin probabilities in the compound binomial model. Insurance: Math Econ 12:133–142

Xiao Y, Guo JY (2007) The compound binomial risk model with time-correlated claims. Insurance: Math Econ 41:124–133

Yuen KC, Guo JY, Ng KW (2005) On ultimate ruin in a delayed-claims risk model. J Appl Prob 42:163–174

# Modeling Spatial Time Series by Graphical Models

**Qifeng Wu and Yuan Li**

**Abstract** We propose the spatial temporal autoregressive models based on graph for spatial time series. With Granger's causal relation, we first define the spatial temporal chain graph for spatial time series. Based on the chain graph, the spatial temporal autoregressive model is constructed. Model building procedures are given by graph selection and Bayesian method.

**Keywords** Causality · Graphical models · Spatial time series

## 1 Introduction

Modeling spatial data associated with space sites and time constitutes one of the most interesting recent rese.arch fields in spatial statistics and spatial time series. Spatial temporal statistical models have been widely used in a number of fields such as econometrics, epidemiology, environmental science, image analysis, oceanography, etc. There are an enormous literature on this subject. For details, we the readers refer to Anselin and Florax (1995), Pfeifer and Deutsch (1980), Cressie (1993), Elhorst (2000), Kamarianakis and Prastacos (2005), Ripley (1981), Rosenblatt (1995), Tj$\phi$ stheim (1987).

Spatial temporal statistical models studied in recent years mainly focuses on parametric, nonparametric and semi-parametric forms. However, because of space complex and the curse of dimensionality, most studies deal with parametric models. A typical representative of parametric models is the space-time autoregressive integrated moving average model (STARIMA model), which was first presented

The research was supported by the National Natural Science Foundation of China (NSFC 10971042).

Q. Wu
Shaozhou Nomal Colledge, Shaoguan University, China

Y. Li (✉)
School of Mathematics & Information Science, Guangzhou University, China

Y. Zhou and D.D. Wu (eds.), *Modeling Risk Management for Resources and Environment in China*, Computational Risk Management,
DOI 10.1007/978-3-642-18387-4_60, 

in the literature in the early eighties last century. Since then this model had obtained wide uses in a variety of disciplines such as river flow, spread of disease and spatial econometrics (Kamarianakis and Prastacos, 2005). As Pfeifer and Deutsch (1980) point out, STARIMA model reflects the dependence between the N-dimensional time series consisted of univariate time series observed at N space locations. By introducing the weight matrices specified by a graph prior to analyzing the data to the model, STARIMA models N time series simultaneously as a linear combination of past observations and disturbances at neighboring sites. STARIMA model incorporates the dependence structure of not only time, but also space.

Before Kamarianakis and Prastacos (2005), as a purely inductive model, STARIMA model was univariate in nature when it was used to model the spatial temporal behavior of traffic flow. Kamarianakis and Prastacos (2005) tailored STARIMA model to model the traffic flow of a road network. However, STARIMA model has two defects. One is that it limits to one variable. The other is that the weight matrices in the model has to be specified by a graph prior to analyzing the data. In practice, the graph is usually unknown and should be specified by data. These defects limit application of STARIMA model greatly. Furthermore, it is difficult to extend STARIMA model to multivariate cases by Kamarianakis and Prastacos (2005) method.

In this paper, we consider the problem of modeling spatial time series. Our model overcomes the drawback of STARIMA model by chain graph, graph selection and Bayesian method. With Granger's causal relation, we first define the spatial temporal chain graph for spatial time series. Based on the chain graph, the spatial temporal autoregressive model is proposed.

The paper is organized as follows. The spatial temporal chain graphs and spatial temporal autoregressive model based on the chain graph are presented in Sect. 2. The model building procedures are given in Sect. 3.

## 2 The Spatial Temporal Autoregressive Model Approach

### *2.1 The Spatial Temporal Chain Graph*

Let $Z^m$ be the set of integer lattice points in the $m$-dimensional Euclidean space where $m \geq 1$ and $Z = \{..., -2, -1, 0, 1, 2, ...\}$. A $d$-dimensional stationary spatial time series $\{X(s,t), s \in Z^m, t \in Z\}$ is a $R^d$ valued time series defined on space domain $Z^m$. Here stationarity is in sense of both time and space. Denote by $\gamma(s,t)$ the covariance function of $\{X(s,t), s \in Z^m, t \in Z\}$. Then stationarity means that $\gamma(s,\tau) = Cov(X(a,t), X(a+s, t+\tau))$.

A graph over $V$ is an ordered pair $(V, E)$ where the elements in $V$ are called the vertices, and $E$ is a set of directed or undirected edges denoted as $a \rightarrow b$, and $a - b$ for distinct vertices $a$ and $b$ in $V$. $a \rightarrow b$ is called directed edges, while $a - b$ is called undirected edge. $Pa(a) = \{b : a \rightarrow b, b \in V\}$ is called the parent set of $a$, while $Neb(a) = \{b : a - b, b \in V\}$ is called the neighbor set of $a$.

Causality is very important in analyzing time series. In the following, we take Granger causality to define spatial temporal graphs, which leads to the spatial temporal chain graph. Let $V_{TS} = \{(i,j,l), i \in V, j \in Z^m, l \in Z\}$ be the set of vertices, i.e., $V_{TS} = V \times Z^m \times Z$ where $V = \{1,2,...,d\}$, $i$ denotes the $i$-th variable, $j$ spatial location and $l$ the time. Hence vertex $(i,j,l)$ represents the value $X_i(j,l)$ of the $i$-th variable of $X_V$ at spatial location $j$ and the time point $l$. Following Dahlhaus and Eichler (2003), we define the spatial temporal chain graph for space time series.

**Definition 1.** *The spatial temporal chain graph, written as STSC-graph, for stationary process $X_V$ is the graph $G_{TS} = (V_{TS}, E_{TS})$ where edge set $E_{TS}$ satisfies*

1. $(a,s_1,t-u) \to (b,s_2,t) \notin E_{TS} \Leftrightarrow u<0$ *or* $X_a(s_1,t-u)^\wedge X_b(s_2,t) \,|\, \bar{X}(s_1,t) - \{X_a(s_1,t-u)\}$
2. $(a,s_1,t-u) - (b,s_2,t) \notin E_{TS} \Leftrightarrow u \neq 0$ *or* $X_a(s_1,t)^\wedge X_b(s_2,t) \,|\, \bar{X}_V(s_1,s_2,t) \cup \{X_{V-\{a,b\}}(t)\}$

*where $\bar{X}(s_1,t) = \{X(s_1,l), l<t\}$ denotes the past of the process $X(s,t)$ at space $s_1$ and time $t$. $X_{V-\{a,b\}}(t) = \{X_j(s,t), s \in Z^m, t \in Z, j \in V, j \neq a,b\}$. Here it involves conditional orthogonality.*

*For random vectors X, Y and Z, we call X and Y are conditionally orthogonal given Z, denoted by $X \perp Y|Z$, if X and Y are uncorrelated after the linear effects of Z have been removed. Let X, Y, U, V be random variables, and $\pi = \{\mu + \alpha U + \beta V | \mu, \alpha, \beta \in R\}$ be a linear subspace spanned by U,V. The best linear predictor, denoted by $\mathrm{Pr}\,oj_\pi X$, of X on U,V is defined as*

$$E(X - \mathrm{Pr}\,oj_\pi X)^2 = \min_{\mu,\alpha,\beta} E(X - \mu - \alpha U - \beta V)^2. \tag{1}$$

*$\mathrm{Pr}\,oj_\pi X$ is called the linear projector of X on $\pi$. The linear property is obviously true for $\mathrm{Pr}\,oj_\pi X$: $\mathrm{Pr}\,oj_\pi(\alpha X + \beta Y) = \alpha \mathrm{Pr}\,oj_\pi X + \beta \mathrm{Pr}\,oj_\pi Y$. The definition of the linear projector $\mathrm{Pr}\,oj_\pi X$ can be generalized to a general linear subspace $\pi$ spanned by random variables $X_1, X_2, ..., X_n, ...$ With $\mathrm{Pr}\,oj_\pi X$, we readily define the partial covariance as follows:*

$$Cov_L(X,Y|\pi) = E(X - \mathrm{Pr}\,oj_\pi X)(Y - \mathrm{Pr}\,oj_\pi Y).$$

*It follows from Kamarianakis and Prastacos (2005) that the spatial temporal autoregressive model STAR(p) with order p, based on weighting matrices, is defined as follows:*

$$X_t = \sum_{k=1}^{p} \sum_{l=0}^{\lambda_k} \Phi_{kl} W_l X_{t-k} + \varepsilon_t \tag{2}$$

*where $X_t = (X(s_1,t), X(s_2,t), ..., X(s_M,t))^\tau$, i.e., $X(s,t)$ is consisted of observations at all space locations $s_1, s_2, ..., s_M$ and time t with $X(s,t)$ being one dimensional, p is the autoregressive order, $\lambda_k$ is the spatial order of the $k^{th}$ autoregressive term,*

*$\Phi_{kl}$ are $N \times N$ parametric matrices, $W_l$ is the known $N \times N$ matrix for spatial order $l$ and $\varepsilon_t$ is the random noise vector with the mean $E\varepsilon_t = 0$, the covariance matrix $E\varepsilon_t\varepsilon_{t-s}^\tau = \Sigma\delta_{0,s}$ where $\delta_{0,s} = 1$, if $s = 0$ and $\delta_{0,s} = 0$ if $s \neq 0$.*

*The characteristic of model (2) is that it expresses each observation at time t and location i as a linear combination of previous observations and innovations lagged both in time and locations. As Kamarianakis and Prastacos (2005) and pointed out, the basic mechanism of this representation is the hierarchical ordering of the neighbors of each site, based on which weight matrices $\{W_l\}$ are constructed.*

*For example, in traffic flow system, $X(s,t)$ denotes the traffic flow at location s and time t. If we know the road network tree structure in advance, then we can model the traffic flow as (2). It is easily seen that the traffic flow at downstream locations only depends on that at upstream locations but not vice versa.*

*Model (2) reflects this influence by introducing a hierarchal ordering for the neighbors of each measurement space point. Here involve the conception of "neighbors". If $a \to b$, then a is called the first order neighbor. If $a \to b \to c$, but not $a \to c$, then a is called the second order neighbor, and so on.*

*Let $A_k = \sum_{l=0}^{\lambda_k} \Phi_{kl} W_l$. Then model (2) can be written as the form of VAR model (3).*

$$X_t = A(1)X_{t-1} + A(2)X_{t-2} + \ldots + A(p)X_{t-p} + \varepsilon_t \tag{3}$$

*where $\varepsilon_t$ are independent and identically distributed with mean 0 and covariance matrix $\Sigma$ and $X_t = (X(s_1,t), X(s_2,t), \ldots, X(s_M,t))^\tau$ with $X(s,t)$ in (2) for $d = 1$, i.e., $X(t)$ is a $M \times 1$ vector composed of observations at M locations, $A(l) = (a_{ij}^{(l)})$, $l = 1, 2, \ldots, p$.*

*Note that the dimensionality $d = 1$. Hence the space time series chain graph becomes $(s_1, t-u) \to (s_2, t)$ or $(s_1, t) - (s_2, t)$.*

**Proposition 1.** *For stationary STAR(p) model (2),*

1.
$$(j, t-u) \to (i,t) \notin E_{TS} \Leftrightarrow u \in \{1, 2, \ldots, p\} \quad and \quad a_{ij}^{(u)} = 0 \tag{4}$$

2.
$$(j,t) - (i,t) \notin E_{TS} \Leftrightarrow k_{ij} = 0, \tag{5}$$

where $a_{ij}^{(u)} = \sum_{l=0}^{\lambda_u} \sum_{m=1}^{N} \phi_{im}^{(ul)} w_{mj}^{(l)}$ with $\Phi_{ul} = (\phi_{ij}^{(ul)})_{N\times N}$, $W = (w_{ij}^{(l)})_{N\times N}$ and $K = (k_{ij}) = \Sigma^{-1}$

## 2.2 G-STAR Model

In model (2), if location j is the $l^{\text{th}}$ of site i, then in graphical words, j is the parent or ancestor of i. Therefore we can consider only those terms j which are ancestry of i in model (2). It follows from Proposition 1 that the influence of ancestry is reflected in the $l^{th}$ neighbors implied in the weight matrix $W_l$, which can be interpreted by $a_{ij}^{(u)} = \sum_{l=0}^{\lambda_u} \sum_{m=1}^{N} \phi_{im}^{(ul)} w_{mj}^{(l)}$. So we can consider only those terms j which are parents of I in model (2). Based on this idea and the TSC graph $G_{TS} = (V_{TS}, E_{TS})$, we propose the following spatial temporal AR(p) model G-STAR(p) models:

$$X_i(s_j,t) = \sum_{l=1}^{p} \sum_{(l_1,l_2,t-l)\in Pa((i,j,t))} \alpha_{l_1,l_2,l}^{(i,j)} X_{l_1}(s_{l_2}, t-l) + \varepsilon_{i,j,t} \tag{6}$$

where $\varepsilon_t = (\varepsilon_t^{(1)'}, \varepsilon_t^{(2)'}, ..., \varepsilon_t^{(M)'})'$ are i.i.d. series with $E\varepsilon_t = 0$, $E\varepsilon_t\varepsilon_t' = \Sigma$ with $\Sigma = (\Sigma_{ij})$ where $\Sigma_{ij} = E\varepsilon_t^{(i)}\varepsilon_t^{(j)'}$ being the covariance matrix which reflects the influence of space locations $s_i$ and $s_j$, $\varepsilon_t^{(i)} = (\varepsilon_{i,1,t}, \varepsilon_{i,2,t}, ..., \varepsilon_{i,d,t})'$. Furthermore, $\alpha_{l_1,l_2,l}^{(i,j)}$ and $p$ are unknown parameters.

Our idea to construct model (6) is as follows: the component $X_j(s_i,t)$ at location $s_i$ generally depends not only its history, but also other components at locations. rather than $s_i$. Because of Proposition 1, we consider the influence of the components at other locations on $X_j(s_i,t)$ at location $s_i$ only by the parents. This looks reasonable in many applications.

## 3 Estimation of the parameters for G-STAR model

We firstly express G-STAR(p) models as VAR(p) models. Let $X_t = Vec(A_t)$, $A = (X_j(s_i,t))_{d\times M} = (X_t^{(1)}, X_t^{(2)}, ..., X_t^{(M)})$ where $X_t^{(j)} = (X_1(s_j,t), X_2(s_j,t), ..., X_d(s_j,t))'$ and $Vec(A)$ represents the vector by stacking columns of matrix $A$ one by one. It follows form (6) that

$$X_t^{(j)} = \sum_{l=1}^{p} \sum_{l_1=1}^{M} B_{l_1,l}^{(j)} X_{t-l}^{(l_1)} + \varepsilon_t^{(j)} \tag{7}$$

where $B_{l_1,l}^{(j)} = (b_{km})_{d\times d}$ with $b_{km} = \alpha_{m,l_1,l}^{(k,j)} w_{m,l_1,l}^{(k,j)}$, satisfying $w_{m,l_1,l}^{(k,j)} = 1$, if $(m,l_1,t-l) \in Pa(k,j,t)$; otherwise $w_{m,l_1,l}^{(k,j)} = 0$, and $\varepsilon_t^{(j)} = (\varepsilon_{1j,t}, \varepsilon_{2j,t}, ..., \varepsilon_{dj,t})'$. It follows from (7) that G-STAR(p) can be written as VAR(p) models as follows:

$$X_t = \sum_{l=1}^{p} A_l X_{t-l} + \varepsilon_t \tag{8}$$

where $A_l$ is a $dM \times dM$ matrix with the (i,j)-th block $B_{l,j}^{(i)}$, i.e., $A_l = (B_{l,j}^{(i)})_{M\times M}$ with the $d \times d$ block matrix $B_{l,j}^{(i)}$.

We know from Proposition 1 that $(a, s_{l_1}, t-u) \to (b, s_{l_2}, t) \notin E_{TS} \Leftrightarrow u \in \{1,2,...,p\}$ and $a_{ij}^{(u)} = 0$ with $A_u = (a_{ij}^{(u)})$ and i=(l_2-1)d + b, j = (l_1-1)d + b. It is easily seen form (7) that $a_{ij}^{(u)} = \alpha_{a,l_1,u}^{(b,l_2)} w_{a,l_1,u}^{(b,l_2)}$. Then $a_{ij}^{(u)} = 0$ implies $w_{a,l_1,u}^{(b,l_2)} = 0$, which is consistent with the definition of model (6) where $(a, s_{l_1}, t-u) \notin Pa((b, s_{l_2}, t))$. We also know from Proposition 1 that

$$(a, s_{l_1}, t) - (b, s_{l_2}, t) \notin E_{TS} \Leftrightarrow k_{ij} = 0$$

where $\mathrm{Y} = (\mathrm{k_{lm}})_{\mathrm{dM \times dM}} = [\mathrm{Cov}(\varepsilon_t, \varepsilon_t)]^{-1}$, and i = (l_2-1)d+b, j = (l_1-1)d+b. Especially for $Cov(\varepsilon_t^{(i)}, \varepsilon_t^{(j)}) = 0$ for $i \neq j$, i.e., noises are uncorrelated for different locations, we have that $[Cov(\varepsilon_t, \varepsilon_t)]^{-1} = diag(\Sigma_1^{-1}, \Sigma_2^{-1}, ..., \Sigma_M^{-1})$ with $\Sigma_i = Cov(\varepsilon_t^{(i)}, \varepsilon_t^{(i)})$, i = 1,2,...,M. Then for $l_1 \neq l_2$, it holds that $\kappa_{ij} = 0$ where i = (l_2-1)d+b, j = (l_1-1)d+b which implies that $(a, s_{l_1}, t) - (b, s_{l_2}, t) \notin E_{TS}$, i.e., $X_a(s_{l_1}, t)$ is not correlated with $X_b(s_{l_2}, t)$ contemporaneously. Hence only at the same location s can $X_a(s, t)$ and $X_b(s, t)$ be correlated contemporaneously.

Model (6) can be written as

$$Y = X\Phi + \varepsilon, \tag{9}$$

where $Y$ and $\varepsilon$ are $(T-p) \times dM$ matrices, $X$ is $(T-p) \times dMp$ matrix, $\Phi$ is a $dMp \times dM$.

**Priors on** $\phi = Vec(\Phi) = (\phi_1, \phi_2, \ldots, \phi_n)^\tau$ with $n = (dM)^2 p$. We know from (8) that $\phi_i = 0$ if and only if there does not exist an edge between some two vertices. Let m be the number of zeros of $\{\phi_i\}$ which implies that graph G has n-m edges, $\phi_m = (\phi_{10}, \phi_{20}, \ldots, \phi_{m0})^\tau$ be parameters from $\phi$, corresponding to those vertices between which no an edge is connected, and $\phi_{non}$ be the n-m-dimensional vector composed of the left components of $\phi$ after deducting $\phi_m$. In other words, $\phi_m$ represents the restricted parameters on graph G, while $\phi_{non}$ is free parameters. Given the graph G, we assume that the prior distribution of $\phi_{non}$ is

$$\phi_{non} | G \sim N_{n-m}(\mu_{non}, \Sigma_{non})$$

and the prior distribution of $\phi_m$ is

$$\phi_m | G \sim N_m(0, \Sigma_{st0}).$$

Note that

$$\begin{aligned} f(Y, \phi, \Sigma, G) &= f(Y|\phi, \Sigma, G) f(\phi, \Sigma, G) \\ &= f(\phi, \Sigma, G|Y) f(Y) \end{aligned}$$

Then we have

$$\begin{aligned} f(\phi, \Sigma, G|Y) &\propto f(Y|\phi, \Sigma, G) f(\phi, \Sigma, G) \\ &= f(Y|\phi, \Sigma, G) f(\phi, \Sigma|G) f(\Sigma|G) f(G) \\ &= f(Y|\phi, K, G) f(\phi, K|G) f(K|G) f(G), \end{aligned} \tag{10}$$

where

$$f(Y|\phi,K,G) \propto |\Sigma|^{-(T-P)/2} etr\left\{-\frac{1}{2}(Y-\Phi X)\Sigma^{-1}(Y-\Phi X)^{\tau}\right\}$$
$$= K|^{(T-P)/2} etr\left\{-\frac{1}{2}(Y-\Phi X)K(Y-\Phi X)^{\tau}\right\} \quad (11)$$

with $etr(A) = \exp\{tr(A)\}$ and $K = \Sigma^{-1}$,

$$f(\phi|K,G) = N_n(\mu,\Sigma_0) \propto |\Sigma_0|^{-n/2} \exp\left\{-\frac{1}{2}(\phi-\mu)^{\tau}\Sigma_0^{-1}(\phi-\mu)\right\} \quad (12)$$

with

$$\mu = \begin{pmatrix} \mu_{non} \\ 0 \end{pmatrix}, \quad \Sigma_0 = \begin{pmatrix} \Sigma_{non} & 0 \\ 0 & \Sigma_{st0} \end{pmatrix}.$$

In order to reduce number of parameters in $\Sigma$, we have to parameterize $K = \Sigma^{-1}$. Following Talih (2003) and Zhang (2008), we parameterize $K = (\kappa_{ij})_{dM\times dM}$ as follows:

$$\kappa_{ii} = \frac{v_i}{\sigma_i^2}, \quad \kappa_{ij} = -\frac{\theta}{\sigma_i\sigma_j} I_{\{(i,j)\in E_{sub}, i\neq j\}}$$

where $I_{\{.\}}$ is a indicator function, (i,j) represents the undirected edge between vertices i and j, $E_{sub}$ is the subgraph of E by deleting directed edges at time t. $E_{sub}$ has at most dM(dM-1)/2 edge.

Let $\sigma = (\sigma_1, \sigma_2, \ldots, \sigma_{dM})^{\tau}$. In order to keep K positive definite, we have to limit $|\theta|<1$ (see Talih 2003; Zhang 2008). Then $f(K|G) = f(\theta)f(\sigma)$. It follows from (3.4) that

$$f(\phi,\Sigma,G|Y) \propto f(Y|\phi,K,G) f(\phi,K|G) f(\theta) f(\sigma) f(G). \quad (13)$$

For the prior distribution of $\theta$ and $\sigma$, we follows Talih (2003) and Zhang (2008) to assume that

$$f(\theta) \propto \frac{\pi/2}{\cos^2(\pi\theta/2)}, \quad f(\sigma_i^2) \propto \frac{1}{\sigma_i^2}.$$

Note the stationarity of space time series $\{X(s,t)\}$. Then graph G, there are at most $(dM)^2(T-p-1)$ directed edges and $dM(dM-1)/2$ undirected edges. Hence there are $2^{n_G}$ possible graphs G for $\{X(s,t)\}$ where $n_G = (dM)^2(T-p-1)+ dM(dM-1)/2$ . We assume that each graph has the same prior probability, i.e., the prior distribution of G is as follows: $f(G) = \frac{1}{n_G}$. Given the sample $Y$, the posterior joint distribution of $\phi, \Sigma, G$ is given by (13). It is not easy to get the solution from (13). We consider to use Monte Carlo simulation to get the sample from $f(\phi, \Sigma, G|Y)$. Then we get the joint posterior distribution $G$, $\phi$ and $\Sigma$ given $Y$ by large simulations. For this purpose, we apply Metropolis-Hastings iterative method.

## References

Anselin L, Florax RGJM (1995) New directions in spatial econometrics. Springer, Berlin

Cressie NAC (1993) Statistics for spatial data. Wiley, New York

Dahlhaus R, Eichler M (2003) Causality and graphical models in time series analysis. In: Green P, Hjort N, Richardson S (eds) Highly structured stochastic systems. University Press, Oxford

Elhorst JP (2000) Space-time ARMA modeling. Dynamic models in space and time. Manuscript

Kamarianakis YI, Prastacos P (2005) Space-time modeling of traffic flow. Computers and Geosciences 31(2):119–133

Pfeifer PE, Deutsch SJ (1980) A three-stage iterative procedure for space-time modeling. Technometrics 22(1):35–47

Ripley B (1981) Spatial statistics. Wiley, New York

Rosenblatt M (1995) Stationary sequences and random fields. Birkhauser, Boston

Talih M (2003) Markov random fields on time-varying graphs, with an application to portfolio selection. Ph.D. dissertation, Yale University

Tjøstheim D (1987) Spatial series and time series: similarities and differences. In: Droesbeke F (ed) Spatial process and spatial time series analysis. FUSL, Brussels, pp 217–228

Zhang F (2008) Detection of structural changes in multivariate models and study of time-varying graphs. M.S. dissertation, Guangzhou University

# Standard Deviation Method for Risk Evaluation in Failure Mode under Interval-Valued Intuitionistic Fuzzy Environment

**Yejun Xu**

**Abstract** Failure mode and effects analysis (FMEA) has been extensively used for examining potential failures in products, process, designs and services. An important issue of FMEA is the determination of risk priories of the failure modes that have been identified. In this paper, we treat the risk factors occurrence (O), severity (S) and detection (D) as interval-valued intuitionistic fuzzy numbers, and then standard deviation method is proposed to determine the weights of the risk factors objectively for prioritization of failure modes. A practical example is given to verify the developed method and to demonstrate its practicality and effectiveness.

**Keywords** Interval-valued intuitionistic fuzzy number · Interval-valued intuitionistic fuzzy weighted averaging(IVIFWA) operator, risk evaluation · Standard deviation

## 1 Introduction

Failure mode and effects analysis(FMEA) is a widely used engineering technique for defining, identifying and eliminating known and/or potential failures, problems, errors and so on from system, design, process, and/or service before they reach the customer (Stamatis 1995; Wang et al. 2009). The traditional FMEA determines the risks priorities of failure modes through the risk priority number, which is the product of the occurrence (O), severity (S) and detection (D) of a failure, where O and S are the frequency and seriousness(effects) of the failure, and D is the ability to detect the failure before it reaches the customer.

Significant efforts have been made in FMEA literature. As a result, fuzzy logic has been extensively used for FMEA. For example, Braglia et al. (2003) proposed

Y. Xu
Business School, HoHai University, Nanjing, Jiangsu 210098, China
e-mail: xuyejohn@163.com

Y. Zhou and D.D. Wu (eds.), *Modeling Risk Management for Resources and Environment in China*, Computational Risk Management,
DOI 10.1007/978-3-642-18387-4_61, 

a multi-attribute decision-making approach called fuzzy TOPSIS approach for FMECA, which is a fuzzy version of the technique for order preference by similarity to ideal solution(TOPSIS). The proposed fuzzy TOPSIS approach allows for the risk factors O, S, and D and their relative importance weights to be evaluated using triangular fuzzy numbers. Chang et al. (1999) used fuzzy linguistic terms such as *Very Low*, *Low*, *Moderate*, *High* and *Very High* to evaluate O, S, and D, and utilized grey relational analysis to determine the risk priorities of potential causes. Bowles and Peláez (1995) described a fuzzy logic based approach for prioritizing failures in a system FMEA, which uses fuzzy linguistic terms to describe O, S, D, and the risks of failure. There are also many methods which have been proposed to deal with the FMEA, the reader could refer (Lertworasirikul et al. 2003; Pillay and Wang 2003; Xu et al. 2002). But in the real situation, the decision making problem is so complex, and the decision information provided by a decision maker is often imprecise or uncertain due to time pressure, lack of data, or the decision maker's limited attention and information processing capabilities. Accordingly, intuitionistic fuzzy sets(IFS) introduced by Atanassov have been found to be well suited to dealing with vagueness.

In this paper, we focus our attention on developing a method objectively named standard deviation method for risk evaluation and prioritization of failure modes. To do so, the rest of the paper is organized as follows. In Sect. 2, we introduce the notion of intuitionistic fuzzy values and interval-valued IFSs. In Sect. 3, we introduce the MADM problem with interval-valued intuitionistic fuzzy information, in which the information about attribute weights is completely unknown, and the attribute values take the form of interval-valued intuitionisitic fuzzy numbers. To determine the attributes, an optimization based on the standard deviation method, is established to determine the weights. In Sect. 4, a practical example is used to illustrate the developed models. In Sect. 5, we conclude the paper and give some remarks.

## 2 Preliminaries

In Atanassov (1986), Atanassov introduced a generalized fuzzy set called intuitionistic fuzzy set, shown as follows:

**Definition 1.** *An IFS in X is given by*

$$A = \{<x, \mu_A(x), v_A(x)>|x \in X\} \tag{1}$$

*which is characterized by a membership function* $\mu_A : X \to [0,1]$ *and a non-membership function* $v_A : X \to [0,1]$, *with the condition*

$$0 \leq \mu_A(x) + v_A(x) \leq 1, \forall x \in X$$

*where the numbers* $\mu_A(x)$ *and* $v_A(x)$ *represent, respectively, the degree of membership and the degree of non-membership of the element x to the set A.*

*In the real world, however, due to the decision maker's limited attention and information-processing capabilities, or a decision should be made under time pressure or lack of data, the decision makers may express their preferences over attributes by means of interval formats, so Atanassov (Atanassov and Gargov 1989) extended the Intuitionistic Fuzzy Sets (IFS) into Interval-Valued Intuitionistic Fuzzy Sets (IVIFS).*

**Definition 2.** *An IVIFS over X is given by*

$$\tilde{A} = \{<x, \tilde{\mu}_{\tilde{A}}(x), \tilde{v}_{\tilde{A}}(x)>|x \in X\} \tag{2}$$

*where* $\tilde{\mu}_{\tilde{A}}(x) \subset [0,1]$ *and* $\tilde{v}_{\tilde{A}}(x) \subset [0,1]$ *are interval values and for every* $x \in X$*, and*

$$\sup \tilde{\mu}_{\tilde{A}}(x) + \sup \tilde{v}_{\tilde{A}}(x) \leq 1, \forall x \in X.$$

*For convenience, we call* $\tilde{\alpha} = (\tilde{\mu}_{\tilde{\alpha}}, \tilde{v}_{\tilde{\alpha}}) = ([a,b],[c,d])$ *an interval-valued intuitionistic fuzzy number (IVIFN) (Xu 2007), where* $\tilde{\mu}_{\tilde{\alpha}} = [a,b] \subset [0,1]$, $\tilde{v}_{\tilde{\alpha}} = [c,d] \subset [0,1], b+d \leq 1$.

**Definition 3.** *(Xu 2007): Let* $\tilde{\alpha} = (\tilde{\mu}_{\tilde{\alpha}}, \tilde{v}_{\tilde{\alpha}}) = ([a,b],[c,d])$*be an interval-valued intuitionistic fuzzy number, a score function S of an intuitionistic fuzzy value can be represented as follows:*

$$S(\tilde{\alpha}) = (a - c + b - d)/2 \tag{3}$$

*where* $S(\tilde{\alpha}) \in [-1,1]$.

*Obviously, for an IVIFN* $\tilde{\alpha} = (\tilde{\mu}_{\tilde{\alpha}}, \tilde{v}_{\tilde{\alpha}}) = ([a,b],[c,d])$*, it is clear that* $S(\tilde{\alpha})$*gets bigger, then the IVIFN* $\tilde{\alpha}$ *gets greater. Especially, if* $S(\tilde{\alpha}) = 1$*, then* $\tilde{\alpha}$ *be the maximum value* $\tilde{\alpha} = ([1,1],[0,0])$*, on the contrary, if* $S(\tilde{\alpha}) = -1$ *be the minimum value* $\tilde{\alpha} = ([0,0],[1,1])$.

**Definition 4.** *(Xu 2007): Let* $\tilde{\alpha} = (\tilde{\mu}_{\tilde{\alpha}}, \tilde{v}_{\tilde{\alpha}}) = ([a,b],[c,d])$ *be an interval-valued intuitionistic fuzzy number, an accuracy function H to evaluate the degree of accuracy of the interval-valued intuitionistic fuzzy number can be represented as follows:*

$$H(\tilde{\alpha}) = (a + b + c + d)/2 \tag{4}$$

*where* $H(\tilde{\alpha}) \in [0,1]$*. The larger the value of* $H(\tilde{\alpha})$*, the higher the degree of accuracy of the degree of membership of the IFVN* $\tilde{\alpha}$.

*Based on the score function S and the accuracy function H, Xu(Xu & Yager 2006) introduced an order relation between two interval-valued intuitionistic fuzzy numbers in the following:*

**Definition 5:** *Let* $\tilde{\alpha} = (\tilde{\mu}_{\tilde{\alpha}}, \tilde{v}_{\tilde{\alpha}})$ *and* $\tilde{\beta} = (\tilde{\mu}_{\tilde{\beta}}, \tilde{v}_{\tilde{\beta}})$ *be two inverval-valued intuitionistic fuzzy numbers,* $S(\tilde{\alpha})$ *and* $S(\tilde{\beta})$ *be the scores of* $\tilde{\alpha}$ *and* $\tilde{\beta}$*, respectively, and let* $H(\tilde{\alpha})$ *and* $H(\tilde{\beta})$ *be the accuracy degrees of* $\tilde{\alpha}$ *and* $\tilde{\beta}$*, then*

- *If* $S(\tilde{\alpha}) < S(\tilde{\beta})$*, then* $\tilde{\alpha}$ *is smaller than* $\tilde{\beta}$*, denoted by* $\tilde{\alpha} < \tilde{\beta}$.

- *If* $S(\tilde{\alpha}) = S(\tilde{\beta})$, *then*
  1. *If* $H(\tilde{\alpha}) = H(\tilde{\beta})$, *then* $\tilde{\alpha} = \tilde{\beta}$;
  2. *If* $H(\tilde{\alpha})<H(\tilde{\beta})$, *then* $\tilde{\alpha}$ *is smaller than* $\tilde{\beta}$, *denoted by* $\tilde{\alpha}<\tilde{\beta}$.

*To aggregate interval-valued intuitionistic preference information, Xu (2007) defined the following operations:*

**Definition 6.** *Let* $\tilde{\alpha}_1 = ([a_1, b_1], [c_1, d_1])$*and* $\tilde{\alpha}_2 = ([a_2, b_2], [c_2, d_2])$ *be two interval-valued intuitionistic fuzzy numbers, then*

1. $\tilde{\alpha}_1 + \tilde{\alpha}_2 = ([a_1 + a_2 - a_1a_2, b_1 + b_2 - b_1b_2], [c_1c_2, d_1d_2])$;
2. $\tilde{\alpha}_1 \times \tilde{\alpha}_2 = ([a_1a_2, b_1b_2], [c_1 + c_2 - c_1c_2, d_1 + d_2 - d_1d_2])$;
3. $\lambda\tilde{\alpha}_1 = ([1 - (1 - a_1)^{\lambda}, 1 - (1 - b_1)^{\lambda}], [c_1^{\lambda}, d_1^{\lambda}])\ , \lambda>0$;
4. $\tilde{\alpha}_1^{\lambda} = ([a_1^{\lambda}, b_1^{\lambda}], [1 - (1 - c_1)^{\lambda}, 1 - (1 - d_1)^{\lambda}])$, $\lambda>0$.

**Definition 7.** $\tilde{\alpha}_1 = ([a_1, b_1], [c_1, d_1])$ *and* $\tilde{\alpha}_2 = ([a_2, b_2], [c_2, d_2])$ *be two interval-valued intuitionistic fuzzy numbers, then we call*

$$d(\tilde{\alpha}_1, \tilde{\alpha}_2) = \|\tilde{\alpha}_1 - \tilde{\alpha}_2\| = \frac{1}{4}(|a_1 - a_2| + |b_1 - b_2| + |c_1 - c_2| + |d_1 - d_2|) \quad (5)$$

*the deviation between* $\tilde{\alpha}_1$ *and* $\tilde{\alpha}_2$.

## 3 Standard Deviation Method for Deriving Attributes Weight Information from Interval-Valued Intuitionistic Fuzzy Numbers

Consider that a decision-maker provides his/her preferences over the attributes $U = \{u_1, u_2, ..., u_n\}$by means of interval-valued intuitionistic fuzzy values, i.e., $\tilde{r}_{ij} = (\tilde{\mu}_{ij}, \tilde{v}_{ij}) = ([a_{ij}, b_{ij}], [c_{ij}, d_{ij}])$, where $\tilde{\mu}_{ij} = [a_{ij}, b_{ij}] \subset [0, 1]$,$\tilde{v}_{ij} = [c_{ij}, d_{ij}] \subset [0, 1]$, $b_{ij} + d_{ij} \leq 1$. $i = 1, 2, ..., m$,$j = 1, 2, ..., n$, then all the preference values of the alternatives consists the decision matrix $\tilde{R} = (\tilde{r}_{ij})_{m\times n}$.

**Definition 8.** *Let* $\tilde{R} = (\tilde{r}_{ij})_{m\times n}$ *be the interval-valued intuitionistic fuzzy decision matrix,* $\tilde{r}_i = (\tilde{r}_{i1}, \tilde{r}_{i2}, ..., \tilde{r}_{in})$ *be the vector of attribute values corresponding to the alternative*$x_i$, $i = 1, 2, ..., m$, *then we call*

$$\tilde{z}_i(w) = \text{IFWA}_w(\tilde{r}_{i1}, \tilde{r}_{i2}, ..., \tilde{r}_{in}) = w_1\tilde{r}_{i1} + w_2\tilde{r}_{i2} + \cdots + w_n\tilde{r}_{in}$$
$$= \left(\left[1 - \prod_{j=1}^{n}(1 - a_{ij})^{w_j}, 1 - \prod_{j=1}^{n}(1 - b_{ij})^{w_j}\right], \left[\prod_{j=1}^{n}(c_{ij})^{w_j}, \prod_{j=1}^{n}(d_{ij})^{w_j}\right]\right) \quad (6)$$

*the overall value of the alternative* $x_i$, *where* $w = (w_1, w_2, ..., w_n)^T$ *is the weighting vector of attributes.*

In the situation where the information about attribute weights is completely known, i.e., each attribute weight can be provided by the expert with crisp numerical value, we can aggregate all the weighted attribute values corresponding to each alternative into an overall one by using (6). Based on the overall attribute values $z_i(w)$ of the alternatives $x_i(i = 1, 2, ..., m)$, we can rank all these alternatives and then select the most desirable one(s). The greater $z_i(w)$, the better the alternative $x_i$ will be. However, in this paper, we consider the attribute weight information about the attribute is incomplete known, thus, we need to determine the attribute weight firstly.

The standard deviation method is proposed by Wang (2003) to deal with MADM problems with numerical information. Its main ideal is as follows. For the MADM problems, we need to compare the collective preference values to rank the alternatives, the larger the ranking value $z_i(w)$, the better the corresponding alternative $x_i$ is. If the performance values of each alternative have little differences under an attribute, it shows that such an attribute plays a small important role in the priority procedure. Contrariwise, if some attribute makes the performance values among all the alternatives have obvious differences, such an attribute plays an important role in choosing the best alternative. So to the view of sorting the alternatives, if one attribute has similar attribute values across alternatives, it should be assigned a small weight; otherwise, the attribute which makes larger deviations should be assigned a bigger weight, in spite of the degree of its own importance. Especially, if all available alternatives score about equally with respect to a given attribute, then such an attribute will be judged unimportant by most experts. In other word, such an attribute should be assigned a very small weight. Wang suggested that zero should be assigned to the attribute of this kind. The difference of attribute values can be measured using standard deviation.

For the attribute $u_j$, the standard deviation of alternative $x_i$ to all the other alternatives can be expressed as follows:

$$S_j = \sqrt{\frac{1}{m}\sum_{i=1}^{m}\left(\left\|\tilde{r}_{ij} - \frac{1}{m}\sum_{t=1}^{m}\tilde{r}_{tj}\right\| w_j\right)^2} = w_j\sqrt{\frac{1}{m}\sum_{i=1}^{m}\left(d(\tilde{r}_{ij},\tilde{r}_j)\right)^2}, j=1,2,...,n, \quad (7)$$

where $\tilde{r}_j = \frac{1}{m}\sum_{t=1}^{m}\tilde{r}_{tj} = \left(\left[1-\prod_{t=1}^{m}(1-a_{tj})^{\frac{1}{m}}, 1-\prod_{t=1}^{m}(1-b_{tj})^{\frac{1}{m}}\right], \left[\prod_{t=1}^{m}(c_{tj})^{\frac{1}{m}}, \prod_{t=1}^{m}(d_{tj})^{\frac{1}{m}}\right]\right)$ denotes the mean value of the attribute $u_j$. $d(\tilde{r}_{ij},\tilde{r}_j) = \frac{1}{4}\left(\left|a_{ij} - 1 + \prod_{t=1}^{m}(1-a_{tj})^{\frac{1}{m}}\right| + \left|b_{ij} - 1 + \prod_{t=1}^{m}(1-b_{tj})^{\frac{1}{m}}\right| + \left|c_{ij} - \prod_{t=1}^{m}(c_{tj})^{\frac{1}{m}}\right| + \left|d_{ij} - \prod_{t=1}^{m}(d_{tj})^{\frac{1}{m}}\right|\right)$ denotes the deviation of mean value $\tilde{r}_j$ to the attribute value $\tilde{r}_{ij}$ of the alternative $x_i$ for the attribute $u_j$. So $S_j$ denotes the standard deviation for the attribute $u_j$.

Based on the aforementioned analysis, we have to choose the weight vector $w$ to maximize all standard deviation values for all the attributes. To do so, we can construct the model as follows:

$$(\text{M-1}) \max\ F(w) = \sum_{j=1}^{n} S_j = \sum_{j=1}^{n} w_j \sqrt{\frac{1}{m}\sum_{i=1}^{m}\left(d(\widetilde{r}_{ij},\widetilde{r}_j)\right)^2}$$

$$\text{s.}t.\ \sum\nolimits_{j=1}^{n} w_j^2 = 1,\ w_j \geq 0,\ j = 1,2,\ldots,n$$

Let

$$\overline{\sigma}_j = \sqrt{\frac{1}{m}\sum_{i=1}^{m}\left(d(\widetilde{r}_{ij},\widetilde{r}_j)\right)^2}, \tag{8}$$

Then, the above model can be transformed into the following model (M-2):

$$(\text{M-2}) \begin{cases} \max F(w) = \sum\limits_{j=1}^{n} w_j \overline{\sigma}_j \\ \text{s.}t. \sum\limits_{j=1}^{n} w_j^2 = 1, w_j \geq 0, j = 1,\ldots,n, \end{cases}$$

We solve the model and get:

$$w_j = \frac{\overline{\sigma}_j}{\sqrt{\sum\limits_{j=1}^{n}\left(\overline{\sigma}_j\right)^2}} \tag{9}$$

Thus (9) is the extreme point of model (M-6).
By normalizing $w_j$ to let the sum of $w_j, j = 1,\ldots,n$ be a unit, we have

$$w_j^* = \frac{w_j}{\sum\limits_{j=1}^{n} w_j} = \frac{\overline{\sigma}_j}{\sum\limits_{j=1}^{n} \overline{\sigma}_j}, j = 1,\ldots,n \tag{10}$$

which also meats the information theory, that is, the attribute providing more information should be assigned a bigger weight.

Based on the above models, we develop a practical method for solving the multiple attribute decision making problems, in which the information about attribute weights is completely unknown, and the attribute values take the form of interval-valued intuitionistic fuzzy numbers. The method involves the following steps:

**Step 1.** Let $X = \{x_1,\ x_2,\ \ldots, x_m\}$ be a discrete set of alternatives, $U = \{u_1, u_2, \ldots, u_n\}$ be a set of attributes, and $w = (w_1, w_2, \ldots, w_n)^T$ be the weight vector of attributes, where $w_j \in [0,1],\ j = 1,2,\ldots,n, \sum_{j=1}^{n} w_j = 1$. For each alternative $x_i \in X$, the decision maker gives his/her preference value $\widetilde{r}_{ij}$ with respect to attribute $u_j \in U$, where $\widetilde{r}_{ij}$ takes the form of interval-valued intuitionistic fuzzy values, that is $\widetilde{r}_{ij}=(\tilde{\mu}_{ij},\tilde{v}_{ij})=([a_{ij},b_{ij}],[c_{ij},d_{ij}])$, $\tilde{\mu}_{ij}=[a_{ij},b_{ij}]\subset[0,1]$, $\tilde{v}_{ij}=[c_{ij},d_{ij}]\subset[0,1]$, $b_{ij}+d_{ij}\leq 1$, $i=1,2,...,m, j=1,2,...,n$, then all the preference values of the alternatives consists the decision matrix $\tilde{R}=(\tilde{r}_{ij})_{m\times n}$.

**Step 2.** If the information about the attribute weights is completely unknown, we solve the model (M-2) to obtain the optimal weight vector $w^* = (w_1^*, w_2^*, ..., w_n^*)^T$;

**Step 3.** Utilize the weight vector $w^* = (w_1^*, w_2^*, ..., w_n^*)^T$ and by Eq.(6), we can obtain the overall values or $\widetilde{z}_i(w^*)(i = 1, 2, ..., m)$ of the alternatives $x_i(i = 1, 2, ..., m)$.

**Step 4.** Calculate $S(\widetilde{z}_i)$ and $H(\widetilde{z}_i)$ of the overall interval-valued intuitionistic fuzzy preference values $\widetilde{z}_i(w^*)(i = 1, 2, ..., m)$ by (3) and (4).

**Step 5.** Rank all the alternatives $x_i(i = 1, 2, ..., m)$ and select the best one(s) in accordance with the $S(\widetilde{z}_i)$ and $H(\widetilde{z}_i)(i = 1, 2, ..., m)$.

**Step 6.** End.

## 4 An Illustrative Example

In this section, we provide a numerical example to illustrate the potential applications of the proposed FMEA.

A FMEA team identifies seven potential failure modes in a system and needs to prioritize then in terms of their failure risks such as probability of occurrence (O), severity (S) and detectability (D) so that high risky failure modes can be corrected with top priorities. Due to the difficulty in precisely assessing the risk factors and their relative importance weights, the FMEA team members reach a consensus to evaluate them using the interval-valued intuitionistic fuzzy numbers. The assessment information of the seven failure modes $x_i(i = 1, 2, ..., 7)$ on each factor is presented in Table 1.

**Step 1.** we first utilize the Eq. (11) to get the weight vector of attributes.

$$w^* = (0.3288, 0.3940, 0.2772)^T$$

**Table 1** Interval-valued intuitionistic fuzzy decision matrix $\tilde{R} = (\tilde{r}_{ij})_{m\times n}$

| | $O$ | $S$ | $D$ |
|---|---|---|---|
| $x_1$ | ([0.7,0.8],[0.1,0.2]) | ([0.8,0.9],[0.1,0.1]) | ([0.7,0.9],[0.1,0.1]) |
| $x_2$ | ([0.5,0.7],[0.1,0.2]) | ([0.6,0.7],[0.1,0.3]) | ([0.4,0.5],[0.2,0.4]) |
| $x_3$ | ([0.3,0.5],[0.1,0.3]) | ([0.4,0.5],[0.1,0.3]) | ([0.3,0.6],[0.3,0.4]) |
| $x_4$ | ([0.6,0.7],[0.1,0.2]) | ([0.7,0.8],[0.1,0.2]) | ([0.5,0.7],[0.1,0.3]) |
| $x_5$ | ([0.5,0.7],[0.2,0.3]) | ([0.5,0.7],[0.1,0.3]) | ([0.4,0.6],[0.2,0.3]) |
| $x_6$ | ([0.3,0.4],[0.4,0.6]) | ([0.2,0.4],[0.5,0.6]) | ([0.4,0.5],[0.4,0.5]) |
| $x_7$ | ([0.3,0.5],[0.3,0.5]) | ([0.4,0.6],[0.3,0.4]) | ([0.4,0.5],[0.2,0.4]) |

We utilize the proposed algorithm to prioritize the seven failure modes.

**Step 2**. Utilize the weight vector $w^* = (w_1, w_2, w_3)^T$ and (6) to calculate the overall values $\widetilde{z}_i(w^*)(i = 1, 2, ..., 7)$ of the alternatives $x_i(i = 1, 2, ..., 7)$.

$\widetilde{z}_1(w^*) = ([0.744, 0.874], [0.1, 0.126])$, $\widetilde{z}_2(w^*) = ([0.518, 0.654], [0.121, 0.284])$,

$\widetilde{z}_3(w^*) = ([0.341, 0.53], [0.136, 0.325])$, $\widetilde{z}_4(w^*) = ([0.62, 0.744], [0.1, 0.224])$,

$\widetilde{z}_5(w^*) = ([0.474, 0.675], [0.152, 0.3])$, $\widetilde{z}_6(w^*) = ([0.293, 0.430], [0.437, 0.570])$,

$\widetilde{z}_7(w^*) = ([0.369, 0.542], [0.268, 0.431])$.

**Step 3.** Calculate the scores $S(\widetilde{z}_i)(i = 1, 2, ..., 7)$ of the overall interval-valued intuitionistic fuzzy values $\widetilde{z}_i(w^*)(i = 1, 2, ..., 7)$.

$$S(\widetilde{z}_1) = 0.696,\ S(\widetilde{z}_2) = 0.3835,\ S(\widetilde{z}_3) = 0.205,\ S(\widetilde{z}_4) = 0.52,\ S(\widetilde{z}_5) = 0.3485,\ S(\widetilde{z}_6) = -0.142,\ S(\widetilde{z}_7) = 0.106$$

thus

$$S(\widetilde{z}_1) \succ S(\widetilde{z}_4) \succ S(\widetilde{z}_2) \succ S(\widetilde{z}_5) \succ S(\widetilde{z}_3) \succ S(\widetilde{z}_7) \succ S(\widetilde{z}_6)$$

**Step 4.** Rank all the alternatives $x_i(i = 1, 2, ..., 7)$ in according with the scores $S(\widetilde{z}_i)$ $(i = 1, 2, ..., 7)$

$$x_1 \succ x_4 \succ x_2 \succ x_5 \succ x_3 \succ x_7 \succ x_6$$

and thus the final conclusion for this example is that failure mode $x_1$ is given the top priority for correction.

## 5 Conclusions

FMEA is a very important safety and reliability analysis tool which has been widely used in many areas and industries. In this paper, standard deviation method is proposed to determine the weights of the risk factors objectively for prioritization of failure modes which allows the risk factors to be evaluated in interval-valued intuitionistic fuzzy numbers rather than a crisp number. For the information about the attribute weights is completely unknown, we establish an optimization model, by solving the model, we get a simple and exact formula which can be used to determine the attribute weights of interval-valued intuitionistic fuzzy numbers. After that, we utilize the interval-valued intuitionistic fuzzy weighted average (IVIFWA) operator to aggregate the interval-valued intuitionistic fuzzy numbers decision information and then select the most desirable alternative according to the score function and accuracy function. Finally, a practical example is given to verify the developed method and to demonstrate its practicality and effectiveness. And also, the method can be extended to the group intuitionisitic fuzzy decision making easily.

## References

Atanassov KT (1986) Intuitionistic fuzzy sets. Fuzzy Sets Syst 20:87–96

Atanassov K, Gargov G (1989) Interval valued intuitionistic fuzzy sets. Fuzzy Sets Syst 31:343–349

Bowles JB, Peláez CE (1995) Fuzzy logic prioritization of failures in a system failure mode, effects and criticality analysis. Reliab Eng Syst Saf 50:203–213

Braglia M, Frosolini M, Montanari R (2003) Fuzzy TOPSIS approach for failure mode, effects and criticality analysis. Qual Reliab Eng Int 19:425–443

Chang CL, Wei CC, Lee YH (1999) Failure mode and effects analysis using fuzzy method and grey theory. Kybernetes 28:1072–1080

Lertworasirikul S, Fang SC, Joines JA, Nuttle HLW (2003) Fuzzy data envelopment analysis (DEA): a possibility appraoch. Fuzzy Sets Syst 139:379–394

Pillay A, Wang J (2003) Modified failure mode and effects analysis using approximate reasoning. Reliab Eng Syst Saf 79:69–85

Stamatis DH (1995) Failure mode and effect analysis: FMEA from theory to execution. ASQC Quality Press, Milwaukee

Wang YM (2003) A method based on standard and mean deviations for determining the weight coefficients of multiple attributes and its applications. Math Stat Manage 22:22–26

Wang YM, Chin KS, Poon GKK, Yang JB (2009) Risk evaluation in failure mode and effects analysis using fuzzy weighted geometric mean. Expert Syst Appl 36:1195–1207

Xu ZS (2007) Methods for aggregating interval-valued intuitionistic fuzzy information and their application to decision making. Control Decis 22:215–219

Xu ZS, Yager RR (2006) Some geometric aggregation operators based on intuitionistic fuzzy sets. Int J Gen Syst 35:417–433

Xu K, Tang LC, Xie M, Ho SL, Zhu ML (2002) Fuzzy assessment of FMEA for engine systems. Reliab Eng Syst Saf 75:17–29

# A Computer Auditing Model of the Balance Sheet Parallel Simulation Based On Data Mining

**Li Zhang, Lu Wang, and Jianping Zhang**

**Abstract** In order to find possible problems in system auditing of the Large enterprise groups, quickly locking the audit doubt, avoid audit risk, we set up a parallel simulation program for single company based on the logic of balance sheet. To compared the differences of report's project, using correlation analysis Rules, combination auditing experience, using statistical model to analyze the unusual data, rapid locked doubts subjects. Examples of application show the effectiveness of the model, we can quickly find the audit trail of many companies, reduce the risk of incorrect acceptance caused by the delay of data acquisition.

**Keywords** Balance sheet · Computer auditing · Data mining · Parallel simulation · Risk

## 1 Introduction

The wide application of the accounting information system has made far-reaching influence for the organizational form of enterprise management and the business process. The environment the auditing facing also has been changed profoundly.

L. Zhang (✉)
School of Information Management, Beijing Information Science & Technology University, Qinghe Xiaoying Donglu.12, Haidian District 100192 Beijing, China
e-mail: yeszhl@163.com

L. Wang
China Audit International Certified Public Accountants Ltd, Fu Cheng Road.73, Haidian District 100142 Beijing, China
e-mail: wlcaf@163.com

J. Zhang
College of Textile Engineering and Art, Taiyuan University of Technology, Welcome Road. 113, Yuci, Jinzhong City 030600, Shanxi Province, China
e-mail: yeszhl@126.com

Y. Zhou and D.D. Wu (eds.), *Modeling Risk Management for Resources and Environment in China*, Computational Risk Management,
DOI 10.1007/978-3-642-18387-4_62, 

Especially in the auditing for the large state-owned enterprise groups: facing the multi-level organizational structure, the diverse accounting forms and the heterogeneous data sources, it has become a difficult problem for the auditors to use the limited resources to carry out the auditing analysis of the authenticity, the accuracy and the integrality of the balance sheet of the single company within the Group, and lock a batch of doubtful subjects and find out the audit trails.

Auditor General Liu Jiayi once highlighted particularly that audit was the "immune system" of the running entire economy. Currently, the environmental changes of the "immune system" have endowed the auditing techniques with the following characteristics: (1) The auditing methods have been adapted to the information conditions. And the auditing techniques are higher than the capacity of the audited unit; (2) In the process of implementation, the monitoring function of auditing is reflected particularly, and the dynamic early warning and control are stressed. Therefore, the auditors need to, and have to adopt the advanced technologies for auditing.

The study of computer auditing began with 1960s. United States and Canada have taken some useful explorations in the guidelines, technologies and the software development. The Chinese National Audit Office has also actively promoted the computer auditing which have achieved certain effects. At present, the research results on the computer auditing mainly focuses on the following aspects: (1) The implications of the computer auditing: The National Auditing Office, the Computing Center of Board of Audit of Japan, Cao Jianxin, Li Xuerou and other scholars all classified the implications of the computer auditing into two: the auditing on the computer system itself and the computer assisted auditing (Cao et al. 2007); (2) The criteria: A trend the global economy brings is the convergence of global accounting and auditing standards. The foreign research on the computer auditing standards began with 20 years ago, mainly focusing on the standards, the specifications and the system changes. What attract the most attention currently are COBIT4.1, ISO27000, ITIL and other frameworks, standards, guidelines and procedures which were developed by ISACA, the International Auditing Association and ISO, etc. (ISACA 2009). (3) The application of the information technology in audit is still at the exploratory stage, while the research results of some scholars have certain reference. For example, Liu Ruzhuo and Wang zhiyu systematically expounded the general controls, the application controls and the testing technology of applied software auditing, opening up the new approach of the domestic computer auditing Geng, Liqiang, Nam and Hun analyzed the integration test, system control, and the continuous and intermittent simulating technology, and established the unstructured mining model of audit log (Park and Lee 2008); Yu and Fe introduced the application of the isolated-point detection in the multi-dimensional analysis of auditing; Allinson and Caroline took the auditing trail as the vector space model of the judicial evidence. In customs auditing, Liao Yi used the analysis of isolated point to develop data mining, detect abnormal importing goods, the fraud valuation reporting and other audit trail (Liao 2007). Other Scholars have searched for a number of rules using data mining to (Chen et al. 2009; Huang 2006; Zhang et al. 2006; Chen 2009); He Yujie introduced the SQL query and the OLAP analysis in the audit; Wang Zhong applied the fuzzy neural network and genetic algorithm to solve the overall analyzing problems of audit data; Zhang Jincheng and Huang Zuoming described the method of inspecting procedure code, the controlled processing and the parallel simulation method, etc., which has played a certain role in the development of the computer auditing technology.

Although these studies have formed a preliminary framework, there are still some limitations in view of the inadequate application of the combination of the information technology and the analysis of accounting statements, and the rare application studies of the computer auditing to the financial statements of the large-scale enterprise group. That how to use information technology to serve the audit of financial statements of the large enterprise group, has become an urgent problem to be solved in current auditing practices. Below are the studies about the computer auditing methods of the balance sheet.

# 2 The Model of the Balance Sheet Audit with Parallel Simulation

In order to rapidly implement the balance sheet audit of the heterogeneous multi-level group, we established the model of the balance sheet audit with parallel simulation (Fig. 1).

## 2.1 *The Definition of the Problem*

By computer auditing, found out the doubtful items in the balance sheet of the single company of large enterprise groups. Then encoded all the companies within the Group in order to generate the company code table.

## 2.2 *Data Pre-processing*

Put up the audit data warehouse server. The data sources of Group generally fall into two categories: the business adopting the uniform ERP software and non-online unit (using different soft wares). Had the data of the same source automatically be taken into the audit data warehouse, and then loaded the non-online unit data into the audit server (DW) after clean-up and conversion.

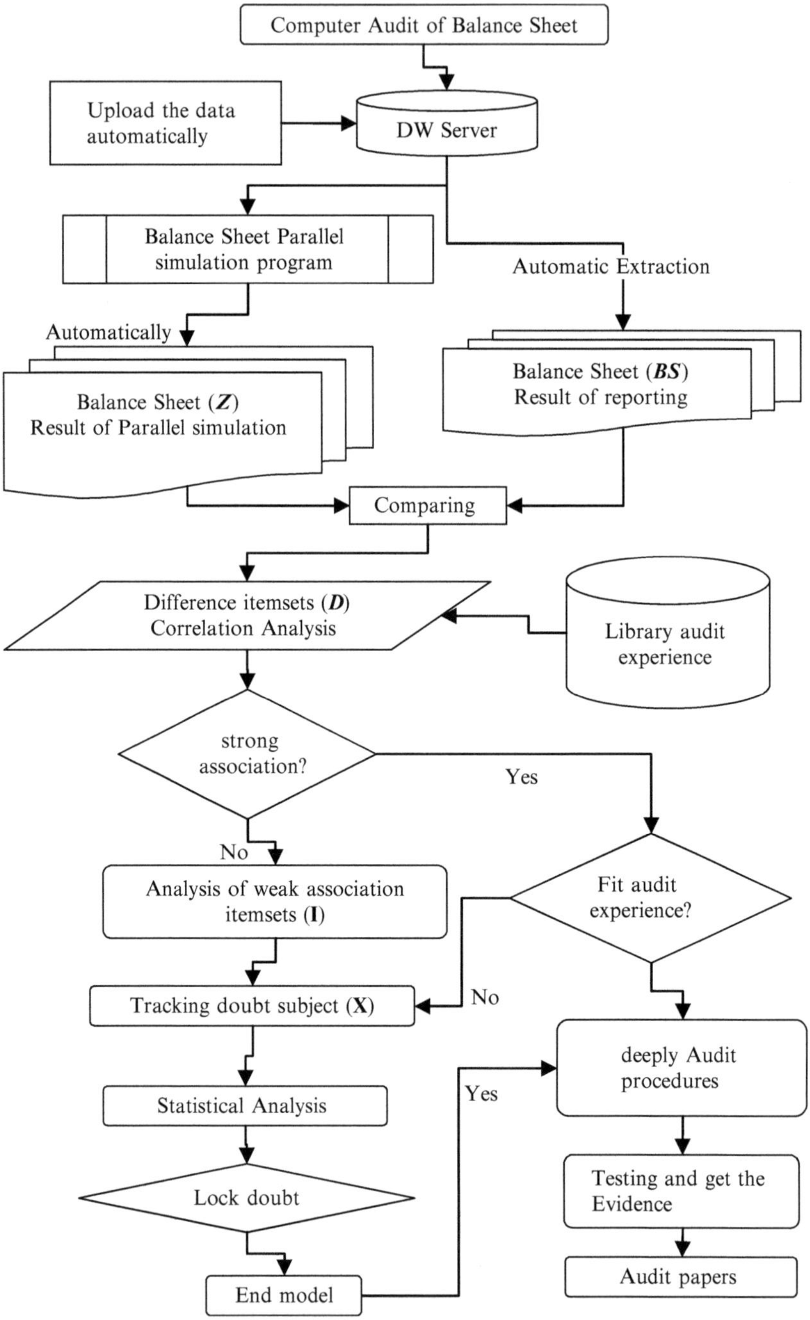


**Fig. 1** The flow chart of parallel simulation for balance sheet audit

## 2.3 The Parallel Simulation Program

According to the fetch logic of the financial statements, designed and developed the report system with parallel simulation, took the numbers from the data warehouse directly to generate the single company's balance sheet (Zi, Zi was on behalf of the i-company's balance sheet which was generated through parallel simulation). The data acquisition by parallel simulation were removed to the data warehouse server of the audited units, which could reduce the subjected risks effectively compared to the other audit softwares that took the balance sheet directly.

## 2.4 Comparative Analysis

Designed the comparative analysis program, compared the paralleling simulated balance sheet ($Z_i$) with the automatically generated one by the system when the enterprise reported ($BS_i$, the i-company's balance sheet submitted by the Enterprise), and then identified the different companies, generating report sets ($R_j$, the variances in the report of k-company). n is the number of companies within the Group.

## 2.5 The Parallel Simulation Program

Established the affair set of association rules (D), discovered all frequent item sets according to the minimum, and then produced the association rules according to the frequent item sets and the smallest confidence.

## 2.6 Analysis of the Doubted Subjects

Compared the audit subjects the strong association rules related with the audit experience library, and then excluded the fit items. Took the items not meeting the audit experience and the weak associated item sets as isolated points, and pulled out doubtful itemsets (I).

According to the fetch logic, took the subject analysis to the doubted item sets of the report, and traced the doubted subject sets (X).

After that, according to characteristics of the debit-and-credit relationship of the debit-and-credit accounting method, analyzed the trend, gather, merge, and made regression analysis, locked auditing doubtful points by audit methods.

Having finish the model, continued to carry out the further audit procedures, took substantive tests to the doubted points, collect evidence, and filled the audit papers to form a basis for the report.

# 3 Data Mining of the Balance Sheet

In the model of balance sheet audit with parallel simulation, the design of data mining algorithms was essential.

## 3.1 The Association Rules and the Apriori Algorithm

Variance sets ($R_j$) was on behalf of the difference between the balance sheet submitted by the companies and the parallel simulated results according to the enterprise accounts by the auditors, showing that there might be not matched items in the corporate accounts table.

**Definition 1.** *The transaction database D is constituted by the companies' variance report table* $R_k$, $D = (R_1, R_2, \ldots R_k)$, $(k = 1, 2, 3 \ldots\ldots n)$ $R_k = (i_1, i_2, \ldots i_m)$, $(m = 1, 2, 3 \ldots\ldots p)$ *was the item set, among which $i_m$ was the item of the report table, such as monetary capital, receivable accounts and other items, each of which was a item.*

Used the Apriori algorithm proposed by Agrawal and others to find the itemsets with the frequent differences. Adopted the layer-by-layer search iterative algorithm, and first calculated the degree of support of each item set at one time to determine the frequent 1 – itemsets. Made use of the frequent itemsets $L_{k-1}$ found in the (k – 1)th iteratary, and used the Apriori-gen to generate candidate itemsets $C_k$ in the two steps of the connection and pruning. Scanned the database to calculate the degree of support of candidate itemsets. Hash tree could be effectively used to determine if a candidate itemsets in $C_k$was included in the transactionR. Stopped when the frequent itemsets generated in the kth is the space algorithm.

In the correlation analysis, the affairs of the difference report form could be focused on the frequent item sets of differences and the degree of support, the quantitative confidence. Combined with the auditing experience, the auditors could determine the doubtful subjects with high- risks.

## 3.2 Statistical Analysis

To determine the root of the differences, summarized, merged and took other statistical analysis to the correlation between the doubtful subjects. Furthermore, took rate analysis and structural analysis to the difference items, determining the priorities for further audit.

# 4 Discovery of the Audit Trail

A large enterprise group had more than 500 subsidiary companies in its merged scope; most of the enterprises shared the ERP system. The Group had the data storehouse server, to which the data of the online subsidiary units could be uploaded two real-time every day. The non-online units uploaded data by adopting the united templates of fetching to the group's server regularly. When auditing, the auditors shared data directly in real-time from the data warehouse DW server of the audited units.

Take the information system auditing of a large enterprise group's balance sheet for example, to verify the accuracy and effectiveness of the model of balance sheet audit with parallel simulation.

## *4.1 The Parallel Simulation*

Developed the balance sheet program of parallel simulation based on VB, and generated the comparing report forms. Compared them with the report forms issued by the company's report form system, and then generated the difference transaction sets which contained 54 variance reports.

## *4.2 Correlation Analysis*

We count frequent item sets using Apriori algorithm, Correlation analysis the difference sets, we get the results (Table 1):

We find five stronger correction rules such as:

Accounts receivable ⇒ Receipts in advance [support = 31.48%, confidence = 82%]

Accounts Payable ⇒ Prepayments [support = 35.19%, confidence = 73.4%]

Other receivables ⇒ other payable [support = 33.33%, confidence = 86.2%]

Investment Property ⇒ Fixed assets ⇒ Accumulated depreciation [support = 40.74%, confidence = 91.3%]

**Table 1** Correlation analysis's results

| Items | Count | Support | confidence |
|---|---|---|---|
| Accounts receivable, receipts in advance | 17 | 31.48% | 82% |
| Accounts payable, prepayments | 19 | 35.19% | 73.40% |
| Other receivables, other payables | 18 | 33.33% | 86.20% |
| Investment property, fixed assets, accumulated depreciation | 22 | 40.74% | 91.30% |
| Investment property, fixed assets | 28 | 51.85% | 97.40% |
| Construction in progress, fixed assets | 6 | 11.11% | 78% |
| Investment property, intangible assets | 3 | 5.56% | 100% |
| Other payables, should pay special | 4 | 7.41% | 85% |

Investment Property ⇒ Fixed assets [support = 51.85%, confidence = 97.4%]

When support is more than 30% and confidence more than 70%, we think the rules are stronger correction rules. Other itemsets are non-frequent itemsets, such as {Construction in progress, fixed assets}, {Investment Property, Intangible assets}, {Other payables, Should pay special}, {Accounts receivable}, {Other receivables}, {inventory}.

## 4.3 Compared the Audit Experience Database

In some cases, some transferred from data of companies could be accepted by auditors, such as the re-classification of the back-and-forth fund. While in some other cases, the transferred item would affect the current quality of the accounting information directly, and make some influence to the following accounting information as well, such as the process of turning to solid. If the time of the process of turning to solid was incorrect, the later accumulated depreciation and profit and loss items would be greatly affected.

The auditors found that, among the above five strong association rules, there were three-classified transfer items. According to the data rules of the audit experience database, the auditors judged that the items of the strong association rules might be the adjustment of the regular re-classification in the items of the report form, which was further verified to be acceptable and would not be concerned as the high-risk areas. While the other two were the identified projects for Investment Property. Further we locked the auditing doubts of the Investment Property; at the same time, focused on the low-correlation projects, such as fixed assets, the projects in construction, other payable accounts, receivable credits and inventories which can be recognized as the key projects of the next step.

## 4.4 Analyzed the Doubtful Subjects

According to the fetch logic, identified the detailed projects from the variance of the inventory projects and determined if the variance of the stock merchandise was bigger. Confirmed the next doubtful subjects.

## 4.5 Statistical Analysis

Taking statistical analysis according to the items of the report form, the auditor found that there are 28 companies had variances in Investment Property, 30 companies with the differences in fixed assets, 6 companies with the differences in the construction and 31 companies with the differences in receivable accounts.

Besides, there were 20 companies with differences in other payment, and one company having differences in stock merchandise. Further tracing the audit annual and the following years, the auditor found that a number of adjustments of audited units were achieved through the transferred table, not having accounting treatment.

After carrying out the ratio analysis and structural analysis to the balance sheet of the companies, and performing regular audit analyzing procedures, the auditor discovered that the 23rd company has product lending activities, which had resulted in the inconsistence between the report tables for the stock merchandise with the reality of the accounts, and the false carry-over of the cost. Therefore, further audit procedures should be designed to trace the doubtful points.

### *4.6 Audit Discovered*

By auditing, we found that the report form system the group used had no direct functions of transferring table. Many subordinate enterprises adjusted the items of the report forms directly through the report form system. However, these problems were not handled correspondingly in the current and next year's accounts system, resulting in the accounts which are not in. We designed the further audit procedures to trace the reasons for differences.

## 5 Conclusion

Our goal is to validate the audited accounting information system authenticity and reliability, quickly found the audit doubt. in auditing large enterprise groups, there many difficulties include the resource cost and time cost. To achieve our aims, firstly, we conduct computer audits in more than 500 subsidiaries on a group of enterprises set up a parallel simulation reporting program of single company based on the logic of balance sheet, We find there are differences in 54 enterprise's reports between the account system and parallel simulation. Secondly, we set up the Correlation analysis model and compare the difference items of balance sheet, we find five strong association rules. According to auditing experience, using statistical analyze the unusual data, rapid lock doubts items of different company, such as fixed assets, construction in progress, other payables, accounts receivable, inventory. Lastly, we find a direct adjust function in enterprise's reporting system, lead to the account, tables, and reality are different. Examples of application show the effectiveness of the model, we can quickly find the audit trail of many companies, reduce the risk of incorrect acceptance caused by the delay of data acquisition.

**Acknowledgments** We are grateful to Beijing Municipal Commission of Education for provided the humanities and social science fund (SM201010772004).

## References

Cao J, Zhu B, Hua F (2007) Review on computer audit developing. J Pioneering Sci Technol 10,165-166 (2007)
Chen D (2009) Data mining in the application of modern audit. J Nanjing Audit Univ 6(1):57–61
Chen D, Wang J, Han B (2009) Computer audit based on outlier mining. J Nanjing Audit Univ 6(1):62–66
Huang Y (2006) Outlier analysis in the application of computer audit. J Audit Res S1:122–125
ISACA (2009) Certified information system auditor review manual. ISACA J 2:1–3
Liao Y (2007) Research of data mining based on outliers and application in IT audit system. Master's degree thesis, Beijing Jiaotong University
Liu R, Wang Z (2006) Computer audit techniques and methods. Tsinghua University Press, Beijing
Park NH, Lee WS (2008) Anomaly detection over clustering multi-dimensional transactional audit streams. In: Proceedings – 1st IEEE International Workshop on Semantic Computing and Applications, IWSCA, IEEE Press, pp 78–80
Zhan Z, Li J, Luo W (2006) Found in the audit trail based on event found. J Audit Res Supplement, 67–70

MIX
Papier aus verantwortungsvollen Quellen
Paper from responsible sources
FSC® C105338

FSC
www.fsc.org

If you have any concerns about our products,
you can contact us on
**ProductSafety@springernature.com**

In case Publisher is established outside the EU,
the EU authorized representative is:
**Springer Nature Customer Service Center GmbH**
**Europaplatz 3, 69115 Heidelberg, Germany**

Printed by Libri Plureos GmbH
in Hamburg, Germany